Synthetic Aperture Radar (SAR) Techniques and Applications

Synthetic Aperture Radar (SAR) Techniques and Applications

Special Issue Editor

Fabio Bovenga

MDPI • Basel • Beijing • Wuhan • Barcelona • Belgrade • Manchester • Tokyo • Cluj • Tianjin

Special Issue Editor
Fabio Bovenga
Italian National Research Council,
Institute for Electromagnetic
Sensing of the Environment
(IREA)
Italy

Editorial Office
MDPI
St. Alban-Anlage 66
4052 Basel, Switzerland

This is a reprint of articles from the Special Issue published online in the open access journal *Sensors* (ISSN 1424-8220) (available at: https://www.mdpi.com/journal/sensors/special_issues/ SAR_techniques_applications).

For citation purposes, cite each article independently as indicated on the article page online and as indicated below:

LastName, A.A.; LastName, B.B.; LastName, C.C. Article Title. *Journal Name* **Year**, *Article Number*, Page Range.

ISBN 978-3-03936-122-9 (Hbk)
ISBN 978-3-03936-123-6 (PDF)

Contents

About the Special Issue Editor

Fabio Bovenga (Ph.D.): Fabio Bovenga received the Laurea degree and the Ph.D. degree in Physics from the University of Bari (Italy) in 1997 and 2005, respectively. He was at the University of Bari until 2007, after which he joined the Italian Research Council, where he is presently a researcher at IREA in Bari. His main research interests include advanced processing techniques for Synthetic Aperture Radar (SAR) imaging and SAR interferometry and the application of multi-temporal/multi-frequency analysis to ground monitoring. He is also involved in the exploitation of a new generation of wide-band SAR data. He has co-authored more than 150 journal papers, book chapters, and conference communications and serves as a reviewer for several international peer-reviewed journals. He currently chairs sessions in international symposia on remote sensing and signal processing (IEEE, SPIE). Since 2002, he has delivered lectures on SAR/InSAR principles and techniques for the Master's program in "Spaceborne Remote Sensing" at Bari University. He has been co-advisor for several degree theses in Physics and Electronic Engineering on SAR processing, SAR interferometry, and their applications. He is also a board member of the IEEE Geoscience and Remote Sensing South Italy Chapter.

Editorial

Special Issue "Synthetic Aperture Radar (SAR) Techniques and Applications"

Fabio Bovenga

Institute for Electromagnetic Sensing of the Environment (IREA), Italian National Research Council, Via Amendola, 122/d, 70126 Bari, Italy; bovenga.f@irea.cnr.it

Received: 24 March 2020; Accepted: 25 March 2020; Published: 27 March 2020

Abstract: This editorial of the special issue titled "Synthetic Aperture Radar (SAR) Techniques and Applications", reviews the nineteen papers selected for publication. The proposed studies investigate different aspects of SAR processing including signal modelling, simulation, image analysis, as well as some examples of applications. The papers are grouped according to homogeneous subjects, then objectives and methods are summarised, and the more relevant results are commented.

Keywords: SAR imaging; multi-angle/wide angle SAR; inverse SAR; ground-based SAR; ionospheric effects; SAR interferometry; SAR image analysis; SAR image change detection; SAR sea applications

1. Introduction

Synthetic Aperture RADAR (SAR) became a well-established and powerful remote sensing technology used worldwide for several applications thanks to the possibility of sensing the Earth's surface at night and day, and in any weather condition. Recent advances have dramatically increased the SAR monitoring potential by improving spatial resolution, revisit time, swath width and polarimetric capability. Moreover, the present and forthcoming spaceborne missions, allow SAR imaging at different bands and acquisition modes (e.g., spotlight, wide swath, bistatic, multi-static and geosynchronous). All these advances stimulated the investigation of new processing algorithms, products, and applications able to fully exploit the new sensor capabilities (e.g., wide spectral band, multi-angle view, short revisit time), as well as the large and continuously updated SAR data archives. The same holds for SAR imaging from ground-based platforms, airplanes and Unmanned Aerial Vehicles (UAVs).

This editorial paper reviews the content of the special issue dedicated to SAR techniques and applications, by presenting advances on SAR signal modelling, SAR simulation, SAR processing, SAR image analysis and SAR-based applications.

2. Contributions

The special issue has collected nineteen papers investigating different aspects of SAR processing, SAR image analysis and SAR applications. The contributions cover topics related to multi-angle/wide angle SAR imaging, Doppler parameter estimation, data-driven focusing, Inverse SAR (ISAR) applied to pulsar signal modelling and detection, Ground-Based SAR (GBSAR), near-field Interferometric ISAR, interaction between SAR signal and infosphere, SAR interferometry for ground displacement monitoring, feature extraction, change detection and SAR-based sea applications. In the following, the papers are grouped according to homogeneous subjects, and are reviewed by summarizing objectives, methods and main contributions. Comments are also provided on those research aspects that are more relevant with respect to the state-of-the-art research on SAR processing and applications.

2.1. SAR Imaging

The works in [1–6] concern several aspects of the SAR imaging. In [1], the authors propose a multi-angle SAR imaging system suited for an ultrahigh speed platform and based on multi-beamforming. By acquiring images at different angles during the same flight, the system allows better characterization of the target on the ground, as well as a simplified motion error compensation. A procedure is proposed aimed at both improving the range migration algorithm and imaging data from different view angles in a unified coordinate system. This provides images with the same resolution, not deformed and scaled, that can be fused quickly and accurately.

The work in [2] also deals with the optimal processing of SAR data acquired with wide aperture, as for circular SAR systems, and in particular tackles the problem of aspect-dependent backscattering. The authors propose an approach based on the least squares of compressed sensing residuals, which is used in video imaging and does not require the isotropic scattering assumption adopted by other methods. This procedure is able to reconstruct time sequences of sparse signals changing slowly with time, and thus it is well suited to processing images derived from SAR sub-apertures, which are highly overlapped. The proposed approach was tested on real data, providing a more accurate estimation of aspect dependent scattering than other methods based on compressed sensing.

The work in [3] concerns the processing of multi-pass squinted SAR data. The proposed algorithm combines images acquired with the same azimuth squint angle on each pass for performing 3D imaging (as in SAR tomography), and data acquired with different azimuth squint angles for refining the suppression of azimuth sidelobes. A performance analysis is carried out by using both simulated point targets and real data acquired by TerraSAR-x satellite mission. The algorithm is able to improve the azimuth suppression while preserving the mail lobe resolution.

The work in [4] considers the problem of Doppler parameter estimation and compensation for SAR data acquired by airborne systems with very high squint geometries. The authors propose an algorithm based on extended multiple aperture mapdrift, which is able to estimate the Doppler phase spatial variation of the third-order. This high order is required for focusing data acquired at high squint angle and at high resolution along very variable aircraft trajectories. In this case, indeed, the inertial navigation system is not able to provide positioning, velocity and angle information accurate enough for a reliable SAR focusing. The method was used for processing both simulated data and real airborne data, and provided accurate targets focusing, thus demonstrating the reliability of high order Doppler parameter compensation.

The imaging of airborne wide-area surveillance (WAS) radar is considered in [5]. For this kind of radar system, Doppler beam sharpening (BDS) imaging is adopted, which, however, suffers from low cross range resolution due to the short dwell time. The authors propose a knowledge-aided DBS processing able to increase the cross-range resolution by a factor of two. The algorithm exploits the strong spatial coherence between adjacent pulses for increasing the number of pulses processed in each coherent interval, thus enhancing the DBS final resolution. A performance analysis was carried out by first using simulated point targets, and then real WAS data. Results demonstrated that the proposed algorithm outperforms other methods adopted in DBS imaging.

The work in [6] proposes an innovative SAR data focusing algorithm, which does not need the knowledge of neither nominal SAR system parameters, nor sensor attitude and trajectory information. The basic idea consists in estimating directly from the data the range and azimuth reference functions needed for SAR focusing, by exploiting, through Singular Value Decomposition, the inherent redundancy present in the SAR raw data. To ensure reliable parameter estimation, strong point scatterers are needed within the imaged scene. This blind focusing could be very useful, for instance, for SAR systems onboard simple aerial unmanned vehicles to be used in real-time and low-cost applications. The algorithm performances were first assessed through simulations, and then SAR data from the ERS mission were processed by using both the proposed algorithm and the standard Range Doppler focusing approach. Results showed a reasonably good quality of the focused image both in amplitude and phase.

The list of papers devoted to SAR imaging concludes with the interesting study in [7], which uses the principles of inverse SAR for modelling signals coming from a pulsar, reflected by space objects (e.g., asteroids), and then detected by a radio telescope on the Earth. Thanks to the coherence of pulsar emissions, pulsar signals can be modelled as monochromatic Gaussian pulses distributed in a time-frequency signal grid. Accordingly, the detected pulsar signals reflected by a moving space object can be considered as a delayed copy of such pulses and modelled within a passive inverse SAR scenario. A range compression approach for this specific ISAR imaging was introduced theoretically and demonstrated through numerical simulations. The Crab Nebula pulsar was considered as emission source. This study can be used for performing space object navigation, localization and imaging based on pulsar emission.

2.2. Ground-Based SAR

The special issue includes also three interesting papers [8–10] dedicated to ground-based SAR (GBSAR) systems.

The first one [8] is a communication describing the Imaging Multiple-Input Multiple-Output (MIMO) ground-based interferometric radar developed in order to overcome the main limitations of traditional GBSAR systems, which are based on the mechanical movement of the antenna. The proposed system reduces data acquisition time, thus both limiting the atmospheric artefacts and extending the application to vibration measurements. Moreover, the use of independent modules and integrated technologies allows reducing production costs and improving both the transportability and the deployment of the system. The authors introduced the concept and the design of the system, and presented first results derived by using the developed prototype.

The work in [9] concerns a GBSAR system consisting of an antenna on a rotating boom that performs a 360-degree scanning of the surrounding scene, namely, an ArcSAR system. In order to improve the quality of the ArcSAR imaging, the authors propose a refinement of the image focusing by using a digital elevation model (DEM) derived through interferometric processing. The DEM of the scene is first generated by processing ArcSAR interferometric images, and then projected on ground range. Finally, this interferometric DEM is used for enhancing the ArcSAR imaging of the targets on the scene. The authors described the procedure, provided an accuracy analysis and presented results obtained by simulating ArcSAR images according to existing radar amplitude data and an external DEM.

Finally, the work in [10] proposes improving the interferometric near-field 3D imaging by using a multichannel joint sparse reconstruction. The basic idea consists in deriving multichannel signals by dividing the two observed full apertures into sub-apertures. Then, by exploiting the sparsity of the target echo in each channel, the imaging problem is set up as a multichannel joint sparse reconstruction, and the 3D target image of each sub-aperture target is obtained through the improved orthogonal matching pursuit method. Finally, the high-resolution 3D image is derived by synthesizing the 3D images from each sub-aperture. The proposed algorithm improves the imaging accuracy of both strong scattering centres and anisotropic targets. The procedure was tested by using both electromagnetic simulations and real data acquired in an anechoic chamber by using a prototype system.

2.3. Ionosphere

This special issue includes also two works investigating the effects of the ionosphere onto SAR imaging from spaceborne platforms.

In [11], the ionospheric scintillation is considered, being the main limiting factor of spaceborne P-band SAR imaging. The sliding spotlight mode, while increasing the azimuth resolution, is particularly affected by scintillation artefacts due to the long integration time. The theoretical analysis of scintillation effects on the P-band spotlight SAR images was performed by introducing a novel scintillation simulator based on the reverse back-projection algorithm. SAR raw data for the sliding spotlight mode were simulated for both pointy and extended targets, under different scintillation conditions. Simulations allowed investigating the image degradation in azimuth induced by scintillation, and deriving threshold values of scintillation strength and spectral index, which guarantee acceptable P-band spotlight imaging.

One of the most relevant characteristics of the ionosphere is the total electron content (TEC), which is investigated in [12] by exploiting spaceborne polarimetric SAR (PolSAR) data. First, the authors assessed the precision of SAR-based TEC retrieval by comparing measurements derived from L-band PolSAR data acquired by the ALOS satellite, and from very accurate incoherent scatter radars. Then, the TEC of the topside profile was estimated by subtracting from the TEC derived by PolSAR, the TEC of the bottom side profile measured by an ionosonde observing the same space of the satellite. This procedure allows refining the topside TEC estimations and thus improves the modelling of the electron density topside profile.

2.4. Ground Displacement Monitoring

The following seven papers are specifically dedicated to applications. The first two [13,14] use SAR interferometry (InSAR) for investigating displacements related to land subsidence and highway deformation. Multi-temporal InSAR is a well-established technique currently applied to displacement monitoring thanks to the availability of reliable processing tools developed during the last two decades, as well as data archives continuously updated by operative satellite missions. Nowadays, there is an increasing need for advanced, application-oriented procedures for analysing the InSAR-based displacement records.

In [13], high-resolution RADARSAT-2 SAR data were processed through the Small BAseline Subset (SBAS) algorithm for deriving 3-year displacement time series over Wuhan city, which suffers from subsidence problems related to urban construction. First, the InSAR results were compared to measurements from levelling benchmarks for a quality check. Then, the mean displacement maps and time series were analysed for studying the subsidence characteristics in space and time. Thanks to the availability of data covering the whole urban area with unprecedented spatial and temporal density, it was possible to provide a reliable assessment of subsidence causes.

The work in [14] proposes an interesting method for improving analysis and interpretation of InSAR displacement products. It consists of modelling the displacement time series of man-made structures by using rheological parameters, namely, viscosity and elasticity, which are engineering properties of the soil allowing to define quantitatively how materials deform in response to forces. The SBAS algorithm was used for processing high-resolution TerraSAR-X data, and then for monitoring a highway segment built on soft clay and affected by deformation. Results were validated by using independent levelling measurements, demonstrating that the proposed modelling method allows improving the estimation of the nonlinear component of the deformation.

2.5. Target Discrimination and Change Detection

Thanks to the information content available under all-weather conditions and also during night-time, SAR is widely used for target detection, classification and change detection. The following two papers [15,16] concern this kind of applications.

The work in [15] proposes the use of aspect entropy for quantifying the degree of anisotropy in SAR backscattering, and thus improving the detection of anisotropic targets and their classification. The proposed method applies to SAR systems acquiring under different look angles, as circular SAR. First, the use of aspect entropy as a reliably index of anisotropy was verified in simulations. Afterwards, the authors described the algorithm, which consists of computing the aspect entropy at pixel level for distinguishing between isotropic and anisotropic scattering mechanisms, and then at target level for target classification. Moreover, a denoising method for the radar cross section curve was proposed. Finally, the procedure was validated by using X-band data acquired by a circular SAR.

The work in [16] experiments with the use of two-colour multiview (2CMV) advanced geospatial information products for detecting changes between SAR images acquired at different times.

The proposed procedure consists of a pre-processing step for denoising SAR amplitude, followed by the change detection step performed by running algorithms of unsupervised feature learning (K-SVD) and clustering (k-means). Moreover, an optical flow algorithm is used for distinguishing changes related to actual target motions (correct detections) from those related to errors in image co-registration (false positives). The procedure was first introduced, and then tested on datasets coming from both an airborne high-resolution X-band SAR, and the spaceborne medium resolution C-band ERS-2 mission. Results were also compared with other methods, showing improvements in presence of co-registration and perspective errors.

2.6. Sea Applications

The last three papers [17–19] of the special issue present SAR imaging algorithms devoted to two sea applications, namely, ship detection and classification, and oceanic eddy detection and analysis.

The work in [17] presents a new method for refocusing moving ships in SAR images. Ship detection is widely required for both civilian and military surveillance; however, SAR images derived by standard focusing suffer from strong blurring in presence of moving ships. Therefore, a further processing step is needed, which performs reliable motion compensation and refocusing. The paper proposes an algorithm based on an inverse SAR technique able to refocus only the portions of SLC matrix containing moving ships, instead of the whole raw image. Motion phase compensation is performed through an iterative procedure based on a fast minimum entropy method. The procedure was presented and validated by using both airborne data and spaceborne images acquired by TerraSAR-X and Gaofeng-3 missions. Results showed improvements with respect to other refocusing methods.

The work in [18] also deals with ship detection. It proposes a method that performs ship classification by processing SAR images through convolutional deep neural networks (CNN). In this application contest, the SAR datasets available for training the network are often limited, whereas CNNs require thousands of examples to avoid overfitting. In order to overcome this problem, the proposed algorithm starts with an augmentation method for enlarging the training dataset. Then, transfer learning is used for improving the classification accuracy. The procedure was tested by processing TerraSAR-X high-resolution images. Results were compared with outcomes coming from other classification methods, showing improved performances.

Spaceborne SAR observations are very promising for identifying and studying the mechanism governing oceanic eddies. The last paper [19] of this special issue concerns SAR imaging of oceanic eddies generated by shear-waves. The authors developed a method for simulating the current field of an ocean eddy according to the Burgers–Rott vortex model. These simulated ocean eddies were then used to generate SAR images through a simulation tool. The developed procedure is able to perform simulations under different geometric and radiometric SAR configurations, and different wind conditions (speed and direction). Results were validated by using real SAR data provided by ERS-2 and ENVISAT satellite missions. This kind of tool results very useful to understand how radar and wind characteristics impact on the eddy features in SAR images, and thus to support their interpretation and study.

3. Conclusions

In this editorial paper, we reviewed the content of the special issue dedicated to SAR techniques and applications. All the selected nineteen papers proposed interesting advances on different aspects of the SAR processing concerning signal modelling, imaging simulation, image analysis and some applicative examples. In particular, issues were addressed related to multi-angle and wide-angle acquisition modes, which, in the last years, have been becoming more and more common. Several SAR systems were considered, including spaceborne and aerial platforms, light unmanned vehicles, as well as ground-based radars. Specific aspects of SAR signal propagation (e.g., in ionosphere) and back-scattering (e.g., anisotropic backscattering, ocean eddies) were also investigated though modelling and simulation. Moreover, an interesting study proposed modelling pulsar signals by using inverse SAR imaging principles. Applicative examples were also presented based on advanced techniques for image analysis and signal modelling, including, among others, convolutional deep neural networks, unsupervised feature learning and clustering, and deformation modelling through rheological parameters.

In conclusion, the proposed studies represent valid examples of the fertile research ongoing in the field of SAR processing and applications, and demonstrate as SAR imaging still presents large margins for investigations.

Funding: This work was partially supported by the Italian Ministry of University and Research (MIUR) in the framework of "CLOSE to the Earth" project (ARS01_00141), PON Ricerca e innovazione 2014–2020.

Acknowledgments: The guest editor would like to thank the authors' contribution to this Special Issue and all reviewers for providing valuable and constructive recommendations.

Conflicts of Interest: The author declares no conflicts of interest.

References

1. Chang, W.; Tao, H.; Sun, G.; Wang, Y.; Bao, Z. A Novel Multi-Angle SAR Imaging System and Method Based on an Ultrahigh Speed Platform. *Sensors* **2019**, *19*, 1701. [CrossRef] [PubMed]
2. Wei, Z.; Zhang, B.; Wu, Y. Accurate Wide Angle SAR Imaging Based on LS-CS-Residual. *Sensors* **2019**, *19*, 490. [CrossRef] [PubMed]
3. Wang, Y.; Yang, W.; Chen, J.; Kuang, H.; Liu, W.; Li, C. Azimuth Sidelobes Suppression Using Multi-Azimuth Angle Synthetic Aperture Radar Images. *Sensors* **2019**, *19*, 2764. [CrossRef] [PubMed]
4. Zhou, Z.; Li, Y.; Wang, Y.; Li, L.; Zeng, T. Extended Multiple Aperture Mapdrift-Based Doppler Parameter Estimation and Compensation for Very-High-Squint Airborne SAR Imaging. *Sensors* **2019**, *19*, 213. [CrossRef] [PubMed]
5. Chen, H.; Wang, Z.; Liu, J.; Yi, X.; Sun, H.; Mu, H.; Li, M.; Lu, Y. Knowledge-Aided Doppler Beam Sharpening Super-Resolution Imaging by Exploiting the Spatial Continuity Information. *Sensors* **2019**, *19*, 1920. [CrossRef] [PubMed]
6. Guaragnella, C.; D'Orazio, T. A Data-Driven Approach to SAR Data-Focusing. *Sensors* **2019**, *19*, 1649. [CrossRef] [PubMed]
7. Lazarov, A. Pulsar Emissions, Signal Modeling and Passive ISAR Imaging. *Sensors* **2019**, *19*, 3344. [CrossRef] [PubMed]
8. Michelini, A.; Coppi, F.; Bicci, A.; Alli, G. SPARX, a MIMO Array for Ground-Based Radar Interferometry. *Sensors* **2019**, *19*, 252. [CrossRef] [PubMed]
9. Wang, Y.; Song, Y.; Lin, Y.; Li, Y.; Zhang, Y.; Hong, W. Interferometric DEM-Assisted High Precision Imaging Method for ArcSAR. *Sensors* **2019**, *19*, 2921. [CrossRef] [PubMed]
10. Fang, Y.; Wang, B.; Sun, C.; Wang, S.; Hu, J.; Song, Z. Joint Sparsity Constraint Interferometric ISAR Imaging for 3-D Geometry of Near-Field Targets with Sub-Apertures. *Sensors* **2018**, *18*, 3750. [CrossRef] [PubMed]
11. Yu, L.; Zhang, Y.; Zhang, Q.; Ji, Y.; Dong, Z. Performance Analysis of Ionospheric Scintillation Effect on P-Band Sliding Spotlight SAR System. *Sensors* **2019**, *19*, 2161. [CrossRef] [PubMed]
12. Wang, C.; Guo, W.; Zhao, H.; Chen, L.; Wei, Y.; Zhang, Y. Improving the Topside Profile of Ionosonde with TEC Retrieved from Spaceborne Polarimetric SAR. *Sensors* **2019**, *19*, 516. [CrossRef] [PubMed]
13. Zhang, Y.; Liu, Y.; Jin, M.; Jing, Y.; Liu, Y.; Liu, Y.; Sun, W.; Wei, J.; Chen, Y. Monitoring Land Subsidence in Wuhan City (China) using the SBAS-InSAR Method with Radarsat-2 Imagery Data. *Sensors* **2019**, *19*, 743. [CrossRef] [PubMed]
14. Xing, X.; Chen, L.; Yuan, Z.; Shi, Z. An Improved Time-Series Model Considering Rheological Parameters for Surface Deformation Monitoring of Soft Clay Subgrade. *Sensors* **2019**, *19*, 3073. [CrossRef] [PubMed]
15. Teng, F.; Hong, W.; Lin, Y. Aspect Entropy Extraction Using Circular SAR Data and Scattering Anisotropy Analysis. *Sensors* **2019**, *19*, 346. [CrossRef] [PubMed]
16. Kanberoglu, B.; Frakes, D. Improving the Accuracy of Two-Color Multiview (2CMV) Advanced Geospatial Information (AGI) Products Using Unsupervised Feature Learning and Optical Flow. *Sensors* **2019**, *19*, 2605. [CrossRef] [PubMed]

17. Huang, X.; Ji, K.; Leng, X.; Dong, G.; Xing, X. Refocusing Moving Ship Targets in SAR Images Based on Fast Minimum Entropy Phase Compensation. *Sensors* **2019**, *19*, 1154. [CrossRef] [PubMed]
18. Lu, C.; Li, W. Ship Classification in High-Resolution SAR Images via Transfer Learning with Small Training Dataset. *Sensors* **2019**, *19*, 63. [CrossRef] [PubMed]
19. Wang, Y.; Yang, M.; Chong, J. Simulation and Analysis of SAR Images of Oceanic Shear-Wave-Generated Eddies. *Sensors* **2019**, *19*, 1529. [CrossRef] [PubMed]

Article

A Novel Multi-Angle SAR Imaging System and Method Based on an Ultrahigh Speed Platform

Wensheng Chang, Haihong Tao, Guangcai Sun *, Yuqi Wang and Zheng Bao

National Laboratory of Radar Signal Processing, Xidian University, Xi'an 710071, China;
victorycws@163.com (W.C.); hhtao@xidian.edu.cn (H.T.); xdwangyuqi@163.com (Y.W.);
zhbao@mail.xidian.edu.cn (Z.B.)
* Correspondence: gcsun@xidian.edu.cn

Received: 20 January 2019; Accepted: 3 April 2019; Published: 10 April 2019

Abstract: Considering the difficulty of pulse repetition frequency (PRF) design in multi-angle SAR when using ultra-high speed platforms, a multi-angle SAR imaging system in a unified coordinate system is proposed. The digital multi-beamforming is used in the system and multi-angle SAR data can be obtained in one flight. Therefore, the system improves the efficiency of data recording. An improved range migration algorithm (RMA) is used for data processing, and imaging is made in a unified imaging coordinate system. The resolution of different view images is the same, and there is a fixed delay between the images. On this basis, the SAR image fusion is performed after image matching. The results of simulation and measured data confirm the effectiveness of the system and the method.

Keywords: SAR imaging; multi-angle SAR; improved RMA; SAR image fusion

1. Introduction

Synthetic Aperture Radar (SAR) imaging is able to work day and night under all weather conditions [1]. Therefore, it has wide applications in topographic mapping, environmental monitoring and information acquisition, but the electromagnetic scattering property of a complex object varies with incidence angle [2]. In order to meet requirements of omnidirectional observation, it is necessary to implement new research on SAR imaging systems. The multi-angle SAR imaging system has attracted considerable attention [2–8].

The electromagnetic scattering property varies with incidence, so the SAR imaging is greatly affected by the incidence angle [3]. When the target is observed from one angle, since it is occluded, or the scattering coefficient of the angle is low, the complete information of the target cannot be obtained, but multi-angle SAR observes the target from different angle, and it can obtain as much information as possible about the target. The current multi-angle SAR includes spotlight SAR [4], wide azimuth beam SAR [5] and multiple flight paths SAR [6] In the spotlight SAR, the antenna is steered to increase extend the synthetic time and to observe targets from different angles. In this mode, the azimuth bandwidth of the signal may greater than the PRF, which causes spectrum ambiguity and makes signal processing more complicated [4]. The spotlight SAR expands observation angle but reduces the imaging scope. When using an ultra-high speed platform, the azimuth bandwidth of the signal becomes large, and a very large PRF is required. The wide beam angle SAR increases the beam width and obtains echoes of targets from different angles. The wide beam SAR increase the imaging scope, but the two-dimensional spectrum is a sector. This means increased range cell migration (RCM) and severe coupling of range and azimuth [5]. The error of range cell migration compensation in the frequency domain will affect the imaging accuracy. The back projection algorithm can completely compensate the RCM in time domain, but it needs a lot of calculations [7]. Multi angle observation can be realized by multiple flight paths [6], and a large imaging scope can be obtained, but the flight efficiency is low and the cost is high.

For the above problems, a new multi-angle SAR imaging system is proposed in this paper. Digital multi-beamforming is used to obtain SAR data from different angles. The digital T/R modules are divided into three groups and three sets of receiving feeders are used to obtain the multi-beam signals in the time domain. On this basis, an improved RMA in a unified coordinate system is proposed. Modified Stolt interpolation was proposed to correct the distorted spectrum in squinted SAR and improve the efficiency of the spectrum. Then, imaging is performed in a uniform coordinate system. The images of different views only have a translational relationship in azimuth, which can achieve fast matching of multi-angle images. This multi-angle SAR is not required to adjust the antenna direction, nor large beam angle, which reduces the equipment requirements. SAR data from different angles can be obtained in one flight, which reduces experimental costs.

2. A Multi-angle SAR Imaging System and Signal Model on a High-Speed Platform

As shown in Figure 1, the multi-angle SAR imaging system proposed in this paper adopts the digital multi-beamforming, which uses the same antenna to form multiple beams. The squint angles of three beams are different. The date of forward-looking beam, side-looking beam and backward-looking beam are recorded simultaneously, and the data received by each channel are independent from each other.

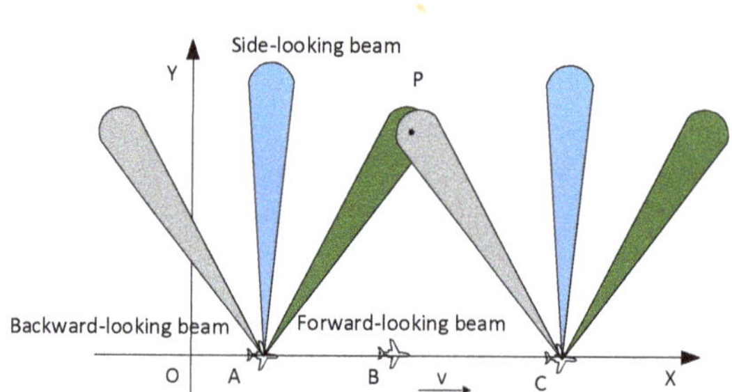

Figure 1. Model of multi-angle SAR imaging system.

2.1. Digital Multi-Beamforming

There are two ways to obtain multi-beam data. One way is using one set of receiving feeders to separate multi-beam data in the Doppler domain, the other way is using multiple sets of feeders to obtain multi-beam data in time domain. When multi-beam data is separated in the Doppler domain, the Doppler bandwidth of the multi-beam signal is large. To prevent spectrum aliasing, a large PRF is required. Reference [8] gives a design to reduce the PRF, and the PRF is the sum of the Doppler bandwidth of multi-beam signals, but the method limits beam pointing. In addition, when using the ultrahigh speed platform, the Doppler bandwidth of the multi-beam signal becomes large. As a result, a large PRF is required. As shown in Figure 2, the multi-angle SAR imaging system proposed in this paper uses three sets of receiving feeders and the digital T/R modules are divided into three groups, each with independent receiving feeder and phase shifter. The multi-beam data are separated in the time domain. Thus, the PRF is equal to the Doppler bandwidth of a single beam. At the same time, there is no restriction on the direction of the beam, and the required beam pointing can be set. When the scattering angles vary from $20°$ to $-20°$, the results of the imaging will be different [3]. In order to get as much information as possible in the scene, the difference in the direction of the three beams is at least $20°$.

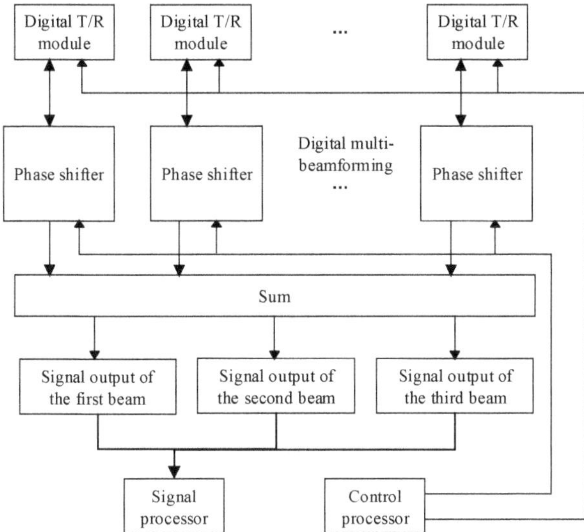

Figure 2. Reception of digital multi-beamforming.

2.2. Signal Model

As shown in Figure 3, the speed of the carrier is v and the wavelength is λ. Taking the three beams as an example, in data collecting, the echo data of target $P_i(X_i, R_s)$ from forward-looking beam is firstly obtained. When the carrier is located at A, the forward-looking beam center points to the target P_i. At this time, the squint angle is θ, and the beam-width of forward-looking beam is θ_{BW1}. When θ_{BW1} is small, the Doppler bandwidth is approximately

$$BW1 = \frac{2v}{\lambda}\left[\sin(\theta + \frac{\theta_{BW1}}{2}) - \sin(\theta - \frac{\theta_{BW1}}{2})\right] \approx \frac{2v}{\lambda}\cos\theta \cdot \theta_{BW1} \tag{1}$$

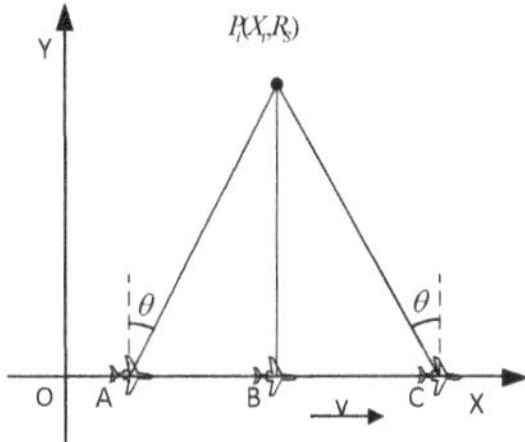

Figure 3. Multi-angle SAR signal model.

Then, the echo data of target P_i from a side-looking beam is obtained. When the aircraft is at B, the center of the side-looking beam points to the target P_i, the beam-width of side-looking beam is θ_{BW2}, and the Doppler bandwidth of side-looking beam is

$$BW2 = \frac{2v}{\lambda}\left[\sin(\frac{\theta_{BW2}}{2}) - \sin(\frac{\theta_{BW2}}{2})\right] \approx \frac{2v}{\lambda}\theta_{BW2} \tag{2}$$

Finally, the echo data of target P_i from backward -looking beam is obtained. When the aircraft is at C, the center of the backward-looking beam points to the target P_i, the backward-looking beam-width is θ_{BW3}, and the Doppler bandwidth of backward-looking beam is

$$BW3 = \frac{2v}{\lambda}\left[\sin(-\theta + \frac{\theta_{BW3}}{2}) - \sin(-\theta - \frac{\theta_{BW3}}{2})\right] \approx \frac{2v}{\lambda}\cos\theta \cdot \theta_{BW3} \tag{3}$$

The beam-width of the phased array antenna is $\theta_{BW} = \theta'_{BW} / \cos\theta$, where θ'_{BW} is the beam-width of the side-looking beam, and substitute it into Equations (1)–(3). It can be obtained that $BW1 = BW2 = BW3$ and then the Doppler bandwidth of three beams is the same. Therefore, the three beam images have the same azimuth resolutions.

The distance between AB is

$$L_{AB} = R_S \tan\theta \tag{4}$$

The distance between the BC and the distance between the AB are the same. The time difference between the forward-looking beam and the side-looking beam is

$$\Delta t = L_{AB}/v \tag{5}$$

The repetition frequency of the transmitted pulse is PRF and the data are received in the strip mode. For a same target, it is located at different azimuth sampling units, and the difference of azimuth sampling units between different beams is

$$\Delta nan = \Delta t \cdot PRF = L_{AB} \cdot PRF/v \tag{6}$$

The data of each beam are processed independently to obtain images. When fusing the images from different beams, it is necessary to ensure the matching of the position of the target. According to (6), the forward-looking image is moved back by $2\Delta nan$ azimuth sampling units, and the side-looking image is moved back by Δnan azimuth sampling units. The images obtained from the three beams are fused in backward-looking image.

3. Multi-Angle SAR Imaging Method Based on a Unified Coordinate

3.1. Problems of Multi-Angle SAR Registrations and Fusion

The fusion objects of current SAR images are various remote sensing images, including the fusion of infrared images and SAR images, the fusion of optical images and SAR images, and the fusion of SAR images. Most current SAR image registrations are performed in the image domain. A heterogeneous-SAR image registration method by normalized cross correlation is proposed in [9]. Frost filtering is implemented on the SAR image and then the Gaussian gradient images of SAR image is used to form two Gabor characteristic matrixes, and then the normalized cross correlation matching is implemented on the two characteristic matrixes to achieve the registration of the image. The edge features of the target and the feature points can be extracted from the SAR image [10–13], and the SAR image registration is performed by the information. A new method is proposed in [10] to detect stable features by intersecting Coherent Scatters. The stable features are used to achieve the coarse registration and the Powell algorithm is used for precise registration. A new method using boundary features of images to achieve SAR image registration is proposed in [12]. A globalized boundary detection algorithm is used for feature extraction and the coherence point drift algorithm is used to match the boundaries. A method for non-homologous SAR image registration is proposed in [14]. The method utilizes multi-look technology to multi resolution images, then uses the coherent phase to deal with multi resolution images, respectively, getting the registration point and achieving image registration.

3.2. Improved RMA Algorithm

RMA [15–21] achieves SAR imaging in the wave number domain. In spite of the squint angle value, it can perfectly focus the whole scene without using any approximate conditions. The range cell migration compensation, secondary range compression and azimuth compression are achieved by Stolt interpolation [15]. In principle, it is the optimum algorithm for SAR imaging [16]. However, the Stolt interpolation needs huge computation. Since the multi-angle SAR imaging system adopts the method of multi-beamforming, the beam squint angle is more than $20°$, and the RMA algorithm can process the data of the squint SAR. It can focus the whole scene by interpolation. For $20°$ squint, the general interpolation formula has low spectrum utilization (Section 3.3. for details.), and the improved RMA algorithm is used to improve the spectrum utilization.

To illustrate the derivation process of the echo signal, the imaging relationship at point A in Figure 3 is drawn separately, as shown in Figure 4. $M_i(X_i, R_b)$ is one point in the scenario and R_b is the closest distance from the point target to the aircraft trajectory. The distance from the aircraft to the point target can be expressed as:

$$R(t_m) = \sqrt{R_b^2 + (vt_m - X_i)^2} \tag{7}$$

where t_m is the azimuth slow-time. Assuming that the transmitted signal is a LFM signal, the received baseband echo signal is [17]:

$$s_0(t_r, t_m) = A_0 w_r\left(t_r - 2\frac{R(t_m)}{c}\right) w_a(t_m - t_{mc}) \exp\left\{-j4\pi f_c \frac{R(t_m)}{c}\right\} \exp\left\{j\pi\gamma\left(t_r - \frac{2R(t_m)}{c}\right)^2\right\} \tag{8}$$

where A_0 is the amplitude of the signal, $w_r(\cdot)$ is the range envelope, t_r is the range fast time, $w_a(\cdot)$ is the azimuth envelope, t_{mc} is the center of synthetic aperture time, f_c is the center frequency of the transmitted signal, and γ is the chirp rate of the chirp signal. A two-dimensional FFT is applied to the echo signal, and the two-dimensional frequency domain expression can be obtained:

$$S_{2DF}(f_r, f_a) = A_1 W_r(f_r) W_a(f_a - f_{ac}) \exp\{j\theta_{2DF}(f_r, f_a)\} \tag{9}$$

where

$$\theta_{2DF}(f_r, f_a) = -\frac{4\pi R_b(f_c + f_r)}{c}\sqrt{1 - \frac{(cf_a)^2}{4(f_c + f_r)^2 v^2}} - \frac{\pi f_r^2}{\gamma} - 2\pi f_a \frac{X_i}{v} \tag{10}$$

$$W_a(f_a) = w_a\left(\frac{-cR_0 f_a}{2(f_c+f_r)v^2\sqrt{1-\frac{c^2 f_a^2}{4v^2(f_c+f_r)^2}}}\right)$$ is the envelope of the azimuth spectrum, and $W_r(f_r) = w_r\left(\frac{f_r}{\gamma}\right)$ is the envelope of the range spectrum.

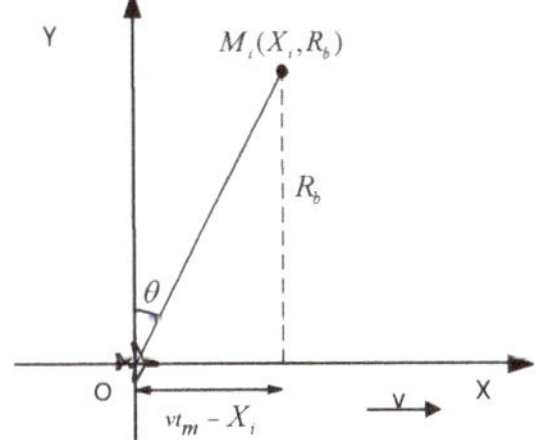

Figure 4. Single beam signal model.

Pulse compression needs to eliminate the quadratic term of f_r in Equation (10), and a matched filter can be constructed in frequency:

$$H_r(f_r) = \exp(j\frac{\pi f_r^2}{\gamma})$$ (11)

After multiplication of Equation (9) and Equation (11) to complete pulse compression, the phase after pulse compression is:

$$\theta(f_r, f_a) = -\frac{4\pi R_b(f_c + f_r)}{c}\sqrt{1 - \frac{(cf_a)^2}{4(f_c + f_r)^2 v^2}} - 2\pi f_a \frac{X_i}{v}$$ (12)

Let $k_r = \frac{4\pi(f_r + f_c)}{c}$, $k_x = \frac{2\pi f_a}{v}$, Formula (12) is rewritten as:

$$\theta(k_r, k_x) = -R_b\sqrt{k_r^2 - k_x^2} - k_x X_i$$ (13)

Since the signal processing of the RMA algorithm is performed in the two-dimensional frequency domain, and R_b represents the time domain, the phase compensation cannot handle the change along the range direction. At this time, a reference range is first selected, and the phase at the reference distance is compensated. Generally, the reference range is set at the center of the scenario. At this time, the matched function of consistent compression is:

$$H_{COMP}(k_r, k_x) = jR_S\sqrt{k_r^2 - k_x^2} + jk_x R_s \tan\theta$$ (14)

where, R_s is the closest distance from the center point of the scenario to the aircraft trajectory. After consistent compression, the point at the center of the scenario is completely focused, and the residual phase at the other range is:

$$\theta_{RFM}(k_r, k_x) = -(R_b - R_S)\sqrt{k_r^2 - k_x^2} - k_x(X_i - R_s \tan\theta)$$ (15)

The RMA algorithm performs range cell migration compensation, secondary range compression and azimuth compression by interpolation $k_y = \sqrt{k_r^2 - k_x^2}$ [1,17]. For 20° squint, the two-dimensional spectrum is distorted, and it needs to extract a rectangular aperture of data adequately in such 2-D support [18], as shown in Figure 5; it needs to discard part of the spectrum due to the squint angle, which reduces the energy of targets after imaging. For each of the determined k_r, the variation of k_y with k_x is shown by the arc in Figure 5.

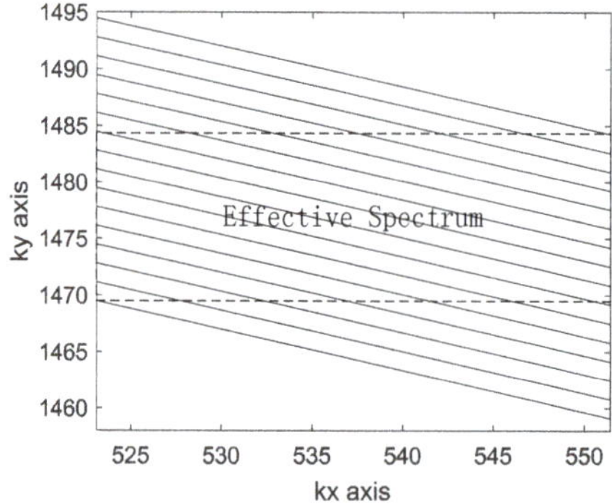

Figure 5. Spectrum of traditional interpolation.

The improved interpolation uses the tangent of each arc instead of the traditional k_y, and corrects the distorted spectrum. Therefore, the method can effectively improve the utilization of the spectrum in squint SAR. The improved interpolation is:

$$k_y = \sqrt{k_r^2 - k_x^2} - \left[\sqrt{k_{rc}^2 - k_{xc}^2} - \frac{k_{xc}}{\sqrt{k_{rc}^2 - k_{xc}^2}} (k_x - k_{xc}) \right] \tag{16}$$

where $k_{rc} = \frac{4\pi f_c}{c}$, $k_{xc} = \frac{2\pi f_{ac}}{c}$ and $f_{ac} = \frac{2v\sin\theta}{\lambda}$ is Doppler center. The residual phase after interpolation is

$$\theta_{STOLT}(k_y, k_x) = -(R_b - R_S) \left[k_y + \left(\sqrt{k_{rc}^2 - k_{xc}^2} - \frac{k_{xc}}{\sqrt{k_{rc}^2 - k_{xc}^2}} (k_x - k_{xc}) \right) \right] - k_x(X_i - R_s \tan\theta) \tag{17}$$

Since the interpolation introduces a linear phase that varies with range, it is necessary to compensate for the introduced linear phase in the Range–Doppler domain. After IFFT along the range, the following is obtained:

$$\begin{aligned} s_{RD}(Y, k_x) &= A_2 \mathrm{sin}\,\mathrm{c}\left(\frac{B_{ky}}{2\pi}Y\right) W_a\left(\frac{vk_x}{2\pi}\right) \exp\{-jk_x(X_i - R_s \tan\theta)\} \\ &\cdot \exp\left\{ -j(R_b - R_S)\left(\sqrt{k_{rc}^2 - k_{xc}^2} - \frac{k_{xc}}{\sqrt{k_{rc}^2 - k_{xc}^2}} (k_x - k_{xc}) \right) \right\} \end{aligned} \tag{18}$$

where B_{ky} is the bandwidth of k_y, $Y = R_b - R_S$, and the second phase in Equation (18) needs to be compensated along azimuth, and the azimuth compensation function is:

$$H_{AZIMUTH}(R_b, k_x) = \exp\left\{ j(R_b - R_S)\left(\sqrt{k_{rc}^2 - k_{xc}^2} - \frac{k_{xc}}{\sqrt{k_{rc}^2 - k_{xc}^2}} (k_x - k_{xc}) \right) \right\} \tag{19}$$

Multiply Equation (18) and Equation (19) and perform IFFT along azimuth to obtain:

$$s_{RX}(Y, X_i) = A_3 \mathrm{sin}\,\mathrm{c}\left(\frac{B_{ky}}{2\pi}Y\right) \mathrm{sin}\,\mathrm{c}\left(\frac{B_{kx}}{2\pi}(X_i - R_s \tan\theta)\right) \tag{20}$$

where B_{kx} is the bandwidth of k_x. The point target $M_i(X_i, R_b)$ is focused at $(X_i - R_s \tan\theta, R_b - R_S)$ in the time domain.

The algorithm processing flow is shown in Figure 6:

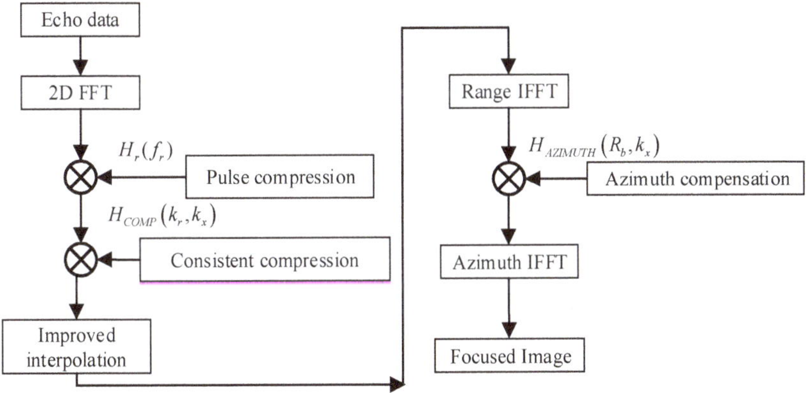

Figure 6. Multi-angle SAR algorithm flow chart.

3.3. Application and Consideration

For many artificial objects, the SAR image is greatly affected by the azimuth angle. Through multi-angle image fusion, we can obtain more detailed information about the target, which improves the target detection and recognition ability of SAR images. SAR image matching fusion can be achieved quickly by imaging in a unified coordinate system. In order to maximize the use of the spectrum, it is necessary to make the interpolated spectrum as rectangular as possible. After interpolation, the original coordinate axis k_r is replaced by the new coordinate axis k_y. Figure 7 is the bandwidth of the spectrum after interpolation, and the effective spectrum is the part within the dashed box. Figure 5 shows the spectrum of the traditional interpolation method, and the spectrum is approximated as a character quadrilateral. The effective spectrum is significantly smaller than the spectrum obtained by the method of this paper.

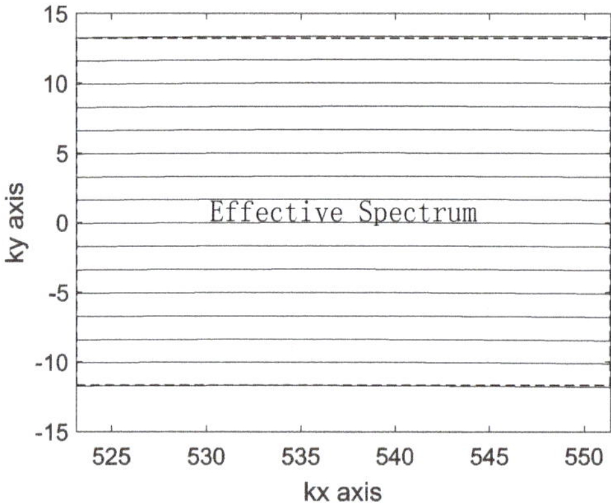

Figure 7. Spectrum of proposed interpolation.

For accurate matching, images need to have a uniform scale. The bandwidth of k_y represents the range bandwidth after interpolation. In order to have the same range resolution of multiple-angle images in the time domain, the bandwidth of k_y is required to be the same. The traditional method is to intercept the largest rectangle in the interpolated spectrum, as shown in Figure 5. The traditional method is used to determine k_{y1}. Let $k_{rL} = \min(kr)$, $k_{rH} = \max(kr)$, $k_{yL} = \max\left(\sqrt{k_{rL}^2 - k_x^2}\right)$, $k_{yH} = \min\left(\sqrt{k_{rH}^2 - k_x^2}\right)$ and N is the number of range sampling units, and then $k_{y1}(i) = k_{yL} + (i-1)(k_{yH} - k_{yL})/N$, $i = 1, 2, \cdots, N$. The result of $k_{y1} - (k_r - k_{rc})$ is shown in Figure 8. The slope greater than 0 represents the bandwidth of k_{y1} is greater than the bandwidth of k_r. In the images of different views, the bandwidth of k_{y1} is inconsistent and there is a slight change in the range resolution of the time domain. In general SAR imaging applications, it can be ignored. However, the change in the range resolution will lead to inaccurate matching and affect the quality of the fusion in image matching. In the proposed method, in order to unify the bandwidth of k_y in different view images, the center value of k_y is first determined, and then the bandwidth of k_y is determined according to the bandwidth of the k_r. The proposed method is used to determine k_{y2} and $k_{y2}(i) = k_{rc} + (i - N/2)(k_{rH} - k_{rL})/N, i = 1, 2, \cdots, N$. As shown in Figure 9, the bandwidth of k_{y2} is smaller than the bandwidth of k_{y1}, which means that the proposed method discards a small portion of the spectrum. The result of $k_{y2} - (k_r - k_{rc})$ is shown in Figure 8. The slope is 0, which represents the bandwidth of k_{y2} is the same as the bandwidth of k_r. In the images of different views, the bandwidth of k_{y2} is consistent, and different images have the same range resolution in the time domain. The advantage of the scale uniformity is obvious in image matching.

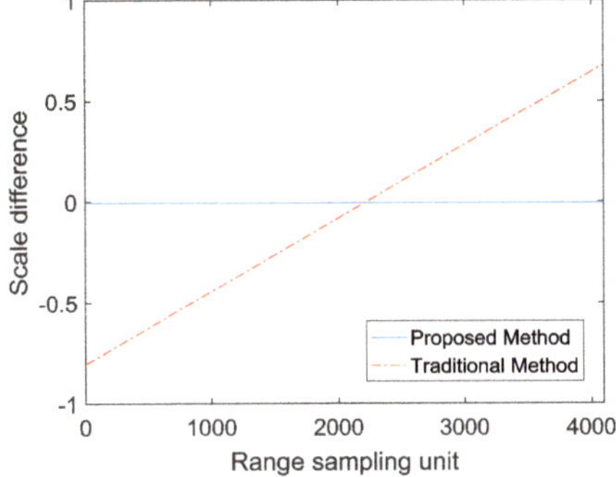

Figure 8. Scale difference of two methods.

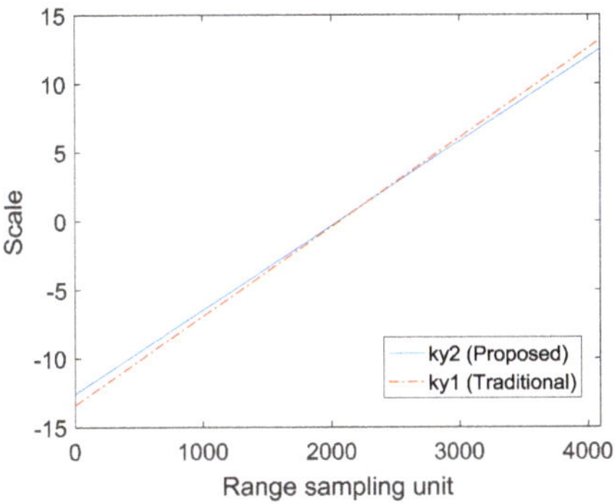

Figure 9. Comparison of two sampling methods.

It is difficult for the aircraft to maintain an ideal state due to factors such as airflow during flight. Therefore, motion compensation is required in data processing. In the mode of multi-flight acquisition for imaging data, the motion compensation of each SAR image is different because of the different motion errors of each flight, which brings difficulties to image matching. When multi-angle SAR data are taken by this system, data of each angle have the same motion error, and the data of multiple angles can be compensated by the motion error of a single view, simplifying the compensation process. It is also possible to jointly perform motion compensation through multiple viewing angles to improve compensation accuracy.

4. Experimental Simulation, Measured Data

4.1. Experimental Simulation

In order to verify the validity of the algorithm, the simulation data are used for explanation. The simulation resolution is 0.3 m × 0.3 m, the wavelength is 3 cm, the center frequency is 10 GHz, the signal bandwidth is 500 MHz, the range sampling rate is 600 MHz, the pulse width is 3.5 μs, the speed of aircraft is 100 m/s, the antenna aperture is 0.6 m, and the pulse repetition frequency is 450 Hz. The closest distance from the center of the scenario to the aircraft route is 30 km. Three beams are used with a beam spacing of 20° and the beam width is 2.86°. There are five points in the scene, and the simulation scenario layout is shown in Figure 10. The center point target is located at (0, 0), and the remaining four points are located at (±30, ±30).

The squint angle of the forward-looking beam is 20°, and the scenario image processed by the above imaging algorithm is shown in Figure 11a, the position of the center point target is (1025, 2050), and the positions of the other four points are (1025 ± 135, 2050 ± 120). The azimuth sampling rate is

1.35 times of the azimuth bandwidth, so the distance between the center point target and the rest of the point target in the azimuth direction is $135/1.35 \times 0.3 = 30$ m, which is consistent with the scenario layout; The range sampling rate is 1.2 times of the bandwidth, and the distance between the center point target and the rest of the point target in the range direction is $120/1.2 \times 0.3 = 30$ m, which is consistent with the scenario layout. Figure 11b is a result of interpolation of the point (1025 − 135, 2050 − 120) in Figure 11a. It can be seen that the point target in forward-looking beam is well focused. The profiles of range and azimuth-spread function of the target are presented in Figure 11c,d. The peak sidelobe ratio (PLSR) along the range direction shown in Figure 11b is −13.2242. The integral sidelobe ratio (ISLR) along the range direction is −9.8468. The PLSR along the azimuth direction shown in Figure 11b is −13.2611. The ISLR along the azimuth direction is −9.8963.

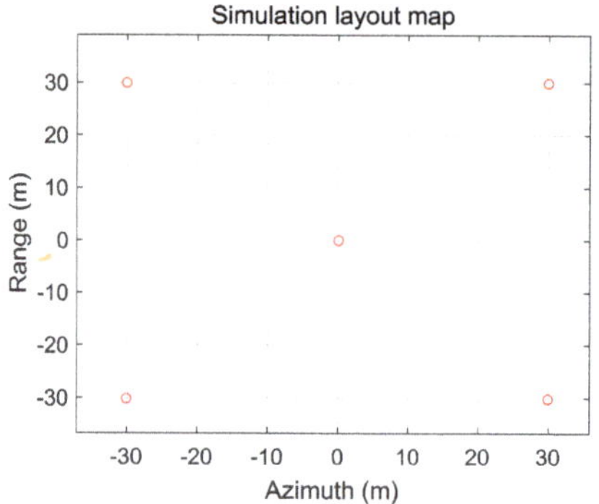

Figure 10. Simulation layout map.

The scenario image of the side-looking beam processed by the above imaging algorithm is shown in Figure 12a, the position of the center point target is (1025, 2050), and the positions of the remaining four points are (1025 ± 135, 2050 ± 120). The distance between the center point target and the rest of the point target in the azimuth direction is $135/1.35 \times 0.3 = 30$ m, and the distance between the center point target and the rest of the point target in the range direction is $120/1.2 \times 0.3 = 30$ m, which is consistent with the scenario layout. Figure 12b is a result of interpolation of the point (1025 − 135, 2050 − 120) in Figure 12a. It can be seen that the point target in side-looking beam is well focused. The profiles of range and azimuth-spread function of the target are presented in Figure 12c,d. The PLSR along the range direction shown in Figure 12b is −13.2231. The ISLR along the range direction is −9.8464. The PLSR along the azimuth direction shown in Figure 12b is −13.2602. The ISLR along the azimuth direction is −9.8962.

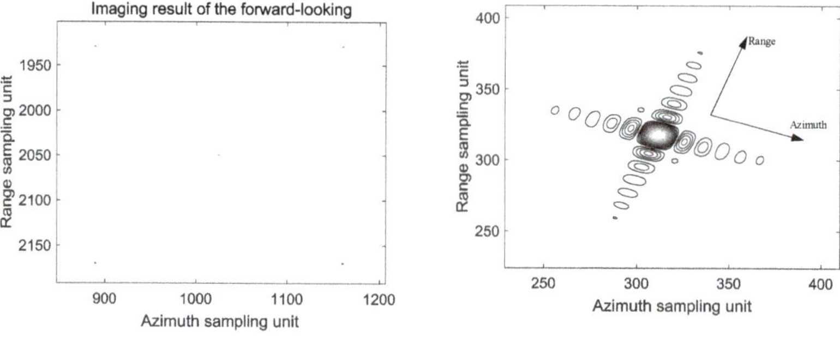

(**a**) Imaging result of targets (**b**) Interpolation of a single point target

Figure 11. *Cont.*

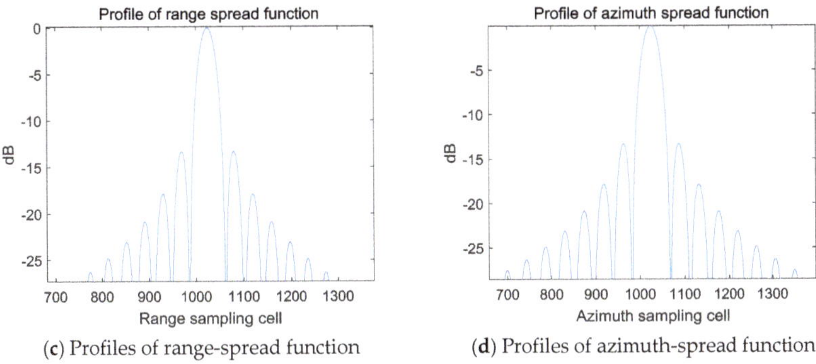

(**c**) Profiles of range-spread function   (**d**) Profiles of azimuth-spread function

Figure 11. Imaging result of the forward-looking beam.

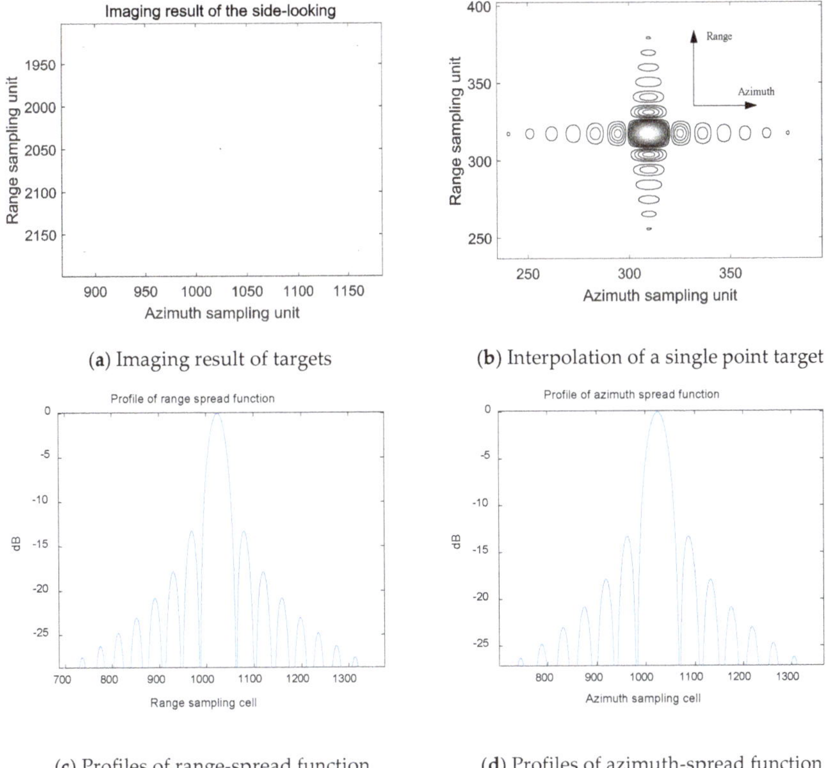

(**a**) Imaging result of targets   (**b**) Interpolation of a single point target

(**c**) Profiles of range-spread function   (**d**) Profiles of azimuth-spread function

Figure 12. Imaging result of the side-looking beam.

The scenario image processed by the above imaging algorithm for the backward-looking beam is shown in Figure 13a. the position of the center point target is (1025, 2050), and the positions of the remaining four points are (1025 ± 135, 2050 ± 120). The distance between the center point target and the rest of the point target in the azimuth direction is $135/1.35 \times 0.3 = 30$ m, and the distance between the center point target and the rest of the point target in the range direction is $120/1.2 \times 0.3 = 30$ m, which is consistent with the scenario layout. Figure 13b is a result of interpolation of the point $(1025 - 135, 2050 - 120)$ in Figure 12a. It can be seen that the point target in backward-looking beam is well focused. The profiles of range and azimuth-spread function of the target are presented in Figure 13c,d. The PLSR along the range direction shown in Figure 13b is −13.2299. The ISLR along the range direction is −9.8458. The PLSR along the azimuth direction shown in Figure 13b is −13.2536. The ISLR along the azimuth direction is −9.8859.

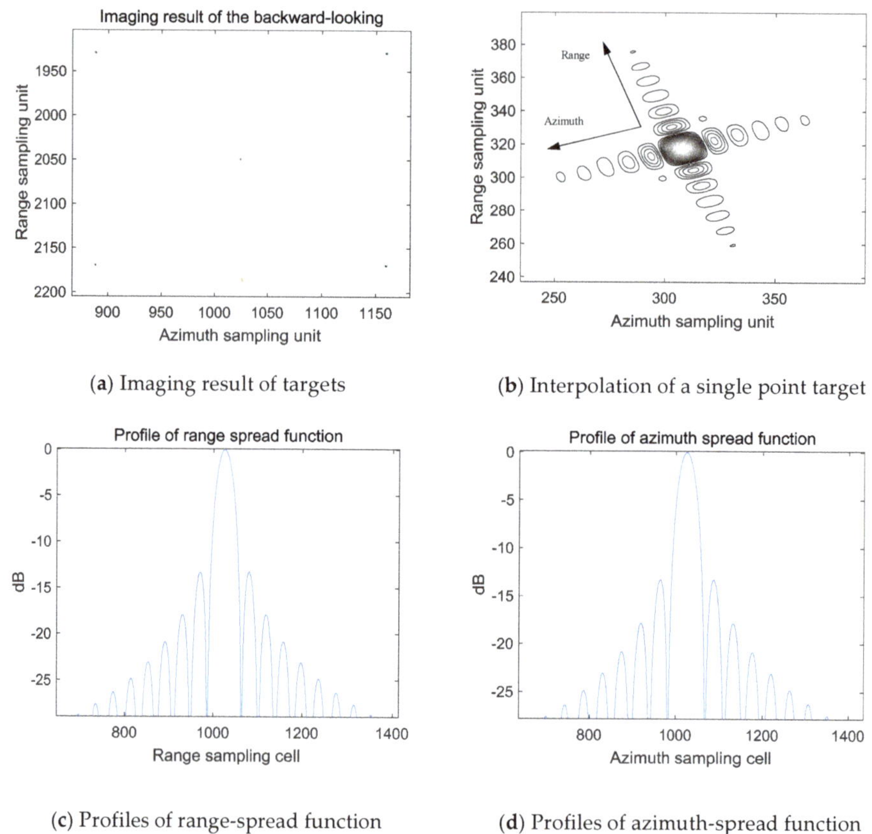

(**a**) Imaging result of targets

(**b**) Interpolation of a single point target

(**c**) Profiles of range-spread function

(**d**) Profiles of azimuth-spread function

Figure 13. Imaging result of the backward-looking beam.

In each beam, the absolute position and relative position of the point target are not changed and matched with the ground point, so the imaging of the same point target on the ground by different beams only has the difference in azimuth time. According to the time difference represented by Formula (5) or the azimuth point difference represented by Formula (6), the image fusion of multi-view SAR can be completed by delaying the forward-looking beam imaging result by $2\Delta t$ and delaying the side-looking beam imaging result by Δt, and then superimposing them into the backward-looking beam imaging result.

According to the time difference represented by the Formula (5), or the difference in the number of azimuth points represented by the Formula (6), the front-view beam imaging result is delayed by $2\Delta t$, the due side-view imaging result is delayed by Δt, and then image fusion of multi-angle SAR is completed after superimposition on back-view beam. The result of the fusion is shown in Figure 14a. Figure 14b is a result of interpolation of the point in Figure 14a. It can be seen from Figure 14a,b that the imaging and fusion of images can be completed in a uniform coordinate system within a viewing angle range of $-20°$ to $20°$. The Range PSLR is -8.31 and the azimuth PSLR is -6.37.

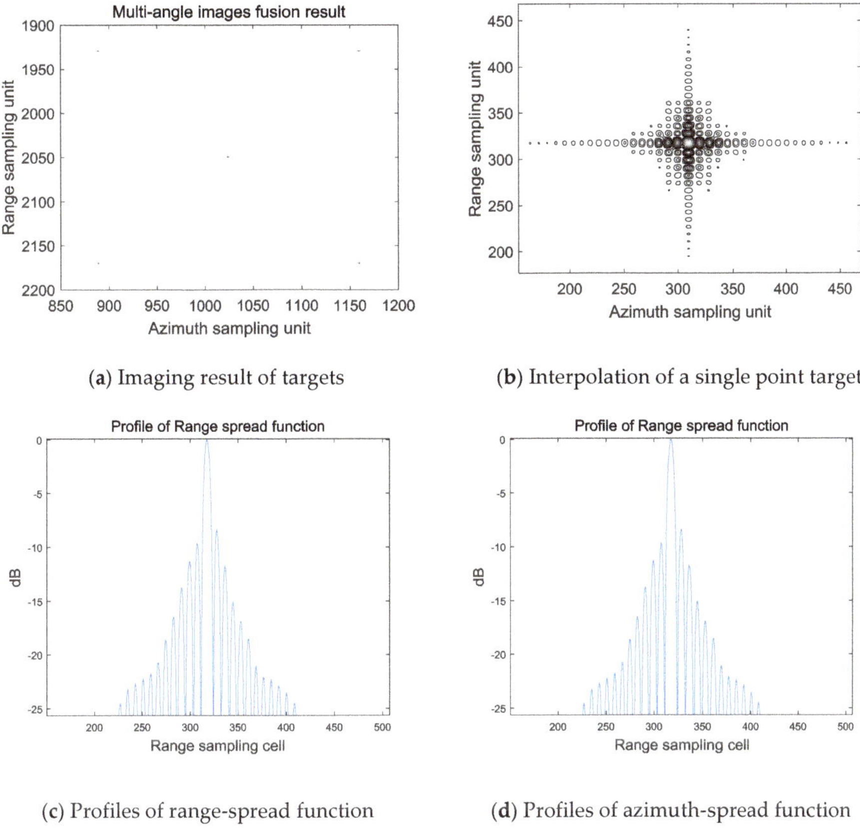

(**a**) Imaging result of targets

(**b**) Interpolation of a single point target

(**c**) Profiles of range-spread function

(**d**) Profiles of azimuth-spread function

Figure 14. Result after image fusion.

4.2. Measured Data

In order to validate the effectiveness of the proposed algorithm, the large-angle spotlight SAR measured data are processed using the proposed algorithm. The large-angle spotlight SAR measured data contains information about multiple perspectives of the target. After dividing the data into two parts according to the two viewpoints of forward-looking and backward-looking, the fusion image of multi-angle SAR is obtained by using the algorithm proposed in this paper. The parameters of the system are shown in Table 1.

Table 1. Parameters of the system.

Parameters	Value
Velocity	102 m/s
Frequency band	9.6 GHz
Bandwidth	600 MHz
PRF	312 Hz
Angle range	$-30°$–$30°$
Reference range	34 km

Figure 15a,b are images of six vehicles with forward-looking and backward-looking views. It can be seen that the target information obtained is not complete because of sheltering of the single-view target. Figure 15c is obtained through the image fusion of two angles of view. From which, complete geometric features of the target can be seen clearly. The information entropy is used to evaluate the effects of image fusion. Information entropy in Figure 15a,b are 6.1819 and 6.1046, and information entropy in Figure 15c is 6.6635. The information entropy in the image increases after fusion. This means the fused image contains more information about the targets.

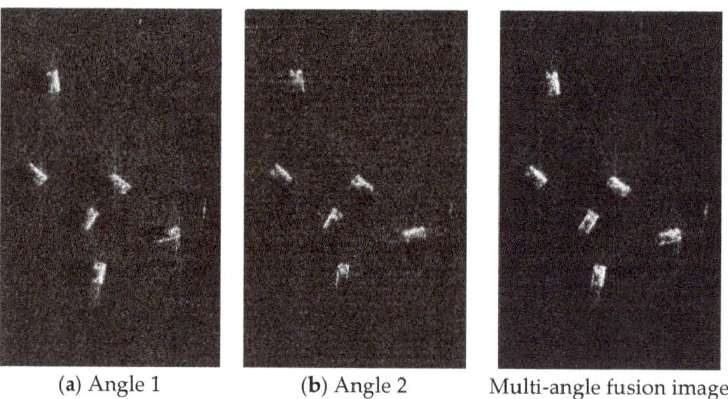

<table>
<tr><td>(a) Angle 1</td><td>(b) Angle 2</td><td>Multi-angle fusion image</td></tr>
</table>

Figure 15. Image fusion results of proposed method.

Figure 16 shows the image fusion results of Range–Doppler algorithm. Different from the proposed method, the result of angle 2 has a deformation, and the image registration needs to be performed after the image is corrected. When the images are fully registered, the images can be well fused as shown in Figure 16c. When the image is not fully registered, part of the target information will be lost as shown in Figure 16d.

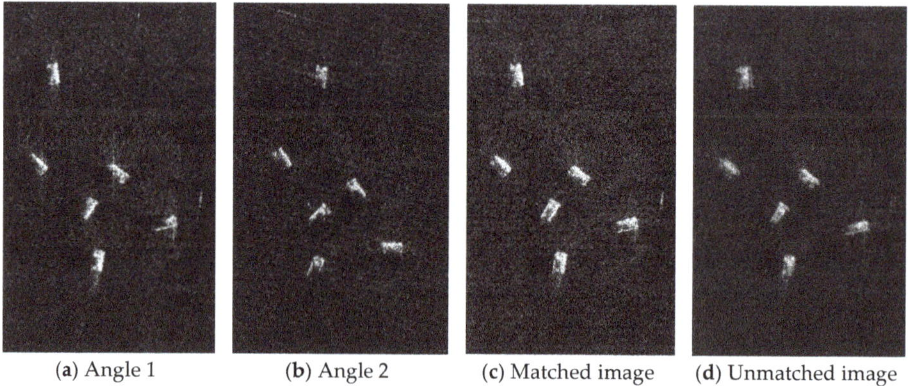

<table>
<tr><td>(a) Angle 1</td><td>(b) Angle 2</td><td>(c) Matched image</td><td>(d) Unmatched image</td></tr>
</table>

Figure 16. Image fusion results of Range–Doppler algorithm.

Compared with the traditional method, the method proposed in this paper does not require additional image registration, which simplifies the process of image fusion. It also avoids the effects of mismatch between images. However, RMA requires interpolation and is computationally intensive, which can cause real-time processing difficulties.

Figure 17 shows a multi-angle fusion result of two views in a large scenario area, in which red represent the components of forward-looking view and green represent the components of backward-looking view. The background is spotlight SAR image, and the segmented portion is forward-looking and backward-looking images.

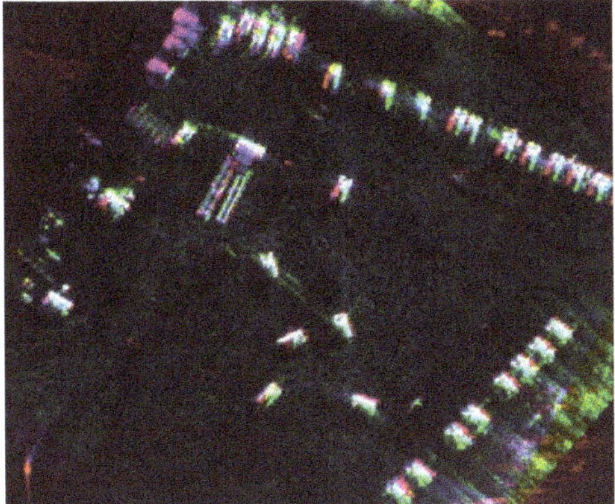

Figure 17. A multi-angle fusion result of a large scenario area.

Figure 18 is an optical picture and enlarged fusion result of the transport vehicle of Figure 18. Different colors represent components of different views. In a single view image, the occluded portion can be supplemented by another view. The geometric characteristics of the transport vehicle are relatively complete, which is beneficial to the identification of the target.

(**a**) Optical image   (**b**) Fusion image by multiple angles

Figure 18. Optical image and fusion image of a vehicle.

5. Conclusions

A multi-angle SAR imaging system is proposed in this paper using multi-beamforming. When using an ultrahigh speed platform, the main issue is an increase in Doppler bandwidth in the signal. As a result, it is difficult to separate signals of multiple beams in the frequency domain. Therefore, this paper separates the multi-beam signal in the time domain using three groups of feeders. In order to achieve accurate matching of multi-view SAR images, an improved RMA in a unified coordinate is proposed. SAR data from different view angles is imaged in a uniform coordinate system. The resolution between images is the same, and the image is not deformed and scaled. There is only a time delay relationship between images of different view angle. Therefore, image fusion does not require additional registration. Multi-angle images can be quickly and accurately fused.

Author Contributions: Conceptualization, W.C. and H.T.; Formal analysis, W.C., H.T. and G.S.; Methodology, W.C. and G.S.; Resources, Z.B.; Software, Y.W.

Funding: This research was funded by the National Key R&D Program of China, grant number 2017YFC1405600.

Conflicts of Interest: The authors declare no conflict of interest.

References

1. Bao, Z.; Xing, M.; Wang, T. *Radar Imaging Technology*; Publishing House of Electronic Industy: Beijing, China, 2005.
2. Sun, G.; Xing, M.; Xia, X.; Wu, Y.; Bao, Z. Beam Steering SAR Data Processing by a Generalized PFA. *IEEE Trans. Geosci. Remote Sens.* **2013**, *51*, 4366–4377. [CrossRef]
3. Tan, L.; Ma, Z.; Zhong, X. Preliminary result of high resolution multi-aspect SAR imaging experiment. In Proceedings of the 2016 CIE International Conference on Radar (RADAR), Guangzhou, China, 10–13 October 2016; pp. 1–3.
4. Sun, G.; Xing, M.; Xia, X.; Yang, J.; Wu, Y.; Bao, Z. A Unified Focusing Algorithm for Several Modes of SAR Based on FrFT. *IEEE Trans. Geosci. Remote Sens.* **2013**, *51*, 3139–3155. [CrossRef]
5. Yang, J.; Sun, G.; Chen, J.; Wu, Y.; Xing, M. A subaperture imaging scheme for wide azimuth beam airborne SAR based on modified RMA with motion compensation. In Proceedings of the 2014 IEEE Geoscience and Remote Sensing Symposium, Quebec City, QC, Canada, 13–18 July 2014; pp. 608–611.
6. Arii, M.; Nishimura, T.; Komatsu, T.; Yamada, H.; Kobayashi, T.; Kojima, S.; Umehara, T. Theoretical characterization of multi incidence angle and fully Polarimetric SAR data from rice paddies. In Proceedings of the 2016 IEEE International Geoscience and Remote Sensing Symposium (IGARSS), Beijing, China, 10–15 July 2016; pp. 5670–5673.
7. Vu, V.T.; Sjögren, T.K.; Pettersson, M.I. Fast factorized backprojection algorithm for UWB SAR image reconstruction. In Proceedings of the 2011 IEEE International Geoscience and Remote Sensing Symposium, Vancouver, BC, Canada, 24–29 July 2011; pp. 4237–4240.
8. Wen, X.; Kuang, G.; Hu, J.; Zhang, J. Simultaneous multi-beam SAR mode using phased array radar. *Sci. Sinica (Inf.)* **2015**, *45*, 354–371. [CrossRef]
9. Jiang, Y. Optical/SAR image registration based on cross-correlation with multi-scale and multi-direction Gabor characteristic matrixes. In Proceedings of the IET International Radar Conference 2013, Xi'an, China, 14–16 April 2013; pp. 1–4.
10. Yu, H.; Liu, Y.; Li, L.; Yang, W.; Liao, M. Stable feature point extraction for accurate multi-temporal SAR image registration. In Proceedings of the 2017 IEEE International Geoscience and Remote Sensing Symposium (IGARSS), Fort Worth, TX, USA, 23–28 July 2017; pp. 181–184.
11. Tao, Z.; Dongyang, A.; Cheng, H. Image registration of SAR and optical image based on feature points. In Proceedings of the IET International Radar Conference 2013, Xi'an, China, 14–16 April 2013; pp. 1–5.
12. Shen, D.; Zhang, J.; Yang, J.; Feng, D.; Li, J. SAR and optical image registration based on edge features. In Proceedings of the 2017 4th International Conference on Systems and Informatics (ICSAI), Hangzhou, China, 11–13 November 2017; pp. 1272–1276.
13. Zhu, S.; Ran, D. Multi-angle SAR image fusion algorithm based on visibility classification of non-layover region targets. In Proceedings of the 2017 International Conference on Security, Pattern Analysis, and Cybernetics (SPAC), Shenzhen, China, 15–17 December 2017; pp. 642–647.
14. Zhang, Y.; Wu, T. An SAR image registration method based on pyramid model. In Proceedings of the 2016 CIE International Conference on Radar (RADAR), Guangzhou, China, 10–13 October 2016; pp. 1–5.
15. Cumming, I.G.; Neo, Y.L.; Wong, F.H. Interpretations of the omega-K algorithm and comparisons with other algorithms. In Proceedings of the 2003 IEEE International Geoscience and Remote Sensing Symposium. Proceedings, Toulouse, France, 21–25 July 2003; Volume 3, pp. 1455–1458.

16. Wu, Y.; Song, H.; Shang, X.; Zheng, J. Improved RMA based on Nonuniform Fast Fourier Transforms (NUFFT's). In Proceedings of the 2008 9th International Conference on Signal Processing, Beijing, China, 26–29 October 2008; pp. 2489–2492.

17. Cumming, I.G.; Wong, F.H. *Digital Signal Processing of Synthetic Aperture Radar Data*; Artech House, Inc.: Norwood, UK, 2005.

18. Xiong, T.; Xing, M.; Xia, X.; Bao, Z. New Applications of Omega-K Algorithm for SAR Data Processing Using Effective Wavelength at High Squint. *IEEE Trans. Geosci. Remote Sens.* **2013**, *51*, 3156–3169. [CrossRef]

19. Li, Z.; Xing, M.; Liang, Y.; Gao, Y.; Chen, J.; Huai, Y.; Zeng, L.; Sun, G.-C.; Bao, Z. A Frequency-Domain Imaging Algorithm for Highly Squinted SAR Mounted on Maneuvering Platforms With Nonlinear Trajectory. *IEEE Trans. Geosci. Remote. Sens.* **2016**, *54*, 4023–4038. [CrossRef]

20. Tang, S.; Zhang, L.; Guo, P.; Zhao, Y. An Omega-K Algorithm for Highly Squinted Missile-Borne SAR with Constant Acceleration. *IEEE Geosci. Remote Sens. Lett.* **2014**, *11*, 1569–1573. [CrossRef]

21. Li, Z.; Liang, Y.; Xing, M.; Huai, Y.; Gao, Y.; Zeng, L.; Bao, Z. An Improved range Model and Omega-K-Based Imaging Algorithm for High-Squint SAR with Curved Trajectory and Constant Acceleration. *IEEE Geosci. Remote Sens. Lett.* **2016**, *13*, 656–660. [CrossRef]

Article

Accurate Wide Angle SAR Imaging Based on LS-CS-Residual

Zhonghao Wei [1,2,3,*], Bingchen Zhang [1,2,3] and Yirong Wu [2,3]

1　Key Laboratory of Technology in Geo-Spatial Information Processing and Application System, Institute of Electronics, Chinese Academy of Sciences, Beijing 100190, China; bczhang@mail.ie.ac.cn (B.Z.); wyr@mail.ie.ac.cn (Y.W.)

2　School of Electronic, Electrical and Communication Engineering, University of Chinese Academy of Sciences, Beijing 101408, China

3　Institute of Electronics, Chinese Academy of Sciences, Beijing 100190, China

*　Correspondence: weizhh@163.com; Tel.: +86-10-5888-7208

Received: 4 December 2018; Accepted: 22 January 2019; Published: 25 January 2019

Abstract: Wide angle synthetic aperture radar (WASAR) receives data from a large angle, which causes the problem of aspect dependent scattering. L_1 regularization is a common compressed sensing (CS) model. The L_1 regularization based WASAR imaging method divides the whole aperture into subapertures and reconstructs the subaperture images individually. However, the aspect dependent scattering recovery of it is not accurate. The subaperture images of WASAR can be regarded as the SAR video. The support set among the different frames of SAR video are highly overlapped. Least squares on compressed sensing residuals (LS-CS-Residuals) can reconstruct the time sequences of sparse signals which change slowly with time. This is to replace CS on the observation by CS on the least squares (LS) residual computed using the prior estimate of the support. In this paper, we introduce LS-CS-Residual into WASAR imaging. In the iteration of LS-CS-Residual, the azimuth-range decoupled operators are used to avoid the huge memory cost. Real data processing results show that LS-CS-Residual can estimate the aspect dependent scatterings of the targets more accurately than CS based methods.

Keywords: wide angle SAR; compressed sensing; LS-CS-Residual; aspect dependent

1. Introduction

Wide angle synthetic aperture radar (WASAR) receives echo data from a large angle. Advances in synthetic aperture radar (SAR) technology have enabled coherent sensing of WASAR. Circular SAR (CSAR) is a specific case of WASAR whose track is circular. With the increase of the synthetic angle, because of the reflector geometry, shadowing, and coherent scintillation, the problem of aspect dependent scattering [1,2] arises. Traditional imaging methods are based on the isotropic assumption. It means that the scattering is constant in the synthetic aperture angle, which is not valid in WASAR.

To accommodate the aspect dependent scattering, there are mainly two approaches, the subaperture approach and full aperture approach [2]. The subaperture approach [1] divides the whole aperture into the subapertures and assumes that the scatterings are constant in the subaperture. Then, the narrow angle imaging methods such as matched filtering [3] and an L_1 regularization based SAR imaging method [4,5] can be adopted for the subaperture imaging. For the full aperture approach, they can be divided into two kinds. The first one assumes that the scattering during one subaperture is isotropic and reconstructs an imaging model with all subapertures included [6–8]. The subaperture images are recovered jointly. The other one is the parametric method [9,10]. It assumes that the scene includes some scattering targets and that their scatterings follow some functions. The scattering functions of the targets are fitting with the whole aperture data included.

In the past decade, compressed sensing (CS) [11,12] has drawn much attention in sparse signal processing, which provides reconstruction guarantees for sparse solutions to linear inverse problems. It is shown that, when the scene is sparse and the measurement matrix satisfies the Restricted Isometry Property (RIP), the signal can be recovered with down-sampled data by solving an L_1 minimization problem [12]. An L_1 minimization problem is also known as Basis Pursuit. The theory of Lagrange multipliers indicates that we can solve an unconstrained problem that will yield the same solution, provided that the Lagrange multipler is selected correctly. The unconstrained problem is known as LASSO or L_1 regularization [13]. For the subaperture reconstruction based on L_1 regularization, it mainly has two drawbacks. Firstly, as a common reconstructed model in compressed sensing (CS), L_1 regularization is a biased estimator [14], which means that the amplitude of the targets would be underestimated. Secondly, the support set of the targets is not accurately estimated with the data of one subaperture and there are some missed detections. For the first drawback, it can be solved via debiased-CS proposed in [15,16]. Debiased-CS is a two-step method, which firstly reconstructs the signal with CS and calculates the least squares (LS) estimates on the support set of the signal. For the second drawback, since the support sets of different aspect subaperture images are highly overlapped across the whole aperture [9], this information can be adopted in the subaperture image reconstruction to avoid it.

The idea of CS is to compressively sense signals that are sparse in some known domains and then use sparse recovery techniques to recover them. Considering the dynamic CS problem, i.e., the problem of recovering a time sequence of sparse signals, CS recovers each sparse signal in the sequence independently without using any information from other frames. Least squares on compressed sensing residual (LS-CS-Residual) [17] is to replace CS on the observation by CS on the least squares (LS) residual computed using the prior estimate of the support. It is suitable for dynamic CS problem. It is proved that LS-CS-Residual can recover the signal better than CS [17].

The subaperture images of WASAR can be regarded as the SAR video. Every frame is the subaperture image indexed by the aspect angle. In WASAR, the backscattering from a complex target at high frequencies can be approximately modeled as a discrete set of the scattering centers [9].The scattering center can be described by the aspect-dependent amplitude and position [9]. The supports of the scattering centers overlap across the whole aperture. This information can be adopted in WASAR imaging.

In this paper, we propose a novel subaperture imaging method based on LS-CS-Residual. The proposed method firstly implements Backprojection (BP) on all of the aperture data. Then, the coarse support set is estimated from the BP image. For every subaperture of WASAR, the least squares estimate on the support set is calculated. Then, the observation residual is calculated. With the residual data, we can solve the residual observation model with L_1 regularization. The accurate supports of subaperture images are estimated from the L_1 regularization image. Finally, the LS estimate on the accurate supports is calculated. Since the structure information and LS estimate on the support set are adopted in the proposed method, it can recover the aspect dependent scattering more accurately than CS and debiased-CS.

In the iteration of LS-CS-Residual, there are matrix-vector products. For large scale scenes, the storage of measurement matrix can cost a huge amount of memory. A common strategy is to adopt the azimuth-range decouple operators in the algorithm. In this paper, the BP based azimuth-range decouple operators are adopted. The memory cost is reduced from $\mathcal{O}(n^2)$ to $\mathcal{O}(n)$.

This paper is organized as follows. In Section 2, we describe the WASAR subaperture imaging model based on CS. Section 3 introduces the WASAR imaging method based on LS-CS-Residual. Section 4 presents the experimental results. The conclusions are presented in Section 5.

2. WASAR Subaperture Imaging Based on Compressed Sensing

WASAR receives data from a large angle. The configuration of WASAR is depicted in Figure 1. The whole aperture can be divided into subapertures. For the data collected from a little subaperture, its scattering can be regarded as constant. Then, the phase history of the i-th ($i = 1, 2 \cdots I$) subaperture is formulated as

$$r_i(f_p, \theta_q) = \sum_{m=1}^{M} \sum_{n=1}^{N} s_i(x_m, y_n) \cdot \exp\{-j\frac{4\pi f_p}{c} \cdot (x_m \cos(\theta_q) + y_n \sin(\theta_q))\} + z_i, \tag{1}$$

where r_i is the phase history data of the i-th subaperture, and m and n are the pixel indexes along x and y. M and N are the pixel numbers along x and y, s is the scattering reflectivity of the i-th subaperture which is located at (x_m, y_n), f_p ($p = 1, 2 \cdots P$) is the frequency, c is the light velocity, θ_q ($q = 1, 2 \cdots Q$) is the aspect angle, and z_i is additive noise.

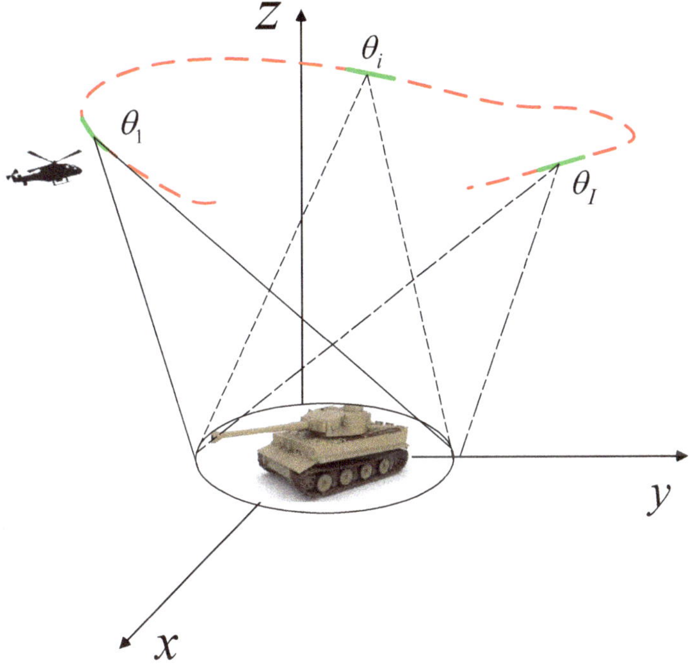

Figure 1. The configuration of WASAR.

We vectorize Equation (1) and express it in a compact form

$$r_i = \mathbf{\Phi}_i \cdot s_i + z_i, \tag{2}$$

where r_i is the history data of i-th subaperture, s_i is the backscattering of i-th subaperture, and z_i is the noise, the measurement matrix $\mathbf{\Phi}_i$ is shown as

$$\mathbf{\Phi}_i = \begin{bmatrix} \phi_i(1,1) & \phi_i(1,2) & \cdots & \phi_i(1,MN) \\ \vdots & \vdots & \ddots & \vdots \\ \phi_i(pq,1) & \phi_i(pq,2) & \cdots & \phi_i(pq,MN) \\ \vdots & \vdots & \ddots & \vdots \\ \phi_i(PQ,1) & \phi_i(PQ,2) & \cdots & \phi_i(PQ,MN) \end{bmatrix}, \tag{3}$$

where $\phi_i(pq, mn) = \exp\{-j\frac{4\pi f_p}{c}(x_m \cos(\theta_q) + y_n \sin(\theta_q))\}$.

The subaperture imaging methods for WASAR imaging assume that the scattering of the targets are not relevant to the aspect angle in a narrow angle. Then, a traditional imaging method can be implemented in subaperture image focusing.

CS has been introduced into SAR imaging [4]. When the scene is sparse and the measurement matrix satisfies the restricted isometry property (RIP) condition, Equation (2) can be solved via L_1 regularization [18]

$$\min_{s_i} \| r_i - \mathbf{\Phi}_i \cdot s_i \|_2^2 + \lambda \| s_i \|_1.$$ (4)

where λ is the regularization parameter.

For Equation (4), the optimality condition is

$$2\mathbf{\Phi}_i^H (\mathbf{\Phi}_i s_i - r_i) + \lambda p = 0,$$ (5)

where $(\cdot)^H$ is the conjugate transpose and

$$p = \partial \| s_i \|_1.$$ (6)

Suppose the oracle support of the s_i is T, and then the solution of (4) is

$$(s_i)_T = \mathbf{\Phi}_{iT}^\dagger r_i - \lambda \left(\mathbf{\Phi}_{iT}^H \mathbf{\Phi}_{iT} \right)^{-1} \text{sign}(s_{iT}), (s_i)_{T^C} = 0,$$ (7)

where $\mathbf{\Phi}_{iT}^\dagger = \left(\mathbf{\Phi}_{iT}^H \mathbf{\Phi}_{iT} \right)^{-1} \mathbf{\Phi}_{iT}^H$, T^C denotes the complement of T. $\text{sign}(\cdot)$ is the signal function formulated as

$$\text{sign}(s_i) = \frac{s_i}{|s_i|}.$$ (8)

If the oracle support is accurate, then the first term of (7) is the exact estimate of the signal. The second term of (7) is the bias that is brought by the regularized term of (4).

In [14], it is shown that L_1 can reconstruct the targets with the underestimated amplitudes. Some missed detections are also introduced in the results of L_1 regularization. In addition, with less azimuth measurements, the resolution of the subaperture is reduced. The underestimation can be reduced via LS on the support. The missed detections can be reduced when more information is adopted. In the next section, we will propose a novel method for WASAR subaperture imaging.

3. WASAR Imaging Based on LS-CS-Residual

L_1 regularization would cause the errors of the amplitude and support set estimation in WASAR imaging. In this section, we propose a novel WASAR imaging based on LS-CS-Residual.

In WASAR, the subaperture images can be regarded as the video indexed on subaperture [2], which is a map of reflectivity as a function of viewing angle. The reflectivities of the targets can be described via their amplitudes and positions. Although they vary with aspect angle, the positions are highly overlapped. Some methods for dynamic scene such as video signal processing and dynamic MRI imaging can be introduced to WASAR imaging.

LS-CS-Residual [17] has been proposed for dynamic CS problems, such as dynamic magnetic resonance imaging (MRI). The idea of LS-CS-Residual is to perform CS not from the observations, but from the least squares residual computed using the previous support estimation. It is shown that it needs fewer samples and the bounded reconstruction error is smaller than the traditional CS. In the model of LS-CS-Residual, the information between different frames are used and there is also a debiasing step in the final to reduce the bias caused by L_1 regularization. It can reconstruct the results much more accurately than CS.

Since the support sets of between different WASAR subaperture images are highly overlapped, which means that WASAR imaging can be regarded as a dynamic problem. Thus, LS-CS-Residual is suitable for WASAR subaperture imaging. In the frame of LS-CS-Residual, the LS estimate is included, which means that the underestimation of L_1 regularization is avoided. In addition, the support information of different subapertures will be used, which will make the results more accurate.

LS-CS-Residual mainly has three steps: initial LS estimation, implementing CS on the residual (CS-Residual) and final LS estimate.

Initial LS Estimate

For Equation (2), if the support set of s_{θ_i} is known, we could simply compute the LS estimate on the support while setting all other values to zeros. The previous support can be estimated from the prior information. Suppose the estimated support is T, to compute and initial LS estimate

$$(s_{i,init})_T = (\Phi_{iT})^\dagger r_i, (s_{i,init})_{T^c} = 0. \tag{9}$$

Then, the LS residual is calculated as

$$r_{i,res} = r_i - \Phi_i s_{i,init}. \tag{10}$$

In WASAR, the scattering of the targets is aspect dependent. However, the support sets of the subaperture images are highly overlapped, which means that a fairly accurate support T can be estimated from the data. T is estimated via

$$T = supp(s_0 : |s_0| > \alpha), \tag{11}$$

which is the support of the elements whose amplitudes are larger than α.

In [17], the threshold α is determined by the $b\%$-Energy support, which means that T contains at least $b\%$ of the signal energy. In WASAR imaging, we set $b\% = 90\%$.

Notice that the LS residual, $r_{i,res}$, can be rewritten as

$$r_{i,res} = \Phi_i \beta_i, \beta_i = s_i - s_{i,init}. \tag{12}$$

CS-Residual

In this step, CS is implemented on the LS residual, i.e., solve (12) with CS in the following model

$$\min_{\beta_i} \|r_{i,res} - \Phi_i \beta_i\|_2^2 + \lambda \|\beta_i\|_1. \tag{13}$$

Iterative shrinkage thresholding algorithm (ISTA) [19] can be used to solve (13). In the iteration of ISTA, there are no matrix inversions involved. ISTA is preferred for its simplicity in implementation for distributed or parallel recovery due to nature of the involved matrix-vector multiplications [20,21]. The iteration is formulated as

$$\hat{\beta}_i^t = \beta_i^t + \mu \left[\Phi_i^H (r_i - \Phi_i \beta_i^t) \right], \tag{14}$$

$$\beta_i^{t+1} = f_{\lambda\mu}\left(\hat{\beta}_i^t\right) = \begin{cases} \text{sgn}(\hat{\beta}_i^t)(|\hat{\beta}_i^t| - \lambda\mu), & \text{if } |\hat{\beta}_i^t| > \lambda\mu \\ 0, & \text{otherwise,} \end{cases} \tag{15}$$

where $\mu \in (0, \|A\|_2^{-2})$ is the step size controlling the convergence, λ is the regularization parameter, and f is the iterative function of ISTA. In the iteration, the value of λ is

$$\lambda = |\hat{\beta}_i^t|_{K+1}/\mu, \tag{16}$$

where $|\hat{\beta}_i^t|_{K+1}$ is the $(K+1)$-th largest element of $\hat{\beta}_i^t$ and $K = \|\hat{\beta}_i^t\|_0$.

The final estimation is

$$\hat{s}_i = \beta_i + s_{i,init}. \tag{17}$$

Final LS Estimation

It is shown that β_i is obtained after L_1 regularization, and the estimation will be biased towards zeros. Thus, a debiasing step is needed

$$T' = \text{supp}(\hat{s}_i), \tag{18}$$

$$s_{iT} = (\mathbf{\Phi}_{iT'})^{\dagger} r_i, \, s_{iT'^c} = 0. \tag{19}$$

After the construction of the subaperture images, the generalized likelihood ratio test (GLRT) [1] can be implemented for the final composite image. GLRT is defined as

$$s(x,y) = \max_i |s^i(x,y)|, \tag{20}$$

where $s^i(x,y)$ is the scattering at pixel (x,y) of i-th subaperture.

The algorithm is summarized in Algorithm 1.

Algorithm 1 LS-CS-Residual based WASAR imaging.

Input: Subaperture echo data r_i $(i = 1 : I)$ and measurement matrix $\mathbf{\Phi}_i$, iterative parameter μ, maxmum iterative step $T_{\max}$.

1: Implement BP on the whole aperture data, estimate T from the BP image. $s_i = 0$ $(i = 1 : I)$, $t = 0$.
2: **for** $i = 1 : I$ **do**
3:    $(s_{i,init})_T = (\mathbf{\Phi}_{iT})^{\dagger} r_i$, $(x_{i,init})_{T^c} = 0$
4:    $r_{i,res} = r_i - \mathbf{\Phi}_i s_{i,init}$
5:    $\beta_i^0 = 0$
6:    $Res = \varepsilon + 1$
7:    **while** $t < T_{\max}$ and $Res > \varepsilon$ **do**
8:       $\hat{\beta}_i^t = \beta_i^t + \mu \left[\mathbf{\Phi}_i^H (r_i - \mathbf{\Phi}_i \beta_i^t) \right]$
9:       $\lambda = |\beta_i^t|_{K+1} / \mu$
10:      $\beta_i^{t+1} = f_{\lambda\mu}(\beta_i^t + \mu \left[\mathbf{\Phi}_i^H \left(r_i - \mathbf{\Phi} s_{i,res}^t \right) \right])$
11:      $Res = \|\beta_i^{t+1} - \beta_i^t\|_2$
12:      $t = t + 1$
13:    **end while**
14:    $\hat{s}_i = \beta_i^{t+1} + x_{i,init}$
15:    $T' = \text{supp}(\hat{s}_i)$
16:    $s_i = \mathbf{\Phi}_{iT'}^{\dagger} y_i$
17: **end for**
18: $s(x,y) = \max_i |s^i(x,y)|$
Output: s(x,y)

In WASAR imaging, it will cost huge amount of memory to store the measurement matrix. The azimuth-range decouple operators can be used to reduce the memory cost [5]. In this paper, we take BP based operators to substitute the measurement matrix and its conjugate transpose in real WASAR imaging. With the BP based operators, the memory cost can be reduced dramatically. If we reconstruct the measurement matrix, the memory cost is $\mathcal{O}(PQMN)$. With the BP based operators, the memory cost is $\mathcal{O}(MN)$. It means that, with the measurement matrix, the memory cost is reduced from $\mathcal{O}(n^2)$ to $\mathcal{O}(n)$.

BP mainly includes two operations: range Fourier transform and azimuth coherent addition. The imaging and raw data generation procedures are formulated as

$$\mathcal{I}\{\cdot\} \cong \mathcal{R}^{-1}\{\mathcal{H}\{\mathcal{F}^{-1}\{\mathcal{R}\{\cdot\}\}\}\}, \tag{21}$$

$$\mathcal{G}\{\cdot\} \cong \mathcal{R}\{\mathcal{F}\{\mathcal{H}^{-1}\{\mathcal{R}^{-1}\{\cdot\}\}\}\}, \tag{22}$$

where $\mathcal{F}$ and $\mathcal{F}^{-1}$ are the the Fourier transform pairs, $\mathcal{H}$ is azimuth coherent addition operator and $(\mathcal{H})^{-1}$ is its inverse operation, $\mathcal{R}$ reshapes the vector into matrix and $\mathcal{R}^{-1}$ reshapes the matrix into a vector.

4. Real Data Experiment

In this section, we will use two datasets to show the effectiveness of the proposed method.

4.1. Turntable Data

The turntable data collected by the Institute of Electronics, Chinese Academy of Sciences will be used to show the effectiveness of the proposed method. The real data of a metal tank model are measured in an anechoic chamber on a turntable, which is in uniform circular motion. The radar is a stepped frequency type and has a center frequency of 15 GHz and bandwidth 6 GHz. The turntable plane and its center are set as the imaging ground plane and the coordinate origin, respectively. The radius of equivalent circular passes is 8.54 m. The 360° whole aperture is divided into 36 subapertures. The pixel size of the SAR image is 0.25 cm × 0.38 cm.

We reconstruct the subaperture images with BP, CS, debiased-CS and LS-CS-Residual. The results of the three methods are shown in Figure 2. Figure 2a is the result of BP, which is used as the referenced image. Compared with the result of the three method, the result of LS-CS-Residual remains less artifacts as shown in the white circle.

To compare the performance of the three methods in the reconstruction of aspect dependent scattering, we select an aspect dependent scattering target P and plot its aspect dependent amplitude curve Figure 3. The result of BP is used as the reference. In Figure 3, we select Area 1 to show the performance of LS-CS-Residual to reduce the underestimation. Area 2 in Figure 3 is selected to show the performance of LS-CS-Residual to reduce the missed detections. As shown in Figure 3 Area 1, CS underestimates the amplitude of the target. The results of Debiased-CS and LS-CS-Residual highly overlap the result of BP. So Debiased-CS and LS-CS-Residual avoid the underestimation caused by CS. In Figure 3 Area 2, CS and debiased-CS fail in reconstructing the weak scattering. Since the support information of the other subapertures is adopted in LS-CS-Residual, the support of weak scattering target is preserved in the subapertures. So with the prior support information and the final debiasing step, LS-CS-Residual can reconstruct the aspect dependent scatterings of the targets more accurately than CS and debiased-CS.

The time taken by the three algorithms is given in the Table 1. Debiased-CS takes more time because of the debias step compared with CS. Compared with the former two algorithms, LS-CS-Residual takes similar amount time.

Table 1. Time taken (in minutes) by the three algorithms.

CS	Debiased-CS	LS-CS-Residual
22.31	23.43	20.08

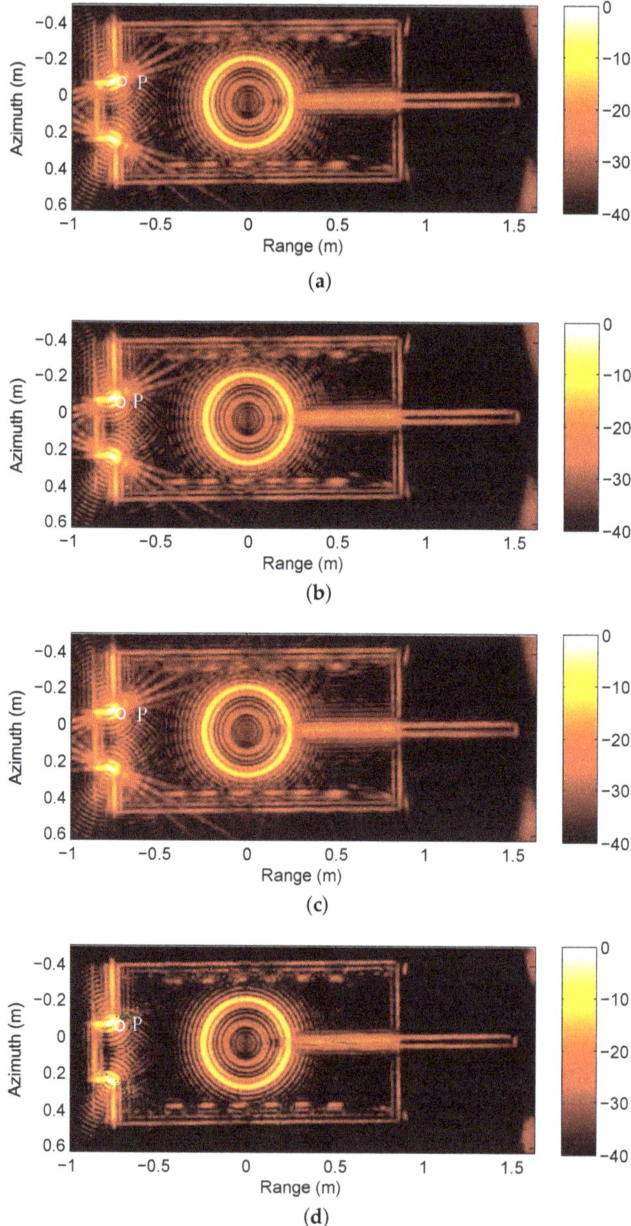

Figure 2. Results of the four methods. (**a**) GLRT result of BP; (**b**) GLRT result of CS; (**c**) GLRT result of debiased-CS; (**d**) GLRT result of LS-CS-Residual.

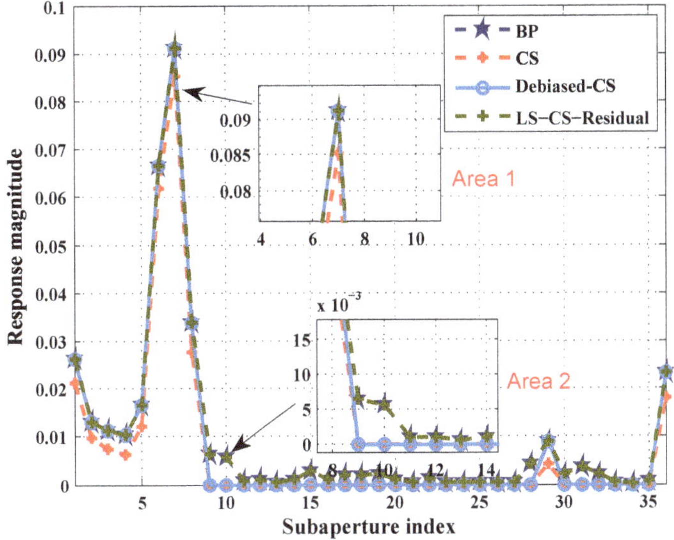

Figure 3. Reconstructed aspect dependent scattering of pixel P via the three methods.

4.2. Gotcha Volumetric SAR DATA

Gotcha volumetric SAR dataset [22] is X-band circular SAR data that consists of CSAR phase history data collected at the X-band with a 640-MHz bandwidth. The spotlighted scene is a parking lot in an urban environment. The scene consists of numerous civilian vehicles and reflectors.

In this experiment, the HH polarization data are used. The whole aperture of $360°$ is divided into 180 subapertures. Every subaperture is $4°$. The apertures overlap every $2°$. The pixel size is 0.2 m $\times$ 0.2 m. The area of reflectors is chosen. We reconstruct the scene with BP, CS, debiased-CS and LS-CS-Residual. The GLRT results of the four methods are shown in Figure 4.

To evaluate the aspect dependent reconstruction performance of different methods, we select an aspect dependent scattering target and plot its reconstructed aspect dependent scattering. The selected target is a reflector that distributes across several pixels. We reconstruct the subaperture images with BP, CS, debiased-CS and LS-CS-Residual. To compare the aspect dependent scattering reconstruction performance of the three methods, we add the intensities of these pixels together and plot the results in Figure 5. Figure 5 is the main valid scattering area of the reflector. BP result is used as the reference. It is shown that the result of LS-CS-Residual is highly overlapped with the results of BP. The intensities of CS and debiased-CS are less than BP and LS-CS-Residual. The underestimation of debiased-CS is mainly caused by the missed detections. Since there are some missed detections in the result of debiased-CS, the intensities of the debiased-CS is less than BP. CS causes bias and missed detections because of the regularizer term. With the bias and missed detections, the peak of CS is lower than the other three methods.

The time taken by the three algorithms is given in Table 2. Table 2 shows that the three algorithms take similar amounts of time.

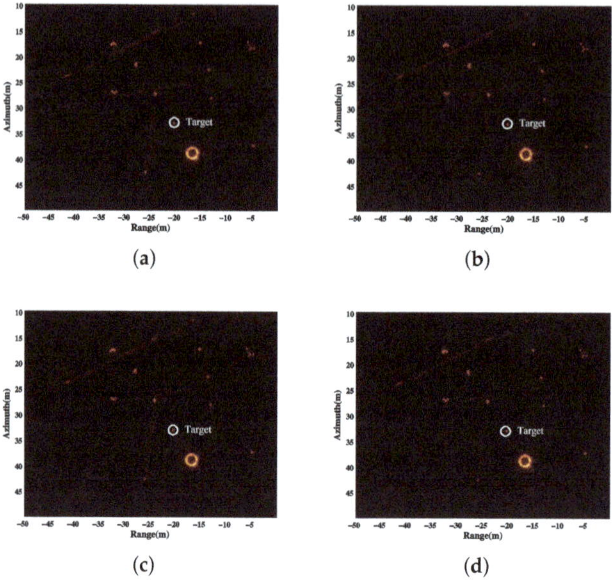

(a)   (b)

(c)   (d)

Figure 4. Results of the four methods. (**a**) GLRT result of BP; (**b**) GLRT result of CS; (**c**) GLRT result of debiased-CS; (**d**) GLRT result of LS-CS-Residual.

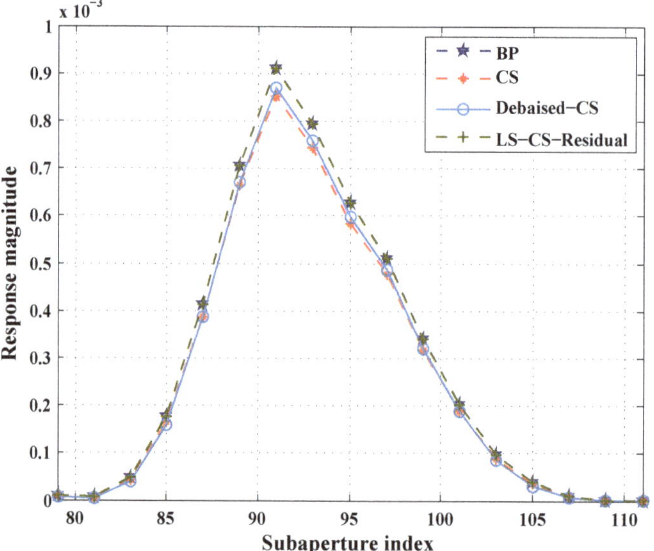

Figure 5. Reconstructed aspect dependent scattering of target via the three methods.

Table 2. Time taken (in minutes) by the three algorithms.

CS	Debiased-CS	LS-CS-Residual
154.94	162.69	139.45

5. Conclusions

In this paper, an accurate WASAR imaging algorithm based on LS-CS-Residual is proposed. The traditional regularized subaperture imaging method based on L_1 regularization introduces the bias and missed detections which will cause inaccurate aspect dependent scattering estimates. To overcome this problem, LS-CS-Residual has been introduced into WASAR imaging. LS-CS-Residual mainly has three steps: initial LS estimate, CS on the residual and final LS estimate. The LS estimate step can be used to reduce the bias. The missed detections are reduced because the support information is adopted in the process of the LS-CS-Residual. The proposed method accommodates aspect dependent scattering better than CS and debiased-CS. The experiment results demonstrate its validity.

Author Contributions: Conceptualization, Z.W.; Data Curation, B.Z.; Methodology, Z.W.; Project Administration, B.Z.; Supervision, B.Z. and Y.W.; Validation, Z.W.; Writing—Original Draft, Z.W.; Writing—Review and Editing, B.Z.

Acknowledgments: This work was supported by the National Natural Science Foundation of China under Grant No. 61571419.

Conflicts of Interest: The authors declare no conflict of interest.

Abbreviations

The following abbreviations are used in this manuscript:

SAR	Synthetic aperture radar
WASAR	Wide angle synthetic aperture radar
CSAR	Circular synthetic aperture radar
LS	Least squares
CS	Compressed sensing
LS-CS-Residual	Least squares on compressed sensing residual
BP	Backprojection
CS-Residual	CS on the residual
ISTA	Iterative shrinkage thresholding algorithm
GLRT	Generalized likelihood ratio test

References

1. Moses, R.; Lee P.; Mujdat, C. Wide-angle SAR imaging. *Proc. SPIE* **2004**, *5427*, 164–175.
2. Ash, J.; Emre, E.; Lee, P.; Edmund, Z. Wide-Angle Synthetic Aperture Radar Imaging: Models and algorithms for anisotropic scattering. *IEEE Signal Process. Mag.* **2014**, *31*, 16–26. [CrossRef]
3. Soumekh, M. *Synthetic Aperture Radar Signal Processing*; Wiley: New York, NY, USA, 1999.
4. Baraniuk, R.; Steeghs, P. Compressive Radar Imaging. In Proceedings of the IEEE Radar Conference, Boston, MA, USA, 17–20 April 2007; pp. 17–20.
5. Zhang B.; Hong W.; Wu, Y. Sparse microwave imaging: Principles and applications. *Sci. China Inf. Sci.* **2012**, *55*, 1722–1754. [CrossRef]
6. Stojanovic, I.; Mujdat, C.; William, K. Joint space aspect reconstruction of wide-angle SAR exploiting sparsity. *Proc. SPIE* **2008**, *6970*. [CrossRef]
7. Cong, X.; Guan, G.; Yong, J.; Li, X.; Wen, G.; Huang, X.; Qun, W. A novel adaptive wide-angle SAR imaging algorithm based on Boltzmann machine model. *Muldimens. Syst. Signal Process.* **2016**, *29*, 119–135. [CrossRef]
8. Wei, Z.; Jiang, C.; Zhang, B.; Bi, H.; Hong, W.; Wu, Y. WASAR imaging with backprojection based group complex approximate message passing. *Electron. Lett.* **2016**, *52*, 1950–1952. [CrossRef]

9. Trintinalia, L.C.; Rajan, B.; Ling, H. Scattering center parameterization of wide-angle backscattered data using adaptive Gaussian representation. *IEEE Trans. Antennas Propag.* **1997**, *45*, 1664–1668. [CrossRef]

10. Gerry, M.J.; Potter, L.C.; Gupta, I.J.; Van Der Merwe, A. A parametric model for synthetic aperture radar measurements. *IEEE Trans. Antennas Propag.* **1999**, *47*, 1179–1188. [CrossRef]

11. Donoho, D. Compressed Sensing. *IEEE Trans. Inf. Theory* **2006**, *52*, 1289–1306. [CrossRef]

12. Candes, E.; Wakin, M. An Introduction to Compressive Sampling. *IEEE Signal Process. Mag.* **2008**, *25*, 21–30. [CrossRef]

13. Tibshirani, R. Regression Shrinkage and Selection via the Lasso. *J. R. Stat. Soc. Ser. B* **1996**, *58*, 267–288. [CrossRef]

14. Osher, S.; Feng, R.; Jiechao, X.; Yuan, Y.; Wotao, Y. Sparse recovery via differential inclusions. *Appl. Comput. Harmon. Anal.* **2016**, *41*, 436–469. [CrossRef]

15. Candes, E.; Tao, T. The Dantzig selector: Statistical estimation when p is much larger than n. *Ann. Stat.* **2007**, *35*, 2313–2351. [CrossRef]

16. Figueiredo, M.A.T.; Nowak, R.D.; Wright, S.J. Gradient Projection for Sparse Reconstruction: Application to Compressed Sensing and Other Inverse Problems. *IEEE J. Sel. Top. Signal Process.* **2007**, *1*, 586–597. [CrossRef]

17. Vaswani, N. LS-CS-Residual (LS-CS): Compressive Sensing on Least Squares Residual. *IEEE Trans. Signal Process.* **2010**, *58*, 4108–4120. [CrossRef]

18. Candes, E.; Romberg, J.; Tao, T. Robust uncertainty principles: Exact signal reconstruction from highly incomplete frequency information. *IEEE Trans. Inf. Theory* **2006**, *52*, 489–509. [CrossRef]

19. Daubechies, I.; Defrise, M.; De Mol, C. An iterative thresholding algorithm for linear inverse problems with a sparsity constraint. *Commun. Pure Appl. Math.* **2004**, *57*, 1413–1457. [CrossRef]

20. Ravazzi, C.; Fosson, S.; Magli, E. Distributed Iterative Thresholding for ℓ_0/ℓ_1-Regularized Linear Inverse Problems. *IEEE Trans. Inf. Theory* **2015**, *61*, 2081–2100. [CrossRef]

21. Fiandrotti, A.; Fosson, S.; Ravazzi, C.; Magli, E. PISTA: Parallel Iterative Soft Thresholding Algorithm for Sparse Image Recovery. In Proceedings of the Picture Coding Symposium (PCS), San Jose, CA, USA, 8–11 December 2013; pp. 325–328.

22. Casteel, C.; Gorham, L.; Minardi, M.; Scarborough, S.; Naidu, K.; Majumder, U. A challenge problem for 2D/3D imaging of targets from a volumetric data set in an urban environment. *Proc. SPIE* **2007**, *6568*. [CrossRef]

Article

Azimuth Sidelobes Suppression Using Multi-Azimuth Angle Synthetic Aperture Radar Images

Yamin Wang [1], Wei Yang [1], Jie Chen [1,*], Hui Kuang [1], Wei Liu [2] and Chunsheng Li [1]

[1]   School of Electronic and Information Engineering, Beihang University, Beijing 100191, China;
      wangyamin@buaa.edu.cn (Y.W.); yangweigigi@sina.com (W.Y.); kuanghui@buaa.edu.cn (H.K.);
      lics@buaa.edu.cn (C.L.)
[2]   Electronic and Electrical Engineering Department, University of Sheffield, Sheffield S1 3JD, UK;
      w.liu@sheffield.ac.uk
*   Correspondence: chenjie@buaa.edu.cn; Tel.: +86-10-8233-7049

Received: 20 March 2019; Accepted: 17 June 2019; Published: 19 June 2019

Abstract: A novel method is proposed for azimuth sidelobes suppression using multi-pass squinted (MPS) synthetic aperture radar (SAR) data. For MPS SAR, the radar observes the scene with different squint angles and heights on each pass. The MPS SAR mode acquisition geometry is given first. Then, 2D signals are focused and the images are registered to the master image. Based on the new signal model, elevation processing and incoherent addition are introduced in detail, which are the main parts for azimuth sidelobes suppression. Moreover, parameter design criteria in incoherent addition are derived for the best performance. With the proposed parameter optimization step, the new method has a prominent azimuth sidelobes suppression effect with a slightly better azimuth resolution, as verified by experimental results on both simulated point targets and TerraSAR-X data.

Keywords: multi-pass squinted (MPS); azimuth sidelobes suppression; synthetic aperture radar (SAR)

1. Introduction

For many applications of synthetic aperture radar (SAR), the images can be adversely affected by sidelobes, especially in the case of strongly scattering targets with weak targets nearby, such as in a harbour with ships and containers. Hence, it is often desirable to suppress the sidelobes in order to improve image quality.

Several methods have been proposed to do this. The most common approach [1–3] imposes a weighting on the signal spectrum, such as the Taylor and Hamming windows, but such methods tend to widen the mainlobe. Another method, known as spatially variant apodization (SVA) [4,5], and its modified versions [6–9], can reduce the sidelobes without degrading mainlobe resolution. However, nonlinear apodization modifies the statistical distribution of the pixel intensities, thus hindering the extraction of information from homogeneous regions [10]. In [11], a dual-Delta factorization method was proposed to suppress sidelobes in squinted and bistatic SAR images, but this iterative method is complex and computationally expensive.

This letter introduces a novel multi-pass squinted (MPS) SAR, whose data can be used to realize 3D imaging and also to produce 2D images with low azimuth sidelobes. Compared with traditional multi-pass SAR for 3D imaging [12] or multi-baseline SAR for interferometry [13], in which the SAR operates in broadside mode, MPS SAR works in squint mode and observes the scene with different azimuth squint angles on each pass. Squint mode increases the difficulty of SAR signal processing and can also provide more possibilities in terms of applications with corresponding imaging methods. In this paper it is a novel application for azimuth sidelobes suppression based on MPS SAR mode; it can suppress azimuth sidelobes significantly and improve the azimuth resolution slightly simultaneously. For azimuth sidelobes suppression, the first-order phase related to the Doppler centroid frequency is specially preserved, which can be utilized in elevation processing. Some existing algorithms have been adjusted to cater for this new signal model in elevation processing, resulting in the azimuth mainlobe and sidelobes separating in elevation. Moreover, through the parameter optimization of elevation integrated range, the performance of the azimuth sidelobes suppression can be better, which is firstly introduced based on MPS SAR mode. The effectiveness of the proposed method is verified by both simulated data and the real TerraSAR-X image data compared with the signal spectrum weighting algorithms.

This letter describes a method for azimuth sidelobes suppression using MPS SAR data. In Section 2, the imaging geometry is introduced. Section 3 builds the signal model and describes the processing of the stack of images along elevation to yield resolution cells at different elevations, as in tomography. However, unlike tomography, in this mode the azimuth sidelobes occur at different elevations and can be eliminated by incoherent addition in elevation. The performance of the proposed method is related to system parameters in Section 4, and a set of design criteria is proposed. Simulations with point targets and real SAR images are performed to test the proposed method in Section 5 and conclusions are drawn in Section 6.

2. Multi-Pass Squinted SAR

The imaging geometry of MPS SAR is shown in Figure 1a, where X, Y, Z, and S represent range, azimuth, height, and elevation coordinates, respectively. The aircraft is in the azimuth-height plane, L_n represents the nth pass of the aircraft, $A_{n,m}$ is the center position of the SAR in the mth acquisition on the nth pass, $2N + 1$ and $2M + 1$ are the numbers of passes and acquisitions, respectively, $\varphi_{n,m}$ is the azimuth squint angle, which is the angle between line-of-sight and broadside, and the heavy lines on each pass represent the synthetic apertures of the acquisitions.

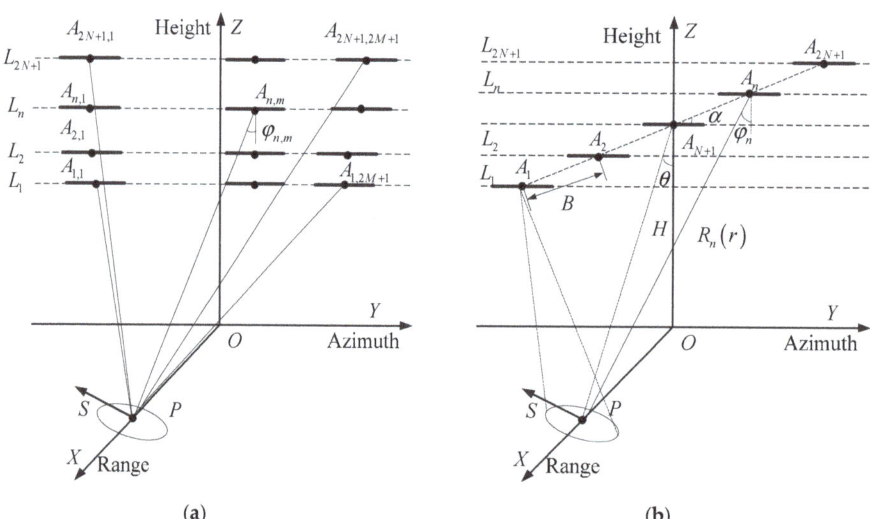

Figure 1. Imaging geometry: (**a**) MPS SAR; (**b**) for azimuth sidelobes suppression.

Based on the imaging geometry in Figure 1a, data acquired with the same azimuth squint angle on each pass can be combined and processed for 3D imaging, which is similar to the traditional TomoSAR. Moreover, the data acquired with different azimuth squint angles on each pass can be selected and used to suppress azimuth sidelobes. However, the acquisition of the data should meet the imaging geometry for azimuth sidelobes suppression, as shown in Figure 1b. Figure 1b represents the $2N + 1$ MPS acquisitions taken from each pass in Figure 1a; A_n is the center position of the acquisition on the nth pass, φ_n is the azimuth squint angle, H is the height of the center pass, θ is the incidence angle and α is the "flight angle", which is the angle between the line of the center positions (assumed collinear and equally spaced) and the azimuth coordinate; this is given by $\alpha = \mathrm{acos}\left(\left(\vec{A_1 A_n} \cdot \vec{y}\right)\big/\left|\vec{A_1 A_n}\right|\right)$.

$B = \left|\vec{A_n A_{n+1}}\right|$ is the distance between two adjacent center positions of the SAR and is referred to as the baseline (assumed to be the same for all adjacent pairs), and $B_{a,n}$, $B_{//,n}$, and $B_{\perp,n}$ are the azimuth, parallel, and orthogonal baselines, respectively, which are the projections of the vector $\vec{A_n A_{N+1}}$ along the azimuth, the line of sight, and elevation, with the following forms:

$$\begin{cases} B_{a,n} = (n - N - 1) \cdot B \cdot \cos\alpha \\ B_{\perp,n} = (n - N - 1) \cdot B \cdot \sin\alpha \cdot \sin\theta \\ B_{//,n} = (n - N - 1) \cdot B \cdot \sin\alpha \cdot \cos\theta \end{cases}, \tag{1}$$

P is a point target in the scene, and represents the distance between the SAR center position A_n and P:

$$R_n(r) \approx \sqrt{\left(r + B_{//,n}\right)^2 + B_{\perp,n}^2 + B_{a,n}^2} \approx \sqrt{\left(r + B_{//,n}\right)^2 + (B_{a,n})^2} + B_{\perp,n}^2 \bigg/ \left(2\sqrt{\left(r + B_{//,n}\right)^2 + (B_{a,n})^2}\right) \tag{2}$$

where r represents the distance between A_{N+1} and P.

3. Signal Processing

The signal model was built and the proposed method for suppressing azimuth sidelobes was derived based on the imaging geometry described in Section 2. It contained four main steps: 2D focusing, image registration, elevation processing, and incoherent addition, as indicated in Figure 2.

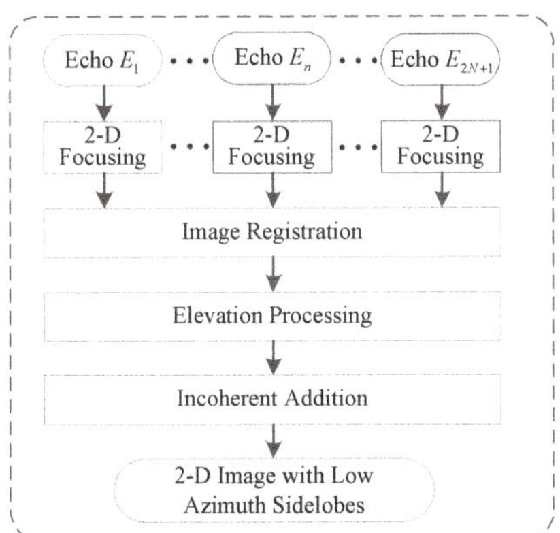

Figure 2. Flowchart of the proposed processing method for azimuth sidelobes suppression.

The data acquired on each pass were first focused to obtain a 2D image. Here, the modified chirp scaling kernel [14], which is suitable for squinted SAR, was used to process the raw data to get focused 2D images. However, not only the energy was focused in this 2D image processing, but also the phase was preserved, which can be utilized in subsequent elevation processing for azimuth sidelobes suppression in the proposed method. The 2D images were then registered to the selected master image

$(n = N + 1)$. This allowed the stack of images to be processed along elevation, where the 2D signal from the target P in the nth image is given by:

$$s_n(y', r') = \sqrt{\sigma_P} \cdot \operatorname{sinc}\left(\frac{r' - r}{\rho_r}\right) \cdot \operatorname{sinc}\left(\frac{y'}{\rho_a}\right) \cdot \exp\left\{-j\frac{4\pi}{\lambda}R_n(r)\right\} \cdot \exp\left\{-j2\pi f_{d,n}y'/v\right\}, \tag{3}$$

where σ_P is the radar cross-section (RCS) of P, λ is wavelength, ρ_a and ρ_r represent azimuth and range resolution, respectively, r' and y' are the variables associated with the range and azimuth position of the focused image, v is the velocity of the SAR, and $f_{d,n}$ is the Doppler centroid frequency of the nth acquisition:

$$f_{d,n} = 2v\sin\varphi_n/\lambda \approx 2vB_{a,n}\left/\left(\lambda\sqrt{\left(r + B_{//,n}\right)^2 + B_{a,n}^2}\right)\right.. \tag{4}$$

In Equation (3), the second exponential term is the specially preserved phase, which indicates that in different acquisitions the phase varies with the Doppler centroid frequency and the target's azimuth position. The 2D focusing and image registration are the basics of the proposed method.

We assumed that the variation in the heights of the targets in the scene was less than the resolution in elevation (see Equation (9)), so the imaging area could be seen as effectively flat. Thus, the targets in the same 2D image cell cannot be separated in elevation after elevation processing.

As a new application, the spectral analysis (SPECAN) [15,16] algorithm was used to process the MPS SAR signal in elevation. The residual phase induced by varying center distance $R_n(r)$ was first compensated by multiplying the signal with its complex conjugate phase:

$$H(r,n) = \exp\left\{j\tfrac{4\pi}{\lambda}R_n(r)\right\} \approx \exp\left\{j\tfrac{4\pi}{\lambda}\left(\sqrt{\left(r + B_{//,n}\right)^2 + B_{a,n}^2} + B_{\perp,n}^2\left/\left(2\sqrt{\left(r + B_{//,n}\right)^2 + (B_{a,n})^2}\right)\right.\right)\right\}. \tag{5}$$

After multiplication, the signal was modeled as:

$$s_n'(y', r') = \sqrt{\sigma_P} \cdot \operatorname{sinc}\left(\frac{r' - r}{\rho_r}\right) \cdot \operatorname{sinc}\left(\frac{y'}{\rho_a}\right) \cdot \exp\left\{-j2\pi f_{d,n}\frac{y'}{v}\right\}, \tag{6}$$

Then, a new variable ξ_n (different from that in [15,16]) was designed to focus the signal in elevation by the single Fourier transform (FFT), to give:

$$ss(y', r', s) = \sum_{n=1}^{2N+1}\left(s_n'(y', r') \cdot \exp\{-j2\pi\xi_n s\}\right) \approx (2N+1)\sqrt{\sigma_P} \cdot \operatorname{sinc}\left(\tfrac{r'-r}{\rho_r}\right) \cdot \operatorname{sinc}\left(\tfrac{y'}{\rho_a}\right) \cdot \operatorname{sinc}\left(\tfrac{s + y'\cot\alpha/\sin\theta}{\rho_e}\right) \tag{7}$$

where,

$$\xi_n = 2B_{\perp,n}\left/\left(\lambda\sqrt{\left(r + B_{//,n}\right)^2 + (B_{a,n})^2}\right)\right. \approx 2B_{\perp,n}/(\lambda r), \tag{8}$$

ρ_e is the resolution in elevation, given by [12]:

$$\rho_e \approx \lambda r\left/\left(2B_{\perp,total}\right)\right.. \tag{9}$$

Here, $B_{\perp,total} = 2N \cdot B\sin\alpha \cdot \sin\theta$ is the total orthogonal baseline. The elevation range after elevation processing is $[-H_{max}/2, H_{max}/2]$, where the maximum ambiguity in elevation H_{max} is:

$$H_{max} \approx \lambda r/(2B\sin\alpha \cdot \sin\theta). \tag{10}$$

From Equations (3) and (7), it can be seen that the signal in $(r', y', 0)$ occurs at $(r', y', -y'\cot\alpha/\sin\theta)$ after elevation processing, which means the azimuth signal of the target will be located at different positions in elevation, i.e., the mainlobe and sidelobes along the azimuth of the focused target will separate in elevation. Moreover, if $\left|-y'\cot\alpha/\sin\theta\right| > H_{max}/2$, the signal will be aliased in elevation; the aliased position is given by:

$$s' = -y'\cot\alpha/\sin\theta - \lfloor -y'\cot\alpha/\sin\theta/H_{\max} + 0.5 \rfloor \cdot H_{\max},\tag{11}$$

where $\lfloor x \rfloor$ is the largest integer not larger than x.

To further explain the effect of elevation processing, Figure 3a shows the azimuth profiles of a target in a 2D image, and Figure 3b is a slice across the 3D image in the azimuth-elevation plane. It can be seen that after elevation processing, the sidelobes at different azimuth positions were shifted to different elevations. Moreover, the F and G parts in Figure 3a were located at the wrong positions due to aliasing in elevation, as given by Equation (11).

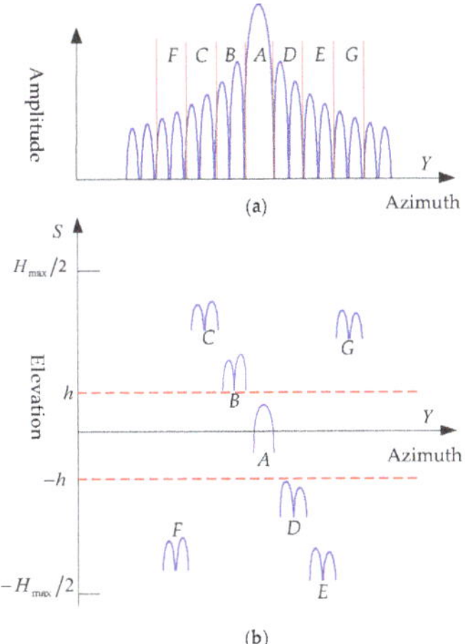

Figure 3. Effect of elevation processing. (**a**) Azimuth profile of a target in the 2D image; (**b**) slice across the 3D image in the azimuth-elevation plane.

Thus, if we integrate the energy of the signal $ss(y',r',s)$ incoherently over the elevation range $[-h,h]$ in which mainly the azimuth mainlobe and maybe several lower sidelobes are distributed, an image, $\widetilde{s}(y',r')$, with low azimuth sidelobes is obtained:

$$\widetilde{s}(y',r') = \int_{-h}^{h} \left|ss(y',r',s)\right|^{2} ds.\tag{12}$$

It can be seen that the selection of h is crucial in determining the performance of the proposed sidelobes suppression method and it will be discussed below in detail.

4. Parameter Design

Optimization of the performance of the proposed method is based on three criteria:

4.1. The Energy of the Azimuth Mainlobe Must be Preserved

From Equation (7), the azimuth signal is distributed in elevation. To preserve the energy of the azimuth mainlobe, the integrating range $[-h,h]$ must contain the position of the azimuth mainlobe. This is distributed in space because a point target can occur anywhere within the resolution cell.

We defined the azimuth mainlobe as the part of the azimuth profile (see Figure 4b) in which the energy exceeds −4 dB (where the maximum value of the azimuth profile is normalized to 0 dB). Setting

$$10 \log\left(\sin c^2\left(y'_0 / \rho_a\right)\right) = -4, \tag{13}$$

yields $y'_0 \approx 0.5\rho_a$. Assuming the center position of the azimuth mainlobe in elevation is s_0 (where $s_0 \in (-0.5\rho_e, 0.5\rho_e)$), the elevation range of the azimuth mainlobe is $[s_0 - y'_0\cot\alpha/\sin\theta, s_0 + y'_0\cot\alpha/\sin\theta]$. So, to preserve the energy of the azimuth mainlobe of the target, we select,

$$h = y'_0\cot\alpha/\sin\theta + 0.5\rho_e = 0.5\rho_a\cot\alpha/\sin\theta + 0.5\rho_e. \tag{14}$$

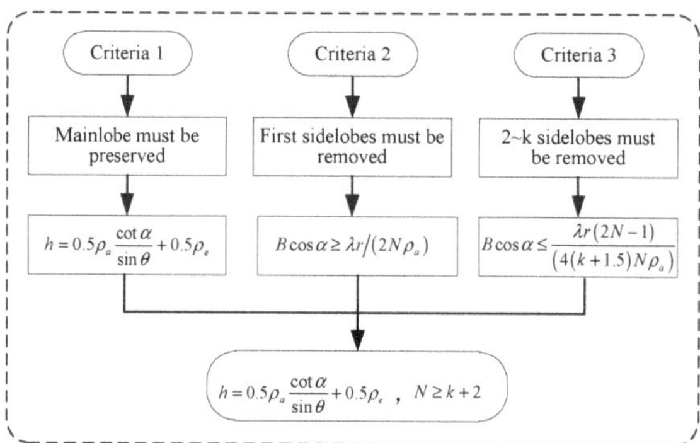

Figure 4. The design of parameter h.

4.2. The First Azimuth Sidelobes in Elevation Must Be Outside the Integrating Range $[-h, h]$

The zeroes of $\sin c(y'/\rho_a)$ occur at the points,

$$y'_m = \pm m\rho_a (m = 1, 2, \cdots). \tag{15}$$

The first zero point is $y'_1 = \rho_a$, so the following relationship must hold:

$$y'_1\cot\alpha/\sin\theta - 0.5\rho_e \geq h. \tag{16}$$

Inserting (9) and (14) into (16), we have:

$$y'_1 = \rho_a B \cos\alpha \geq \lambda r/(2N\rho_a). \tag{17}$$

4.3. The 2nd to the Kth Azimuth Sidelobes in Elevation Must Be Outside the Integrating Range $[-h, h]$

k is a number we select. With the increase of k, the performance of azimuth sidelobes suppression is better. Two cases need to be considered: (a) if the kth azimuth sidelobes is not aliased in elevation, it will be outside $[-h, h]$ when (17) is satisfied, since $y'_k\cot\alpha/\sin\theta - 0.5\rho_e \geq y'_1\cot\alpha/\sin\theta - 0.5\rho_e$, where $y'_k = k\rho_a$; (b) if the kth azimuth sidelobes are aliased in elevation, the following conditions must be satisfied:

$$\begin{cases} y'_{k+1} = (k+1)\rho_a \\ y'_{k+1}\cot\alpha/\sin\theta + 0.5\rho_e \leq H_{\max} - h \end{cases}, \tag{18}$$

Hence,

$$B \cos\alpha \leq \lambda r(2N-1)/(4(k+1.5)N\rho_a). \tag{19}$$

The sidelobes beyond kth with lower energy will contribute little, whether they are relocated in or out the $[-h, h]$ in elevation.

The first to kth azimuth sidelobes up to and including the kth will be suppressed if (17) and (19) are met. Furthermore, based on (17) and (19), we have:

$$N \geq k + 2, \tag{20}$$

which means that more flight passes allow more azimuth sidelobes to be removed. Figure 4 illustrates the process of the parameter selection.

In summary, the integrating range can be computed based on (14), and the baseline B, the flight angle α, and the number of passes $2N + 1$, can be optimized using (17), (19), and (20).

5. Performance Simulation

This section shows the performance based on simulations, using the parameters listed in Table 1 where PRF represents the pulse repetition frequency. The associated elevation resolution is about 55 m, and the maximum ambiguity in elevation (see (10)) is 1654 m. The integrating range in elevation is $[-92\,\text{m}, 92\,\text{m}]$. Moreover, from (17) and (19), the first to the 10th azimuth sidelobes will be suppressed after incoherent addition.

Table 1. List of simulation parameters.

Parameters	Value	Parameters	Value
Aircraft Height	20 km	Bandwidth	80 MHz
Incidence Angle	30°	Sample Rate	100 MHz
Wavelength	0.03 m	PRF	70 Hz
Velocity	100 m/s	Antenna Length	4.0 m
Flight Angle	2°	Pulse Duration	10
Baseline	12 m	Flight passes	31

5.1. Point Target Simulations

Simulations for a point target were first performed to test the proposed method, as shown in Figure 5. After elevation processing, a 3D image was obtained, and Figure 5a is a slice across this image in the azimuth-elevation plane. It can be seen that the azimuth sidelobes were compressed to different positions in elevation, and some sidelobes were aliased. The energy of the elevation signals between the two red lines was then incoherently integrated to form a 2D image of the impulse response function (IRF), as shown in the contour plot of Figure 5b. As can be seen, the target was well focused with low azimuth sidelobes. The azimuth profile obtained using the proposed method was compared with the $(N + 1)$th 2D azimuth profile obtained by a classical 2D focusing algorithm, using a rectangular window (Figure 5c) and a Taylor window with parameters 0.25 (Figure 5d). The proposed method is seen to suppress azimuth sidelobes without degrading the azimuth resolution. The first sidelobes were preserved partly because they were in the integrating region, as shown in Figure 5a, but were less than -30 dB. This was much lower than that for the rectangular window (Figure 5c), and was achieved without the loss of resolution suffered when using the Taylor window (Figure 5d).

To further illustrate the performance of the proposed method, azimuth resolution, peak sidelobe ratio (PSLR), and integrated sidelobe ratio (ISLR) [17] were given as follows. As shown in Table 2, the azimuth resolution using the proposed method was 7.04% slightly higher than that for the rectangular window, at about 1.85 m. PSLR of the images weighted by rectangular window and Taylor window were -13.28 dB and -25.41 dB. ISLR of these two images were -10.11 dB and -20.18 dB, correspondingly. Through processing with the proposed method, the PSLR and ISLR of the image reached -31.07 dB and -29.36 dB, respectively, which were much lower than the other two images. Overall, it could be concluded that the proposed method can suppress azimuth sidelobes splendidly with the maintained azimuth resolution.

Simulations with three point targets with different RCSs illustrate the advantages of the proposed method for detecting weak scatterers. The targets A, B, and C were located at -3 m, 0 m, and 3 m along the azimuth, with RCS −20, −10, and 0 dB, respectively, and they had the same height and range positions. Figure 6 compared the azimuth profiles using the proposed method and the classical 2D focusing algorithm with a Taylor window. The dashed lines in Figure 6 indicated the true target positions. It could be seen that they were clearly separated and had the correct RCS when the proposed algorithm was used. In contrast, under the classical 2D focusing algorithm, the weaker scatterers (A and B) were seriously affected by the sidelobes of the strong scatterer C and could not be detected. Moreover, the mainlobe of C was widened and affected by the sidelobes of B.

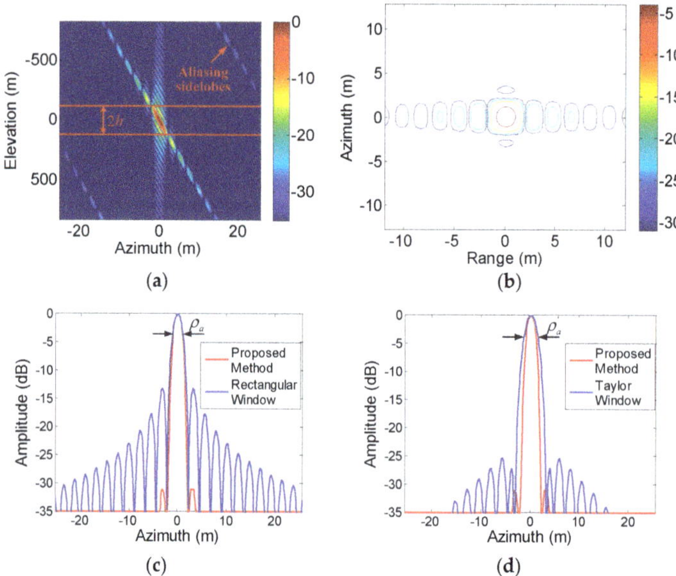

Figure 5. Imaging results for a point target. (**a**) Profile in the azimuth-elevation plane after elevation processing; (**b**) contour plots of the impulse response function (IRF) with the proposed method; (**c**) comparison of azimuth profiles using the proposed method (red) and the classical 2D focusing algorithm with a rectangular window (blue); (**d**) as for (**c**) but with a Taylor window (blue).

Table 2. Imaging quality indicators for a single point target along azimuth.

Single Point Target	Resolution	PSLR	ISLR
Rectangular Window	1.99 m	−13.28 dB	−10.11 dB
Taylor Window	2.49 m	−25.41 dB	−20.18 dB
Proposed Method	1.85 m	−31.07 dB	−29.36 dB

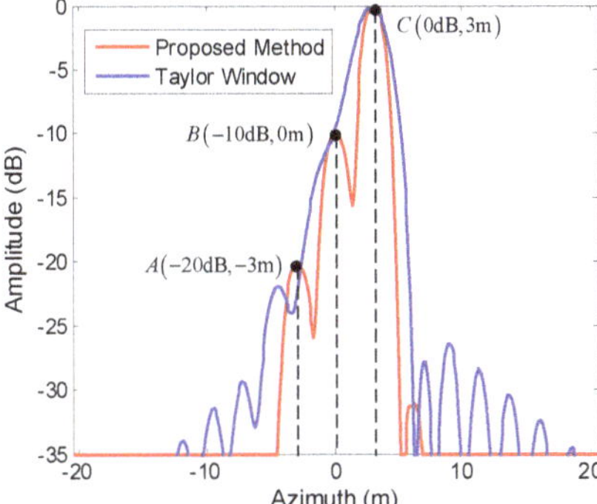

Figure 6. Azimuth profiles of three point targets using the proposed method and the classical 2D focusing algorithm.

Furthermore, simulations with three point targets at different heights were performed to verify the proposed method in a scene with height variations. The targets D, E, and F had the same range position and RCS and were located at (−10 m, −10 m), (0 m, 0 m), and (10 m, 10 m), where (y, z) are the (azimuth, height) coordinates. Figure 7a,b showed the contour plots of the IRF using the classical 2D focusing algorithm with a Taylor window in azimuth and the proposed method. It could be seen that the targets were focused at different range positions due to their different heights. The targets in Figure 7b were all well focused, and had lower azimuth sidelobes and better azimuth resolutions than those in Figure 7a.

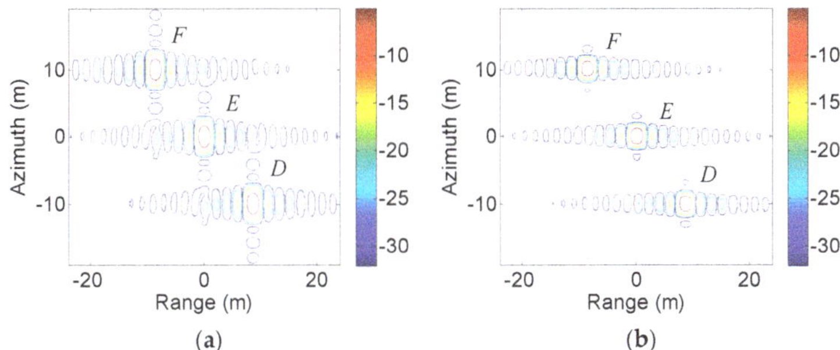

Figure 7. Contour plots of a scene with three point targets using: (**a**) classical 2D focusing algorithm with a Taylor window; (**b**) the proposed method.

5.2. Real SAR Image Simulations

Simulations were also performed on a TerraSAR-X image of a part of Dingxing airport, China, which contains 500 × 300 (azimuth × range) pixels. The original image from TerraSAR-X was used as RCS of the extended target to simulate MPS SAR data with the imaging geometry in Figure 1 and the simulation parameters listed in Table 1. It should be noted that the height of the targets in this real SAR scene is less than the elevation resolution 55 m. The focused 2D images were then processed by the proposed method to suppress the azimuth sidelobes.

Figure 8 showed the resulting images, which were normalized to the same total energy. Figure 8a, b were formed using the classical 2D focusing algorithm with a rectangular window and a Taylor window in azimuth, respectively. High azimuth sidelobes can be seen in Figure 8a, which can be suppressed with a Taylor window at the expense of resolution (Figure 8b). Figure 8c was the image obtained using the proposed method. The targets now had low azimuth sidelobes and the mainlobes had not been widened, compared with those in Figure 8a,b (for example, see the target in the red circle).

As a measure of image focusing quality we used image contrast, γ, defined as the ratio of standard deviation and mean of image intensity. Typically, a larger contrast means better image quality. The values of contrast in the images in Figure 8 were $\gamma_a = 0.1867$, $\gamma_b = 0.1963$, and $\gamma_c = 0.2492$, respectively; as expected, the image in Figure 8c had the highest contrast.

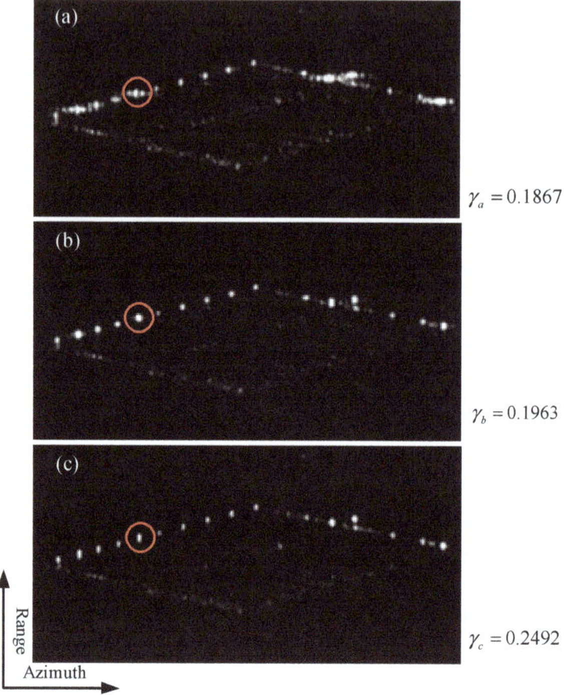

Figure 8. 2D images obtained by processing the raw data simulated with a TerraSAR-X image of part of Dingxing airport, China, with: (**a**) classical 2D focusing algorithm with a rectangular window; (**b**) classical 2D focusing algorithm with a Taylor window; (**c**) the proposed method. The copyright of the original SAR image belongs to Airbus.

6. Conclusions

This article proposed a novel MPS SAR mode and a method to process MPS SAR data together with parameter selection criteria that can be used to optimize system design. Based on the MPS SAR mode, this is a novel application to suppress azimuth sidelobes using some of the existing algorithms, which were already adjusted to meet the new mode. Simulations indicated that it provided 2D images with lower azimuth sidelobes compared with some existing azimuth suppression methods. The analysis presented here is idealized, since it assumes flight passes whose center positions are collinear and equally spaced, which would in practice be difficult to satisfy. Future work will analyze the effects of relaxing these conditions.

Author Contributions: The work presented here was carried out in collaboration among all authors. conceptualization, Y.W. and W.Y.; formal analysis, Y.W. and W.Y.; methodology, Y.W., W.Y. and H.K.; validation, Y.W., W.Y. and H.K.; writing—original draft preparation, Y.W.; writing—review and editing, W.Y.; supervision, J.C., W.L. and C.L.; funding acquisition, W.Y. and J.C.

Funding: This work was supported by the National Natural Science Foundation of China (NSFC) under Grant No. 61701012. National Natural Science Foundation of China (NSFC) under Grant No. 61671043 and Fundamental Research Funds for the Central University under Grant No.YWF-19-BJ-J-304.

Sensors **2019**, *19*, 2764

Conflicts of Interest: The authors declare no conflict of interest.

References

1. Prabhu, K.M.M. *Window Functions and Their Applications in Signal Processing*; CRC Press: Boca Raton, FL, USA, July 2014.
2. Cumming, I.G.; Wong, F.H. *Digital Processing of Synthetic Aperture Radar Data: Algorithms and Implementation*; Artech House: Norwood, MA, USA, 2005.
3. Cohen, I.; Levanon, N. Weight Windows—An Improved Approach. In Proceedings of the 2014 IEEE 28th Convention of Electrical & Electronics Engineers in Israel (IEEEI), Eilat, Israel, 3–5 December 2014.
4. Stankwitz, H.C.; Dallaire, R.J.; Fienup, J.R. Spatially variant apodization for sidelobe control in SAR imagery. In Proceedings of the 1994 IEEE National Radar Conference, Atlanta, GA, USA, 29–31 March 1994; pp. 132–137.
5. Stankwitz, H.C.; Dallaire, R.J.; Fienup, J.R. Nonlinear apodization for sidelobe control in SAR imagery. *IEEE Trans. Aerosp. Electron. Syst.* **1995**, *31*, 267–279. [CrossRef]
6. Smith, B.H. Generalization of spatially variant apodization to noninteger Nyquist sampling rates. *IEEE Trans. Image Process.* **2000**, *9*, 1088–1093. [CrossRef] [PubMed]
7. Castillo-Rubio, C.F.; Llorente-Romano, S.; Burgos-Garcia, M. Spatially variant apodization for squinted synthetic aperture radar images. *IEEE Trans. Image Process.* **2007**, *16*, 2023–2027. [CrossRef] [PubMed]
8. Iglesias, R.; Mallorqui, J.J. Side-lobe cancelation in DInSAR pixel selection with SVA. *IEEE Geosci. Remote Sens. Lett.* **2013**, *10*, 667–671. [CrossRef]
9. Xiong, T.; Wang, S.; Hou, B.; Wang, Y.; Liu, H. A resample-based SVA algorithm for sidelobe reduction of SAR/ISAR imagery with noninteger Nyquist sampling rate. *IEEE Trans. Geosci. Remote Sens.* **2015**, *53*, 1016–1028. [CrossRef]
10. Pastina, D.; Colone, F.; Lombardo, P. Effect of apodization on SAR image understanding. *IEEE Trans. Geosci. Remote Sens.* **2007**, *45*, 3533–3551. [CrossRef]
11. Jun, S.; Yang, L.; Xiaoling, Z.; Ling, F. A novel SAR sidelobe suppression method via Dual-Delta factorization. *IEEE Geosci. Remote Sens. Lett.* **2015**, *12*, 1576–1580. [CrossRef]
12. Fornaro, G.; Serafino, F.; Soldovieri, F. Three-dimensional focusing with multipass SAR data. *IEEE Trans. Geosci. Remote Sens.* **2003**, *41*, 507–517. [CrossRef]
13. Fornaro, G.; Guarnieri, A.M.; Pauciullo, A.; De-Zan, F. Maximum likelihood multi-baseline SAR interferometry. *IEE Proc. Radar Sonar Navig.* **2006**, *153*, 279–288. [CrossRef]
14. Chen, J.; Kuang, H.; Yang, W.; Liu, W.; Wang, P. A novel imaging algorithm for focusing high-resolution spaceborne SAR data in squinted sliding-spotlight mode. *IEEE Geosci. Remote Sens. Lett.* **2016**, *13*, 1577–1581. [CrossRef]
15. Reale, D.; Fornaro, G.; Pauciullo, A.; Zhu, X.; Bamler, R. Tomographic imaging and monitoring of buildings with very high resolution SAR data. *IEEE Geosci. Remote Sens. Lett.* **2011**, *8*, 661–665. [CrossRef]
16. Sack, M.; Ito, M.R.; Cumming, I.G. Application of efficient linear FM matched filtering algorithms to synthetic aperture radar processing. Communications, Radar and Signal Processing. *IEE Proc. F Commun. Radar Signal Process.* **1985**, *132*, 45–57. [CrossRef]
17. Zhu, X.; He, F.; Ye, F.; Dong, Z.; Wu, M. Sidelobe Suppression with Resolution Maintenance for SAR Images via Sparse Representation. *Sensors* **2018**, *18*, 1589. [CrossRef] [PubMed]

Article

Extended Multiple Aperture Mapdrift-Based Doppler Parameter Estimation and Compensation for Very-High-Squint Airborne SAR Imaging

Zhichao Zhou [1,2], Yinghe Li [3], Yan Wang [1,2,*], Linghao Li [1,2] and Tao Zeng [1,2]

[1] School of Information and Electronics, Beijing Institute of Technology, Beijing 100081, China; zcz1024@foxmail.com (Z.Z.); lilinghaodhd@foxmail.com (L.L.); zengtao@bit.edu.cn (T.Z.)
[2] Beijing Key Laboratory of Embedded Real-Time Information Processing Technology, Beijing 100081, China
[3] Beijing Institute of Radio Measurement, Beijing 100081, China; liyinghe@bit.edu.cn
*   Correspondence: yan_wang@bit.edu.cn; Tel.: +86-10-6891-8550

Received: 31 October 2018; Accepted: 3 January 2019; Published: 8 January 2019

Abstract: Doppler parameter estimation and compensation (DPEC) is an important technique for airborne SAR imaging due to the unpredictable disturbance of real aircraft trajectory. Traditional DPEC methods can be only applied for broadside, small- or medium-squint geometries, as they at most consider the spatial variance of the second-order Doppler phase. To implement the DPEC in very-high-squint geometries, we propose an extended multiple aperture mapdrift (EMAM) method in this paper for better accuracy. This advantage is achieved by further estimating and compensating the spatial variation of the third-order Doppler phase, i.e., the derivative of the Doppler rate. The main procedures of the EMAM, including the steps of sub-view image generation, sliding-window-based cross-correlation, and image-offset-based Doppler parameter estimation, are derived in detail, followed by the analyses for the EMAM performance. The presented approach is evaluated by both computer simulations and real airborne data.

Keywords: Doppler parameter estimation and compensation (DPEC); extended multiple aperture mapdrift (EMAM); very-high-squint airborne SAR imaging; spatial variance; the derivative of the Doppler rate

1. Introduction

Airborne synthetic aperture radar (SAR) [1–5] is an all-weather and all-day microwave imaging sensor that can provide two-dimensional high-resolution images of illuminated regions. High-squint airborne SAR [6] is necessary for inverting target electromagnetic scattering characteristics in one track of observation and, therefore, is significant for accurate target identification [7–10]. The larger the squint angle, the more flexible the data acquisition and hence the more information a single observation can achieve. For the very-high-squint (VHS) airborne SAR imaging, targets at different positions have spatially-variant Doppler histories, as shown in Figure 1. While in range (along the direction of electromagnetic wave propagation), the spatial variance can be easily estimated and compensated by range blocking, it is not that convenient to estimate and compensate the azimuth spatially-variant Doppler parameters (along the direction perpendicular to the direction of electromagnetic wave propagation). Note that the spatial variance used below refers to the azimuth spatial variance if without additional denotations.

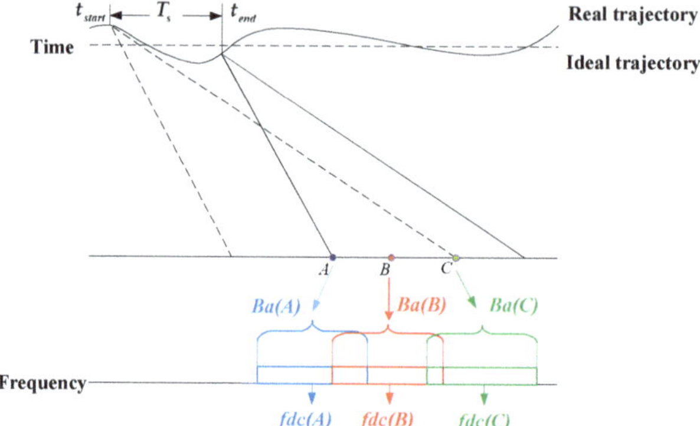

Figure 1. The illumination of the spatially-variant Doppler histories of targets at different positions for the very-high-squint (VHS) airborne SAR imaging. T_s is the aircraft motion time. t_{start} and t_{end} are the time start and end, respectively. B_a is the Doppler bandwidth. f_{dc} is the Doppler centroid.

One of the main challenges of the airborne SAR imaging is the Doppler parameter estimation and compensation (DPEC) because the positioning, velocity, and angle information provided by the onboard inertial navigation system are generally not accurate enough for the high-squint high-resolution imaging [11]. Moreover, the real aircraft trajectories often deviate from the ideal trajectories due to unexpected disturbances [12–15] as shown in Figure 1, which leads to the Doppler parameter errors. If the spatially-variant Doppler parameter estimation (DPE) is not considered and left compensated, the SAR image quality will be seriously deteriorated. Thus, it is necessary to perform echo-based DPE to ensure good focusing performance [16–21]. For the VHS airborne SAR, the DPEC is more challenging because of its complex spatially-variant characteristics.

Traditionally, the DPE can be implemented by the multiple aperture mapdrift (MAM) via the azimuth multi-view processing. The basic MAM method (as shown in Figure 2a) [22,23] assumes that the Doppler parameters do not change with respect to target positions. In this case, the estimated Doppler parameters are the averaged results of the real ones. Although such approximation is valid for the broadside or small-squint SAR imaging, it is no longer valid for the high-squint cases because the spatially-dependent components of the DPE will seriously degrade the image quality if left uncompensated. Although there exist some methods for the spatially-variant DPE, such as the improved MAM (IMAM) method [24,25], their accuracy is limited as they only deal with the spatial variance of the second-order Doppler phase. In the VHS case, for instance, with a 70-degree squint angle [26], the less accurate DPEC methods will lead to serious image quality degradation.

Aiming at implementing accurate enough DPEC for the VHS airborne SAR imaging, we propose an extended MAM (EMAM) method, as shown in Figure 2b. Compared with the IMAM method, the EMAM method realizes higher accuracy by further estimating and compensating the second-order component of the spatially-dependent Doppler rate and the first-order component of the spatially-dependent derivative of the Doppler rate. The former is to avoid the azimuth sidelobe lifting, and the latter is to get rid of the azimuth sidelobe asymmetry. Specifically, the new EMAM method firstly achieves sub-view images via multi-looking processing. Then, a sliding-window-based cross-correlation is implemented to achieve image offsets. Based on the unique mapping between such offset and the Doppler parameters, the DPEC can be accurately implemented.

The paper is arranged as follows. Section 2 introduces the basic MAM method. Section 3 derives the new EMAM method. Section 4 discusses the performance of the proposed method. In Section 5, the validity of the proposed method is verified based on the computer simulations and the real airborne data. Section 6 summarizes this study.

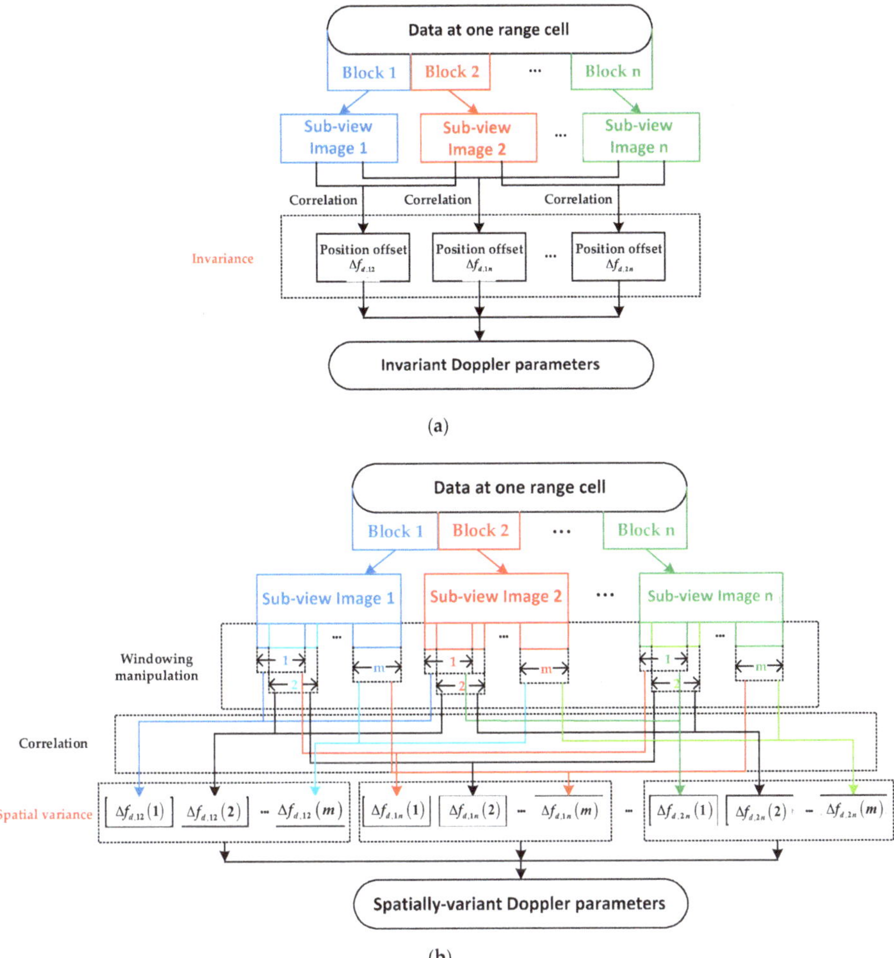

Figure 2. The illuminations of the basic MAM method and the EMAM method. (**a**) The basic MAM method; (**b**) the EMAM method. $\Delta f_{d,ij}$ is the position offset between the sub-*i* image and sub-*j* image.

2. Basic Multiple Aperture Mapdrift Method

The core strategy of the MAM method divides the data along the azimuth into multiple blocks and generates multiple sub-view images. By searching the offsets between two different sub-view images, it is possible to estimate the higher order Doppler parameters. The MAM methods can be implemented via either azimuth frequency-domain blocking [1] or time-domain blocking [22]. Specifically, the data are divided into several parts in azimuth in the time domain after multiplying the deramping function. Then, the azimuth fast Fourier transform (FFT) is carried out individually for each part to achieve multiple sub-view images. As the presented EMAM method is an extension of the basic MAM method, it is necessary to firstly give a brief introduction to the basic MAM method as follows.

Assume that the signal at a certain range cell is expressed as (1) (ignoring the azimuth four-order and higher order terms of the phase).

$$s(t_a) = \mathrm{rect}\left(\frac{t_a}{T_s}\right) \exp\left(\mathrm{j}2\pi f_{dc} t_a + \mathrm{j}\pi f_{dr,a} t_a^2 + \mathrm{j}\pi f_{3rd,a} t_a^3\right), \tag{1}$$

where f_{dc} is the Doppler centroid. T_s is the azimuth accumulation time. $f_{dr,a}$ and $f_{3rd,a}$ represent the real Doppler rate and the derivative of the Doppler rate, respectively. t_a is the azimuth slow time.

The deramping function is described as (2).

$$s(t_a) = \text{rect}\left(\frac{t_a}{T_s}\right) \exp\left(-j\pi f_{dr,b} t_a^2 - j\pi f_{3rd,b} t_a^3\right),\tag{2}$$

where $f_{dr,b}$ and $f_{3rd,b}$ represent the calculated Doppler rate and the derivative of the Doppler rate, respectively, which are inaccurate.

After being multiplied by the deramping function, the data are as follows.

$$s(t_a) = \text{rect}\left(\frac{t_a}{T_s}\right) \exp\left(j2\pi f_{dc} t_a + j\pi e_{dr} t_a^2 + j\pi e_{3rd} t_a^3\right),\tag{3}$$

where e_{dr} and e_{3rd} represent the errors of the Doppler rate and the derivative of the Doppler rate, respectively.

Then, the data are divided into three equal long sub-segments as (4).

$$s_i(t_a) = \text{rect}\left(\frac{t_a - t_{ac,i}}{T_s/3}\right) \exp\left(j2\pi f_{dc} t_a + j\pi e_{dr} t_a^2 + j\pi e_{3rd} t_a^3\right),\tag{4}$$

where $t_{ac,i}$ is the azimuth time for the center of each sub-segment and can be expressed as follows.

$$t_{ac,i} = -\frac{T_s}{3} + (i-1)\frac{T_s}{3}, \quad i = 1,2,3.\tag{5}$$

The data in (4) can be translated to the position where $t_a = 0$ and t_a is replaced by $t_a + t_{ac,i}$.

$$s_i(t_a) = \text{rect}\left(\frac{t_a}{T_s/3}\right) \exp\left\{ \begin{array}{l} j2\pi f_{dc}(t_a + t_{ac,i}) \\ +j\pi e_{dr}(t_a + t_{ac,i})^2 \\ +j\pi e_{3rd}(t_a + t_{ac,i})^3 \end{array} \right\}.\tag{6}$$

By performing the phase derivative of the upper formula and letting $t_a = 0$, the coefficient of the first-order phase can be obtained as (7).

$$f_{d,i} = f_{dc} + e_{dr} t_{ac,i} + \frac{3}{2} e_{3rd} t_{ac,i}^2.\tag{7}$$

Three sub-segments are subjected to the azimuth FFT to obtain three sub-view images, respectively. The center of sub-i image is located at $f_{d,i}$, and the position offset between sub-i image and sub-j image can be expressed as (8).

$$\Delta f_{d,ij} = f_{d,i} - f_{d,j} = e_{dr}(t_{ac,i} - t_{ac,j}) + \frac{3}{2} e_{3rd}\left(t_{ac,i}^2 - t_{ac,j}^2\right).\tag{8}$$

Then, three pairs of sub-view images can be formed to get three position offsets. The system of equations is as follows.

$$\Delta \mathbf{f} = \mathbf{T_{ac}} \begin{bmatrix} e_{dr} \\ \frac{3}{2} e_{3rd} \end{bmatrix},\tag{9}$$

where:

$$\Delta \mathbf{f} = [\Delta f_{d,12}\ \Delta f_{d,13}\ \Delta f_{d,23}]^{\text{T}}$$

$$\mathbf{T_{ac}} = \begin{bmatrix} t_{ac,1} - t_{ac,2} & t_{ac,1}^2 - t_{ac,2}^2 & t_{ac,1}^3 - t_{ac,2}^3 \\ t_{ac,1} - t_{ac,3} & t_{ac,1}^2 - t_{ac,3}^2 & t_{ac,1}^3 - t_{ac,3}^3 \\ t_{ac,2} - t_{ac,3} & t_{ac,2}^2 - t_{ac,3}^2 & t_{ac,2}^3 - t_{ac,3}^3 \end{bmatrix}.\tag{10}$$

After the cross-correlation of two sub-view images is computed and the position of the correlation peak is searched, the estimated value of the offset $\Delta \hat{f}_{d,ij}$ between two sub-view images can be obtained, then $\Delta \hat{f}_{d,ij}$ is taken into (9) to estimate the errors of the Doppler parameters by the least squares principle.

$$\begin{bmatrix} \hat{e}_{dr} \\ \frac{3}{2}\hat{e}_{3rd} \end{bmatrix} = \left(\mathbf{T}_{ac}^{T} \mathbf{T}_{ac} \right)^{-1} \mathbf{T}_{ac}^{T} \Delta \hat{f}. \tag{11}$$

In practice, since the sub-view images are defocused and the defocus conditions of the different sub-view images are not exactly the same, there is certain error in the position of correlation peak of the sub-view image, so the MAM methods often require multiple iterations to achieve better estimate accuracy.

It can be seen that the basic MAM method only compensates the spatially-invariant Doppler phases and hence can be only applied for the broadside or small-squint cases. Although the IMAM methods have partly overcome this disadvantage by estimating and compensating the spatial reliance of the Doppler phase up to the second-order, they still suffer from the problem of insufficient accuracy for the VHS SAR imaging. In this study, this problem is solved by further estimating and compensating the spatial variance of the third-order Doppler phase, resulting in the new EMAM method.

3. Extended Multiple Aperture Mapdrift Method

The spatial variance of the Doppler parameters refers to the fact that these parameters change with the azimuth position of target and can be represented as the functions of f_{dc}. Thus, the errors of the Doppler rate e_{dr} and the derivative of the Doppler rate e_{3rd} in (3) become the functions of f_{dc}, i.e., $e_{dr}(f_{dc})$ and $e_{3rd}(f_{dc})$. The offset of two sub-view images in (8) also becomes the function of f_{dc} as shown in (12).

$$\Delta f_{d,ij}(f_{dc}) = e_{dr}(f_{dc})\left(t_{ac,i} - t_{ac,j}\right) + \frac{3}{2}e_{3rd}(f_{dc})\left(t_{ac,i}^{2} - t_{ac,j}^{2}\right). \tag{12}$$

Then, an additional sliding windowing manipulation for the sub-view image correlation is employed to obtain the corresponding image offset $\Delta f_{d,ij}(f_{dc})$. Specifically, the sliding windowing manipulation is implemented by the short time Fourier transform (STFT). The two sub-view images at the same range cell are individually processed by the STFT, followed by the conversion of the data dimension from one to two, where one denotes the original Doppler frequency and the other denotes the newly-generated frequency. After the conjugate multiplication of the data, the IFFT is generated along the new frequency axis. Then, the offset of the sub-view images can be obtained based on the peak position. Figure 3 shows the flowcharts of the basic MAM method and the EMAM method. It can be seen that the use of STFT can achieve the sliding windowing manipulation, and the Doppler parameters changing with the azimuth frequency can be obtained. In order to improve the efficiency in practical applications, the intervals between windows can be appropriately increased, and the offset of each azimuth frequency can be obtained by the curve fitting. Then, the spatially-variant $\hat{e}_{dr}(f_{dc})$ and $\hat{e}_{3rd}(f_{dc})$ can be obtained based on the estimated $\Delta \hat{f}_{d,ij}(f_{dc})$.

After obtaining $\hat{e}_{dr}(f_{dc})$ and $\hat{e}_{3rd}(f_{dc})$, the operation of the curve fitting is performed. Here, the quadratic curve fitting is taken as an example.

$$\begin{aligned} \hat{e}_{dr}(f_{dc}) &= e_{dr0} + e_{dr1}(f_{dc} - f_{dc,cen}) + e_{dr2}(f_{dc} - f_{dc,cen})^{2} \\ \hat{e}_{3rd}(f_{dc}) &= e_{3rd0} + e_{3rd1}(f_{dc} - f_{dc,cen}) + e_{3rd2}(f_{dc} - f_{dc,cen})^{2}, \end{aligned} \tag{13}$$

where $f_{dc,cen}$ is the Doppler centroid of the azimuth center of the scene, the first terms of the two expressions are the fixed errors, the second terms are the first-order spatial variance errors, and the third terms are the second-order spatial variance errors. In general, the second-order spatial variance error of the derivative of the Doppler rate is too small to be ignored. The first- and second-order spatial variance errors of the Doppler rate and the first-order spatial variance error of the derivative of the Doppler rate should be estimated and compensated.

Use the fitting coefficients in (13) to correct the corresponding Doppler parameters in the high-squint airborne SAR imaging algorithm [6] so as to achieve the focus improved image. In order to improve the accuracy of the Doppler parameter estimation, multiple iterations are performed. The flowchart of the azimuth compression combined with the EMAM method in the high-squint SAR imaging algorithm is shown in Figure 4.

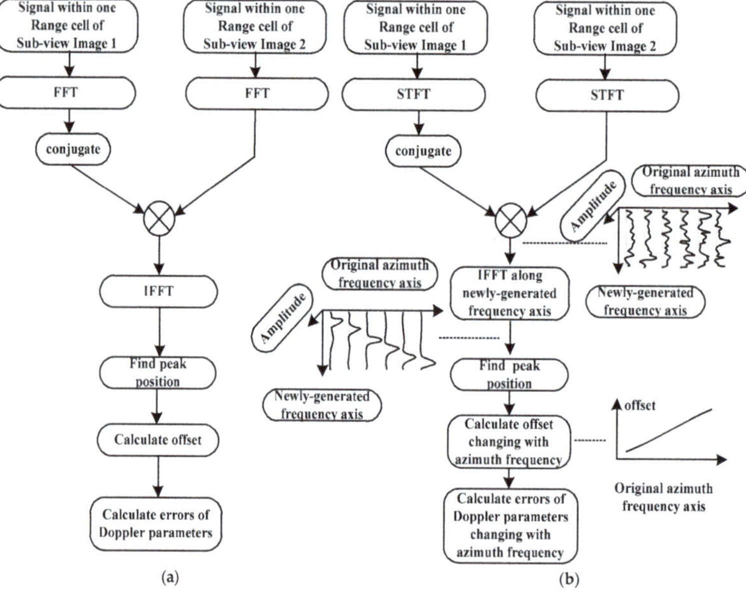

Figure 3. The flow charts of the basic MAM method and the EMAM method. (**a**) The basic MAM method; (**b**) the EMAM method.

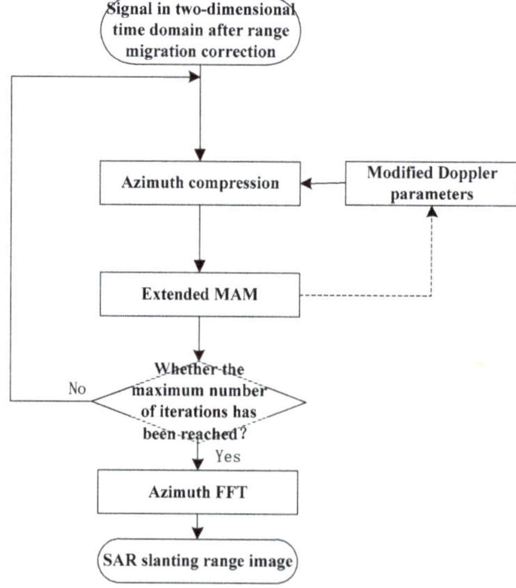

Figure 4. The flowchart of the azimuth compression combined with the EMAM method in the high-squint airborne SAR imaging algorithm.

4. Performance Analysis

4.1. Spatial Variance of Doppler Parameters

The spatial variance of the Doppler parameters is analyzed based on a typical VHS airborne SAR geometry, as shown in Figure 5. The XOY plane is the ground plane. V and A are the velocity and acceleration of the aircraft. The velocity vector is in the YOZ plane. H is the aircraft altitude. R_{ref} is the corresponding slanting distance. γ and γ_A are the velocity dive angle (between the velocity vector and the horizontal plane) and the acceleration dive angle (between the acceleration vector and the horizontal plane), respectively. α and α_A are the velocity azimuth angle (between the projections of the slanting distance vector and the velocity vector to the ground) and the acceleration azimuth angle (between the projections of the slanting distance vector and the acceleration vector to the ground), respectively. θ is the squint angle (between the velocity vector and the slanting distance vector). The aircraft motion time is 4 s.

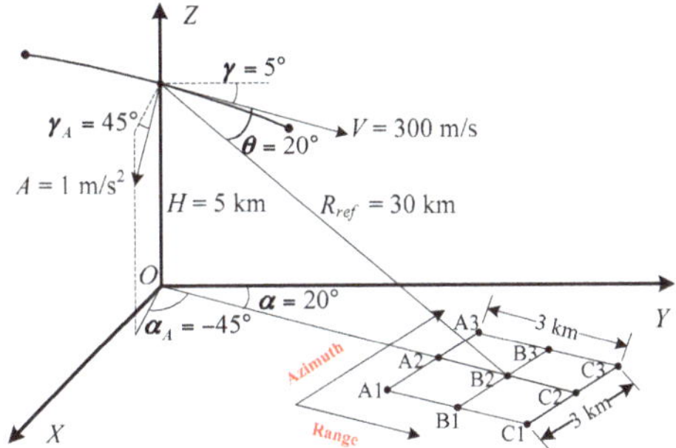

Figure 5. Typical VHS airborne SAR geometry.

Figure 6a,b shows the spatial variance of the Doppler rate and the derivative of the Doppler rate, respectively. The center of the figure represents the beam irradiation position B2. It can be clearly seen that the spatial variations of the Doppler rate and the derivative of the Doppler rate are about 8 Hz/s and 0.08 Hz^3, respectively. Figure 7a,b shows the spatially-variant phase errors caused by the spatially-variant Doppler parameters, which are about 100 rad (larger than $\frac{\pi}{4}$) and 4 rad (larger than $\frac{\pi}{8}$), respectively. If the phase error caused by the Doppler rate is larger than $\frac{\pi}{4}$, or the phase error caused by the derivative of the Doppler rate is larger than $\frac{\pi}{8}$, it will seriously degrade the image quality. Thus, the spatial variance of the Doppler rate and the derivative of the Doppler rate should be estimated and compensated.

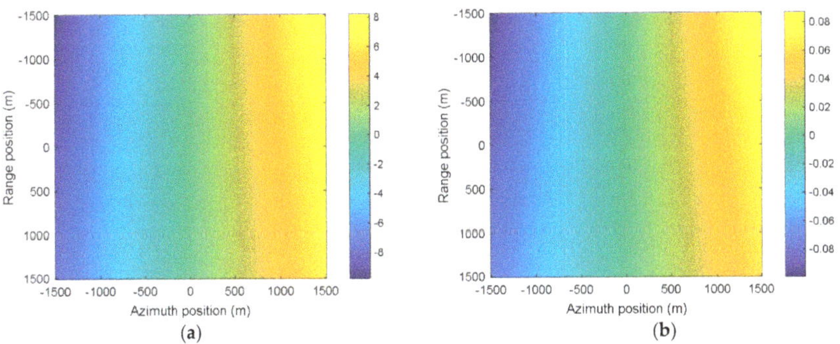

Figure 6. The spatial variance of the Doppler parameters. (**a**) The Doppler rate; (**b**) the derivative of the Doppler rate.

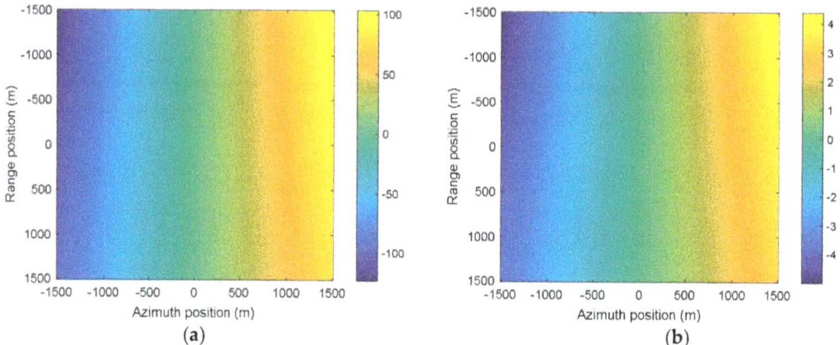

Figure 7. The spatially-variant phase errors caused by the spatially-variant Doppler parameters. (**a**) The phase error caused by the Doppler rate; (**b**) the phase error caused by the derivative of the Doppler rate.

The data at the range time domain and the azimuth frequency domain in the step of the azimuth compression for the high-squint airborne SAR imaging algorithm is as follows.

$$s(f_a, t_r; R, \theta) = \operatorname{sinc}\left(\frac{t_r - 2R/c}{1/B_r}\right) \operatorname{rect}\left(\frac{f_a - f_{ac}}{f_{dr}T_s}\right) \exp\left\{j\varphi_0 + j\pi a_2(f_a - f_{ac})^2 + j\pi a_3(f_a - f_{ac})^3\right\}, \quad (14)$$

where the first sinc function is the result of the range pulse compression, and the latter two are the azimuth envelope and phase modulations. t_r and f_a are the range time and the azimuth frequency, respectively. f_{ac} and f_{dr} are the Doppler centroid and the Doppler rate of the target with the slant range R and the squint angle θ (which is the angle between the velocity vector and range vector), respectively. B_r is the signal bandwidth. T_s is the azimuth accumulation time. In the phase modulation, the constant phase φ_0 does not affect focus, and the spatial variance of a_2 and a_3 (related to the spatial variance of the Doppler parameters) is analyzed below. a_2 and a_3 can be expressed as the functions of (R, f_{dc}). Then, these functions can be further expanded at $f_{dc} = f_{dc,cen} = (2V\cos\theta_{cen})/\lambda$ as the Taylor series shown in (15) [6].

$$\begin{aligned} a_2 &\approx a_{20} + a_{21}(f_{dc} - f_{dc,cen}) + a_{22}(f_{dc} - f_{dc,cen})^2 \\ a_3 &\approx a_{30} + a_{31}(f_{dc} - f_{dc,cen}), \end{aligned} \quad (15)$$

where a_{20} and a_{30} are the constant coefficients, a_{21} and a_{31} are the first-order spatial variance coefficients, and a_{22} is the second-order spatial variance coefficient of a_2.

When the target is at the azimuth center of the distance-isoline of the illuminated scene, θ and f_{ac} become θ_{cen} and $f_{dc,cen}$, respectively. V is the aircraft velocity. λ is the wavelength.

In order to get a well-focused image, the absolute values of the phase errors caused by the spatial variance of a_2 and a_3 should be less than $\frac{\pi}{4}$ and $\frac{\pi}{8}$, as expressed by (16) and (17), respectively.

$$\left|a_{21}(f_{dc} - f_{dc,cen})\left(\frac{B_a}{2}\right)^2 \pi\right| < \frac{\pi}{4}, \quad \left|a_{22}(f_{dc} - f_{dc,cen})^2\left(\frac{B_a}{2}\right)^2 \pi\right| < \frac{\pi}{4}, \quad (16)$$

$$\left|a_{31}(f_{dc} - f_{dc,cen})\left(\frac{B_a}{2}\right)^3 \pi\right| < \frac{\pi}{8}, \quad (17)$$

where B_a is the Doppler width of the target.

4.2. Complexity

The computational complexity (floating-point operation) of the EMAM method is analyzed in detail. For the signal at a certain range cell, the number of azimuth points is N_a. The complexity of the main steps of the EMAM method is as shown in Table 1.

Table 1. The complexity of the main steps of the EMAM method.

Main Step	Operation	Complexity
S_I: Achieving three sub-view images	FFT	$15N_a \log_2 N_a$
S_{II}: Estimating $\hat{e}_{dr}(f_{dc})$ and $\hat{e}_{3rd}(f_{dc})$ (L operations of sliding windowing manipulation, window width: N_w)	STFT Complex conjugate multiplication IFFT Modulus	$30LN_w \log_2 N_w$ $21LN_w$ $15LN_a \log_2 N_a$ $27LN_a$
S_{III}: Estimating the fitting coefficients of $\hat{e}_{dr}(f_{dc})$ and $\hat{e}_{3rd}(f_{dc})$	Curve fitting	$32L$
S_{com}: Total		$15N_a \log_2 N_a + (30N_w \log_2 N_w + 21N_w + 15N_a \log_2 N_a + 27N_a + 32)L$

Therefore, the computational complexity of the main steps for the EMAM method can be written as (18).

$$S_{com} = S_I + S_{II} + S_{III}. \tag{18}$$

In order to improve the accuracy of the Doppler parameters, multiple iterations are performed. Therefore, if the number of iterations is K, the computational complexity of the EMAM method can be written as:

$$S_{EMAM} = KS_{com}. \tag{19}$$

The complexity of the main steps of the basic MAM method and the IMAM method are shown in Tables 2 and 3, respectively.

Table 2. The complexity of the main steps of the basic MAM method.

Main Step	Operation	Complexity
S_I: Achieving three sub-view images	FFT	$15N_a \log_2 N_a$
S_{II}: Estimating $\hat{e}_{dr}$ and $\hat{e}_{3rd}$	FFT Complex conjugate multiplication IFFT Modulus	$30N_a \log_2 N_a$ $21N_a$ $15N_a \log_2 N_a$ $27N_a$
S_{com}: Total		$15N_a \log_2 N_a + 30N_a \log_2 N_a + 21N_a + 15N_a \log_2 N_a + 27N_a$

Table 3. The complexity of the main steps of the improved MAM (IMAM) method.

Main Step	Operation	Complexity
S_I: Achieving two sub-view images	FFT	$10N_a \log_2 N_a$
S_{II}: Estimating $\hat{e}_{dr}(f_{dc})$ (L operations of sliding windowing manipulation, window width: N_w)	STFT Complex conjugate multiplication IFFT Modulus	$10LN_w \log_2 N_w$ $7LN_w$ $5LN_a \log_2 N_a$ $9LN_a$
S_{III}: Estimating the fitting coefficients of $\hat{e}_{dr}(f_{dc})$	Curve fitting	$16L$
S_{com}: Total		$10N_a \log_2 N_a +$ $(10N_w \log_2 N_w + 7N_w + 5N_a \log_2 N_a + 9N_a + 16)L$

Assuming that the number of iterations K is three, the window width N_w in the IMAM method and the EMAM method is 100, and the number of sliding windowing manipulations L is $N_a/50$, then the complexity of the different methods can be compared as shown in Figure 8. It can be clearly seen that the complexity of the EMAM method is larger than the basic method and the IMAM method due to further estimating and compensating the first-order component of the spatially-dependent derivative of the Doppler rate, which increases the data processing time. When N_a is 4096, the complexities of the basic method, the IMAM method, and the EMAM method are 9.437×10^6, 7.196×10^7, and

2.136×10^8, respectively. However, the computational complexity of the proposed method does not change qualitatively, and the real-time implementation of the EMAM onboard could be achieved after evaluating the existing hardware systems.

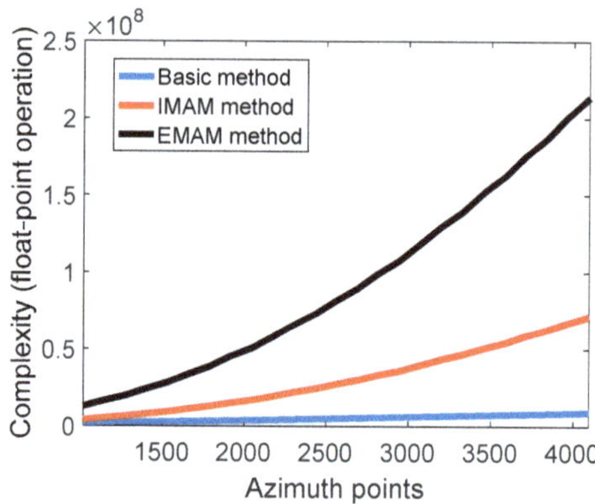

Figure 8. The computational complexity of the different methods.

5. Results

5.1. Simulation

The point target simulations are performed based on the geometry in Figure 5. The point targets are distributed as a 3×3 matrix on the ground plane with both 3 km in range and azimuth. In order to illustrate the advantages of the EMAM method, the imaging results of the basic MAM method and the IMAM method are given. The velocity error $\Delta V(10 \text{ m/s})$ and the acceleration error $\Delta A\left(-0.1 \text{ m/s}^2\right)$ are added in the imaging process. Here are the examples of point targets C1, C2, and C3 in Figure 5 to illustrate and compare the estimation results of the Doppler parameters of the different methods.

Figures 9 and 10 show the two-dimensional imaging results and the azimuth impulse responses of targets by the different methods with the velocity and the acceleration errors, respectively. In Figure 9, the horizontal axis and the vertical axis represent the azimuth samples and range samples, respectively. In Figure 10, the horizontal axis represents the azimuth frequency and the vertical axis refers to the corresponding amplitude of the target (converted to dB). The sub-images from left to right represent C1, C2, and C3 in turn. Figures 9a and 10a show the two-dimensional imaging results and the azimuth impulse responses based on the Doppler parameters with errors, respectively, and there is no Doppler parameter estimation. It can be clearly seen that the images are seriously defocused, and the Doppler bandwidths of the three points after the deramping are still about 10 Hz, indicating that there are still significant secondary phases, and the errors of the Doppler rates are very large. Figures 9b and 10b show the two-dimensional imaging results and the azimuth impulse responses by the basic MAM method, respectively. It can be seen that the focus of point target C2 at the azimuth center is better, but point targets C1 and C3 at the azimuth edges are noticeably defocused. The main reason is that the estimated Doppler parameters by the basic MAM method are the averaged results of the real ones, and their spatial variance is not considered, resulting in the fact that the point targets at the azimuth edges still have significant secondary phase errors. Figures 9c and 10c show the results of the IMAM method. It can be seen that the sidelobes of the three point targets are asymmetrical because the IMAM method only deals with the spatial variance of the Doppler rate. The peak sidelobe ratios are about -10 dB, as shown in Table 4. Figures 9d and 10d show the results of the EMAM method. It can be seen that the targets both at the azimuth center and edges are well-focused, indicating that the EMAM method can estimate the spatial variance of the Doppler rate and the derivative of the Doppler rate well. The peak sidelobe ratios are about -13 dB, as shown in Table 4, indicating that the EMAM method has achieved higher estimation accuracy.

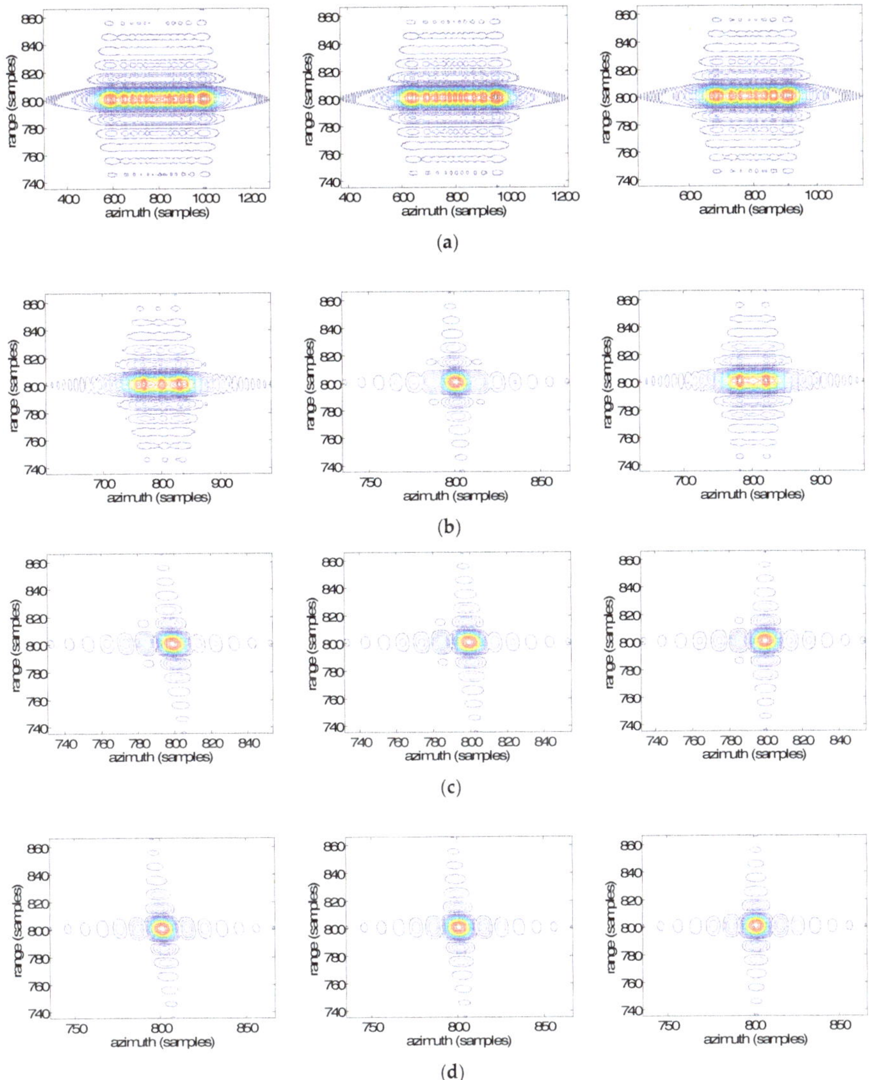

Figure 9. The two-dimensional imaging results of targets by the different methods with the velocity and acceleration errors. (**a**) No Doppler parameter estimation; (**b**) the basic MAM method; (**c**) the IMAM method; (**d**) the EMAM method.

In order to further illustrate the accuracy of the Doppler parameter estimation, Table 5 shows the estimation results of the errors of the Doppler parameters based on the basic MAM method, the IMAM method, and the EMAM method. It can be seen that the estimation results of $\hat{e}_{dr0}$ based on the three methods are relatively close to the real values, and likewise for the estimation results of $\hat{e}_{3rd0}$ by the basic method and the EMAM method. The IMAM method and the EMAM method can estimate $\hat{e}_{dr1}$ well. However, the errors of the estimation results of $\hat{e}_{dr2}$ based on the IMAM method and the EMAM method are relatively large. The reason is that the phase error caused by this term is very small and has little effect on the image focus based on the specific geometry in Figure 5. Moreover, the EMAM method can further estimate $\hat{e}_{3rd1}$ well.

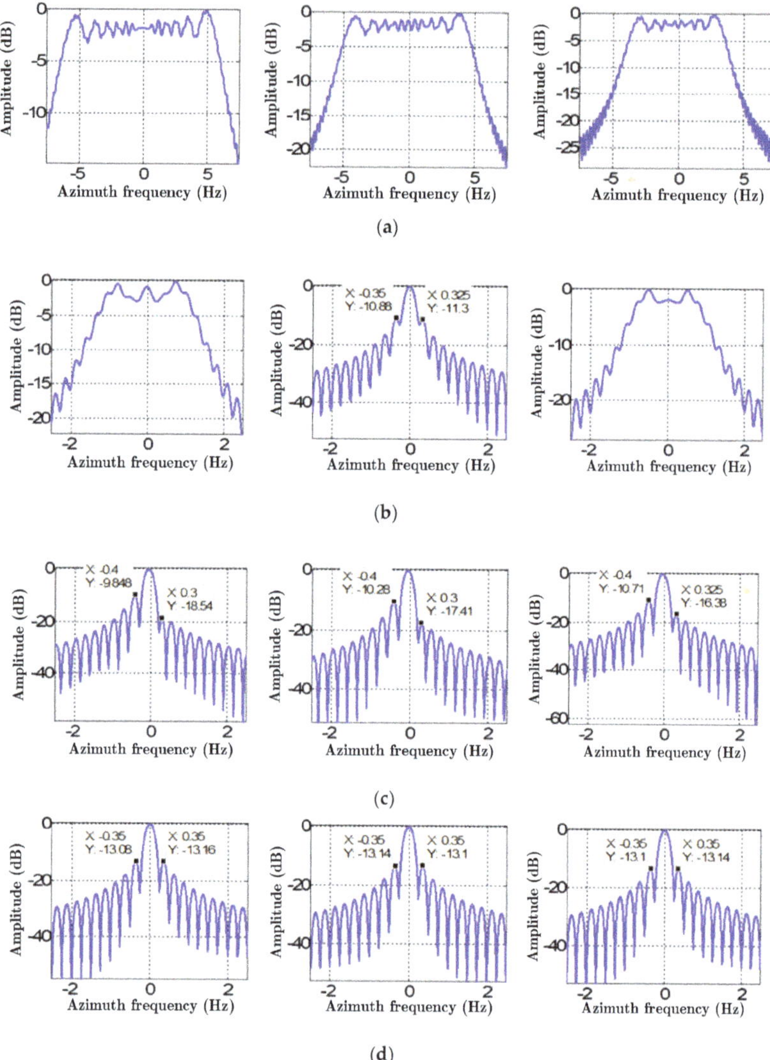

Figure 10. The azimuth impulse responses of targets by the different methods with the velocity and acceleration errors. (**a**) No Doppler parameter estimation; (**b**) the basic MAM method; (**c**) the IMAM method; (**d**) the EMAM method.

Table 4. The azimuth performance analysis of the three point targets C1, C2, and C3 based on the basic MAM, the IMAM method, and the EMAM method.

Method	Index	Point Target C1	Point Target C2	Point Target C3
Basic MAM	PSLR (dB)	−4.72	−10.88	−6.12
	ISLR (dB)	−8.25	−7.87	−8.72
	Azimuth resolution (m)	5.67	0.59	4.35
IMAM	PSLR (dB)	−9.85	−10.28	−10.71
	ISLR (dB)	−8.69	−8.95	−9.18
	Azimuth resolution (m)	0.59	0.58	0.58
EMAM	PSLR (dB)	−13.08	−13.10	−13.10
	ISLR (dB)	−9.63	−9.64	−9.63
	Azimuth resolution (m)	0.57	0.57	0.57

Note: PSLR represents the peak sidelobe ratio (the peak strength ratio of the highest side-lobe to the main-lobe), and ISLR represents the integral sidelobe ratio (the energy radio of all side-lobes to the main-lobe). The theoretical azimuth resolution of the three point targets is 0.57 m.

Table 5. The estimation results of the errors of the Doppler parameters by the different methods.

Error Coefficient	Real Value	Basic MAM	IMAM	EMAM
$\hat{e}_{dr0}$ (Hz/s)	−2.6426	−2.7048	−2.6485	−2.6464
$\hat{e}_{dr1}$ (Hz)	0.0012	-	0.0012	0.0012
$\hat{e}_{dr2}$	1.2575×10^{-7}	-	1.4614×10^{-7}	1.5608×10^{-7}
$\hat{e}_{3rd0}$ $\left(\mathrm{Hz}^3\right)$	−0.0360	−0.0396	-	−0.0390
$\hat{e}_{3rd1}$ $\left(\mathrm{Hz}^2\right)$	1.2540×10^{-5}	-	-	1.2630×10^{-5}

Note: "-" indicates that the basic MAM method or the IMAM method cannot estimate this error coefficient. $\hat{e}_{dr2}$ is non-dimensional.

5.2. Real Data

To validate the EMAM method in practical applications, this section gives the results of real airborne SAR data based on the different methods. The velocity of the aircraft is about 105 m/s, and the acceleration is about 0.26 m/s^2. The aircraft altitude is about 5 km. The squint angle is about 30°. The azimuth width of the image is about 1.2 km, and the range width is about 500 m. The data are processed based on the inertial navigation information (inaccurate), the basic MAM method, the IMAM method, and the EMAM method, respectively, and the results of the slanting distance image are shown in Figure 11. In the figures, the horizontal direction represents the azimuth frequency domain, and the vertical direction refers to the range time domain.

Figure 11a is the image based on the inertial navigation information. It can be seen that the defocus condition of the image is more and more serious from left to right, indicating that the spatial variance of the Doppler parameters is very obvious. The Doppler parameters calculated from the inertial information are closer to the real ones of the left scene. Figure 11b is the image based on the basic MAM method. The azimuth center of the scene is well-focused, but there is still obvious defocus at the azimuth edges, indicating that the estimation results of the Doppler parameters by the basic MAM method are the averages of the real Doppler parameters of the whole scene, which are close to the real ones of the central scene. Figure 11c,d shows the images based on the IMAM method and the EMAM method, respectively. It can be seen that the focus of the image has been significantly improved compared with the basic MAM method, but the comparison between these two methods is not obvious. Therefore, a strong scatterer in the small red square is chosen as shown in Figure 11c,d to further compare the two methods. Figure 12 is the azimuth impulse responses of the chosen strong scatterer based on the IMAM method and the EMAM method. It can be clearly seen that the azimuth sidelobe asymmetry exists in the IMAM result, while for the EMAM result, the main-lobe is narrower and the side-lobe is lower and basically symmetrical, which explains that the EMAM method is better than the IMAM method.

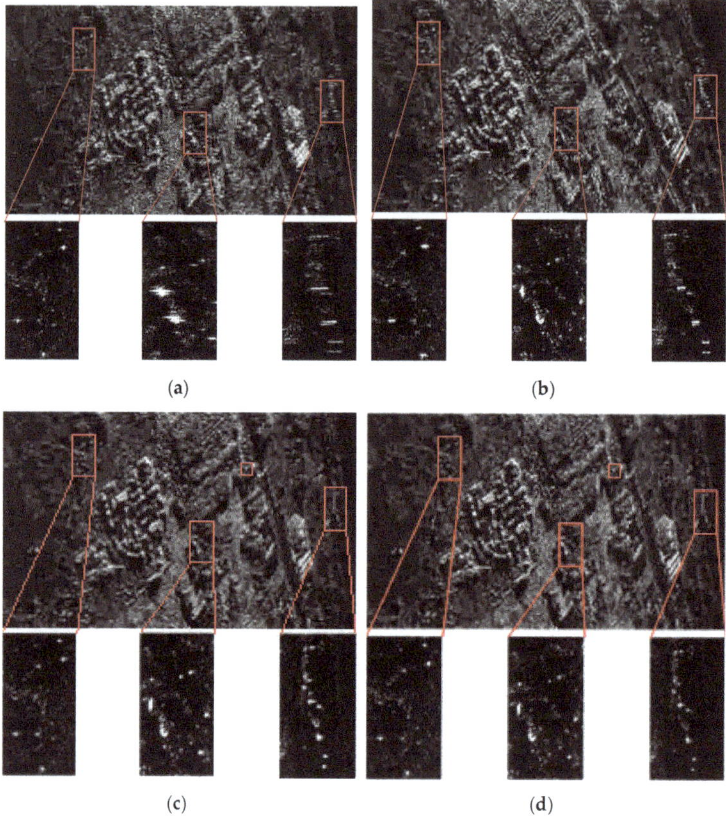

Figure 11. The images of real airborne data based on the different methods. (**a**) The inertial navigation information; (**b**) the basic MAM method; (**c**) the IMAM method; (**d**) the EMAM method.

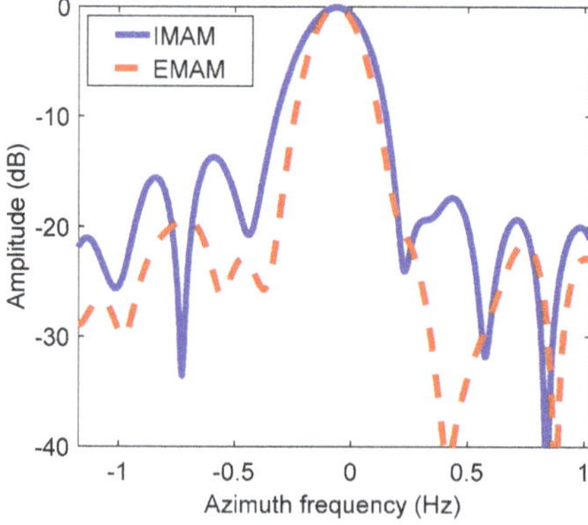

Figure 12. The azimuth impulse responses of the chosen strong scatterer based on the IMAM method and the EMAM method.

Figure 13 shows the estimation curves of the spatially-variant Doppler parameters based on the different methods. It can be seen from the figures that the Doppler parameters obviously change with the azimuth frequency. The estimation results of the Doppler rate and the derivative of the Doppler rate by the basic MAM method are basically the averages of the EMAM method. The blue solid lines represent the estimation results of the Doppler parameters by the IMAM method or the EMAM method, and the red dotted lines refer to the curve fitting values of the estimated Doppler parameters. As shown in Figure 13a, the IMAM method can estimate the spatially-variant Doppler rate, and the estimation result is basically consistent with the EMAM method; while the EMAM method can further estimate the spatially-variant derivative of the Doppler rate as shown in Figure 13b.

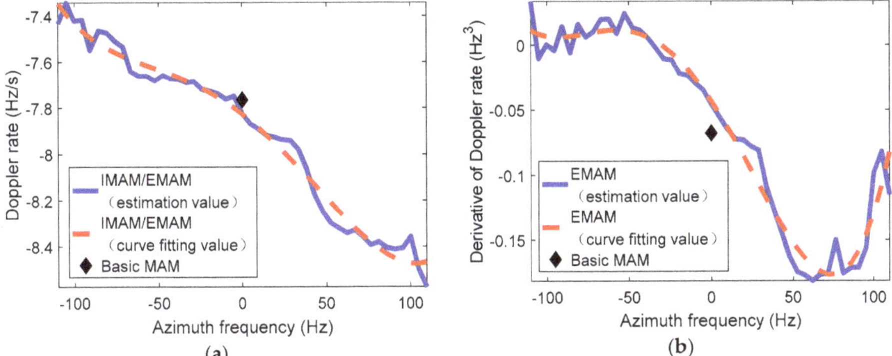

Figure 13. The estimation curves of the spatially-variant Doppler parameters based on the different methods. (**a**) The Doppler rate; (**b**) the derivative of the Doppler rate.

Figure 14 shows the residual spatial variance of the Doppler parameters after compensation based on the estimation results of the EMAM method. It can be seen that the first- and second-order spatial variance of the Doppler rate and the first-order spatial variance of the derivative of the Doppler rate are basically eliminated; only the higher order spatial variance is left, which does not affect the focus of the image.

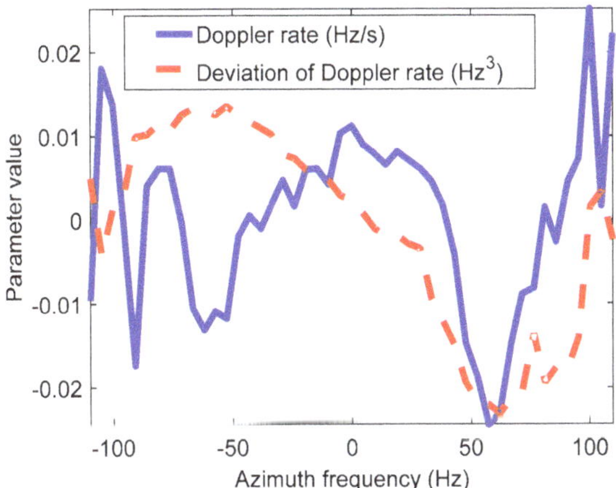

Figure 14. The residual spatial variance of the Doppler parameters after compensation based on the estimation results of the EMAM method.

6. Conclusions

In this study, an EMAM method has been proposed for DPEC of the VHS airborne SAR imaging. Comparing with the existing MAM-based DPEC methods, the EMAM is superior in achieving higher accuracy benefiting from the additional estimation and compensation for the spatial dependence of the third-order Doppler phase, corresponding to the derivative of the Doppler rate. The EMAM method not only avoids the azimuth sidelobe lifting, but also gets rid of the azimuth sidelobe asymmetry. Specifically, the EMAM method firstly achieves sub-view images via multi-looking processing. Then, a sliding-window-based cross-correlation is implemented to achieve image offsets. Based on the unique mapping between such offsets and the Doppler parameters, the DPEC can be accurately implemented. By showing that the EMAM outperforms the existing DPEC methods in both the computer simulations and the real airborne data processing experiments, the effectiveness of the presented approach has been validated. Both the computer simulations and the real airborne data processing experiments show that based on the EMAM method, the targets both at the azimuth center and edges are well focused, indicating that the EMAM method can accurately estimate and compensate the spatial variance of the Doppler rate and the derivative of the Doppler rate. Further research may focus on the real-time implementation of the EMAM onboard.

Author Contributions: Z.Z., Y.L., and Y.W. designed the study. Z.Z., Y.L., and L.L. developed the computer simulations and performed the real airborne data analysis. Z.Z., Y.L., and T.Z. prepared the figures and tables. Z.Z. and Y.L. wrote the final manuscript. All authors read and approved the final manuscript.

Funding: This research was funded by the National Key R&D Program of China Grant Number 2017YFC0804700, by the National Science Fund for Distinguished Young Scholars Grant Number 61625103, and by the Key Program of National Natural Science Foundation of China Grant Numbers 91738302, 91438203. The APC was funded by the Beijing Institute of Technology Research Fund Program for Young Scholars.

Conflicts of Interest: The authors declare no conflict of interest.

References

1. Cumming, I.; Wong, F. *Digital Processing of Synthetic Aperture Radar Data: Algorithms and Implementation*; Artech House: Norwood, MA, USA, 2005.
2. Soumekh, M. *Synthetic Aperture Radar Signal Processing with MATLAB Algorithms*; Wiley: New York, NY, USA, 1999.
3. Cantalloube, H.; Dubois-Fernandez, P. Airborne X-band SAR imaging with 10 cm resolution: Technical challenge and preliminary results. *IET Radar Sonar Navig.* **2006**, *153*, 163–176. [CrossRef]
4. Ballester-Berman, J.; Lopez-Sanchez, J.; Fortuny-Guash, J. Retrieval of biophysical parameters of agricultural crops using polarimetric SAR interferometry. *IEEE Trans. Geosci. Remote Sens.* **2005**, *43*, 683–694. [CrossRef]
5. Curlander, J.; McDonough, R. *Synthetic Aperture Radar: Systems and Signal Processing*; Wiley: New York, NY, USA, 1991.
6. Zeng, T.; Li, Y.; Ding, Z.; Long, T.; Yao, D.; Sun, Y. Subaperture Approach Based on Azimuth-Dependent Range Cell Migration Correction and Azimuth Focusing Parameter Equalization for Maneuvering High-Squint-Mode SAR. *IEEE Trans. Geosci. Remote Sens.* **2015**, *53*, 6718–6734. [CrossRef]
7. Steele-Dunne, S.; Mcnairn, H.; Monsivais-Huertero, A.; Judge, J.; Liu, P.; Papathanassiou, K. Radar Remote Sensing of Agricultural Canopies: A Review. *IEEE J. Sel. Top. Appl. Earth Obs.* **2017**, *10*, 2249–2273. [CrossRef]
8. Guo, H.; Li, X. Technical characteristics and potential application of the new generation SAR for Earth observation. *Chin. Sci. Bull.* **2011**, *56*, 1155–1168. [CrossRef]
9. Hu, C.; Li, Y.; Dong, X.; Wang, R.; Cui, C. Optimal 3D deformation measuring in inclined geosynchronous orbit SAR differential interferometry. *Sci. China (Inf. Sci.)* **2017**, *60*, 060303. [CrossRef]
10. Yin, W.; Ding, Z.; Lu, X.; Zhu, Y. Beam scan mode analysis and design for geosynchronous SAR. *Sci. China (Inf. Sci.)* **2017**, *60*, 060306. [CrossRef]
11. Kennedy, T. Strapdown inertial measurement units for motion compensation for synthetic aperture radars. *IEEE Trans. Aerosp. Electron. Syst. Mag.* **1988**, *3*, 32–35. [CrossRef]
12. Bamler, R. Doppler frequency estimation and the Cramer-Rao bound. *IEEE Trans. Geosci. Remote Sens.* **1991**, *29*, 385–390. [CrossRef]

13. Madsen, S. Estimating the Doppler centroid of SAR data. *IEEE Trans. Aerosp. Electron. Syst.* **1989**, *25*, 134–140. [CrossRef]

14. Cantalloube, H.; Nahum, C. Multiscale Local Map-Drift-Driven Multilateration SAR Autofocus Using Fast Polar Format Image Synthesis. *IEEE Trans. Geosci. Remote Sens.* **2011**, *49*, 3730–3736. [CrossRef]

15. Li, X.; Liu, G.; Ni, J. Autofocusing of ISAR images based on entropy minimization. *IEEE Trans. Aerosp. Electron. Syst.* **1999**, *35*, 1240–1252. [CrossRef]

16. Li, F.; Held, D.; Curlander, J.; Wu, C.; Curlander, J. Doppler Parameter Estimation for Spaceborne Synthetic-Aperture Radars. *IEEE Trans. Geosci. Remote Sens.* **1985**, *1*, 47–56. [CrossRef]

17. Long, T.; Lu, Z.; Ding, Z.; Liu, L. A DBS Doppler Centroid Estimation Algorithm Based on Entropy Minimization. *IEEE Trans. Geosci. Remote Sens.* **2011**, *49*, 3703–3712. [CrossRef]

18. Zeng, T.; Wang, R.; Li, F. SAR Image Autofocus Utilizing Minimum-Entropy Criterion. *IEEE Geosci. Remote Sens. Lett.* **2013**, *10*, 1552–1556. [CrossRef]

19. Samczynski, P.; Kulpa, K. Coherent mapdrift technique. *IEEE Trans. Geosci. Remote Sens.* **2010**, *48*, 1505–1517. [CrossRef]

20. Morrison, R.; Do, M.; Munson, D. SAR image autofocus by sharpness optimization: A theoretical study. *IEEE Trans. Image Process.* **2007**, *16*, 2309–2321. [CrossRef] [PubMed]

21. Berizzi, F.; Corsini, G. Autofocusing of inverse synthetic aperture radar images using contrast optimization. *IEEE Trans. Aerosp. Electron. Syst.* **1996**, *32*, 1185–1191. [CrossRef]

22. Carrara, W.; Goodman, R.; Majewski, R. *Spotlight Synthetic Aperture Radar: Signal Processing Algorithms*; Artech House: Norwood, MA, USA, 1995.

23. Xing, M.; Jiang, X.; Wu, R.; Zhou, F.; Bao, Z. Motion Compensation for UAV SAR Based on Raw Radar Data. *IEEE Trans. Geosci. Remote Sens.* **2009**, *47*, 2870–2883. [CrossRef]

24. Tang, Y.; Zhang, B.; Xing, M.; Bao, Z.; Guo, L. The Space-Variant Phase-Error Matching Map-Drift Algorithm for Highly Squinted SAR. *IEEE Geosci. Remote Sens. Lett.* **2013**, *10*, 845–849. [CrossRef]

25. Pu, W.; Li, W.; Wu, J.; Huang, Y.; Yang, J.; Yang, H. An Azimuth-Variant Autofocus Scheme of Bistatic Forward-Looking Synthetic Aperture Radar. *IEEE Geosci. Remote Sens. Lett.* **2017**, *14*, 689–693.

26. Wang, Y.; Li, J.; Chen, J.; Xu, H. A Parameter-Adjusting Polar Format Algorithm for Extremely High Squint SAR Imaging. *IEEE Trans. Geosci. Remote Sens.* **2014**, *52*, 640–650. [CrossRef]

Article

Knowledge-Aided Doppler Beam Sharpening Super-Resolution Imaging by Exploiting the Spatial Continuity Information

Hongmeng Chen [1], Zeyu Wang [2], Jing Liu [1], Xiaoli Yi [1], Hanwei Sun [1], Heqiang Mu [1], Ming Li [3] and Yaobing Lu [1,*]

[1] Beijing Institute of Radio Measurement, Beijing 100854, China; chenhongmeng123@163.com (H.C.); jingliu_sp2306@163.com (J.L.); 13466798179@163.com (X.Y.); 13488682636@126.com (H.S.); 13611258792@163.com (H.M.)

[2] China Academy of Electronics and Information Technology, China Electronic Technology Group Corporation, Beijing 100846, China; beidou13579@163.com

[3] National Laboratory of Radar Signal Processing, Xidian University, Xi'an 710071, China; liming@xidian.edu.cn

* Correspondence: luyaobing65@163.com; Tel.: +86-10-8852-7368

Received: 16 January 2019; Accepted: 11 April 2019; Published: 23 April 2019

Abstract: This paper deals with the problem of high cross-range resolution Doppler beam sharpening (DBS) imaging for airborne wide-area surveillance (WAS) radar under short dwell time situations. A knowledge-aided DBS (KA-DBS) imaging algorithm is proposed. In the proposed KA-DBS framework, the DBS imaging model for WAS radar is constructed and the cross-range resolution is analyzed. Since the radar illuminates the imaging scene continuously through the scanning movement of the antenna, there is strong spatial coherence between adjacent pulses. Based on this fact, forward and backward pulse information can be predicted, and the equivalent number of pulses in each coherent processing interval (CPI) will be doubled based on the autoregressive (AR) technique by taking advantage of the spatial continuity property of echoes. Finally, the predicted forward and backward pulses are utilized to merge with the initial pulses, then the newly merged pulses in each CPI are utilized to perform the DBS imaging. Since the number of newly merged pulses in KA-DBS is twice larger than that in the conventional DBS algorithm with the same dwell time, the cross-range resolution in the proposed KA-DBS algorithm can be improved by a factor of two. The imaging performance assessment conducted by resorting to real airborne data set, has verified the effectiveness of the proposed algorithm.

Keywords: wide-area surveillance; super-resolution; Doppler beam sharpening

---

1. Introduction

Airborne or spaceborne wide-area surveillance (WAS) radar [1,2] can acquire a wide-area surveillance scene at a very short time, which is usually accomplished by steering the antenna beam from one azimuth angle to another. In the scanning movement of the antenna, the dwell time at each azimuth angle is very short to guarantee a high revisit radio. Accordingly, airborne or spaceborne WAS radar is widely applied in civilian and military fields [3–7], and Doppler beam sharpening (DBS) technique is a very effective tool to accomplish the WAS ability [8–11]. However, the large surveillance region in WAS radar is at the expense of low cross-range resolution. The low cross-range resolution limits its further application. Therefore, it is definitely essential to study the high cross-range resolution further for airborne or spaceborne WAS radar in short dwell time situations.

Several researchers have studied this issue of WAS imaging in previous works, Scan-synthetic aperture radar (Scan-SAR) [1,12,13] can acquire a wide-region compared with conventional TOPs SAR mode [14,15], strip SAR mode [16–18] and spotlight SAR mode [19], and it is an effective means for WAS imaging. However, the synthetic time is usually as large as 1~10 s, which limits the high revisit ratio. Since DBS is the non-focused form of SAR [1,8–11], the imaging time for DBS is usually as small as 0.05~0.1 s, the revisit ratio is very high. Therefore, our attention in this paper is mainly paid to the DBS imaging. For DBS imaging, Fourier transform (FT) [1,8,9], Relax [10], APES [11] are used to increase the cross-range resolution. However, the performance of these existing methods is generally not satisfactory in the engineering applications.

In this paper, we propose an efficient DBS cross-range resolution enhancement architecture, namely knowledge-aided DBS (i.e., KA-DBS), to increase the DBS imaging performance. For the airborne WAS radar, the antenna usually works in a scanning mode, where the antenna illuminates the surveillance region continuously by steering the antenna beam from one azimuth viewing angle to another. Therefore, the echoes reflected from the scatterers on the ground may be coherent in the space. The space coherence property means that more spatial information about the echoes may be mined if proper means are used. Based on this fact, the knowledge of the spatial coherence property is fully exploited in KA-DBS. And then, the spatial continuity model of the radar echo is constructed. In order to well estimate the pulses information outside the observed coherent processing interval (CPI), the forward prediction pulses and the backward prediction pulses are estimated based on the autoregressive (AR) technique [20–22], respectively. Accordingly, the number of pulses at one azimuth angle can be equivalently increased by merging the forward prediction pulses and the backward prediction pulses with the original pulses. Finally, the "merged pulses" are utilized to perform the DBS imaging. The number of "merged pulses" in the proposed KA-DBS algorithm is twice larger than that in the conventional DBS algorithm with the same dwell time. Therefore, the cross-range resolution in KA-DBS is doubled compared with the conventional DBS imaging algorithm. Real-data results show that the proposed algorithm performs well with short dwell time.

The rest of this paper is organized as follows: in Section 2, the DBS architecture is discussed. In Section 3, we introduce the novel KA-DBS algorithm in detail. The performance of the proposed algorithm is verified by real measured data in Section 4. Finally, some conclusions are given in Section 5.

2. DBS Imaging Model

For airborne WAS radar system, the radar dwells in a particular beam position continuously with a set of coherent processing intervals (CPIs). The surveillance region is searched by sequentially looking in all azimuth angles, the working mode of the WAS radar is illustrated in Figure 1.

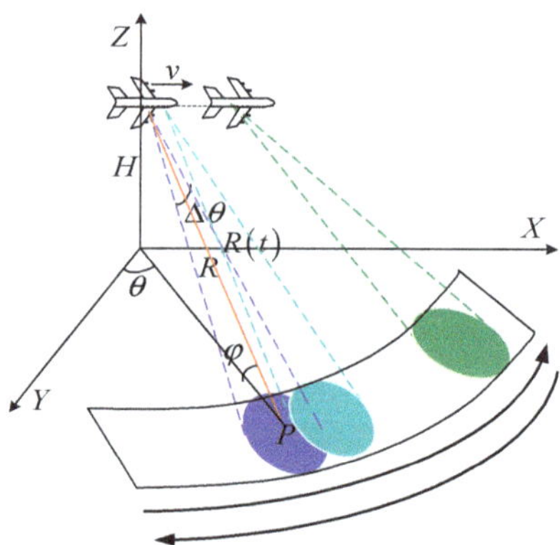

Figure 1. Geometry of DBS imaging.

Suppose that the aircraft flies along to the X-axis with velocity v, and with the flight altitude, H, and the initial slant range between the target, R_0. The azimuth angle and the elevation angle are θ and ϕ, respectively. It is assumed that a linear frequency-modulated (LFM) signal is transmitted, and it can be written as:

$$s(\tau) = \text{rect}\left(\frac{\tau}{T_p}\right)\exp\left[j2\pi\left(f_c\tau + \frac{\gamma}{2}\tau^2\right)\right]$$ (1)

where τ denotes the fast time, T_p denotes the pulse width, f_c is the carrier frequency of the transmitted signal, and rect($\cdot$) stands for the unit rectangular function. γ is the chirp rate. The echoed signal reflected from a point target can be expressed as:

$$s(\tau,t) = \sigma \cdot \text{rect}\left(\frac{\tau - \frac{2R(t)}{c}}{T_p}\right)\exp\left\{j2\pi\left[f_c\left(\tau - \frac{2R(t)}{c}\right) + \frac{\gamma}{2}\left(\tau - \frac{2R(t)}{c}\right)^2\right]\right\}$$ (2)

where t denotes the slow time, $R(t)$ is the instantaneous slant range history from the radar to the point target at time t. σ is the radar cross-section (RCS) of the target, and c is the velocity of light.

According to the DBS geometry illustrated in Figure 1, the instantaneous slant range history between the target and the radar can be written as:

$$R(t) = \sqrt{R_0^2 + (vt)^2 - 2R_0vt\sin\theta\cos\varphi}$$ (3)

Equation (3) can be expanded into a Taylor series, and it can be expressed as:

$$R(t) = R_0 - vt\sin\theta\cos\varphi + \frac{(vt)^2\left(1 - \sin^2\theta\cos^2\varphi\right)}{2R_0} + O(t^3)$$ (4)

Since the dwell time in each fixed azimuth viewing angle is very short, the value of the airplane travels vt is far less the slant range R_0 (i.e., $vt \square R_0$), then the instantaneous slant range history can be approximately expressed as:

$$R(t) \approx R_0 - vt\sin\theta\cos\varphi$$ (5)

The Doppler centroid can be estimated by the follows:

$$f_d = -\frac{2}{\lambda}\frac{dR(t)}{dt} = \frac{2v\sin\theta\cos\varphi}{\lambda}$$ (6)

It can be known that different azimuth angles corresponds to different Doppler frequencies. Therefore, the problem of distinguishing different scatterers at different azimuth angles can be transformed into the problem of distinguishing different scatterers at different Doppler frequencies. Suppose a scatterer is located at the azimuth angle θ_0, and the azimuth angle θ_0 is also the center of the antenna beam. The boundaries of the antenna beam can be denoted as $\theta_0 - \Delta\theta/2$ and $\theta_0 + \Delta\theta/2$, which correspond to the Doppler frequencies f_{dh} and f_{dl}, respectively. $\Delta\theta$ is the 3 dB width of the radar beam. The Doppler bandwidth of the scatterer can be derived as:

$$\Delta f_d = \left|f_{dh} - f_{dl}\right| = \left|f_d\left(\theta_0 - \frac{\Delta\theta}{2}\right) - f_d\left(\theta_0 + \frac{\Delta\theta}{2}\right)\right| = \frac{2v\cos\theta_0\cos\varphi}{\lambda}\Delta\theta$$ (7)

The Doppler bandwidth illustrated in Equation (7) is the frequency excursion experienced by the scatterer during the dwell time in which the scatter is illuminated by the 3 dB width of the antenna. Assuming that pulse compression and range migration correction [23] are performed, then the echoed signal can be written as:

$$s(\tau,t) = \sigma \cdot \text{rect}\left(\frac{t}{T_a}\right)\text{sin c}\left[B\left(\tau - \frac{2R_0}{c}\right)\right]\exp\left(-j4\pi\frac{R(t)}{\lambda}\right)$$ (8)

where $T_a = N_a T_r$, is defined as the CPI in one look direction, and T_r is the pulse repetition interval (PRI). N_a is the coherent pulse number in one CPI. Assuming that there are K scattering centers in one range cell, and fast Fourier transform (FFT) is performed to get the DBS imaging result [1,8,9]:

$$S(\tau, f) = \sin c\left[B\left(\tau - \frac{2R_0}{c}\right)\right]\text{sinc}[T_a(f - f_k)] \tag{9}$$

From Equation (9), it can be seen that the Doppler resolution in the conventional FFT-based method is approximately determined by:

$$\delta f_d = 1/T_a \tag{10}$$

where δf_d is the Doppler resolution.

One important parameter is the sharpening ratio of a DBS image, K_a, as the ratio of the Doppler bandwidth to the Doppler resolution, is given by [1]:

$$K_a = \Delta f_d / \delta f_d = \Delta f_d T_a \tag{11}$$

For DBS imaging, the velocity of the aircraft and the antenna beam width are always fixed, which makes the Doppler bandwidth is fixed at the given azimuth viewing angle. Since the CPI in one look direction satisfies $T_a = N_a T_r$, then Equation (11) can be rewritten as:

$$K_a = \Delta f_d N_a T_r \tag{12}$$

From Equation (11), we can know that increasing the Doppler bandwidth, the coherent pulse number as well as the pulse repetition interval is a useful way to increase the sharpening ratio K_a. However, the sharpening ratio K_a cannot be increased infinitely.

To estimate the maximum value of the sharpening ratio K_a, we should derive the maximum aperture length in DBS imaging. Assuming that the mean slant range in Figure 2 is R_m, then we can get the maximum effective aperture length L_s in one CPI [1], which satisfies:

$$\sqrt{R_m^2 + \left(\frac{L_s}{2}\right)^2} - R_m \leq \frac{\lambda}{8} \tag{13}$$

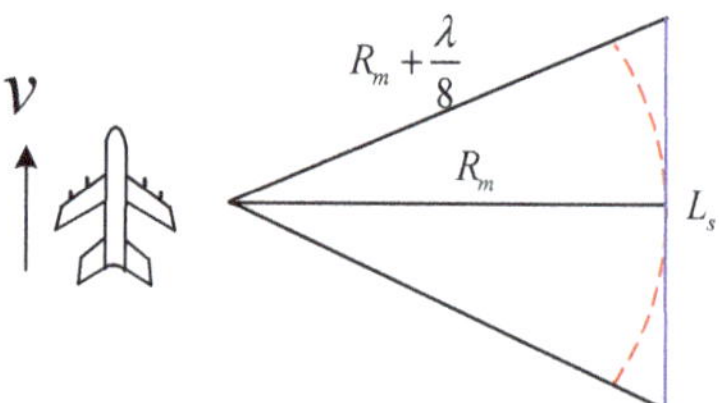

Figure 2. Illustration of the maximum aperture length in DBS imaging.

After some mathematical derivation, the maximum aperture length in DBS imaging is given by:

$$L_s \leq \sqrt{R_m \lambda} \tag{14}$$

and the maximum number of pulses in one CPI can be calculated as:

$$N_{\max} = \frac{\sqrt{R_m \lambda}}{v t_r} \tag{15}$$

Therefore, the sharpening ratio has the upper boundary:

$$K_a \leq \Delta f_d N_{\max} T_r \tag{16}$$

In order to guarantee the wide-area surveillance ability and high revisit ratio, the number of pulses in one CPI is bounded by the maximum value $N_{\max}$. The conflict between short CPI and high resolution in cross-range motivates the study of super-resolution approaches for DBS imaging.

Inspecting Equation (16), we can find that increasing the equivalent number of pulses at each CPI in one look direction may be effective to improve the cross-range resolution. In the following section, we will consider an alternative strategy to increase the sharpening ratio K_a to improve the DBS imaging performance.

3. Knowledge-Aided DBS Super-Resolution Imaging Algorithm

3.1. Spatial Continuity Property of the Echoed Signal

When the antenna of the airborne WAS radar scans the surveillance region, the radar illuminates the imaging scene continuously through the scanning movement of the antenna beam. Since the antenna beam is steered from one azimuth viewing angle to another, a target may be illuminated by many pulses in one CPI in the very short dwell time. Therefore, the received echoes are spatially coherent, and additional pulse information about the target may be acquired with the observed pulses. By exploiting this spatial continuity information, we can try to extrapolate or predict the echo information outside the observed CPI.

In order to demonstrate the assumption that the echoed signal is continuous in the spatial space with a short dwell time, two slow moving targets modeled with Swerling I [1] are injected into the real airborne radar data set. The Doppler frequencies of the two targets are 195 Hz and 215 Hz, respectively. The azimuth angle of the antenna is corresponding to 40 degree. The airborne radar parameters are listed in Table 1.

Table 1. Radar System Parameter.

Parameters	Value
Time width	10 s
Band width	12 MHz
Pulse repetition frequency	2500 Hz
Azimuth beam width	3.2°
Coherent pulses	128
Range gate number	2048

The injected signal just exist about 0.05 s in the slow time domain, which corresponds to about 128 pulses in one CPI with the given dwell time. We predict the forward and backward pulse by exploiting the observed echo information. Detailed forward and backward prediction algorithm will be given in the Section 3.2. The echoed signal of one range cell in the slow time domain is shown in Figure 3.

Based on the spatial continuity assumption, the predicted forward and backward echo information is colored in red, while the initial echo pulse in blue. Detailed pulse information prediction method will be introduced in the Section 3.2, and the prediction factor is set as 0.5 in the experiment. Because the forward and backward pulse lengths are half of the length of one CPI, the equivalent CPI length (i.e., about 0.1 s) in the newly merged signal is doubled than the original CPI length (i.e., about 0.05 s) as shown in Figure 3b. Performing the Fourier analysis to these two data sets shown in Figure 3a,b, respectively.

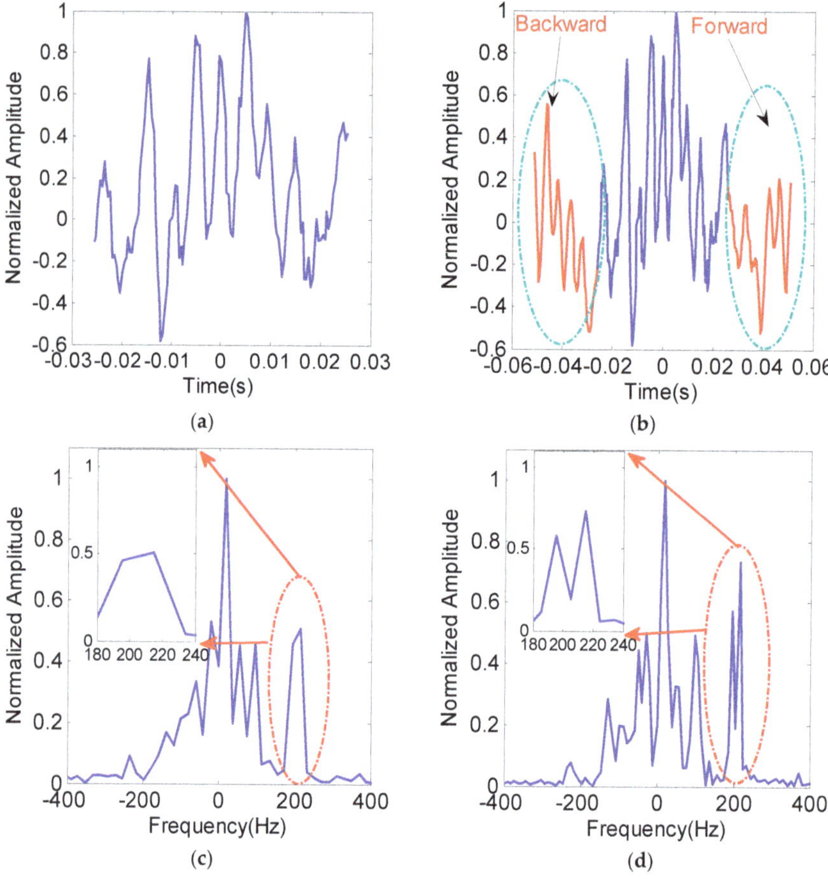

Figure 3. The original signal and the newly merged signal of one range cell in the slow time and the spectrum domain (**a**) Original echoed signal in the slow time domain; (**b**) The newly merged (predicted signal plus original signal) signal based on the spatial continuity property; (**c**)Spectrum of the original echoed signal (**d**); Spectrum of the newly merged signal.

The corresponding spectrum information can be found in Figure 3c,d, respectively. The zoomed in spectrum information about this two targets is left-top of the Figure. Since the Doppler frequency difference between the two targets is 20 Hz, conventional FFT method cannot distinguish them in the spectrum. The spectrum of the two targets is aliasing together and has just one peak in Figure 3c. However, by exploiting the prior knowledge of the spatial continuity, we can distinguish them well in the frequency domain as shown in Figure 3d. This experiment demonstrates that the assumption that the echoed signal is spatially continuous in the spatial space, and the cross-range resolution can be increased if the spatial continuity property is well exploited in DBS imaging.

3.2. KA-DBS Imaging Algorithm

From above analysis, it can be known that the Doppler resolution is proportional to the CPI T_a in one azimuth angle. Therefore, we emphasize on increasing the equivalent number of pulses in each CPI by exploiting the spatial continuity information.

By exploiting the spatial continuity information, the forward pulses and the backward pulses outside the measured CPI can be predicated by taking the advantage of the autoregressive (AR) technique [20,21]. AR is a technique that can forecast the future values on the basis of past values of a time series data, which has well been used in ISAR imaging [22]. In this section, we will introduce the AR technique into the DBS imaging. Firstly, the spatial continuity model of the echoed signal should be constructed. Suppose range compression is performed, the forward and backward predicated pulses can be expressed as:

$$s^f(\tau,n) = -\sum_{i=1}^{P} a_P(i)\cdot\sigma\cdot\text{rect}\left(\frac{(n-i)\cdot t_r}{T_a}\right) \cdot\sin c\left[B\left(\tau - \frac{2R((n-i)t_r)}{c}\right)\right]\exp\left(-j4\pi\frac{R((n-i)t_r)}{\lambda}\right) \tag{17}$$

$$s^b(\tau,n-P) = -\sum_{i=1}^{P} a_P^H(i)\cdot\sigma\cdot\text{rect}\left(\frac{(n-P+i)\cdot t_r}{T_a}\right) \cdot\sin c\left[B\left(\tau - \frac{2R((n-P+i)t_r)}{c}\right)\right]\exp\left(-j4\pi\frac{R((n-P+i)t_r)}{\lambda}\right) \tag{18}$$

Since our emphasis is on how to improve the cross-range resolution. For clear expression, Equations (17) and (18) can be simplified as:

$$s^f(\tau,n) = -\sum_{i=1}^{P} a_P(i)s(\tau,n-i) \tag{19}$$

$$s^b(\tau,n-p) = -\sum_{i=1}^{P} a_P^H(i)s(\tau,n-p+i) \tag{20}$$

Then, the forward and backward prediction error of the echoed signal can be calculated as:

$$e^f(\tau,n) = s^f(\tau,n) + \sum_{i=1}^{P} a_P(i)s(\tau,n-i) \tag{21}$$

$$e^b(\tau,n-p) = s^b(\tau,n-p) + \sum_{i=1}^{P} a_P^H(i)s(\tau,n-p+i) \tag{22}$$

where $s^f(\tau,n)$ and $s^b(\tau,n-p)$ denote the forward and backward prediction pulse, respectively, and $e^f(\tau,n)$ and $es^b(\tau,n-p)$ denote the forward and backward prediction error, respectively. $a(i)$ represents the AR model coefficient. P is the AR model order, and $[\cdot]^H$ denotes the conjugate transpose.

In order to obtain the AR model coefficient $a(k)$, the criterion that to minimize the sum of the forward and backward prediction errors in each iterative procedure is utilized [20,22], which is given as:

$$E_P = \frac{\sum_{n=P+1}^{N} \left|e^f(\tau,n)\right|^2 + \left|e^b(\tau,n)\right|^2}{2} \tag{23}$$

To solve Equation (23), the Levinson recursion algorithm [24–26] is used, which can be solved as follows:

$$a_P(i) = a_{P-1}(i) + a_P(P)a_{P-1}^H(P-i), i = 1,2,\ldots,P-1 \tag{24}$$

By substituting Equation (24) into Equation (19) and Equation (20), respectively, we can have:

$$e_P^f(\tau,n) = e_{P-1}^f(\tau,n) + a_P(p)e_{P-1}^b(\tau,n-1) \tag{25}$$

$$e_P^b(\tau,n) = e_{P-1}^b(\tau,n-1) + a_P^H(p)e_{P-1}^f(\tau,n) \tag{26}$$

Since the forward and backward predictions are known in each iteration, then the AR coefficients $a_p(p)$ can be calculated as:

$$a_p(p) = \frac{-2 \sum\limits_{n=P+1}^{N} e^{f}_{p-1}(\tau,n)\left(e^{b}_{p-1}(\tau,n-1)\right)^{H}}{\sum\limits_{n=P+1}^{N} \left|e^{f}_{p-1}(\tau,n)\right|^{2} + \left|e^{b}_{p-1}(\tau,n-1)\right|^{2}} \tag{27}$$

Based on the calculated AR coefficients $a_p(p)$,, the forward and the backward signals (i.e., $s^f(\tau,n)$ and $s^b(\tau,n\text{-}p)$) and can be calculated as:

$$\begin{aligned} s^f(\tau,n) &= -\sum\limits_{i=1}^{P} a_P(i) \cdot \sum\limits_{k=1}^{K} \sigma_k \cdot \text{rect}\left(\frac{(n-i)\cdot t_r}{T_a}\right) \\ &\cdot \sin c\left[B\left(\tau - \frac{2R((n-i)t_r)}{c}\right)\right] \exp\left(-j4\pi \frac{R((n-i)t_r)}{\lambda}\right) \end{aligned} \tag{28}$$

$$\begin{aligned} s^b(\tau,n-P) &= -\sum\limits_{i=1}^{P} a_P^{H}(i) \cdot \sum\limits_{k=1}^{K} \sigma_k \cdot \text{rect}\left(\frac{(n-P+i)\cdot t_r}{T_a}\right) \\ &\cdot \sin c\left[B\left(\tau - \frac{2R((n-P+i)t_r)}{c}\right)\right] \exp\left(-j4\pi \frac{R((n-P+i)t_r)}{\lambda}\right) \end{aligned} \tag{29}$$

where K is the number of scatters in one range cell. In order to fully utilize the forward and the backward signal, we will merge them with the original signal in the azimuth direction (i.e., the slow time domain). Then the newly merged signal in one range cell can be expressed as:

$$\hat{s}(\tau,1:N+2P) = \left[s^b(\tau,N-P),\ldots,s^b(\tau,N-1),s(\tau,1)\ldots,s(\tau,N),s^f(\tau,N+1),\ldots,s^f(\tau,N+P)\right] \tag{30}$$

where $\hat{s}(\tau,1:N+2P)$ is the newly merged signal in the azimuth direction. One important problem is how to determine the order of the AR model. Shall we increase P infinitely if we want to acquire more predicted signal? Of course not. The selection criteria of the AR model order is detailed discussed in [25,26], and the AR model order is set as one third of the data length in our experiment. Figure 4 gives the predicted energy ratio curve with the predictor factor. In Figure 4, the predicted energy ratio is defined as the energy amount of the predicted signal to the energy amount of the range profile.

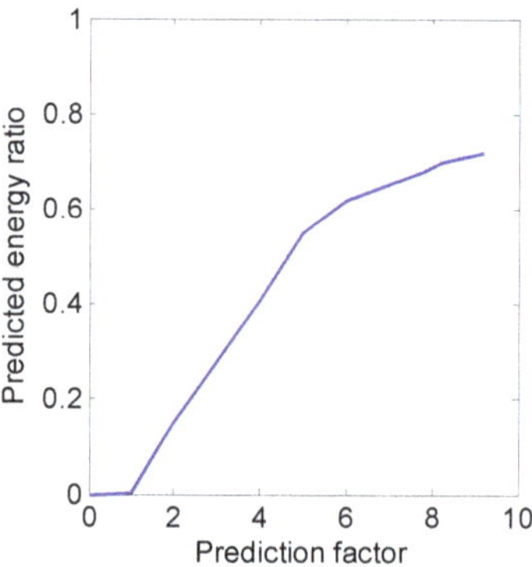

Figure 4. Predicted energy ratio curve with the prediction factor.

From Figure 4, we can know that a higher predictor factor corresponding to a higher predicted energy ratio, which can be explained that more predicted signal contribute more energy. Higher energy radio means a higher uncertainty. Typically, predictor factors less than 3 always lead to useful results (detailed information can be found in [24]). Therefore, the predictor factor is set as 0.5 in our proposed KA-DBS algorithm.

3.3. Performance of the Cross-Range Resolution in KA-DBS

Assuming that the prediction factor is 0.5 in the proposed KA-DBS, then the CPI length in each azimuth angle can be estimated as:

$$\hat{T}_a = 2NT_r = 2T_a \tag{31}$$

Now, the Doppler resolution in the proposed KA-DBS method is approximately given as:

$$\delta \hat{f}_d = \frac{1}{\hat{T}_a} = \frac{\delta f_d}{2} \tag{32}$$

Since the equivalent number of pulse (i.e., the CPI length) is doubled in the proposed KA-DBS imaging algorithm, a finer Doppler frequency resolution can be achieved. Therefore, the sharpening ratio in KA-DBS is given as:

$$\hat{K}_a = \Delta f_d / \delta \hat{f}_d = 2K_a \tag{33}$$

As illustrated in Equation (33), the sharpening ratio $\hat{K}_a$ is theoretically improved by a factor of 2 compared to the conventional DBS imaging algorithm. Therefore, a high cross-range resolution and large sharpening ratio can be achieved in the proposed KA-DBS imaging framework.

3.4. Super-Resolution Imaging Algorithm Based on KA-DBS

So far, the implementation of the proposed KA-DBS super-resolution imaging approach for WAS radar has been described. Figure 5 below shows the whole imaging procedure.

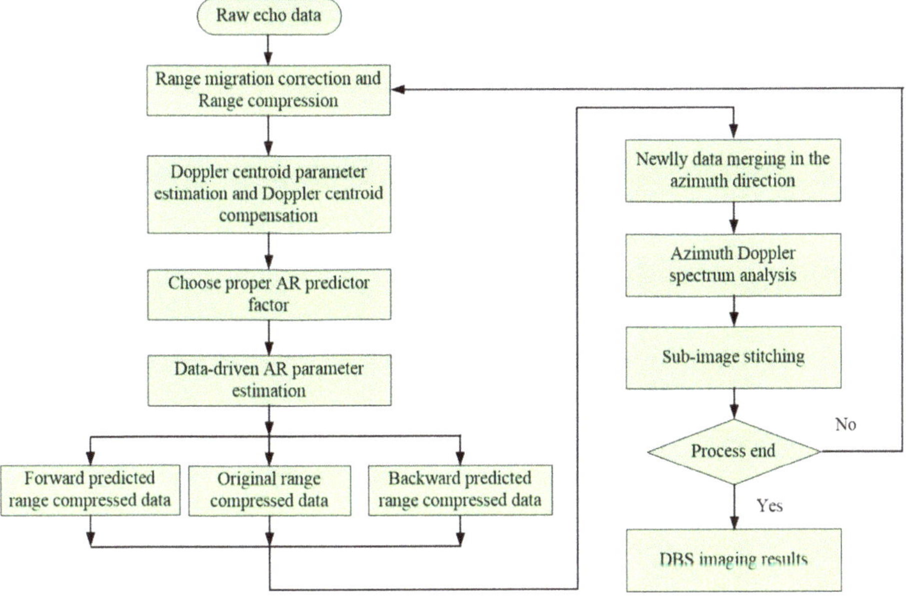

Figure 5. Process of the proposed KA-DBS processing approach.

In Figure 5, first, range migration corrections and range compression are utilized to process the raw echo data. Second, the Doppler centroid estimation parameter is estimated, and the Doppler centre of the signal is modulated to zero frequency, which is quite useful for the latter sub-image stitching process. Third, the proper AR predictor factor is set, and the AR parameter can be data-driven based on the Levinson recursion algorithm. Fourth, the forward range compressed data prediction and the backward range compressed data prediction are performed simultaneously. Then, the forward predicted range compressed data, the backward predicted range compressed data and the original range compressed data are merged together to form a newly merged data. After that, the Doppler analysis is performed to form the DBS sub-image. Finally, the final fan DBS image is formed by stitching all the DBS sub-images based on the affine transformation [9]. The comparison of different super-resolution methods is demonstrated in the following section.

4. Experimental Results

4.1. Simulation

The point target simulations are performed in Figure 6. The point targets are distributed as the outline of an airplane. The simulation parameters are illustrated in Table 1.

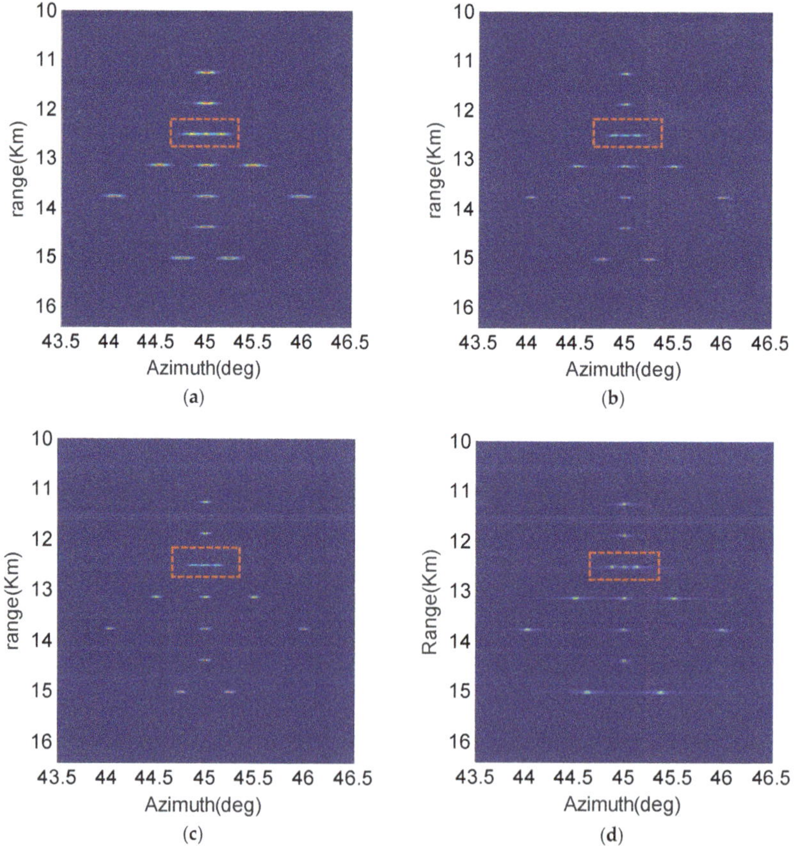

Figure 6. Simulation results in the case of SNR=10 dB; (**a**) FFT algorithm; (**b**) Relax algorithm; (**c**) APES algorithm; (**d**) KA-DBS algorithm.

Figure 6 shows the imaging results with FFT, Relax, APES and the proposed KA-DBS algorithm. The SNR is set as 10 dB. From Figure 6, it can be seen that the imaging results based on the FFT algorithm is a little blurred, especially for the closely spaced scatters in the red rectangle, while the closely spaced scatters can be well distinguished in the imaging results based on Relax, APES and the proposed KA-DBS algorithm. Moreover, the image generated by the proposed KA-DBS algorithm is much clear than that of the other algorithms.

The entropy curve of different imaging algorithms under different SNRs is shown in Figure 7.

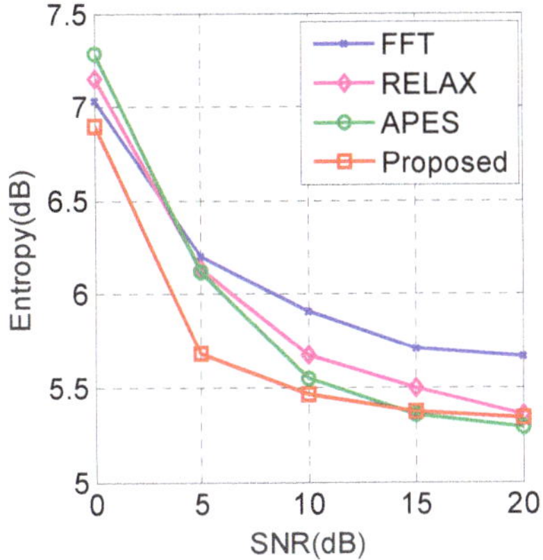

Figure 7. Entropy curves with different algorithms.

From Figure 7, it can be seen that the entropies of all the algorithms are decreasing with the increase of SNR. The proposed algorithm has the smallest entropy compared with the other algorithms when the SNR is less than 15 dB, which is in accordance with the imaging results in Figure 6. An interesting phenomenon can be seen form Figure 7 that the entropy of the proposed algorithm is a little higher than that of the APES algorithm when the SNR is higher than 15 dB. This may be explained that the APES algorithm has lower sidelobes than the proposed algorithm, and the lower sidelobes contribute more in the process of entropy computing under the high SNR situations for the simple simulation scene. The simulation results demonstrate that the proposed KA-DBS imaging algorithm outperforms the other algorithms.

4.2. Real Data

We study the performance of the newly proposed KA-DBS algorithm by resorting to real data set in this section. The experimental data collected on the wide-area surveillance mode of an airborne radar system is selected. The experiment radar parameters are illustrated in Table 2.

Table 2. Radar System Parameter.

Parameters	Value
Time width	24 us
Band width	40 MHz
Pulse repetition frequency	2500 Hz
Scanning area	$45°\sim135°$
Coherent pulses	128
Range gate number	4096

The DBS imaging results using different algorithms (i.e., the conventional FFT-based, Relax-based, APES-based and the proposed KA-DBS algorithm) are given in Figure 8. In KA-DBS, the predicted forward and backward pulse number is half of the pulse number in one CPI. The SNR is about 18 dB in the experiment. All the sub-images are stitched together based on the affine transformation algorithms [9]. In Figure 8a, the image in the conventional FFT-based algorithm suffers from blur. Moreover, it is obvious that the imaging results become clear and clear from the upper-left part of Figure 8 to the lower-right part of Figure 8. And the imaging result based on KA-DBS in Figure 8d performs the best. To further analyze the imaging results, Figure 9 shows the same zoomed in patch of Figure 8.

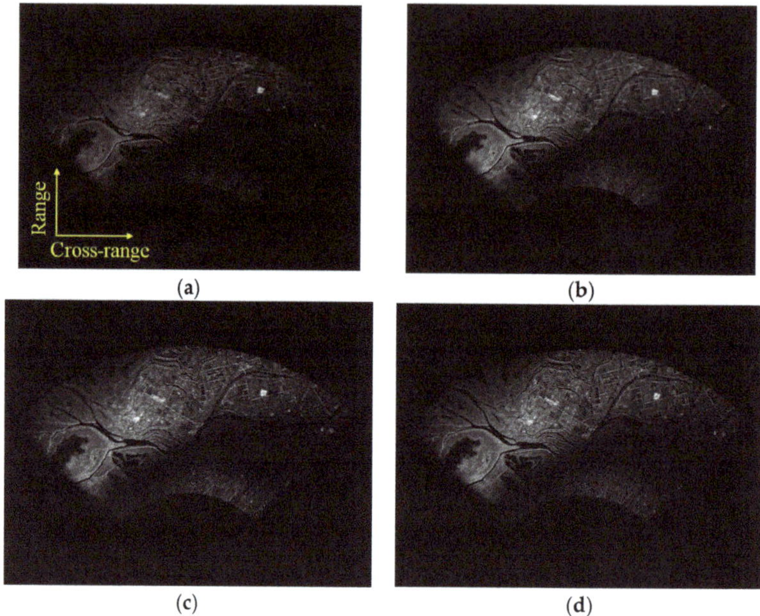

Figure 8. Imaging fan results; (**a**) FFT algorithm with 128 pulses; (**b**) Relax algorithm with 128 pulses; (**c**) APES algorithm with 128 pulses; (**d**) KA-DBS algorithm with 128 pulses.

From Figure 9, it can be found that the proposed KA-DBS algorithm can focus the scene much clear than the other algorithms, and the texture information can be easily distinguished. For example, the roads in Figure 9d are much thinner than that in Figure 9a–c. This can be explained that the forward and backward predicted pulses in each CPI are well utilized to improve the cross-range resolution in KA-DBS imaging.

In order to further verify the proposed algorithm, the coherent pulse number is changed from 32 to 64, and the selected pulse is the central part of the echoed signal. For example, the 49th to the 80th pulse in each azimuth angle is selected to form the 32 pulses data set, and the 33rd pulse to the 96th pulse in each azimuth angle is selected to form the 64 pulses data set. Different algorithms with different pulses are compared in Figure 10. Figure 10a1–d1 are the imaging results with 32 pulses, and Figure 10a2–d2 are results with 64 pulses. From the upper-left to the lower-right, FFT-based, Relax-based, APES-based and the proposed KA-DBS algorithm are given. It can be seen that the imaging results become better and better from the upper-left part of Figure 10 to the lower-right part of Figure 10.

The zoomed in patch of the same zone in Figure 10 is shown in Figure 11. From Figure 11, we can see that the imaging results with 64 pulses are much clear than that with 32 pulses for the same imaging algorithm. Moreover, the image based on FFT method under 32 pulses is the most blurred while the image based on the proposed KA-DBS method under 64 is the most clear. The road, farmland and other detailed information can be well distinguished in the proposed KA-DBS imaging results. Therefore, the proposed KA-DBS algorithm can improve the cross-range resolution at the situation of short dwell time.

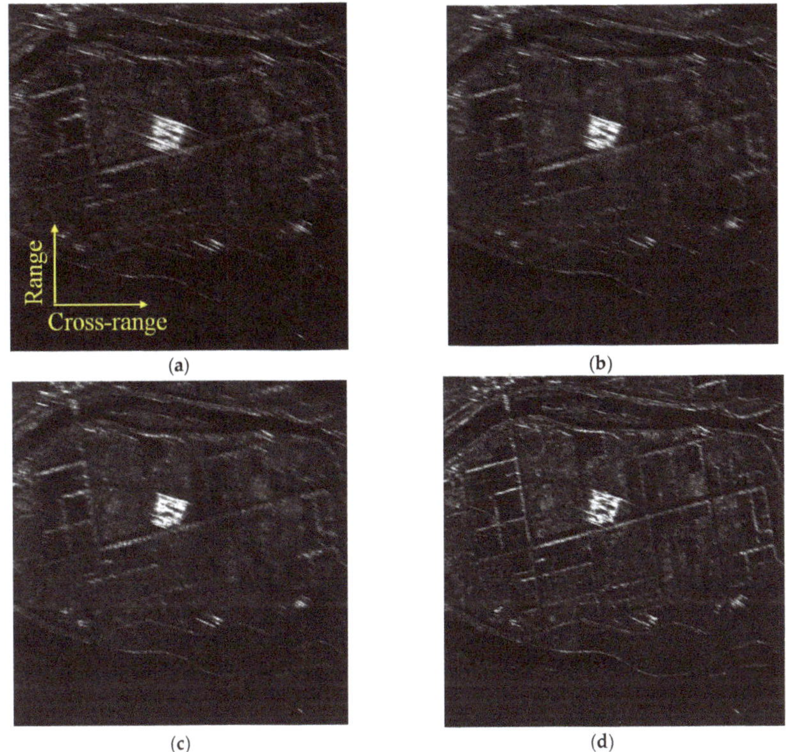

Figure 9. Locally imaging results; (**a**) FFT algorithm with 128 pulses; (**b**) Relax algorithm with 128 pulses; (**c**) APES algorithm with 128 pulses; (**d**) KA-DBS algorithm with 128 pulses.

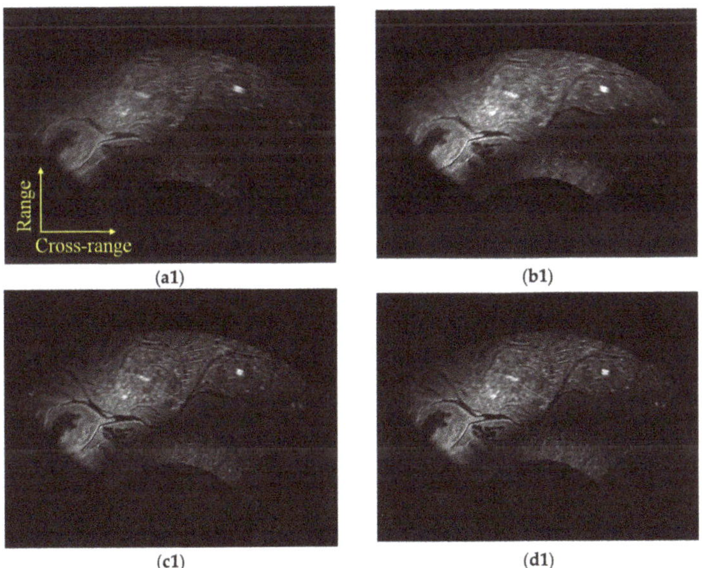

Figure 10. *Cont.*

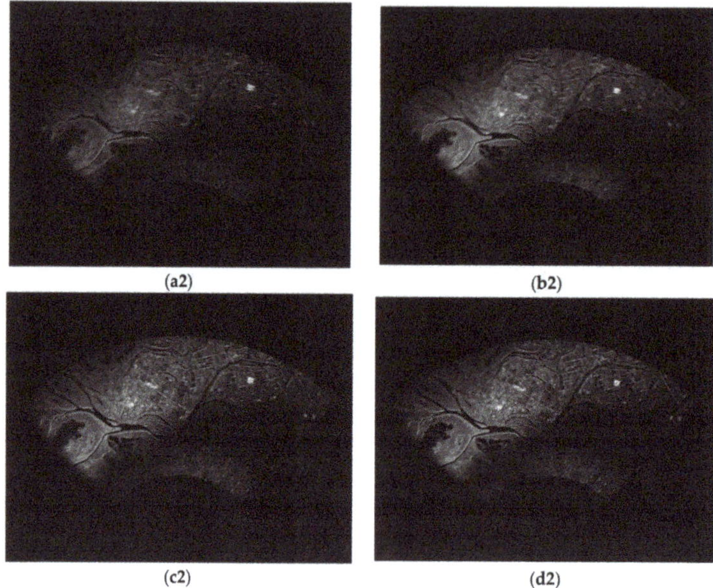

Figure 10. Imaging fan results; (**a1**) FFT algorithm with 32 pulses; (**b1**) Relax algorithm with 32 pulses; (**c1**) APES algorithm with 32 pulses; (**d1**) KA-DBS algorithm with 32 pulses; (**a2**) FFT algorithm with 64 pulses; (**b2**) Relax algorithm with 64 pulses; (**c2**) APES algorithm with 64 pulses; (**d2**) KA-DBS algorithm with 64 pulses.

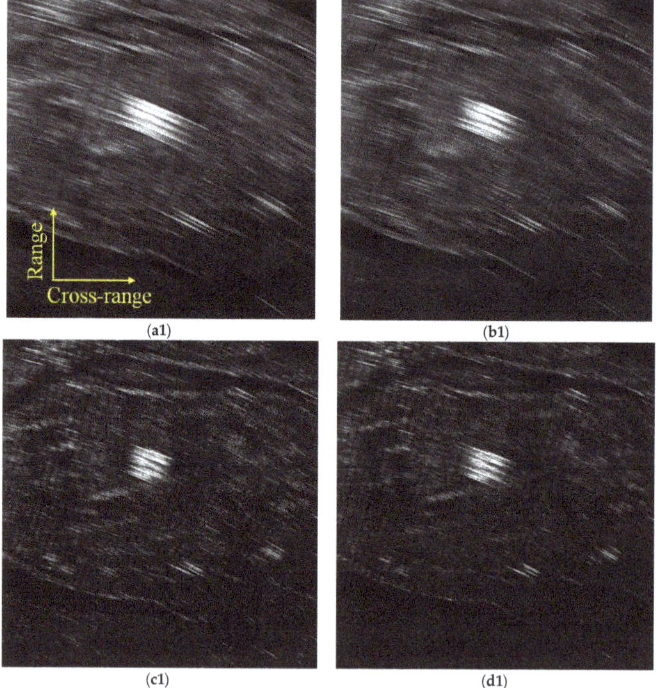

Figure 11. *Cont.*

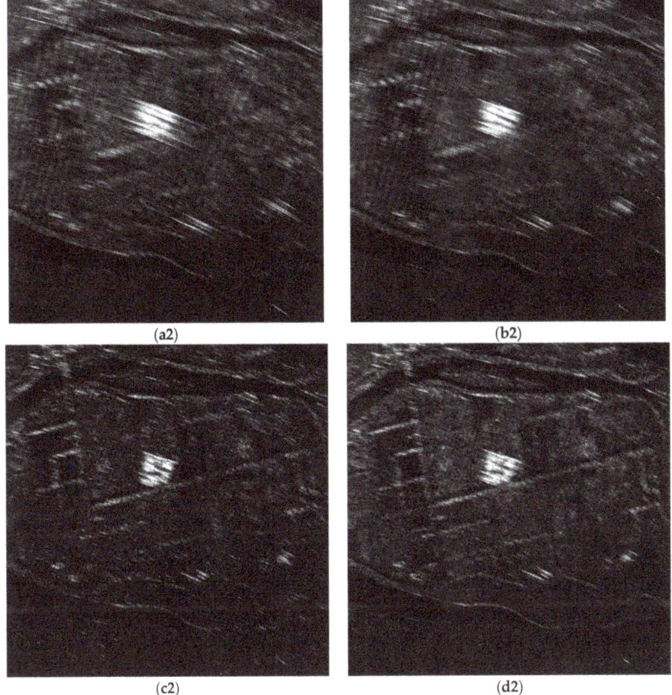

Figure 11. Locally imaging results; (**a1**) FFT algorithm with 32 pulses; (**b1**) Relax algorithm with 32 pulses; (**c1**) APES algorithm with 32 pulses; (**d1**) KA-DBS algorithm with 32 pulses; (**a2**) FFT algorithm with 64 pulses; (**b2**) Relax algorithm with 64 pulses; (**c2**) APES algorithm with 64 pulses; (**d2**) KA-DBS algorithm with 64 pulses.

5. Discussion

In this part, we will evaluate the imaging performance of the KA-DBS algorithm with the other imaging algorithms under different CPI lengths. To estimate the imaging quality, entropy is always utilized, which can be defined as [8]:

$$E = -\sum_{m=1}^{M}\sum_{n=1}^{N} p_{m,n} \log(p_{m,n}) \tag{34}$$

where the probability distribution function is:

$$p_{m,n} = \frac{I_{m,n}^2}{\sum\limits_{m=1}^{M}\sum\limits_{n=1}^{N} I_{m,n}^2} \tag{35}$$

In Equation (35), $I_{m,n}$ ($m = 1,2, \ldots ,M$, $n = 1,2, \ldots ,N$) is the concerned image, which is an $M{\times}N$ matrix.

The entropy curves of the wide-area image and the local image are shown in Figure 12a,b, respectively.

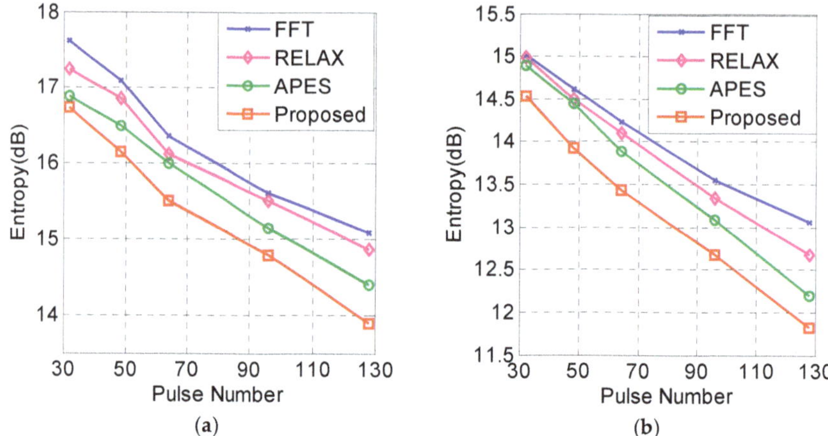

Figure 12. Entropy curves with different algorithms (**a**) the fan images (**b**) the local images.

From Figure 12, we can find that the entropy of the KA-DBS algorithm is the lowest, which means that a higher cross-range resolution can be acquired. Based on above experimental results, it proves that the proposed KA-DBS algorithm outperforms the other algorithms.

6. Conclusions

In this paper, we have considered the problem of high cross-range resolution DBS imaging for airborne WAS radar in short dwell time situations. A knowledge-aided Doppler beam sharpening (i.e., KA-DBS) imaging algorithm is proposed. We have investigated the spatial property of the echoed signal, and the spatial continuity model of the airborne radar system is constructed. By exploiting this spatial continuity knowledge, the forward and backward pulse information outside the observed CPI is well predicted based on the AR technique. Then the predicted pulses are merged together with the original pulses to form the newly merged pulses. Finally, DBS imaging is performed. The number of newly merged pulses in the proposed KA-DBS algorithm is twice larger than that in the conventional DBS algorithm with the same dwell time. Therefore, the cross-range resolution is improved by a factor of two in KA-DBS. Real airborne experiments have demonstrated that the proposed KA-DBS algorithm performs well with short dwell time.

Author Contributions: Conceptualization, H.C., Z.W. and J.L.; Methodology, H.C., Z.W. and J.L.; Software, H.C. and Z.W.; Validation, X.Y., H.S., H.M. and Y.L.; Formal Analysis, H.M. and Y.L.; Investigation, H.C.; Resources, X.Y., H.S., H.M., M.L., and Y.L.; Data Curation, M.L., H.S. and H.M.; Writing—Original Draft Preparation, H.C.; Writing—Review & Editing, H.C., Z.W. and H.S.; Visualization, H.C., Z.W., J.L. and X.Y.; Supervision, M.L. and Y.L.; Project Administration, M.L. and Y.L.; Funding Acquisition, M.L. and Y.L.

Funding: This work was supported by the National Natural Science Foundation of China (No.61271297; No.61272281) and the Postdoctoral Science Foundation of China (No. 2017M610966).

Acknowledgments: The authors would like to thank all the anonymous reviewers for their valuable comments to improve the quality of this paper.

Conflicts of Interest: The authors declare no conflict of interest.

References

1. Cumming, I.G.; Wong, F.H. *Digital Processing of Synthetic Aperture Radar Data: Algorithms and Implementation;* Artech House: Norwood, MA, USA, 2005.
2. Radant, M.E. The evolution of digital signal processing for airborne radar. *IEEE Trans. Aerosp. Electron. Syst.* **2002**, *38*, 723–733. [CrossRef]

3. Brenner, A.R.; Ender, J.H.G. Demonstration of advanced recon-naissance techniques with the airborne SAR/GMTI sensor PAMIR. *IEE Proc.-Radar Sonar Navig.* **2006**, *153*, 152–162. [CrossRef]
4. Cerutti-Maori, D.; Klare, J.; Brenner, A.R.; Ender, J.H.G. Wide-area traffic monitoring with the SAR/GMTI system PAMIR. *IEEE Trans. Geosci. Remote Sens.* **2008**, *46*, 3019–3030. [CrossRef]
5. Yan, H.; Wang, R.; Li, F.; Deng, Y.; Liu, Y. Ground moving target extraction in a multichannel wide-area surveillance SAR/GMTI system via the relaxed PCP. *IEEE Geosci. Remote Sens. Lett.* **2013**, *10*, 617–621. [CrossRef]
6. Wang, C.H.; Liao, G.S.; Zhang, Q.J. First spaceborne SAR-GMTI experimental results for the Chinese Gaofen-3 dual-channel SAR sensor. *Sensors* **2017**, *17*, 2683. [CrossRef] [PubMed]
7. Zheng, M.; Yan, H.; Zhang, L.; Yu, W.; Deng, Y.; Wang, R. Research on strong clutter suppression for Gaofen-3 dual-channel SAR/GMTI. *Sensors* **2018**, *18*, 978. [CrossRef] [PubMed]
8. Long, T.; Lu, Z.; Ding, Z.G.; Liu, L.S. A DBS Doppler centroid estimation algorithm based on entropy minimization. *IEEE Trans. Geosci. Remote Sens.* **2011**, *49*, 3703–3712. [CrossRef]
9. Chen, H.; Li, M.; Lu, Y.; Wu, Y. A DBS image stitching algorithm based on affine transformation. In Proceedings of the IET Radar Conference, Xi'an, China, 14–16 April 2013; pp. 1–4.
10. Cheng, Y.; Sun, C. Applications of superresolution signal estimators to Doppler beam sharpened imaging. *J. Electron.* **2000**, *22*, 392–397.
11. Chen, H.M.; Li, M.; Lu, Y.L.; Zuo, L.; Zhang, P. Novel supper-resolution wide area imaging algorithm based on APES. *Syst. Eng. Electron.* **2015**, *37*, 6–11.
12. Bamler, R.; Eineder, M. ScanSAR processing using standard high precision SAR algorithms. *IEEE Trans. Geosci. Remote Sens.* **1996**, *34*, 212–218. [CrossRef]
13. Liang, C.; Fielding, E.J. Interferometry with ALOS-2 full-aperture ScanSAR data. *IEEE Trans. Geosci. Remote Sens.* **2017**, *55*, 2739–2750. [CrossRef]
14. Prats, P.; Scheiber, R.; Mittermayer, J.; Meta, A.; Moreira, A. Processing of sliding spotlight and TOPS SAR data using baseband azimuth scaling. *IEEE Trans. Geosci. Remote Sens.* **2010**, *48*, 770–780. [CrossRef]
15. Xu, W.; Huang, P.; Wang, R.; Deng, Y.; Lu, Y. TOPS-mode raw data processing using chirp scaling algorithm. *IEEE J. Sel. Top. Appl. Earth Obs. Remote Sens.* **2014**, *7*, 235–246. [CrossRef]
16. Raney, R.K.; Runge, H.; Bamler, R.; Cumming, I.G.; Wong, F.H. Precision SAR processing using chirp scaling. *IEEE Trans. Geosci. Remote Sens.* **1994**, *32*, 786–799. [CrossRef]
17. Chen, J.; Xing, M.; Sun, G.; Li, Z. A 2-D Space-variant motion estimation and compensation method for ultrahigh-resolution airborne stepped-frequency SAR with long integration time. *IEEE Trans. Geosci. Remote Sens.* **2017**, *55*, 6390–6401. [CrossRef]
18. Wang, Y.; Li, J.W.; Sun, B.; Yang, J. A novel azimuth super-resolution method by synthesizing azimuth bandwidth of multiple tracks of airborne stripmap SAR data. *Sensors* **2016**, *16*, 869. [CrossRef] [PubMed]
19. Mittermayer, J.; Moreira, A.; Loffeld, O. Spotlight SAR data processing using the frequency scaling algorithm. *IEEE Trans. Geosci. Remote Sens.* **1999**, *37*, 2198–2214. [CrossRef]
20. Kay, S.M. *Modern Spectral Estimation: Theory and Application*; Prentice-Hall: Englewood Cliffs, NJ, USA, 1988.
21. Kay, S.M.; Marple, S.L., Jr. Spectrum analysis-A modern perspective. *Proc. IEEE* **1981**, *69*, 1380–1419. [CrossRef]
22. Gupta, I.J.; Beals, M.J.; Moghaddar, A. Data extrapolation for high resolution radar imaging. *IEEE Trans. Antennas Propag.* **1994**, *42*, 1540–1545. [CrossRef]
23. Li, W.; Yang, J.; Huang, Y. Keystone transform-based space-variant range migration correction for airborne forward-looking scanning radar. *Electron. Lett.* **2012**, *48*, 121–122. [CrossRef]
24. Moore, T.G.; Zuerndorfer, B.W.; Burt, E.C. Enhanced imagery using spectral-estimation-based techniques. *Linc. Lab. J.* **1997**, *10*, 171–186.
25. Khorshidi, S.; Karimi, M.; Nematollahi, A.R. New autoregressive (AR) order selection criteria based on the prediction error estimation. *Signal Process.* **2011**, *91*, 2359–2370. [CrossRef]
26. Giurcaneanu, C.; Saip, F.A.A. New insights on AR order selection with information theoretic criteria based on localized estimators. *Digit. Signal Process.* **2014**, *32*, 37–47. [CrossRef]

Article

A Data-Driven Approach to SAR Data-Focusing

Cataldo Guaragnella [1,*] and Tiziana D'Orazio [2]

[1] DEI—Department of Electrical and Information Engineering, Politecnico di Bari, 70126 Bari, Italy
[2] STIIMA—Institute of Intelligent Industrial Technologies and Systems for Advanced Manufacturing, CNR—Italian National Research Council, 70124 Bari, Italy; tizianarita.dorazio@cnr.it
* Correspondence: cataldo.guaragnella@poliba.it; Tel.: +39-080-596-3655

Received: 6 February 2019; Accepted: 4 April 2019; Published: 6 April 2019

Abstract: Synthetic Aperture RADAR (SAR) is a radar imaging technique in which the relative motion of the sensor is used to synthesize a very long antenna and obtain high spatial resolution. Several algorithms for SAR data-focusing are well established and used by space agencies. Such algorithms are model-based, i.e., the radiometric and geometric information about the specific sensor must be well known, together with the ancillary data information acquired on board the platform. In the development of low-cost and lightweight SAR sensors, to be used in several application fields, the precise mission parameters and the knowledge of all the specific geometric and radiometric information about the sensor might complicate the hardware and software requirements. Despite SAR data processing being a well-established imaging technique, the proposed algorithm aims to exploit the SAR coherent illumination, demonstrating the possibility of extracting the reference functions, both in range and azimuth directions, when a strong point scatterer (either natural or manmade) is present in the scene. The Singular Value Decomposition is used to exploit the inherent redundancy present in the raw data matrix, and phase unwrapping and polynomial fitting are used to reconstruct clean versions of the reference functions. Fairly focused images on both synthetic and real raw data matrices without the knowledge of mission parameters and ancillary data information can be obtained; as a byproduct, azimuth beam pattern and estimates of a few other parameters have been extracted from the raw data itself. In a previous paper, authors introduced a preliminary work dealing with this problem and able to obtain good-quality images, if compared to the standard processing techniques. In this work, the proposed technique is described, and performance parameters are extracted to compare the proposed approach to RD, showing good adherence of the focused images and pulse responses.

Keywords: SAR system; efficient focusing of SAR data; inverse problem; remote sensing; SAR data-focusing; synthetic aperture radar; Singular Value Decomposition; blind deconvolution; signal processing; parameter estimation; computational modeling

---

1. Introduction

The Synthetic Aperture Radar (SAR) [1–4] can acquire very-high-resolution images of the inspected area using high bandwidth of the transmitted coherent illumination signal by means of an accurate processing of the ground-received returns. In a standard structure, the system is composed of a platform (i.e., airborne or satellite) using the same antenna both for the transmitting and receiving phases; the target scene is repeatedly illuminated with pulses of radio waves. The signal echoes are emitted at equispaced positions along the satellite track, and their returns are received in the band of the transmitted pulse, converted, and IQ-sampled to produce the baseband complex raw data. A baseband algorithm implements the synthetic aperture to produce the equivalent return of a very narrow azimuth antenna beam. Three main algorithms are available to obtain such high-quality images, namely Range Doppler [5] algorithm, Ω-K [6] algorithm and Chirp Scaling algorithm [7]. All such algorithms, which perform equally in terms of focused image quality [5], require precise geometric acquisition parameters and radiometric parameters. Such parameters are always available as a side documentation of each acquired image.

Figure 1 reports the acquisition geometry of a SAR system. The synthesis procedure in focusing the acquired data is carried out by coherent integration. Each target on the ground contributes to the radar return on several subsequent transmitted pulses. In SAR two main directions are important to focus the data: slant range direction, in which transmitted pulses travel, and azimuth direction, i.e., the direction of sensor movement. The precise knowledge of the geometry of the acquisition allows to add in phase all the contributes of each single point scatterer on the ground present in the data to obtain the focused image. The wider the beam, the smaller the detail acquired by any return, but the larger the integration size of the track contributions to synthesize the image. The practical azimuth resolution is limited by the PRF choice (the Pulse Repetition Frequency of transmitted pulses used for coherent illumination of the target area), i.e., the azimuth sampling frequency.

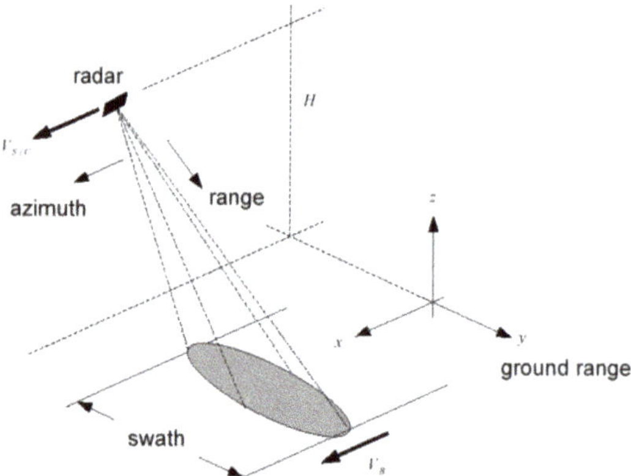

Figure 1. SAR Data acquisition geometry.

In several cases, due to imprecise knowledge of the satellite or aerial vehicle acquisition geometry or due to the presence of motion in the scene (ships, cars, etc.) a bad quality of the image is obtained (defocusing); many authors addressed the problem of a post processing procedure able to exploit the residual correlation present in data to perform accurate focusing of the image. Motion compensation is very important to achieve high resolution in SAR imagery. The phase errors that may be present on the focused image can be compensated through Inertial Measurement Unit (IMU) and Global Positioning System (GPS) side information. However, the need to measure and add such information to the ancillary data complicates the burden on any motion compensation system. In this cases SAR autofocus algorithms [8–12] are used to solve the problem in a blind mode. SAR autofocusing algorithms are categorized into three types: sub-aperture-based algorithms, prominent point-based algorithm, and metric-based autofocus. Most of the traditional autofocus algorithms assume there are strong scatters in the scene. Compared to the other conventional autofocus methods, the metric-based methods can work well without prominent points but deal with an already focused image.

The defocused image can be considered as the perfect focused image convoluted with the point spread function (PSF) caused by the phase error. Only recently and with the advent of lightweight and cheap SAR systems and UAV and drones, the problem of high cost of the system has pushed research to find new solutions in the development of such systems that imply in some cases the development of new techniques trying to focus the acquired raw data matrix using a subset of the ancillary data parameters and in presence of strong geometry anomalies that occur in such cases [13–18]. As the availability of the required parameters has always been guaranteed, at our knowledge, none has been done to develop algorithms able to estimate the reference function to be used as the focusing operator from the data itself and develop a completely blind focusing procedure.

While all the available algorithms to solve the SAR data-focusing problem are model driven, as they use ancillary parameters information to model the inverse problem in radar soil backscatter, in this paper a data-driven approach to develop a totally blind SAR data-focusing, based on the use of the Singular Values Decomposition (SVD) and LMS fitting of the phase information extracted from singular vectors is presented, able to obtain good image quality working on the complex SAR raw data matrix in absence of any information about the sensor. The proposed approach, at the state of the art, works well in the presence of a strong point scatterer in the scene.

The main idea is to exploit all the inherent information intrinsically stored in the data itself to extract the focusing reference functions to be used in a Ω-K or RD algorithm to obtain the Single Look Complex of any SAR sensor without even knowing important ancillary data parameters, needed by all the SAR focusing processors, such as the distance at the center of the beam, the radar sampling frequency, the transmitted chirp bandwidth, the chirp rate, and the chirp duration, the radar wavelength, the PRF, the sensor speed and the off nadir angle, used in data acquisition.

The proposed approach has been tested on several images of ERS raw data, made accessible for the scientific purpose from the Italian Space Agency (ASI) and the cross comparison with the state of the art focusing algorithm were carried out. The results indicate a good accordance to the standard focusing of obtained images with respect to the officially distributed ones. Also, the algorithm can reveal interesting and convenient in several application fields such as local zones monitoring by SAR systems carried by small lightweight and low-cost aerial unmanned vehicles. Modern hardware technology permits to reduce the size and weight of SAR systems into small and cheap flying platforms that can be conveniently used with low-cost platforms and flying drones. High-resolution microwave images of the observed scene can be obtained under various environmental conditions. Thus, UAV-SAR attracts growing interest in recent years [14]. The possibility of developing commercial low-cost systems is anyway still limited by the complication of the development of SAR due to the precise need for mission parameters to obtain good-quality images. Such parameters are very unstable for this kind of applications and the logging system introduces a complication, increasing their cost.

With respect to the reference [19] in which authors have preliminary proposed the blind technique to focus SAR data in the presence of a point scatterers in the scene, in this paper a more complete discussion about the quality of the focused image is carried out. To define the resolution in the range and azimuth directions, point scatterers responses are extracted from the image and azimuth and range cuts are compared with Range Doppler focusing of the same image. Also, an interferometric pair has been processed and the interference fringes have been extracted to show the good performance and phase stability of the proposed technique.

2. Materials and Methods

The proposed approach aims to show that the information needed to obtain the focused image can be extracted from the raw data itself when a point scatterer (either natural or manmade) is present in the scene. The problem is addressed with pattern recognition techniques.

In this section, the information exploitation method is presented. Initially the SVD approach used to extract the maximally correlated information inherently present in the image is introduced. The useful information of chirp in range and in azimuth are shown as extracted from the raw data. The SVD decomposition is discussed to explain the correlation exploitation mechanism. Once defined, the two reference functions extracted by the SVD, are used as inputs of a procedure needed to build clean versions, by LMS fitting procedure on the unwrapped phases both in range and azimuth. Clean versions of such signals are used to define the reference functions that are used to obtain the focused image. The reference functions are finally applied to obtain the focused image. To validate the algorithm, the proposed approach is then tested both on simulated data and on real data images. Focusing results have been carried out on several ERS1/2 raw data images showing the feasibility of the approach.

2.1. Blind SAR Data-Focusing Algorithm

The acquired data are the result of backscattering contribution of the ground at the SAR frequency. As a coherent illuminating source is used, the received data refer to several observations of the same scene taken in different points along the sensor platform trajectory. The radar transmitted pulses are stable in time, so the received returns show a strong azimuth correlation; several subsequent range lines in the raw data matrix contain roughly the same information so that the exploitation of coherence of the received signal can be attempted. This hypothesis allows the use correlation-based algorithm to extract useful information from data. The use of SVD technique can give us information about the reference functions to focus the image.

2.2. SVD—Signal Processing

SVD [20], in its economy formulation, is a standard algorithm able to decompose a given rectangular matrix into the product of three matrices, U, S, and V.

X is the data matrix of size $M \cdot N$, U, and V are orthonormal matrices, respectively named the *left* and *right* singular vectors matrices; each singular vector, either left or right, is represented by the generic column of the matrix U or V, respectively. In analytic form, The SVD decomposition can be written simply as:

$$X = U \cdot S \cdot V^H \tag{1}$$

where the superscript $(\cdot)^H$ represents the transpose and conjugate operator (Hilbert operator). U has the same size of the matrix X while S is a real valued diagonal matrix of order N and V is a complex orthonormal square matrix of order N:

$$U^H \cdot U = I_N \tag{2}$$

and

$$V^H \cdot V = I_N \tag{3}$$

S is the matrix containing the singular values of the matrix decomposition, sorted along the diagonal from the highest value to the lowest.

If a right multiplication for matrix V of both terms in (3) is applied, it can be usefully restated in another form:

$$X \cdot V = U \cdot S = E \tag{4}$$

The V matrix is the matrix performing linear combinations of the columns of the data matrix X to obtain E, an orthogonal matrix. The right multiplication of the matrix U with S only scales the vector columns in U not affecting the orthogonality property.

As a simple example, the SVD decomposition of the matrix X made of only two columns requires the right singular vector matrix V an orthonormal matrix of size 2. The V matrix in this simple case represents a complex Gibbs rotation matrix. In this very simple case, the operation carried out by the decomposition becomes clear:

$$\begin{bmatrix} \vec{x}_1 & \vec{x}_2 \end{bmatrix} \cdot \begin{bmatrix} c & s \\ -s & c \end{bmatrix} = \begin{bmatrix} \vec{e}_1 & \vec{e}_2 \end{bmatrix} \tag{5}$$

The E matrix is thus obtained as a simple linear combination of columns of X. V is the Gibbs rotation matrix and:

$$|c|^2 + |s|^2 = 1 \tag{6}$$

In a geometrical representation we can consider c and s parameters able to scale, rotate and phase shift vectors they multiply so that their sum and difference mixtures give the two orthogonal vectors in E. The modula of such vectors represent the singular values; it is easy to demonstrate that they are proportional to the estimates of the standard deviations of the resulting signals in E. When applied to a multicolumn matrix, this procedure tends to accumulate all the strongly correlated information on columns in the first left singular vector.

2.3. SAR Raw Data SVD Decomposition

When the X matrix to be decomposed is the raw SAR data matrix, the row index is the along track direction (azimuth) while the column index represents the slant range. The coherent illumination due to the transmission of the chirp produces very correlated information; in particular, in the azimuth direction the correlated information is the doppler phase history due to the scanning process. The expected result is that the first left singular vector should closely be related to the doppler history of the SAR system. As in E the transformed vectors are independent, the mostly correlated information in the signals in the columns of X is stored in the first singular vector in E, while the remaining part is in the others. Also, accordingly to the SVD decomposition scheme, the right singular vectors contain the mixing coefficients able to orthogonalize the raw data matrix.

This information is closely related to the transmitted information in the slant range direction. All the rows in the data matrix contain the same information, i.e., the transmitted chirp, delayed and phase shifted of an amount depending on the SAR geometry. The data are, again, strongly correlated, so the SVD decomposition will store in the rotation matrix V all the coefficients able to phase shift the subsequent transmitted chirps in a way that their sum convey the most important part of the global transmitted energy of the pulse. The coefficient must then be closely related to both the geometrical properties of the SAR acquisition and the characteristics of the transmitted chirp. This consideration will be illustrated in the next paragraph. Basing on these observations, a simple and direct scheme to obtain a good focusing of the raw data in a blind mode has been devised.

3. Discussion

3.1. B-SAR—Blind SAR Data-Focusing Algorithm

The focusing algorithm acts as a simple SVD decomposition. Once obtained, the first left and first right singular vectors can be used to define properly the reference functions to be used in the focusing procedure. Figure 2 shows the plot of the elements of the diagonal matrix S. As the matrix E contains orthogonal components, the energy of the whole image can be computed as the sum of the squares of the singular values. In the matrix a large part of the energy remains confined in the first left and first right singular vectors of matrices U and V.

An example of the first right and first left singular vectors of the SVD decomposition are shown in Figure 3.

From the simple observation of the first left singular vector of the matrix U two parameters can be extracted: the antenna beam pattern in azimuth and the phase history of the azimuth reference to be used in the focusing procedure. To select the proper parts of the two singular vectors for the focusing procedure a threshold was selected aiming at retaining the portion of the signal with amplitude greater than a given percentage of its peak value. In this paper, the 10% was selected. This allows to define the reference functions both in range (working on the right singular vectors) and in azimuth (left singular vectors).

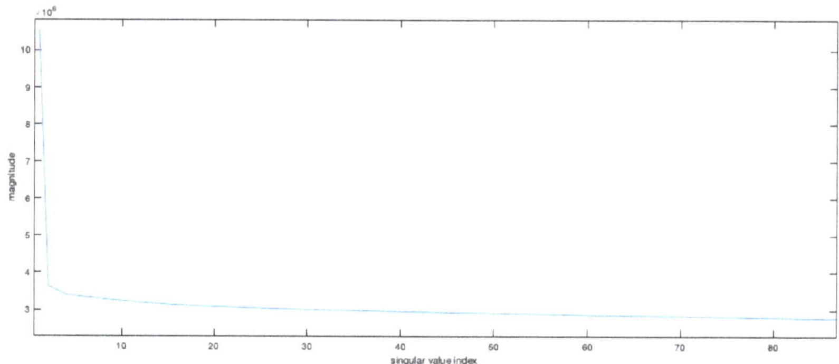

Figure 2. Singular values for the SAR raw data matrix decomposition. On abscissa the indexes of the singular values, on the vertical axis the magnitude. The first singular value is very high compared to the others showing that a large correlation was present in the raw data matrix and exploited in the first left and first right singular vectors.

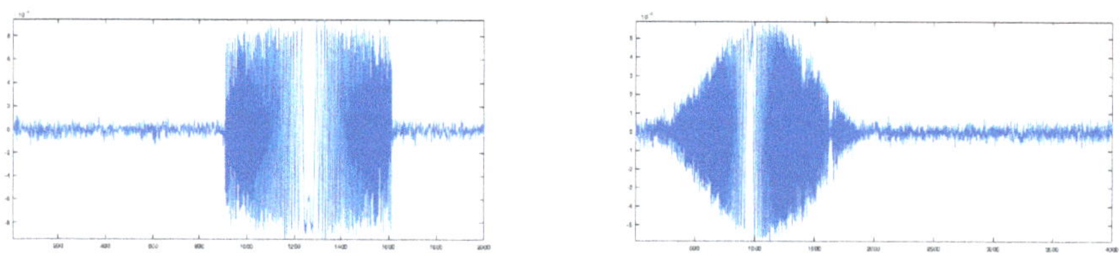

Figure 3. Real parts of first right and first left singular vectors for the SAR raw data matrix decomposition. The range amplitude results almost constant, as the transmitted chirp is assumed to be. The azimuth singular vector is shaped by the azimuth antenna beam pattern.

In real cases, a noisy appearance of the two signals is expected (see Figure 3); for this reason, instead of using the two so extracted reference functions in the focusing procedure, the unwrapped phases of such functions were extracted to define their clean versions and reduce the noise effects on the focused image.

Starting from the extreme points of such estimated phase functions a *LMS* polynomial fitting was used to construct a clean version of the phase histories to be used in the description of the two reference functions, both in azimuth and in range. The unwrapped phase of the reference function in either range or azimuth was approximated by a polynomial function and used to clean up the phase of the reference functions.

Stated y the polynomial function to be used to approximate the unwrapped phase of the reference function, the set of coefficients was computed via a *LMS* fitting procedure.

$$y = \sum_{i=1}^{N} a_i \cdot x^i \tag{7}$$

Here, a_i represent the polynomial coefficients and x the index of samples of the reference function.

Figure 4 represents the three diagrams of the unwrapped phase of the reference function, the polynomial approximation, and the phase error.

The reference function unwrapped phase estimated with the described procedure was selected in an interval of values in which the phase difference between subsequent samples is not larger than π. This choice allows to contain the aliasing effect due to the sampled phase history. To further reduce the aliasing effect at borders of the reference functions, raised cosine tunable length windowing was used.

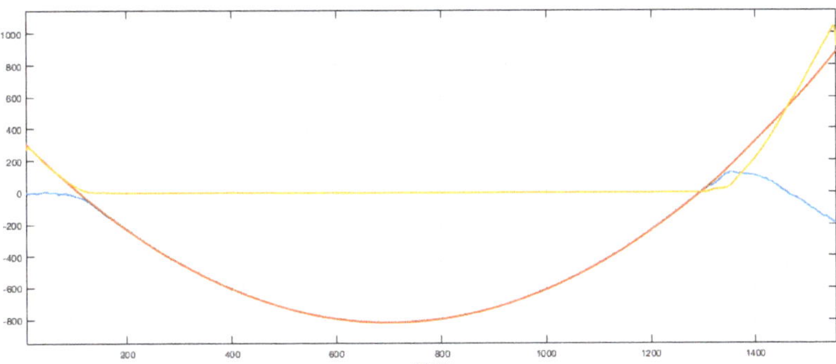

Figure 4. Phase as extracted from the first left singular vector (blue), the phase polynomial LMS approximation (red) and the phase error (yellow).

3.2. Experimental Results

In this section, a simulated experiment of a point target in Additive White Gaussian Noise (AWGN) is presented to show the results that can be obtained using the SVD decomposition; here, all the information inherently present in the raw data file is exploited; in the successive subsections, results of the focusing procedure are presented for real SAR raw data.

3.2.1. Simulated Experiment

To test the performance of the proposed algorithm a simulated point target on a noise floor has been used. The simulated azimuth antenna beam pattern was shaped by a Hanning window. The simulated raw data matrix was affected by AWGN with a low SNR. The ERS mission parameters are used in the simulation and reported in the Table 1.

Table 1. Parameters used to simulate raw SAR data.

	Parameters	Units
Carrier Frequency	$f_c = 5.300$	GHz
Chirp Duration	$T_c = 37.12$	µs
Chirp Bandwidth	$B_{ch} = 15.50829$	MHz
Sampling Frequency	$f_s = 18.962$	MHz
Satellite height	$h = 700$	km
Satellite speed	$v_s at = 11.75$	km/s
Off Nadir Angle	$\theta = 23$	deg
Squint Angle	$\phi = 0$	deg

The obtained raw data matrix was decomposed using the SVD algorithm. A large part of the energy in the data matrix is concentrated in the first singular value, clearly stating that the first left singular vector (i.e., the first column of matrix U) should contain the orthogonal signal with maximum energy in the data.

To select the proper length of the range chirp history and reconstruct a clean reference in range, a thresholding procedure was used; after the SVD decomposition, the reference signal power is much higher than the background noise, so this procedure reveals effective. In this case, the selected range time duration by the choice of a threshold at 10% of the signal peak value was efficient. After the procedure described in the previous subsection, a clean range reference function was obtained.

The same procedure was carried out for the azimuth reference, with a slight more care: the antenna beam pattern estimated in this way is not always effective due to the growing attenuation and the joint influence of azimuth and range beam patterns with the slant range and the possible presence of wide strong scatterers that can reduce the quality of the estimated pattern. Also, it is not clear, in this case, where the azimuth phase history should be stopped. The main objective is to limit the phase history in a proper way to avoid azimuth aliasing. The phase history was selected in a generally asymmetric interval around the peak of the unwrapped phase of the reference function: the possible presence of a squint angle in the true acquired data forces to select the proper doppler phase history to cover all the useful part of the extracted signal phase. As for the range, the thresholding procedure used in the azimuth reference definition was limited to the interval in which the signal was above the 10% of the estimated peak of the beam pattern. Once extracted, anyway, the selection of the proper portion of the phase history to be used in the construction of a clean reference was made basing on the phase, as discussed previously (cfr. Figure 4). Tunable length tapered tails were used to control aliasing effects without reducing severely the azimuth resolution of the focused image, both in the range and azimuth reference functions.

Figure 5 reports the obtained reference functions used to focus on range and azimuth the raw data. Figure reports the range (left column) and azimuth (right column). On the rows, the phase, the real part, and the spectrum of each reference function is presented.

Figure 6 reports the obtained focusing results for the simulated case. Specifically, the figure shows the real part of the simulated (with additive gaussian noise) raw data the first left and right singular vectors as extracted from the SVD decomposition, the 3D version of the ($\times$ 10 interpolated) pulse of the simulated raw data focused with the B-SAR algorithm, the two cuts, in range and azimuth of the interpolated focused pulse. The proposed approach is thus simple and direct and allows to extract useful information to focus the received data.

It should be pointed out that the possibility of obtaining good estimates of the range and azimuth chirp responses is due to the clear presence of a point scatterer with a sufficiently high signal to noise ratio that conveys a large part of the data matrix energy.

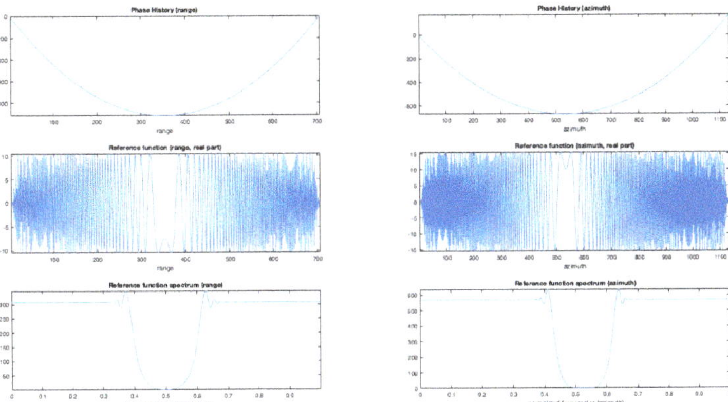

Figure 5. Range and Azimuth references to be used in the focusing procedure. Left range, right azimuth. The three rows present extracted phase histories, the real parts of the reference functions and their spectra.

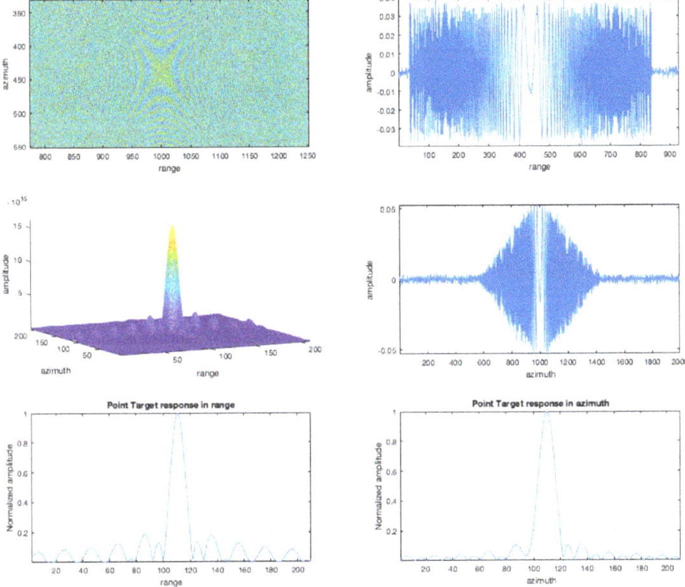

Figure 6. Results of simulated experiment. Top left: real part of the simulated (with additive gaussian noise) raw data; top right: First left singular vector as extracted from the SVD decomposition. center left: The 3D version of the ($\times$ 10 interpolated) focused pulse detail of the simulated raw data with the B-SAR algorithm; center right: First right singular vector as extracted from the SVD decomposition. It refers to the azimuth direction, showing a (hamming simulated) shape of the antenna beam pattern; bottom left: the range cut of the interpolated focused pulse; bottom right: the azimuth cut of the interpolated focused pulse.

Once the azimuth and range reference functions have been defined, focusing can become simple and can be carried out with either RD or Ω-K algorithms. In this paper, the frequency approach has been used to obtain the focused image.

A block diagram of the complete algorithm is reported in Figure 7.

3.2.2. Real SAR Raw Data-Focusing

The SVD decomposition was then applied as a test of the proposed approach to several ERS raw data matrices without any knowledge of the mission ancillary data.

The thresholding applied on the range reference function estimated a length of the range chirp of 704 samples. In Figure 8, the focused ERS image obtained with the proposed algorithm (B-SAR) and the focusing obtained by a standard Range Doppler SAR processor are compared.

The better appearance on the focused image is due to the normalization process. The images have been normalized with the same algorithm that amplifies the focused revealed image to a fixed value after normalization to the standard deviation of the whole image. The appearance of the image brightness is different because of the circular convolution in the Range Doppler procedure due to the frequency domain processing (its effect can be noted on the horizontal axis due to the presence of a periodic structure in the image that folds around the image). No zero padding was used in this procedure in the Range Doppler focusing software, while in the proposed algorithm the zero padding was used both in range and azimuth directions to avoid the circular convolution distortion. This reduces the amplitudes of large portions of the image, giving rise to a lower standard deviation and in the normalization a higher factor. This higher factor enhances very much the point scatterer that appears larger on the image, but the range and azimuth cuts in Figures 9 and 10 allow a better comparison of the performances.

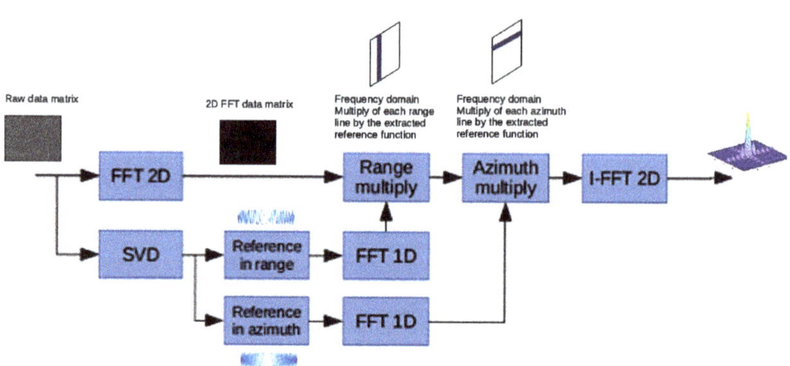

Figure 7. B-SAR Data Processing block diagram.

3.3. Semi-Quantitative Evaluation of Focusing Performance

In this paragraph the comparison between the range and azimuth cuts of the proposed B-SAR and RD algorithm is carried out.

In particular, in Figures 9 and 10 the range and azimuth cuts and the contour shaping of the same point scatterer on respective images obtained focusing the raw SAR data matrix are presented, showing how the range focusing obtained by the proposed procedure seems to adhere more precisely to the theoretical one than the RD focusing.

On the other side, at a lower resolution achieved by RD in the range direction, a higher rejection of side lobes is obtained, showing that the clean aspect of the focused image is due to a higher smoothing of range lines.

For the two focused images, a comparison of the amplitude statistics for the images focused with standard RD approach and B-SAR are plotted in Figure 11. What is evident is a close adherence of the statistics of the B-SAR image to the RD one, showing anyway a more compactness in the values, symptom of an imperfect focusing of highest peaks for this approximated method. The possibility of using the proposed algorithm in several applications in the field of Earth observation, interferometry and multi-temporal interferometry should be a goal to pursue even when mission parameters are unknown. To assess the performance obtained by the proposed approach with respect to the specific parameters of the ERS missions, a comparison of some specific radiometric parameters was carried out. The radiometric parameters table for ERS mission is reported in the Table 2.

Table 2. ERS radiometric parameters.

	Parameter	Units
Carrier Frequency	$f_c = 5.300$	GHz
Chirp Duration	$T_c = 37.12$	µs
Chirp Bandwidth	$B_{ch} = 15.50829$	MHz
Sampling Frequency	$f_s = 18.962$	MHz

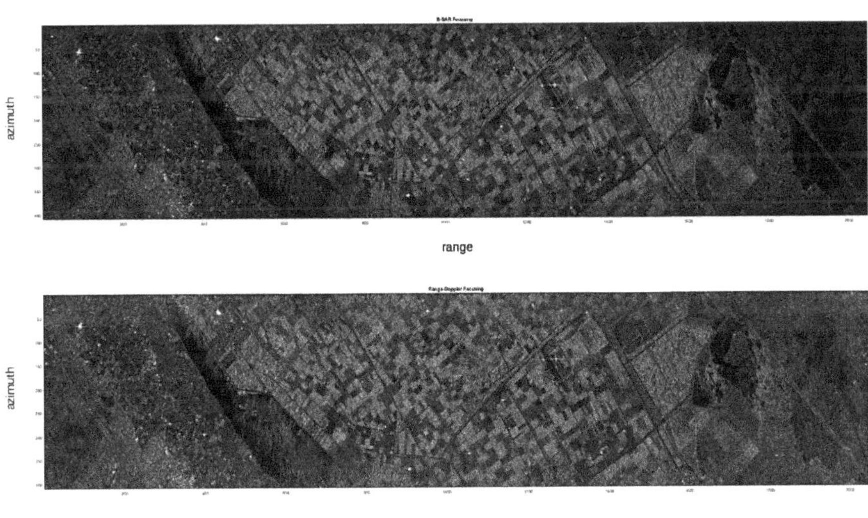

Figure 8. Sample of focused image with the proposed algorithm (upper) compared with the standard focusing obtained via Range Doppler algorithm (lower). ERS 1—Matera. The red circle indicates the point scatterer.

Using these parameters, the theoretical number of samples of the transmitted chirp can be computed as the product of the ERS sampling frequency and the chirp duration, with a close adherence with our estimate of 704 samples of the proposed technique. Also, the theoretical relative bandwidth of the chirp can be obtained as the ratio between the chirp bandwidth and the sampling frequency. For our algorithm, the estimate of this parameter can be computed as the equivalent bandwidth of the reference in range function. Also, in this case a close adherence of the theoretical and experimental relative bandwidth has been obtained, as reported in the Table 3.

Table 3. Comparison of Theoretical and Estimated Chirp Relative Bandwidth.

	Value
Theoretical Relative Bandwidth	0.8178
Estimate of the Relative Bandwidth	0.8164

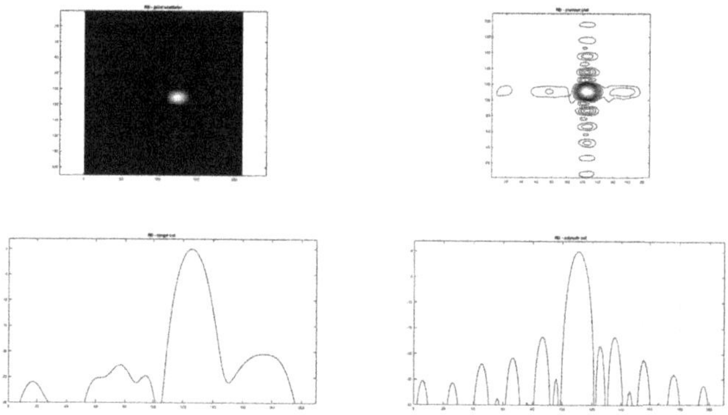

Figure 9. Range Doppler algorithm. Point scatterer image, contour plot, and range and azimuth cuts.

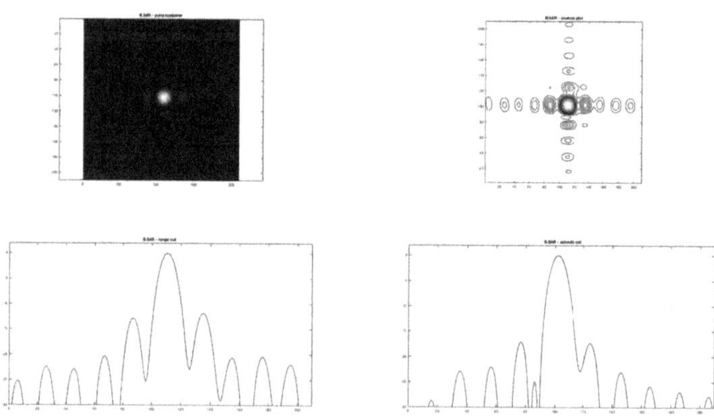

Figure 10. B-SAR algorithm. Point scatterer image, contour plot, and range and azimuth cuts.

3.3.1. B-SAR Phase Stability

An interesting parameter for the proposed approach for blind SAR data-focusing is its phase stability: the interferometric image has been computed on real ERS1/2 tandem pair pass over the Fucino region in Italy. Figure 12 shows one of the focused images of the tandem pass pair, while Figure 13 shows the (5 looks, slope corrected) interferometric image obtained by B-SAR focusing algorithm with superimposed the intensity image. A close correspondence between the flat zones in the valley with the smooth variations of the phase seem to assess the good behavior and the phase stability of the proposed processing technique.

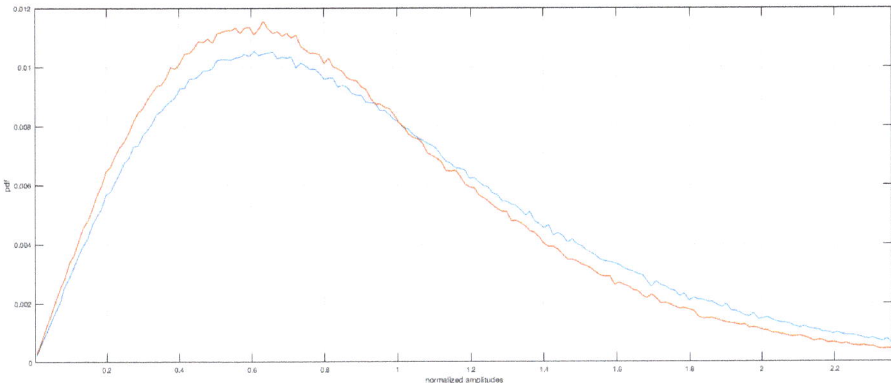

Figure 11. Histogram of focused images amplitudes comparison.

3.3.2. Processing Time

The description of the time processing required by the proposed approach to obtain the focused image is reported in Table 4 with respect to the original size of the raw data matrices. The experiments have been conducted on a desktop pc equipped by an Intel Core i7 Processor with clock speed of 3.4 GHz and a total number of four cores, and a memory of 16 GB.

Table 4. Processing times.

Image Name	[Rows, Cols] (Complex)	Proc. Time [s]
Caramanico 1	[2001, 4001]	21.049431
Caramanico 2	[2001, 4001]	20.528043
Flevoland	[1101, 5000]	13.358860
Matera	[2048, 2048]	10.970114
Simulated Pulse	[776, 2001]	2.473084

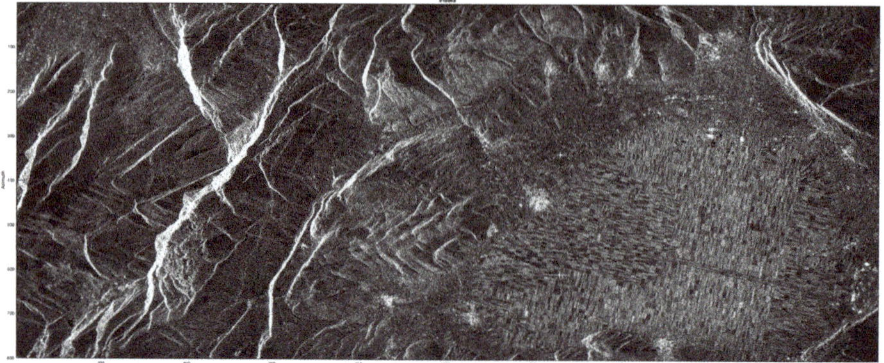

Figure 12. B-SAR focusing. Caramanico site, Fucino Valley, Italy.

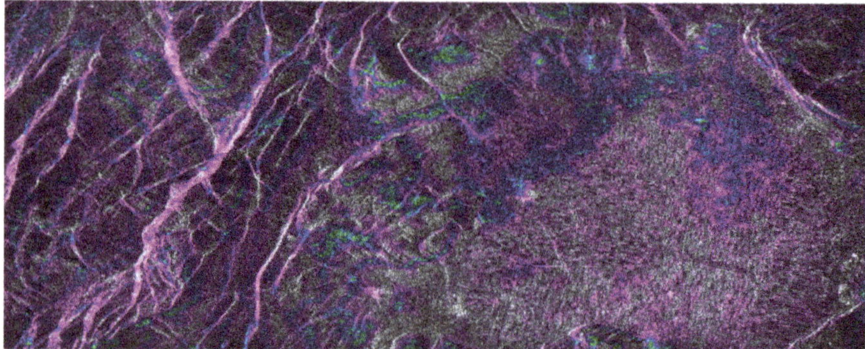

Figure 13. B-SAR, 5 looks, slope corrected interferometric image obtained by B.SAR focusing algorithm with superimposed the intensity image. Caramanico site, Fucino Valley, Italy.

The processing times needed to obtain the SLC/PRI images reported in the table are comparable to the standard processing times of other algorithms.

4. Conclusions

In this work the SVD decomposition has been used to extract correlated information from SAR raw data on scenes where a strong point scatterer is present. The use of the SVD is a sufficient information allowing the development of a simple and direct procedure to focus the acquired data without the need for information about the sensor attitudes, path, and SAR system parameters. The aim of this paper is to define a simple and direct method to obtain good focused images for several application, such as aerial archaeology inspection, agriculture, change detection for land usage and so on. The availability of a blind focusing algorithm can allow the development of simpler SAR systems to be used in low-cost applications in which the highest precision in the focused image is not a strict requirement.

To deal with the problem of precise focusing of the entire image taking care the range space variant system impulse response can be addressed, in a post processing way, using one of the several available autofocusing techniques available in the scientific literature.

The proposed algorithm, at the state of the art, is sufficient to obtain a fair focused image. The appearance of the focused image obtained is comparable with standard RD focusing, as shown in Figure 9. To assess the performance of the proposed approach, point scatterer responses have been compared between the RD and B-SAR focused images, showing a pretty good correspondence.

Also, the problem of phase stability of the algorithm has been addressed, computing the interference fringes corresponding to an ERS 1-2 tandem mission, showing also good adherence of the specific local orography. The main limitations of the proposed algorithm depend on its need for the presence of a strong point scatterer in the imaged zone. This limitation is payed back by its simplicity and the lack of need for the ancillary parameters file in the focusing procedure, aspect that simplifies both the processing and the development of simple and cheap SAR systems to be used in local monitoring also with the recourse to simple aerial unmanned vehicles such as drones.

The focused image is obtained, at the state of the art, by SVD analysis. This algorithm performs correlation exploitation of the contributes of the several azimuth lines. This leaves some room for further analysis as the residual correlation is inherently present in the lower singular vectors and not only on the first one, meaning that better image quality can be addressed using all the correlated components in the SVD decomposition. Of course, the problem is crucial, and attention is being taken on this subject. A precise focusing algorithm taking care of the Range Cells Migration Compensation is the goal of our future work. Recently, some authors have presented good results for SAR With Nonlinear FM Chirp Waveforms [21]. This specific case has not been addressed in this paper and will be the goal for future research. The computational aspects to obtain good-quality focused images are also important: recently some studies about efficiency have been presented [22] exploiting the multicore-based architectures of modern processors. Also, this aspect needs further research, as the proposed approach pays the cost of no information available for the SAR sensor with an increase of computational complexity. Also, the possibility of blind focusing SAR raw data, here addressed only in the presence of a point scatterer (e.g., a corner reflector or a transponder), in the general case of SAR strip map data-focusing represents the field of application for future work.

The proposed algorithm, developed in MATLAB, is distributed under the Noncommercial—Share Alike 4.0—International Creative Common license by the authors.

Author Contributions: Conceptualization, C.G.; Methodology, C.G.; Software, C.G. and T.D.; Validation, C.G. and T.D.; Formal analysis, C.G.; Data curation, C.G.; writing–original draft preparation, C.G. and T.D.; Writing–review and editing, C.G. and T.D.

Funding: This research received no external funding.

Acknowledgments: The authors wish to thank Fabio Bovenga for the interest shown in this research and Davide Palmisano for his help in SAR raw data extraction and data formatting.

Conflicts of Interest: The authors declare no conflict of interest.

References

1. Ridenour, L.N. *Radar System Engineering*; McGraw-Hill Book Co.: New York, NY, USA, 1945.
2. Skolnik, M.I. *Introduction to Radar Systems*; McGraw-Hill Book Co.: New York, NY, USA, 1962.
3. Klauder, J.R.; Price, A.C.; Darlington, S.; Albersheim, W.J. The Theory and Design of Chirp Radars. *Bell Syst. Tech. J.* **1960**, *39*, 745–808. [CrossRef]
4. Cutrona, L.J. Synthetic aperture radar. In *Radar Handbook*; McGraw-Hill: New York, NY, USA, 1970; pp. 23.1–23.25.

Sensors **2019**, *19*, 1649

5. Bamler, R. A comparison of range-Doppler and wavenumber domain SAR focusing algorithms. *IEEE Trans. Geosci. Remote Sens.* **1992**, *30*, 706–713. [CrossRef]

6. Cafforio, C.; Prati, C.; Rocca, F. SAR data focusing using seismic migration techniques. *IEEE Trans. Aerosp. Electron. Syst.* **1991**, *27*, 194–207. [CrossRef]

7. Amein, A.S.; Soraghan, J.J. Fractional Chirp Scaling Algorithm: Mathematical Model. *IEEE Trans. Signal Process.* **2007**, *55*, 4162–4172. [CrossRef]

8. Chen, L.; An, D.; Huang, X. Extended Autofocus Backprojection Algorithm for Low-Frequency SAR Imaging. *IEEE Geosci. Remote Sens. Lett.* **2017**, *14*, 1323–1327. [CrossRef]

9. Ran, L.; Liu, Z.; Zhang, L.; Li, T.; Xie, R. An Autofocus Algorithm for Estimating Residual Trajectory Deviations in Synthetic Aperture Radar. *IEEE Trans. Geosci. Remote Sens.* **2017**, *55*, 3408–3425. [CrossRef]

10. Li, D.; Lin, H.; Liu, H.; Liao, G.; Tan, X. Focus Improvement for High-Resolution Highly Squinted SAR Imaging Based on 2-D Spatial-Variant Linear and Quadratic RCMs Correction and Azimuth-Dependent Doppler Equalization. *IEEE J. Sel. Top. Appl. Earth Obs. Remote Sens.* **2017**, *10*, 168–183. [CrossRef]

11. Ma, J.; Tao, H.; Huang, P. Subspace-based super-resolution algorithm for ground moving target imaging and motion parameter estimation. *IET Radar Sonar Navig.* **2016**, *10*, 488–499. [CrossRef]

12. Torgrimsson, J.; Dammert, P.; Hellsten, H.; Ulander, L.M.H. An Efficient Solution to the Factorized Geometrical Autofocus Problem. *IEEE Trans. Geosci. Remote Sens.* **2016**, *54*, 4732–4748. [CrossRef]

13. Brian D. Rigling, Flying blind: A challenge problem for SAR imaging without navigational data. In Proceedings of the SPIE, Baltimore, MD, USA, 2–3 May 2012; Volume 8394. [CrossRef]

14. Zhang, L.; Hu, M.; Wang, G.; Wang, H. Range-Dependent Map-Drift Algorithm for Focusing UAV SAR Imagery. *IEEE Geosci. Remote Sens. Lett.* **2016**, *13*, 1158–1162. [CrossRef]

15. Guo, P.; Tang, S.; Zhang, L.; Sun, G.C. Improved focusing approach for highly squinted beam steering SAR. *IET Radar Sonar Navig.* **2016**, *10*, 1394–1399. [CrossRef]

16. Li, Z.; Liang, Y.; Xing, M.; Huai, Y.; Zeng, L.; Bao, Z. Focusing of Highly Squinted SAR Data With Frequency Nonlinear Chirp Scaling. *IEEE Geosci. Remote Sens. Lett.* **2016**, *13*, 23–27. [CrossRef]

17. Noviello, C.; Fornaro, G.; Martorella, M. Focused SAR Image Formation of Moving Targets Based on Doppler Parameter Estimation. *IEEE Trans. Geosci. Remote Sens.* **2015**, *53*, 3460–3470. [CrossRef]

18. Tang, S.; Zhang, L.; Guo, P.; Liu, G.; Zhang, Y.; Li, Q.; Gu, Y.; Lin, C. Processing of Monostatic SAR Data with General Configurations. *IEEE Trans. Geosci. Remote Sens.* **2015**, *53*, 6529–6546. [CrossRef]

19. Guaragnella, C.; D'Orazio, T. B.SAR—Blind SAR Data Focusing. In Proceedings of the SPIE—International Conference on Image and Signal Processing for Remote Sensing, Berlin, Germany, 10–13 September 2018.

20. Golub, G.H.; Reinsch, C. Singular value decomposition and least squares solutions. *Numer. Math.* **1970**, *14*, 403–420. [CrossRef]

21. Wang, W.; Wang, R.; Zhang, Z.; Deng, Y.; Li, N.; Hou, L.; Xu, Z. First Demonstration of Airborne SAR With Nonlinear FM Chirp Waveforms. *IEEE Geosci. Remote Sens. Lett.* **2016**, *13*, 247–251. [CrossRef]

22. Imperatore, P.; Pepe, A.; Lanari, R. Spaceborne Synthetic Aperture Radar Data Focusing on Multicore-Based Architectures. *IEEE Trans. Geosci. Remote Sens.* **2016**, *54*, 4712–4731. [CrossRef]

Article

Pulsar Emissions, Signal Modeling and Passive ISAR Imaging

Andon Lazarov

Department of Information Technologies, Naval Academy, 9026 Varna, Bulgaria; lazarov@bfu.bg;
Tel.: +359-887-262-478

Received: 11 June 2019; Accepted: 29 July 2019; Published: 30 July 2019

Abstract: The present work addresses pulsar Crab Nebula emissions from point of view of their modeling and applications for asteroid detection and imaging by applying inverse synthetic aperture radar (ISAR) principles. A huge value of the plasma's effective temperature is a reason for pulsar emission coherency, a property of great practical meaning for a space objects navigation, localization and imaging. Based on measurement data obtained by Goldstone-Apple Valley and Arecibo radio telescopes, an original time frequency grid mathematical model of pulsar emissions is created. Passive ISAR scenario, a space object's geometry and a model of pulsar signals reflected from the space object's surface are also described and graphically illustrated. A new range compression approach for ISAR imaging is suggested and demonstrated. In order to reduce the level of additive white Gaussian noise in signals and enlarge the signal to noise ratio in the final image, coherent summation of multiple complex images is applied. To prove the correctness of the geometry, signal models and theoretical analysis, results of numerical experiments are provided.

Keywords: passive ISAR asteroid imaging; pulsar emission signal modeling

---

1. Introduction

The pulsars are rotating neutron stars formed due to the collapse of massive stars core. They are the densest form of matter in the Universe. During the collapse stage the preservation of angular momentum causes the star to "spin-up" to a rotation period of order 10ms, whereas the preservation of magnetic flux drives the magnetic field strength at the stellar surface up to 10^{12} gauss or higher, with magnetic moments up to 10^{26} gaus-m^3. Typical dimensions (radiuses) of white dwarfs which pulsate with period $\tau = 1$ s, and neutron stars with pulsation period 10^{-2} s are in the interval $(1-5) \times 10^6$ m [1].

The pulsar can be considered as a massive freely spinning top and a powerful particle accelerator since the rotating magnetic field generates enormous electric fields that accelerate charged particles. These accelerated particles emit electromagnetic waves at spin frequency (across the spectrum, from radio waves to gamma-rays). Due to their enormous mass and relatively simple structure, pulsars are exceptionally stable rotators whose timing stability rivals that of conventional atomic clocks. It is the reason pulsar emissions to be used for spacecrafts radio navigation, asteroids detection, and imaging [2]. A navigation system based on celestial sources will be an independent positioning system and available in any Earth orbit as well as in interplanetary and interstellar space [3–5]. The stellar navigation accuracy depends on the emission's time period, which for pulsars is with the high stability and accuracy.

Pulsar navigation tracking system based on the Doppler frequency measurement model and pulsar timing and an interplanetary navigation and positioning system using pulsar signals is discussed in [6,7]. The pulsar signal processing algorithm that consists of epoch-folding, matched filtering and detection is presented and evaluated in [8]. A hybrid detection algorithm based on energy and entropy analysis as an approach for spectrum sensing is considered in [9]. Principles of the pulsar navigation in the solar system are described in [10].

Pulsars are located on thousands of light years from Earth and can be considered as uninterruptible sources of electromagnetic waves. It allows the object illuminated by the pulsar beam as an asteroid or spacecraft, to be considered as a secondary emitter in compliance with Babinet's principle. The diffracted electromagnetic field is a superposition of partial fields reemitted by all illuminated parts of the asteroid and carries information concerning its geometry and kinematics as a moving rigid body. The asteroid has an average orbital speed of 25 km/s. However, asteroids orbiting closer to a sun will move faster than asteroids orbiting between Mars and Jupiter and beyond. The average orbital speed of a main-belt asteroid is 17.9 km/s, the orbital speed of Ceres. If the asteroid comes from far away (e.g., the Oort Cloud), it will be accelerated by the Sun and achieve a velocity equal to the escape velocity from the Sun at the location of Earth, which is 30 km/s. The orbital speed and the escape velocity differ by a factor of $2^{0.5}$. That is, an object falling from infinity toward the Sun, will have a speed equal to $30 \times 2^{0.5} = 42$ km/s. Then the asteroid's speed in the solar system varies from 17.9 km/s to 42 km/s.

In order to analyze and realize localization, navigation and imaging properties of pulsar emissions, adequate mathematical models need to be built, which is one of the goals of the present research. From point of view of pulsar emissions utilization, including localization, navigation and imaging, the most appropriate emissions are those from Crab Nebula pulsar. In this sense, the attention of the present work is on the signal modeling of the pulsar Crab Nebula emission and its passive inverse synthetic aperture radar application for asteroid imaging. The passive ISAR scenario, signal structure, power budget, image reconstruction and numerical experimental results are also discussed.

The remainder of the paper is organized as follows. In Section 2, a general description of Crab Nebula pulsar emission based on radio telescope measurements is given. In Section 3, an overview of synthetic and inverse synthetic aperture radar issues is suggested. In Section 4, passive ISAR scenario and asteroid's geometry are described. In Section 5, a pulsar signal model as a time frequency grid and an algorithm of ISAR signal formation are described. In Section 6, a power budget, ISAR signal processing and asteroid image reconstruction are analytically described. In Section 7, results of a numerical experiment are provided and discussed. In Section 8, conclusion remarks are drawn.

2. Crab Nebula Pulsar Emission Based on Radio Telescope Measurements

It is known that powerful celestial sources with small angular diameters as quasars, sources of hydroxyl radical (OH)-emission and pulsars, are characterized by huge values of effective temperature, the main preposition for a coherent radiation mechanism [1]. The relativistic, magnetized plasma of pulsars radiates energy of $(10^{36}$–$10^{38})/k_B$, where k_B is the Boltzmann constant. This enormous value can be achieved only based on the coherency, the most important feature of pulsar electromagnetic emissions.

There exist measurement data obtained by radio telescopes that can be used for signal modeling purposes [11,12]. For instance, the mean profiles from the Crab pulsar consist of two frequency-dependent components, the main pulse (MP) and inter-pulse (IP). They appear at 70 and 215 degrees of the pulsar's rotation phase, and can be identified from low radio frequencies to hard X-rays. The Main Pulses of the pulsar Crab emission are of particular interest for localization, navigation and imaging purposes. It is due to the coherency and stability of the Main Pulse's repetition period with time duration 0.033 s.

The results of the observation show that most giant main pulses consist of several microbursts. The integrated intensity of pulsar flux is measured with a time resolution of 6.4 ns, whereas the dynamic spectrum is plotted with 19.5 MHz spectral resolution and 25.6 ns time resolution. It is worth noting that pulsar Crab emissions observed by Arecibo radio telescope are preliminary and coherently de-dispersed.

In Figure 1a typical main pulse with three microbursts, registered between 2 and 10 GHz with 125 ns time resolution and 8 MHz frequency spectrum resolution using the Goldstone-Apple Valley radio telescope is presented. The bar at the top right illustrates the generalized, over all frequency channels, root mean square (RMS) power level in the time-frequency coordinates. The left panel shows on-pulse (red) and off-pulse (blue) power as a function of frequency, integrated in time across the pulse.

The high spikes in the on-pulse power correspond to strong, narrowband spikes within each microburst. On the right panel the pulse shape at eight frequencies within the pulse is depicted. The selected eight pulses are over-plotted in the top figure, where the dotted lines separate the on-pulse and off-pulse regions used for the left panel. All powers are given in terms of off-pulse noise (RMS) [12].

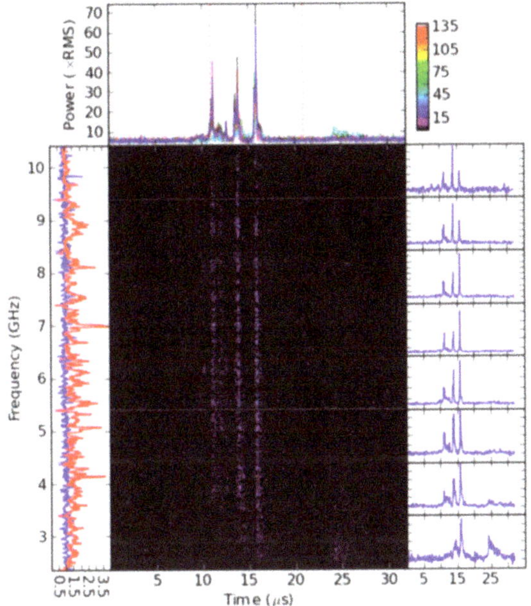

Figure 1. Typical main pulse time-frequency structure of Crab Nebula's pulsar emission registered in Goldstone-Apple Valley radio telescope [12].

Based on characteristics of the main pulse defined in [5,11,12], it is worth noting that the main pulse consists of several microbursts, the time dimension of which is ≤1 μs long at 8–10 GHz, with a bandwidth ≥2 GHz. From measurement data depicted in Figure 1, the following inferences regarding the structure of the pulsar emission can be made.

1. In frequency channels from 4 to 9 GHz, the structure of the main pulse is preserved. The main pulse consists of microbursts almost with the same form. In each frequency channel three monochromatic microbursts can be revealed, which can be described as Gaussian pulses with different amplitudes.

2. The entire time-frequency record of the pulsar's main pulse can be presented as a wideband signal. Each frequency component preserves time delay and, respectively, phase components induced by the object crossing the pulsar emission beam. The algebraic sum of all monochromatic components in frequency channels determines a sinc function with maximum in the position of a particular reemitting scattering point of the space object.

3. Synthetic and Inverse Synthetic Aperture Radar Issues

Problems of synthetic aperture radar imaging are widely discussed in the recent literature. Different modeling and imaging techniques including dominant scattering points and facets models, Fourier transforms, and phase compensation methods are in the focus of the authors' attention. An adaptive ISAR imaging of maneuvering objects based on a modified Fourier transform and distributed ISAR sub-image fusion of a nonuniform rotating target based on matching Fourier transform are discussed in [13,14]. To mitigate blurring and defocusing effects induced by maneuvering targets on the process of ISAR imaging an original method based on modified chirp Fourier transform is suggested in [15]. A Fourier-based image formation algorithm for GNSS-based bistatic forward-looking synthetic aperture radar is presented in [16]. Returns form objects with complex motion in an ISAR imaging system are modeled as multicomponent quadratic frequency modulation (QFM) signals. QFM signals' parameter estimation based on two-dimensional product modified parameterized chirp rate-quadratic chirp rate distribution is discussed in [17].

Non trivial ISAR image reconstruction methods are developed. An accurate method to extract ISAR images of multiple targets by applying Hough transform and particle swarm optimization (PSO) to find residual high order coefficients and to achieve better quality of ISAR imaging and moving target separation is discussed in [18]. A method including a genetic algorithm, PSO and PSO with an island model, to compensate for the inter-pulse phase errors caused by the target movement in stepped-frequency ISAR imaging is proposed in [19]. Residual motion error correction with back-projection multi-squint algorithm for airborne synthetic aperture radar interferometry is applied in [20].

An object's movement of higher order as different kinds of accelerations and range displacements causes the ISAR image to be blurred, which requires focusing the object. An autofocusing method for improving synthetic aperture radar image quality and modified fractal signature image classification technique are described in [21]. An algorithm based on keystone transform and time-domain chirp scaling to deal with the space-variant range cell migration in ISAR imaging with ultrahigh range resolution is proposed in [22]. Phase compensation and image autofocusing algorithms using randomized stepped frequency emitted ISAR signals are described in [23].

A problem of coherent integration for detecting high-speed maneuvering targets, involving range migration, quadratic range migration, and Doppler frequency migration within the coherent processing interval, and a coherent integration algorithm based on the frequency-domain second-order phase difference approach are discussed in [24]. Multi-sensor ISAR radar imaging and phase adjustment based on a combination of signal sparsity and total variation is considered in [25].

To improve the azimuth resolution of ISAR images a fractional sparse energy representation method combined with fractional Fourier transform is proposed in [26]. The clock jitter influence on the signal to noise ratio (SNR) of an analog-to-digital-converter of the ISAR signal acquired from the space object is analyzed in [27].

The abilities of the astronomical radars as unique and powerful information tools to measure physical properties and orbital parameters of asteroids are thoroughly analyzed and illustrated in [28].

Based on the aforementioned, the present work will focus on the passive ISAR scenario—kinematics and geometry—as well as signal modeling and special solutions in the asteroid's imaging algorithm.

4. Passive ISAR Scenario and Asteroid Kinematics and Geometry

Consider a 3D regular grid, where the asteroid's geometry is described. The grid is defined in Cartesian system $O\prime XYZ$ and moves on a rectilinear trajectory at a constant speed in the coordinate system $Oxyz$. The mass-center of the object, the geometric center of the 3D grid and the origin of the coordinate system $O\prime XYZ$ coincide in point O' Figure 2. The distance vector from the radio telescope placed in the origin of the 3D observation coordinate system $Oxyz$ to the g-th generic point of the object space, measured at the p-th moment is described by the vector equation

$$\mathbf{R}_g(p) = \mathbf{R}_{0\prime}(p) + \mathbf{A}\mathbf{R}_g \tag{1}$$

where $\mathbf{R}_{0\prime}(p) = \mathbf{R}_{0\prime}(0) + \mathbf{V}T_p \cdot p$ is the mass center position vector; $\mathbf{R}_g(p) = \left[x_g(p), y_g(p), z_g(p)\right]^T$ is the generic point's distance vector; $x_g(p), y_g(p)$, and $z_g(p)$ are the current coordinates of the generic point, $p = \overline{0, N-1}$; N the number and full number of emitted pulses, respectively, during time of observation (aperture synthesis); $\mathbf{R}_{0\prime}(0) = \left[x_{0\prime}(0), y_{0\prime}(0), z_{0\prime}(0)\right]^T$ is the line-of-sight vector of the object's geometric center at the moment $p = 0$; $\mathbf{V} = \left[V \cos \alpha, V \cos \beta, V \cos \delta\right]^T$ is the object's linear vector velocity; $\mathbf{R}_g = \left[X_g, Y_g, Z_g\right]^T$ is the distance vector to the g-th generic point in the coordinate system $O\prime XYZ$; $X_g = g_X \cdot (\Delta X)$, $Y_g = g_Y \cdot (\Delta Y)$, and $Z_g = g_Z \cdot (\Delta Z)$ are the discrete coordinates of the g-th generic point in the coordinate system $O\prime XYZ$; g_X, g_Y, g_Z are the coordinate indexes of the generic point; ΔX, ΔY, and ΔZ are the spatial dimensions of the 3D grid cell; $\cos \alpha$, $\cos \beta$, and $\cos \gamma = \mathrm{mod}\left(\sqrt{1 - \cos^2 \alpha - \cos^2 \beta} \right)$ are the guiding cosines; and V is the module of the linear velocity vector. The elements of transformation-rotation matrix A are defined by Euler's expressions

$$
\begin{aligned}
a_{11} &= \cos \psi \cos \varphi - \sin \psi \cos \theta \sin \varphi; \\
a_{12} &= \cos \psi \sin \varphi + \sin \psi \cos \theta \cos \varphi; \\
a_{13} &= \sin \psi \sin \theta; \\
a_{21} &= -\sin \psi \cos \varphi - \cos \psi \cos \theta \sin \varphi; \\
a_{22} &= -\sin \psi \sin \varphi + \cos \psi \cos \theta \cos \varphi; \\
a_{23} &= \cos \psi \sin \theta; \\
a_{31} &= \sin \theta \sin \varphi; \\
a_{32} &= -\sin \theta \cos \varphi; \\
a_{33} &= \cos \theta.
\end{aligned}
\tag{2}
$$

where in case of a rotating object, ψ, φ, and θ are time dependent angles, yaw, pitch and roll, respectively.

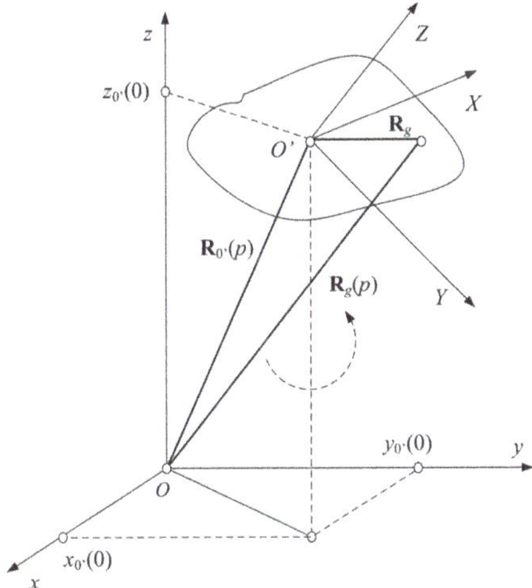

Figure 2. Inverse synthetic aperture radar (ISAR) scenario: asteroid's kinematics and geometry.

5. Pulsar Signal Formation

Models of pulsars' signals can be created based on the structure of their continuous electromagnetic emissions. The pulsar signal detected by a radio telescope is registered in multiple frequency channels Figure 1 as a set of sinusoids (waveforms) having almost the same Gaussian envelopes [5,8,11,12]. This allows the pulsar signal model to be described as a time-frequency grid.

5.1. Time-Frequency Grid Pulsar Signal Model

For each frequency channel the pulsar signal model can be expressed as

$$
s_r(t) = \sum_{p=0}^{N-1} \sum_{l=0}^{L-1} a_l \cdot \exp\left[-\frac{\left[t - T_{pl}\right]^2}{2\sigma_p^2} \right] \cdot \exp\left[j \cdot 2\pi \cdot f_r \cdot (t - T_{pl}) \right]
\tag{3}
$$

where f_r is the frequency of the r-th frequency channel, defined by the expression $f_r = f_0 + r.\Delta f$, where $r = \overline{0, \ R-1}$ is the frequency channel's number, R is the number of the frequency channels the signal is registered in, f_0 is the frequency of the 0-th channel, Δf is the difference between frequency channels (spectral resolution), a_l is the amplitude of the Gaussian microburst, σ_l is the time width of the l-th Gaussian microburst, $l = \overline{0, \ L-1}$ is the index of the Gaussian microburst, L is the number of microbursts inside the main Pulse, p is the index of the main Pulse or the index of the Main Pulse's repetition period, T_{pl} is the composed repetition period defined by $T_{pl} = p \cdot T_p + l \cdot T_l$, where T_p is the main pulse repetition period, T_i is the Gaussian microburst repetition period inside the main pulse, t is the current time, which in discrete form can be presented as $t = T_{pl} + (k-1) \cdot \Delta T$, where k is the time sample index inside the microburst, ΔT is the sample's time duration inside the microburst.

A generalized model of the main pulse with three microbursts with Gaussian envelopes registered in a particular frequency channel is presented in Figure 3.

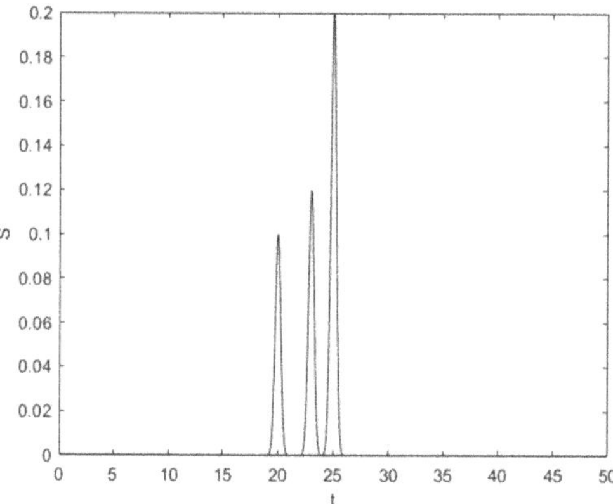

Figure 3. Main pulse with three microbursts measured in a particular frequency channel.

5.2. A Model of a Pulsar Signal Reflected from a Space Object

Signals reflected from the space object preserves the frequency characteristics of the pulsar emission except the amplitude and time delay from a particular generic point of the object surface or from the entire object. The signal amplitudes can be instrumentally measured whereas the time delay can be defined by the correlation of the received real signal (signal plus additive Gaussian noise) reemitted by a pseudo stationary object, slow moving in short integration time interval, with a priory known pulsar signal with time displacement. The signal reemitted by a particular generic point of the asteroid or from the entire surface is a delayed copy of a pulsar signal and can be interpreted as a time-frequency grid signal model with time delay. The pulsar's main pulse train with L microbursts reemitted by the g-th asteroid's generic point and registered in the r-th frequency channel is modeled by the expression

$$\hat{s}_{r,g}(t) = \sum_{p=0}^{N-1} \sum_{l=0}^{L-1} a_g \cdot \exp\left[-\frac{\left[t - T_{pl} - t_g\right]^2}{2\sigma_l^2}\right] \cdot \exp\left[j \cdot 2\pi \cdot f_r \cdot (t - T_{pl} - t_g)\right], \tag{4}$$

where a_g is the amplitude of the reemitted l-th microburst from the g-th generic point; $\sigma_l = T/2$ is the time dispersion of the Gaussian microburst; T is the microburst's time width; $t_g = t_g(p) = R_g(p)/c$ is the time delay of a reemitted signal by the g-th generic point from the object's surface; $R_g(p) = \left[x^2(p) + y^2(p) + z^2(p)\right]^{\frac{1}{2}}$ is the current distance to the particular generic point from the object; c is the speed of light in vacuum.

The time delays $t_g(p)$ for p-th main pulse and l-th microburst are arranged in ascending order, i.e., $g = 0, 1, 2, 3, \ldots, G\text{-}1$, where G is the full number of asteroid's generic points, $g = 0$ is the index of the minimum time delay from the nearest generic point, i.e., $t_0(p) = t_{g,\min}(p)$, $g = G$ -1 is the index of the maximum time delay from the furthest generic point, i.e., $t_{G-1}(p) = t_{g,\max}(p)$.

For the p-th main pulse and l-th microburst in the sequence of pulsar emissions the signal reemitted by g-th generic point limited by the microburst's time width, T, can be rewritten as

$$s_{r,g}(t) = a_g \cdot \text{rect}\left(\frac{t - T_{pl} - t_g}{T}\right) \cdot \exp\left[-\frac{2 \cdot \left[t - T_{pl} - t_g\right]^2}{T^2}\right] \times \exp\left[j \cdot 2\pi \cdot f_r \cdot (t - T_{pl} - t_g)\right], \tag{5}$$

where $\text{rect}\left(\frac{t - T_{pl} - t_g}{T}\right) = \begin{cases} 1, if\ 0 \leq \frac{t - T_{pl} - t_g}{T} < 1 \\ 0, \text{othewise} \end{cases}$.

The following substitutions are made $t = T_{pl} + t_{kp}$, $t_{kp} = t_{g,\min}(p) + k.\Delta T$, $\hat{t}_g(p) = t_{kp} - t_g(p)$, where $k = \overline{0,\ K + (k_{\max} - k_{\min}) - 1}$ is the current signal sample's number in the microburst and/or the signal sample's number measured on the range direction, $K = \text{int}(T/(\Delta T))$ is the full range signal samples' number, $k_{\max} = \text{int}[t_{g,\max}/(\Delta T)]$ is the index of the range bin where the signal reemitted by the furthest generic point with time delay $t_{g,\max}$ is recorded, $k_{\min} = \text{int}[t_{g,\min}/(\Delta T)]$ is the index of the range bin where the signal reemitted by the nearest generic point with time delay $t_{g,\min}$ is recorded, the difference $(k_{\max} - k_{\min})$ denotes the relative object's time width, measured on range direction. The analytical discrete model of the ISAR signal reemitted from the entire surface of the asteroid for each k,p, and r can be written as

$$S(k,p,r) = \sum_{g \in G} a_g \cdot \text{rect}\left(\frac{\hat{t}_g(p)}{T}\right) \cdot \exp\left[-\frac{2 \cdot \left[\hat{t}_g(p)\right]^2}{T^2}\right] \cdot \exp\left[j \cdot 2\pi \cdot f_r \cdot \hat{t}_g(p)\right] \tag{6}$$

where **G** is the asteroid's object space.

5.3. ISAR Signal Modeling Algorithm

The flow chart of asteroid's ISAR signal modelling is presented in Figure 4. The algorithm consists of two parts. In the first part, calculations of time delays, $t_g(p) = R_g(p)/c$, for each $g \in \mathbf{G}$, of signals reemitted by asteroid's scattering points and their arrangement in ascending order, $g = \overline{0,\ G-1}$, are performed. In the second part, the asteroid's ISAR signal modeling in accordance with the expression (7) is accomplished. The summation is correct if and only if the inequality $0 \leq \frac{\hat{t}_g(p)}{T} < 1$ holds, otherwise for particular r and p, k increases, and the procedure is repeated until $k = K + (k_{\max} - k_{\min}) - 1$, then for particular r and k, p increases, and the procedure is repeated until $p = N - 1$, then for particular p and k, r increases, and the procedure is repeated until $r = R - 1$, which is the end of the asteroid's ISAR signal formation.

Only one microburst is considered. The ISAR signal microburst in the main pulse sequence, reemitted by the asteroid, after preliminary signal processing (signal detection and de-dispersion) is recorded in two-dimensional coordinates, time t and frequency f_r. The signal time record is divided into two coordinates: fast time, measured on the range direction with index k (range sample number), and slow time measured in cross-range direction with index p (azimuth sample number). Thus, the ISAR signal microburst is registered in a three-dimensional array with discrete coordinates $[k, p, r]$. In case all microbursts are used for aperture synthesis, the procedure is repeated for all remained microbursts inside the main pulse.

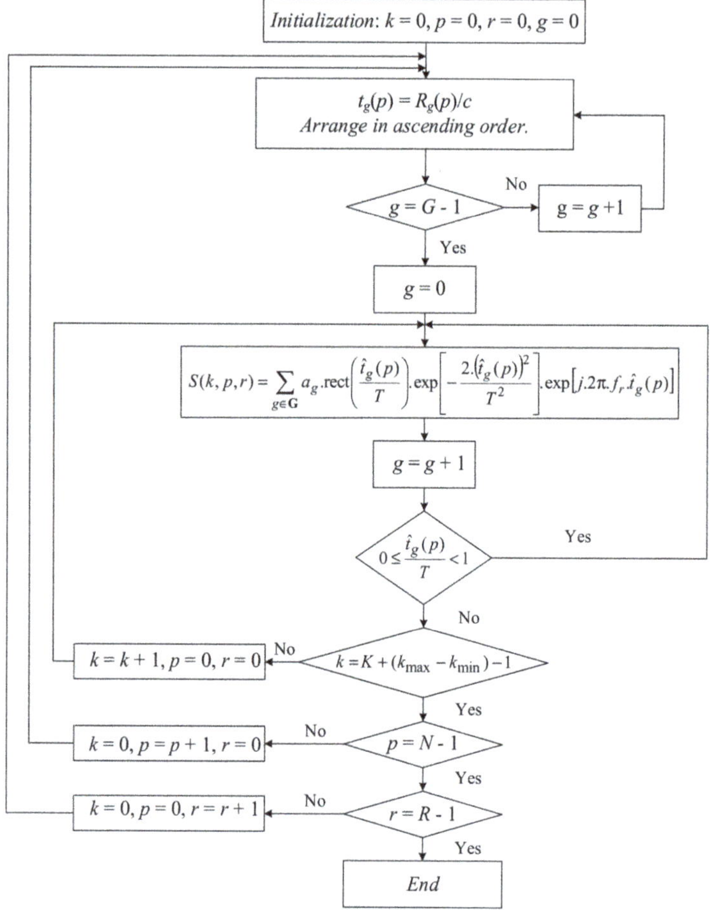

Figure 4. Flow chart of asteroid's ISAR signal modelling.

6. ISAR Signal Processing and Asteroid Image Reconstruction

6.1. Power Budget

Since the asteroid's ISAR signal distinguishes with a low signal power density on the Earth's surface, long observation times will essentially enhance the signal to noise ratio (SNR), which can be defined by the modified radar equation [29]

$$\text{SNR} = \frac{S \cdot \Delta F \cdot G \cdot \lambda^2 \cdot \sigma \cdot T_{\text{int}} \cdot n \cdot N}{4 \cdot \pi^2 (R_{0\prime})^2 k_{\text{B}} \cdot T_s} \tag{7}$$

where S is the spectral flux density of the asteroid's signal on the Earth measured in $\left[\frac{W}{m^2 \cdot Hz}\right]$; $G = q\left(\frac{\pi \cdot D_{pr}}{\lambda}\right)^2$ is the parabolic reflector antenna gain; D_{pr} is the diameter of the parabolic reflector; q is the efficiency factor which is around 0.5 to 0.6; $\sigma = \frac{\pi^3 \cdot D^4}{4 \cdot \lambda^2}$ is the asteroid's radar cross section; $k_{\text{B}} = 1.38 \times 10^{-23} \left[\frac{W}{Hz \cdot K}\right]$ is the Boltzmann constant; D is the asteroid's diameter; $R_{0\prime}$ is the distance to the asteroid's mass center at the moment of imaging; $T_s = 410$ K is the receiver noise temperature;

$T_{int} = T$ is the integration time equal to the microburst time width; N is the number of main pulses for aperture synthesis in one imaging segment; n is the number of imaging segments. In case the spectral flux density of the direct pulsar signal is $10^{-23} \left[\frac{W}{m^2 \cdot Hz} \right]$, assume the spectral flux density of the asteroid's signal $S = 10^{-26} \left[\frac{W}{m^2 \cdot Hz} \right]$. To evaluate SNR assume $N = 128$, $\Delta F = 2 \times 10^9$ Hz; $D_{pr} = 300$ m; $\lambda = 0.03$ m; $D = 100$ m; $T_{int} = 2 \times 10^{-6}$ s; $R_{0,} = 10^9$ m. In case $n = 10^2$; 10^3; 10^4; 10^5; 10^6, the enhancement (reducing of negative values) of the SNR in dB with the number of imaging segments is presented in Table 1.

Table 1. Enhancement of the SNR with the number of imaging segments.

n	10^2	10^3	10^4	10^5	10^6
$10.\log_{10}$ (signal to noise ratio (SNR))	−119	−96	−73	−50	−27

The negative value of SNR (−27 dB) can be further reduced by multiple coherent summations of the current complex image with the previous one after each accomplishment of the image reconstruction procedure.

6.2. Range Compression of the ISAR Signal

Considering that the ISAR signal reemitted by the asteroid is registered in a time-frequency grid, the range compression can be performed by the algebraic summation of signals from all frequency channels. To prove this statement the range compressing will be illustrated assuming that the frequency difference between channels, Δf tends to zero. It allows discrete time and continuous frequency approach to be applied in the ISAR signal's range compressing. The range compressed signal can be expressed as a frequency integration of a pulsar ISAR signal, i.e.,

$$\hat{S}(k,p) = \int_{f_c - \frac{\Delta F}{2}}^{f_c + \frac{\Delta F}{2}} \sum_{g \in G} \widetilde{a}_g[\hat{t}_g(p)] \cdot \exp\left[j \cdot 2\pi \cdot f \cdot \hat{t}_g(p) \right] df, \tag{8}$$

where $\widetilde{a}_g[\hat{t}_g(p)] = a_g \cdot \text{rect}\left(\frac{\hat{t}_g(p)}{T} \right) \exp\left[-\frac{2 \cdot [\hat{t}_g(p)]^2}{T^2} \right]$ is the time dependent amplitude, $f_c = f_0 + (\Delta F/2)$ is the central channel frequency, $\Delta F = R.\Delta f$ is the frequency bandwidth.

After simple mathematical manipulations the solution of the integral can be written as

$$\hat{S}(k,p) = \sum_{g \in G} \widetilde{a}_g[t_{kp} - t_g(p)] \cdot \Delta F \cdot \exp\left[j2\pi[t_{kp} - t_g(p)] \cdot f_c \right] \cdot \frac{\sin\left(\pi[t_{kp} - t_g(p)] \cdot \Delta F \right)}{\pi[t_{kp} - t_g(p)] \cdot \Delta F} \tag{9}$$

Thus, the range compressed ISAR signal is a time displaced copy of a sinc function defining the position of g-th scattering point from the space object, the asteroid. In Figure 5, a range compressed signal with time delay $t_1 = 0.5$ μs is presented.

In Figure 6, three range compressed signals RCS1, RCS2, RCS3 and their sum RCS, is depicted with the following parameters: microburst time width 2 μs, frequency band width $\Delta F = 100$ MHz, signals' time delays: $t_1 = 0.5$ μs, $t_2 = 0.515$ μs, $t_3 = 0.52$ μs, and amplitudes: $a_1 = 1.2, a_2 = 1.8, a_3 = 1.2$.

It is worth noting that in case Δf differs from zero (i.e., has a finite value) an unambiguous time interval $\Delta \tau$ of the ISAR compressed signal registration has to be defined (i.e., $\Delta \tau - 1/\Delta f$).

In a discrete time-frequency grid of the asteroid's signal registration, the range compression can be expressed as

$$\hat{S}(k,p) = \sum_{r=0}^{R-1} \sum_{g \in G} \widetilde{a}_g[\hat{t}_g(p)] \cdot \exp\left(j \cdot 2\pi \cdot f_r \cdot \hat{t}_g(p) \right) \tag{10}$$

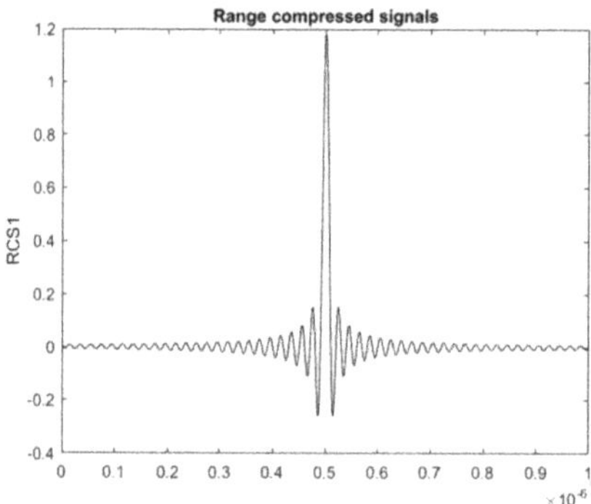

Figure 5. A range compressed signal RCS1 with time delay 0.5 µs.

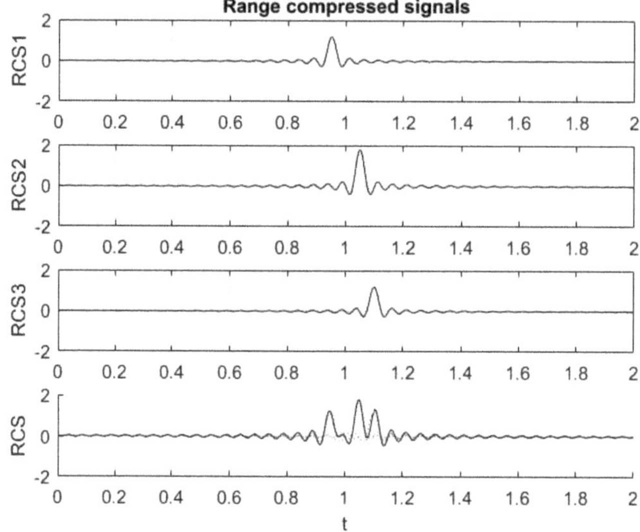

Figure 6. Three range compressed signals RCS1, RCS2, RCS3 and their sum RCS.

6.3. Azimuth Compression of the Range Compressed ISAR Signal and Complex Imaging

Pulsar ISAR signals reemitted by the asteroid are registered in a far field zone of electromagnetic waves propagation. It means that a plane wave approximation can be applied and, hence, an inverse Fourier transform can be used in order to perform azimuth compression of the range compressed ISAR signal, i.e.,

$$\hat{S}(k,\hat{p}) = \frac{1}{N} \sum_{p=0}^{N-1} \hat{S}(k,p) \cdot \exp\left(j\frac{2\pi \cdot p \cdot \hat{p}}{N}\right),\tag{11}$$

where $\hat{p} = \overline{0,\ N-1}$ is the discrete coordinate of the asteroid's generic point at the moment of imaging.

The inverse Fourier transform (11) is a correlation procedure, searching for all Doppler components $\exp\left(-j\frac{2\pi \cdot p \cdot \hat{p}}{N}\right)$ inside the spatial spectrum $\hat{S}(k,\hat{p})$, i.e., searching for that $\hat{p}$ in the interval from 0 to $(N-1)$ that reveal amplitudes by maximizing their intensities $\mathrm{mod}\big[\hat{S}(k,\hat{p})\big]$, and compensate for all phases $\arg[\hat{S}(k,\hat{p})]$ induced by the radial velocities, the motion of first order, except phases proportional to the radial velocities (Doppler frequencies) of scattering points at the moment of imaging. The coordinate

$\hat{p}$ is proportional to a constant radial velocity (Doppler frequency), that corresponds to the azimuth position of a particular scattering point at the moment of imaging. Thus, the expression (11) defines the asteroid's complex image with amplitude $\text{mod}[\hat{S}(k,\hat{p})]$, and phase $\text{arg}[\hat{S}(k,\hat{p})]$. It is worth noting that the Doppler bandwidth and, respectively, Doppler resolution or cross range (azimuth) resolution of the ISAR signal are limited by the fixed main pulse repetition period equal to 0.033 s, and apparent rotation angle between the vector velocity and mass-center's line-of-sight vector in case rectilinear movement of the asteroid.

7. Numerical Experiment

In order to validate the mathematical derivation of asteroid's ISAR geometry, kinematics and signal models, a numerical experiment with a signal model of one microburst is carried out based on the following pulsar's signal parameters: central channel frequency $f_c = 10.075$ GHz; frequency channels bandwidth $\Delta F = 150$ MHz; spectral resolution $\Delta f = 1$MHz; $f_{\min} = 10$ GHz, $f_{\max} = 10.15$ GHz; number of frequency channels $R = 150$; main pulses time repetition period $T_p = 0.033$ s, microburst time width $T = 10^{-6}$ s; number of range samples in microburst $K = 128$; sample's time width $\Delta T = 0.8 \times 10^{-8}$ s; number of azimuth measurements $N = 128$ in one imaging segment defined by the number of main pulses used for aperture synthesis; initial coordinates of asteroid's detection $x_{0'} = 1$ km, $y_{0'} = 350$ km, $z_{0'} = 10^3$ km; distance to the mass-center $R_{0'} = 1.06 \times 10^6$ km; asteroid's velocity $V = 35$ km/s; velocity angles $\alpha = \pi/6$, $\beta = \pi/4$, $\gamma = \pi/2$. The geometry of the asteroid is depicted in a 3-D grid with dimensions $64 \times 64 \times 64$, and grid cell dimensions $\Delta X = \Delta Y = \Delta Z = 0.5$ m, Relative intensity of scattering points is $a_g = 10^{-3}$. In order to obtain an asteroid's image of high resolution the apparent rotation angle between the asteroid's vector velocity and mass-center's line-of-sight vector has to be no less than 10^0. It guaranties the Doppler displacement in the spectrum of the ISAR signal to realize the necessary azimuth resolution of the asteroid's image.

Considering that the ISAR signal from the asteroid is very weak in comparison with the thermal noise, the experiment is carried out assuming the signal is obscured by additive white Gaussian noise, an appropriate model of signal disturbances in deep space navigations and communications. Assume as follows: number of imaging segments $n = 100$, $S = 10^{-26}\left[\frac{W}{m^2 \cdot Hz}\right]$, $\Delta F = 0.15 \times 10^9$ Hz, $D_{pr} = 300$ m, $\lambda = 0.03$ m, asteroid's diameter $D = 29$ m, $T_{\text{int}} = 10^{-6}$ s, then the signal to noise ratio calculated by (7) is equal to -15 dB. To reduce the level of noise, coherent summation of multiple complex ISAR images obtained after multiple applications of the image reconstruction procedure is applied. The number of complex images' sums mitigating the level of the additive white Gaussian noise is 10.

The ISAR signal from the asteroid is modelled in accordance with the algorithm presented by the flow chart in Figure 4. In each frequency channel additive white Gaussian noise is added to the signal using a standard procedure. The complex range compressed ISAR signal is modelled by expression (10). The real and imaginary parts of the range compressed ISAR signal obtained after summation of the signals from 150 frequency channels and registration in range (k) and azimuth (p) coordinates are depicted in Figure 7a,b respectively.

The ISAR complex image as ISAR amplitude and ISAR phase is extracted from the range compressed signal by applying azimuth compression with inverse Fourier transform (11) realized by inverse fast Fourier transform. The ISAR image amplitude and the ISAR image phase, the complex image with (-15) dB signal to noise ratio, just after azimuth compression of the range compressed ISAR signal, are presented in Figure 8a,b, respectively. The asteroid's image is obscured by noise.

A standard additive coherent summation (overlaying) of consecutive complex images is applied to reduce the level of the additive white Gaussian noise. The process of the noise depression and image quality improving by additive coherent summations of 3, 8, and 10 complex images is illustrated in Figures 9–11, respectively.

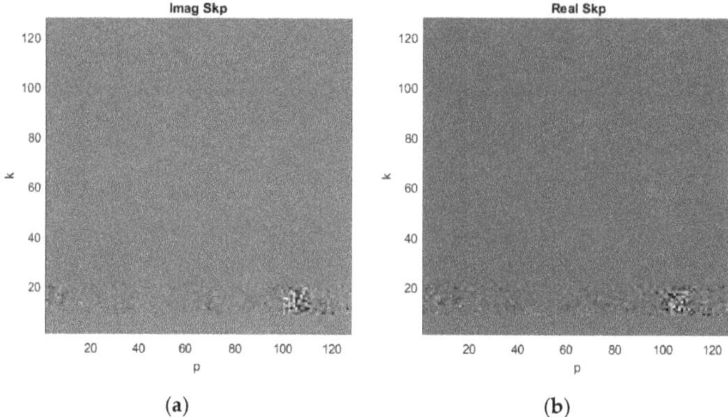

(a)   (b)

Figure 7. Range compressed ISAR signal obtained after summation of signals in all frequency channels: (**a**) real part; (**b**) imaginary part.

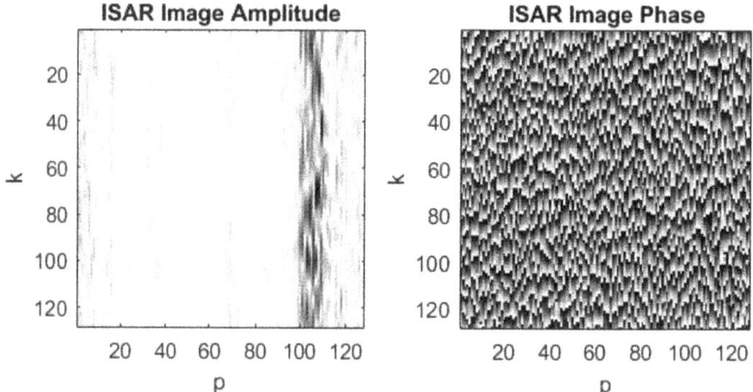

Figure 8. Complex image after azimuth compression of the range compressed ISAR signal with −15 dB signal to noise ratio: (**a**) ISAR image amplitude; (**b**) ISAR image phase.

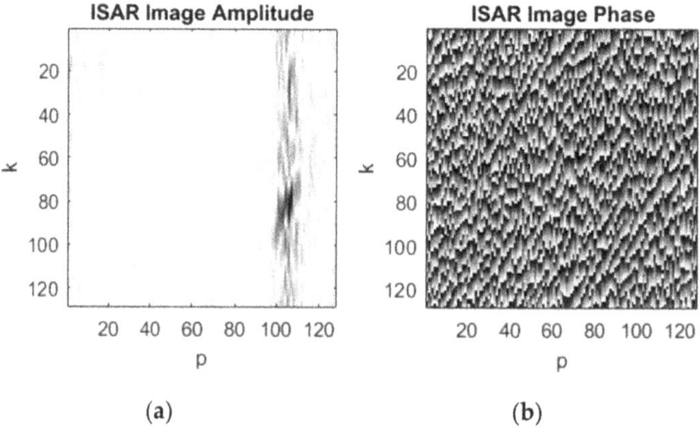

(a)   (b)

Figure 9. Resulting complex image after 3rd additive summation of complex images: (**a**) ISAR image amplitude; (**b**) ISAR image phase.

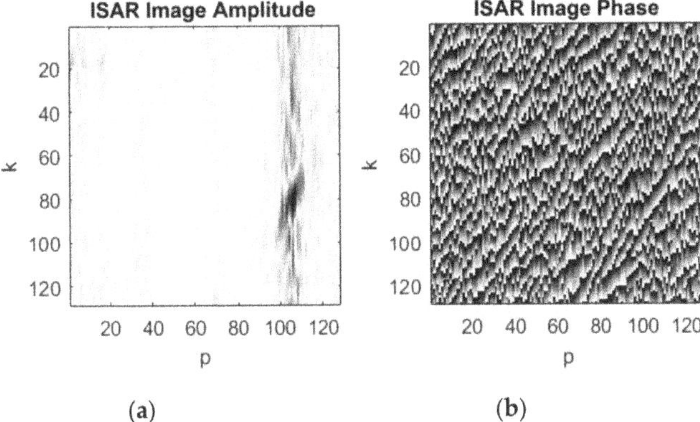

(a)   (b)

Figure 10. Resulting complex image after 8th additive summation of complex images: (**a**) ISAR image amplitude; (**b**) ISAR image phase.

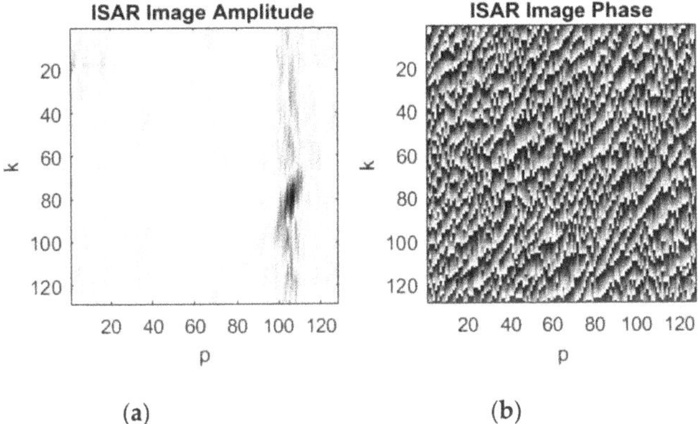

(a)   (b)

Figure 11. Resulting complex image after 10th additive summation of complex images: (**a**) ISAR image amplitude; (**b**) ISAR image phase.

The final complex image, amplitude and phase, obtained after additive Gaussian noise depression by coherent summation of 10 consecutive complex images of the asteroid is presented in Figure 11a,b. As can be seen, the asteroid's ISAR amplitude image is of satisfactory quality, but noise still remains, the asteroid's silhouette is satisfactorily depicted. Further improving of the ISAR image quality can be achieved by increasing the number of coherent summations which is limited by a huge processing time and graphical properties of the software on which the experiment is carried out.

8. Conclusions

In the present work, on the base of real astrophysical measurements by radio telescopes Goldstone-Apple Valley and Arecibo an analytical description of the pulsar Crab emission is suggested. The structure of the Crab pulsar emission has been interpreted as multiple monochromatic Gaussian pulses, distributed in a time-frequency signal grid. Models of pulsar signals reflected from the space object are presented as a time delay copy of time frequency distributed monochromatic Gaussian signals. A new range compression technique is applied to the ISAR complex signal through summation of time recorded ISAR signals in all frequency channels. In case the registration spectral resolution is not satisfactory, which limits the unambiguous time interval of coherency, interpolation in the frequency domain is required.

In the present work only one microburst was applied for aperture synthesis. The author states all microbursts can be used for inverse aperture synthesis in order to improve the image quality through a superposition of all images obtained for each microburst inside the main pulse. In addition, a new ISAR procedure can be developed based on the whole structure of the main pulse. It is supposed that it will further improve the quality of the ISAR image. Based on the sparsity of the ISAR signal due to the limited number of frequency channels in which the ISAR signal is registered, a compressed sensing approach, as l_0 and l_1 norm minimization, can be applied in restoring the pulsar signal structure and ISAR image reconstruction.

Future research works will be focused on other properties and applications of pulsar emissions in the area of stellar navigation and early warning systems for asteroid detection and imaging. From a theoretical point of view, new mathematical structures of pulsar signal models and space object's imaging algorithms based on the atomic clock's stability and wide bandwidth of pulsars' emissions will be developed. From a practical point of view, this work will motivate the development of new highly sensitive technologies to detect pulsars reemissions of asteroids and other nonidentified objects. The usage of steady and stable pulsar emissions for object navigation and imaging purposes is not limited by time and space which is the main advantage of this kind of stellar technology. The problem is the weakness of pulsar emissions, especially reemissions by asteroids and their reliable detection, which requires the development of highly sensitive sensors. From an astronomical point of view, it is very tough to detect the asteroid in the space by the narrow antenna beam of the radio telescope that is not able to cover the whole visible space. A network of multiple synchronized radio-telescopes located on the Earth's surface and directed on different parts of space, or giant phase arrays antennas scanning the space, are needed in order increase the asteroid's detectability.

Funding: This research received no external funding.

Acknowledgments: The author is grateful for the scientific support given by Jean Eilek from New Mexico Institute of Mining and Technology and the anonymous reviewers for their fruitful comments, remarks and suggestions to improve the quality of the paper.

Conflicts of Interest: The author declares no conflicts of interest.

References

1. Ginzburg, V.L.; Zheleznyakov, V.V.; Zaitsev, V.V. *Coherent Mechanisms of Radio Emission and Magnetic Models of Pulsars*; Provided by the NASA Astrophysics Data System; Kluwer Academic Publisher: Dordrecht, The Netherlands, 1969; pp. 464–508. Available online: http://iopscience.iop.org/article/10.1086/311945/fulltext/985796.text.html (accessed on 24 June 2019).
2. Ray, P.S.; Wood, K.S.; Phlips, B.F. Spacecraft navigation using X-ray pulsars. *J. Guid. Control Dyn.* **2006**, *29*, 43–63.
3. Gilster, P. Pulsar Navigation for Deep Space. 2010. Available online: https://www.centauri-dreams.org/?p=15475 (accessed on 30 November 2017).
4. Szondy, D. How to Navigate Deep Space by Pulsar 2016. Available online: https://newatlas.com/pulsars-gps-space-navigation/44978/ (accessed on 30 November 2017).
5. Grootjans, R. Detection of Dispersed Pulsars in a Time Series by Using a Matched Filtering Approach. Master's Thesis, August 2016. Available online: https://essay.utwente.nl/71435/1/GROOTJANS_MA_EWI.pdf (accessed on 30 November 2017).
6. Zhang, X.; Shuai, P.; Huang, L. Phase tracking for pulsar navigation with Doppler frequency. *Acta Astronaut.* **2016**, *129*, 179–185.
7. MIT Technology Review. An Interplanetary GPS Using Pulsar Signals 2013. Available online: https://www.technologyreview.com/s/515321/an-interplanetary-gps-using-pulsar-signals/ (accessed on 1 December 2017).
8. Kabakchiev, C.; Behar, V.; Buist, P.; Heusdens, R.; Garvanov, I.; Kabakchieva, D.; Gaubitch, N.; Bentum, M. Detection and estimation of pulsar signals for navigation. In Proceedings of the IRS 2015, Dresden, Germany, 24–26 June 2015.

9.   Nikonowicz, J.; Kubczak, P.; Matuszewski, Ł. Hybrid detection based on energy and entropy analysis as a novel approach for spectrum sensing, Signals and Electronic Systems (ICSES). In Proceedings of the International Conference, Krakow, Poland, 5–7 September 2016.

10. Dong, J. Pulsar navigation in the solar system. *arXiv* **2011**, arXiv:0812.2635v3.

11. Eilek, J.A.; Hankins, T.H. Radio emission physics in the Crab pulsar. *J. Plazma Phys. arXiv* arXiv:1604.02472, 2016. Available online: https://arxiv.org/pdf/1604.02472.pdf (accessed on 24 June 2019). [CrossRef]

12. Hankins, T.H.; Eilek, J.A.; Jones, G. The Crab pulsar at centimetre wavelengths, II. Single pulse. *Astrophys. J.* **2016**, *833*, 47. [CrossRef]

13. Wang, B.; Xu, S.H.; Wu, W.; Hu, P.; Chen, Z. Adaptive ISAR imaging of maneuvering targets based on a modified Fourier transform. *Sensors* **2018**, *18*, 1370. [CrossRef] [PubMed]

14. Li, Y.; Fu, Y.; Zhang, W. Distributed ISAR sub-image fusion of nonuniform rotating target based on matching Fourier transform. *Sensors* **2018**, *18*, 1806. [CrossRef] [PubMed]

15. Lv, Y.; Wang, Y.; Wu, Y.; Wang, H.; Qiu, L.; Zhao, H.; Sun, Y. A novel inverse synthetic aperture radar imaging method for maneuvering targets based on modified chirp Fourier transform. *Appl. Sci.* **2018**, *8*, 2443. [CrossRef]

16. Zeng, Z.; Shi, Z.; Xing, S.; Pan, Y. A Fourier-Based Image Formation Algorithm for Geo-Stationary GNSS-Based Bistatic Forward-Looking Synthetic Aperture Radar. *Sensors* **2019**, *19*, 1965. [CrossRef] [PubMed]

17. Qu, Z.; Qu, F.; Hou, C.; Jing, F. Quadratic Frequency Modulation Signals Parameter Estimation Based on Two-Dimensional Product Modified Parameterized Chirp Rate-Quadratic Chirp Rate Distribution. *Sensors* **2018**, *18*, 1624. [CrossRef] [PubMed]

18. Choi, G.G.; Park, S.H.; Kim, H.T.; Kim, K.T. ISAR Imaging of Multiple Targets Based on Particle Swarm Optimization and Hough Transform. *J. Electromagn. Waves Appl.* **2009**, *23*, 1825–1834. [CrossRef]

19. Park, S.-H.; Kim, H.-T.; Kim, K.-T. Stepped-frequency ISAR motion compensation using particle swarm optimization with an island model. *Prog. Electromagn. Res.* **2008**, *85*, 25–37. [CrossRef]

20. Xie, P.; Zhang, M.; Zhang, L.; Wang, G. Residual motion error correction with backprojection multisquint algorithm for airborne synthetic aperture radar interferometry. *Sensors* **2019**, *19*, 2342. [CrossRef] [PubMed]

21. Malamou, A.; Pandis, C.; Karakasiliotis, A.; Stefaneas, P.; Kallitsis, E.; Daras, N.; Frangos, P. SAR imaging: An autofocusing method for improving image quality and MFS image classification technique. In *Applications of Mathematics and Informatics in Science and Engineering*; Daras, N., Ed.; Hellenic Military Academy: Vari Attikis, Greece, 2014; pp. 199–215.

22. Gao, Y.; Xing, M.; Zhang, Z.; Guo, L. Ultrahigh range resolution ISAR processing by using KT-TCS algorithm. *IEEE Sens. J.* **2018**, *18*, 8311–8317. [CrossRef]

23. Wang, L.; Huang, T.; Liu, Y. Phase compensation and image autofocusing for randomized stepped frequency ISAR. *IEEE Sens. J.* **2019**, *19*, 3784–3796. [CrossRef]

24. Jin, K.; Lai, T.; Wang, Y.; Li, G.; Zhao, Y. Coherent integration for radar high-speed maneuvering target based on frequency-domain second-order phase difference. *Electronics* **2019**, *8*, 287. [CrossRef]

25. Yang, J.; Su, W.; Gu, H. Multi-sensor inverse synthetic aperture radar imaging and phase adjustment based on combination of sparsity and total variation. *J. Appl. Remote Sens.* **2018**, *12*, 025011. [CrossRef]

26. Zhao, Z.; Tao, R.; Li, G.; Wang, Y. Fractional sparse energy representation method for ISAR imaging. *IET Radar Sonar Navig.* **2018**, *12*, 988–997. [CrossRef]

27. Li, J.; Hu, P.; Zhang, Y.; Chen, Z. Analysis of clock jitter effects on LFM-signal pulse compression based on matched filter. In Proceedings of the 2018 2nd IEEE Advanced Information Management Communicates, Electronic and Automation Control Conference (IMCEC), Xi'an, China, 25–27 May 2018.

28. Ostro, S.J.; Hudson, R.S.; Benner, L.A.M.; Giorgini, J.D.; Magri, C.; Margot, J.-L.; Nolan, M.C. *Asteroid Radar Astronomy*; 2002; pp. 151–168. Available online: https://echo.jpl.nasa.gov/asteroids/ast3_ostro+.pdf (accessed on 30 November 2017).

29. Cherniakov, M.; Saini, R.; Zuo, R.; Antoniou, M. Space-Surface Bistatic Synthetic Aperture Radar with Global Navigation Satellite System Transmitter of Opportunity—Experimental Results. *IET Radar Sonar Navig.* **2007**, *1*, 447–458. [CrossRef]

Communication

SPARX, a MIMO Array for Ground-Based Radar Interferometry

Alberto Michelini [1,*], Francesco Coppi [1], Alberto Bicci [1] and Giovanni Alli [2]

[1] IDS GeoRadar, 56121 Pisa, Italy; francesco.coppi@idsgeoradar.com (F.C.);
   alberto.bicci@idsgeoradar.com (A.B.)
[2] IDS Ingegneria Dei Sistemi S.p.A., 56121 Pisa, Italy; g.alli@idscorporation.com
* Correspondence: a.michelini@idsgeoradar.com

Received: 13 November 2018; Accepted: 5 January 2019; Published: 10 January 2019

Abstract: Ground-Based SAR Interferometry (GB-InSAR) is nowadays a proven technique widely used for slope monitoring in open pit mines and landslide control. Traditional GB-InSAR techniques involve transmitting and receiving antennas moving on a scanner to achieve the desired synthetic aperture. Mechanical movement limits the acquisition speed of the SAR image. There is a need for faster acquisition time as it plays an important role in correcting rapidly varying atmospheric effects. Also, a fast imaging radar can extend the applications to the measurement of vibrations of large structures. Furthermore, the mechanical assembly put constraints on the transportability and weight of the system. To overcome these limitations an electronically switched array would be preferable, which however faces enormous technological and cost difficulties associated to the large number of array elements needed. Imaging Multiple-Input Multiple Output (MIMO) radars can be used as a significant alternative to usual mechanical SAR and full array systems. This paper describes the ground-based X-band MIMO radar SPARX recently developed by IDS GeoRadar in order to overcome the limits of IDS GeoRadar's well-established ground based interferometric SAR systems. The SPARX array consists of 16 transmit and 16 receive antennas, organized in independent sub-modules and geometrically arranged in order to synthesize an equally spaced virtual array of 256 elements.

Keywords: GB-SAR; MIMO radar; radar imaging

1. Introduction

Nowadays, thanks to its distinguishing features [1,2] GB-InSAR technology has become a consolidated technique for measure ground displacements in many geophysical applications [2–4] and has proved to be particularly suitable in environments where continuous and real-time monitoring is required [5,6]. Despite the high number of technological advances seen in the last decade, some typical limitations are still present in the standard GB-InSAR systems, and therefore many improvements can be performed with respect to the current technique.

Some of the persisting GB-InSAR limitations are related to its mechanical scanning. In fact, traditional GB-InSAR techniques require a radar sensor equipped with transmitting and receiving antennas, moved by a mechanical scanner to achieve the desired synthetic aperture [7]. This approach, even if it has proved to be simple and effective, can have a considerable impact on important operational aspects such as scanning times, maintenance and installation, which will be discussed briefly below.

The data acquisition time is one of the most important parameters in the evaluation of a remote sensing monitoring system; since the first GB-InSAR introduction, the scanning times have been significantly reduced, and currently the fastest systems can scan the entire 360° circular sector in just

40 s [8,9]. However, further reducing this time by using a mechanical scan, could result too demanding both in terms of power consumption and system operation.

Installation and maintenance are other important aspects to consider in the overall assessment of a monitoring system. For example, even if the use of a large mechanical scanning system does not entail particular problems in easily accessible installations, or in well-equipped environments such as open-pit mines, it could turn out a strong limitation in remote regions operations.

Therefore, within the field of GB-InSAR systems, one of the most interesting lines of research is the replacement of mechanical scanning with some kind of Electronically Scanned Array (ESA). To this scope, Multiple Input Multiple Output (MIMO) radar are promising systems in the evolution of the GB-InSAR technology [10,11]. The MIMO principle of operation is to transmit and receive the radar signal alternately from various appropriately located elements. This strategy allows to synthesize an arbitrary antenna array, using a relatively small number of physical elements, thus limiting the complexity and cost of the whole system with respect to standard ESA [12].

Despite the innovations introduced by the MIMO strategy, the prototypes developed so far [10,11] still seem to suffer from major disadvantages in terms of production cost and ease of installation. To let GB-InSAR MIMO become a feasible and easily installable technology, IDS GeoRadar aimed to develop a system with a modular and integrated architecture. Modularity will help the installation procedure, allowing the sequential assembly of the various modules, rather than the whole system at the same time. Furthermore, the exploitation of highly integrated technologies such as microstrips and patch antennas, will lead to a cost reduction compared with other technologies such as coaxial cables and horn antennas [10].

In this paper, we present the SPARX array, a MIMO system developed by IDS GeoRadar, composed by 16 transmit and 16 receive antennas, organized in independent and integrated sub-modules and geometrically arranged in order to synthesize an equally spaced virtual array of 256 elements.

2. MIMO Imaging System

A generic MIMO imaging system [13] is composed by n_{TX} transmitting antennas placed in the positions x_m and by n_{RX} receiving antennas placed in the positions y_l. Given a MIMO configuration consider a target located in $r = re$ at a distance r far away respect to the system extent; the time of flight τ_{ml} that an electromagnetic signal takes to go from the m-th transmitter to the target and come back to the l-th receiver is approximately:

$$\tau_{ml} \cong \frac{2}{c} \left[r - \left(\frac{x_m + y_l}{2} \right) \cdot e \right]; \tag{1}$$

that is equivalent to transmit and receive a signal from a unique antenna placed in the virtual phase center placed in $(x_m + y_l)/2$. Therefore, transmitting alternately from every transmitter and receiving alternately from every receiver, it can be generated a virtual array with $N = n_{TX} \cdot n_{RX}$ elements placed in the MIMO virtual phase centers. It can be noticed that, given a MIMO configuration, the dual one with transmitter and receiver swapped, generates the same phase centers positions. One of the main advantages of MIMO technique is that the number of virtual antennas grows as the square of the number of physical antennas; it is therefore possible to generate a large number of virtual elements, exploiting a relatively small number of physical antennas, with a significant cost and complexity reduction of the imaging system hardware.

2.1. MIMO Array Factor

Consider a MIMO array working with central wavelength λ, the corresponding array factor $F_{MIMO}(e)$ is given by the product of the transmitting and the receiving array factors:

$$F_{MIMO}(e) = F_{TX}(e)F_{RX}(e) = \frac{1}{N} e^{i\frac{4\pi}{\lambda}r} \sum_{m,l} e^{i\frac{4\pi}{\lambda}(\frac{x_m + y_l}{2}) \cdot e} \tag{2}$$

To avoid grating lobes in the MIMO array factor, it is necessary that F_{TX} has a null in correspondence of every grating lobe of F_{RX} and vice versa. In Figure 1 is shown an example of a typical MIMO array factor resulting from the product of a transmitting and receiving array factors, it can be noticed that all the grating lobes inside the receiving array factor are compensated by the nulls of the transmitting one.

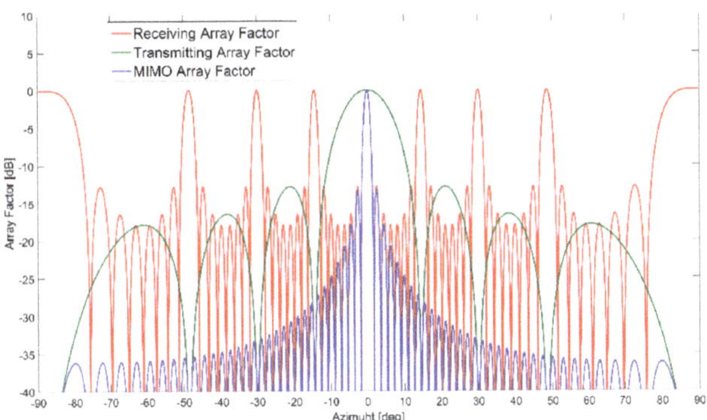

Figure 1. MIMO array factor (in blue), obtained from the product of a transmitting array factor (in green) and a receiving array factor (in red).

In a physical array, different sources of errors cause deviations from the ideal model; this fact limits the array performances, typically degrading the SideLobe Level (*SLL*) of the point spread function. Assuming that, for a real system, the array factor F can be modeled as a random variable whose expectation value $\langle F \rangle$ is equal to the ideal one, then the *SLL* distribution can be read from the power of the statistical deviation $\delta F = F - \langle F \rangle$:

$$SLL = \langle |\delta F|^2 \rangle. \tag{3}$$

In a MIMO array, the physical imperfections cause statistical deviations δF_{TX}, δF_{RX} of the transmitting and the receiving array factors from the ideal ones; ignoring second order terms, the total deviation δF_{MIMO} from the ideal MIMO array factor can be expressed as:

$$\delta F_{\mathrm{MIMO}} \simeq \delta F_{TX} \cdot F_{RX} + F_{TX} \cdot \delta F_{RX}. \tag{4}$$

Thus, in a MIMO array the *SLL* distribution is strongly dependent on the transmitting and the receiving array factors:

$$SLL_{\mathrm{MIMO}} \simeq |F_{RX}|^2 \langle |\delta F_{TX}|^2 \rangle + |F_{TX}|^2 \langle |\delta F_{RX}|^2 \rangle + \Re[\langle \delta F_{TX} \cdot \delta F_{RX}^* \rangle \delta F_{RX} \cdot \delta F_{TX}^*]. \tag{5}$$

If the errors on the transmitter and the receiver are uncorrelated then the third term of this expansion vanishes. As a simple example, consider small and uncorrelated phase and amplitude errors on every transmitting and receiving element. From array theory it is well known that the effect of this kind of errors is to add to the sidelobes a uniform power level proportional to the mean square error:

$$\langle |\delta F_{TX}|^2 \rangle = \sigma_{TX}^2 / n_{TX}; \ \langle |\delta F_{RX}|^2 \rangle = \sigma_{RX}^2 / n_{RX}. \tag{6}$$

Using these relations to compute the MIMO SideLobe Level it is possible to gather that, in a MIMO array, small and uncorrelated phase and amplitude errors generate a non-uniform SideLobe Level:

$$SLL_{\mathrm{MIMO}} \simeq |F_{RX}|^2 \frac{\sigma_{TX}^2}{n_{TX}} + |F_{TX}|^2 \frac{\sigma_{RX}^2}{n_{RX}}. \tag{7}$$

In particular, it can be noticed from the previous formula that, for a MIMO array, small errors produce strong sidelobes in the correspondence of transmitting and receiving grating lobes. To compensate this undesired effect, an efficient and reliable calibration procedure should be applied on the MIMO system. This reasoning can be easily generalized to other sources of errors like inaccuracy in the antennas placement, deformation of the system geometry, etc. The discussed behavior in the MIMO SideLobe Level has already been noticed in various experimental tests with MIMO arrays [14,15].

2.2. MIMO Uniform Linear Configurations

The simplest MIMO array configuration is composed by a uniformly spaced linear array of transmitting antennas parallel to a uniformly spaced linear array of receiving antennas [13,16]. Denoting with w the array axis direction, the antennas positions can be expressed as:

$$x_m = m \cdot p_{TX} \cdot w + x_0 m = 1, \cdots, n_{TX};$$
$$y_l = l \cdot p_{RX} \cdot w + y_0 l = 1, \cdots, n_{RX}. \tag{8}$$

where p_{TX} and p_{RX} are the transmitting and receiving spacing, respectively. The corresponding virtual phase centers are located in:

$$\frac{m \cdot p_{TX} + l \cdot p_{RX}}{2} \cdot w + \frac{x_0 + y_0}{2}. \tag{9}$$

Starting from this configuration, in order to generate a uniformly spaced array with $N = n_{TX} \cdot n_{RX}$ virtual elements it is necessary that the array spacing satisfy the relation $p_{RX} = n_{TX} \cdot p_{TX}$, or the dual one $p_{TX} = n_{RX} \cdot p_{RX}$. If one of these relations is satisfied, then the resulting linear array is uniformly spaced with a virtual spacing equal to $p_{TX}/2$, or $p_{RX}/2$ in the dual configuration (Figure 2).

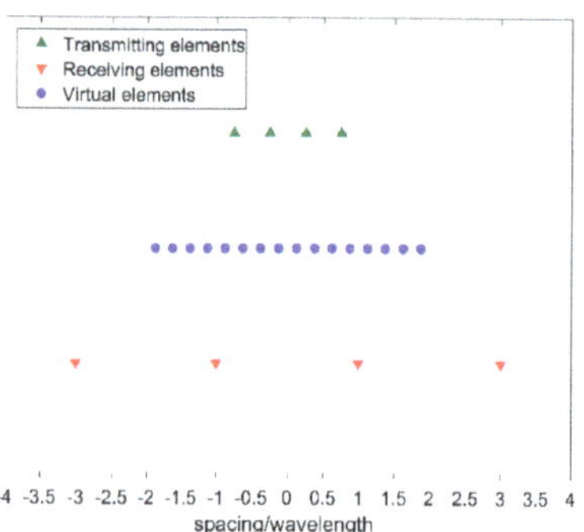

Figure 2. 16 elements MIMO uniform linear array (in blue) obtained from 4 elements transmitting linear array and 4 elements receiving linear array.

In this case the smallest spacing between physical antennas is double the spacing of virtual elements, yet MIMO technique allows to create configuration with arbitrary large ratio between real spacing and the virtual spacing; to see this consider a MIMO configuration composed by a linear array of $n_{TX} = 2k + 1$ transmitting antennas uniformly spaced by p_{TX}, and a linear array of $n_{RX} \geq 2$ receiving antennas uniformly spaced by $n_{TX} \cdot p_{TX}/2$:

$$x_m = m \cdot p_{TX} \cdot w + x_0 m = 1, \cdots, 2k + 1;$$
$$y_l = (2k + 1) \cdot l \cdot \frac{p_{TX}}{2} \cdot w + y_0 l = 1, \cdots, n_{RX}. \tag{10}$$

The corresponding virtual array is linear and contains a $N = n_{TX} \cdot (n_{RX} - 1) + 1$ elements sub-array uniformly spaced by $p_{TX}/4$. The uniformly spaced sub-array in Figure 3 is equivalent to the virtual array in Figure 2, however it has been obtained with five transmitting elements spaced by

p_{TX} instead of four transmitting elements spaced by $p_{TX}/2$; the second MIMO configuration although is require more physical antennas than the first one, allows to space apart the radiating elements decreasing mutual coupling effects.

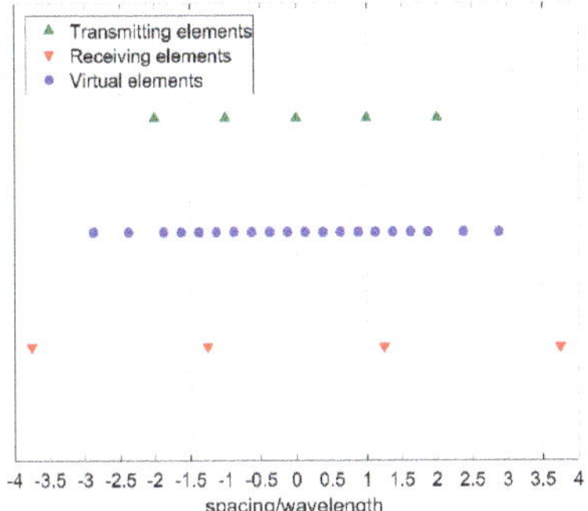

Figure 3. Twenty elements MIMO linear array (in blue) obtained from a five elements transmitting linear array and a four elements receiving linear array.

3. SPARX Design

Recently, in order to overcome the current limits of the ground based interferometric SAR systems, IDS GeoRadar developed SPARX: an X-band MIMO array. The SPARX array consists of 16 transmit and 16 receiver antennas, organized in independent sub-modules and geometrically arranged in order to synthesize a uniformly spaced virtual array of 256 elements. In Figure 4 is shown the SPARX block diagram: a radar sensor generates an X-band RF signal with a central frequency of 9.7 GHz and an instantaneous bandwidth of 275 MHz. The RF signal is transmitted to a Single Pole Double Throw (SPDT) switch stage and then switched between two transmitting antenna modules, each one consisted of an eight radiating elements switched array. The reflected signal is received by four receiving antenna modules each one composed by four radiating elements switched array and then collected by the Radar sensor through a Single Pole 4 Throw (SP4T) switch stage.

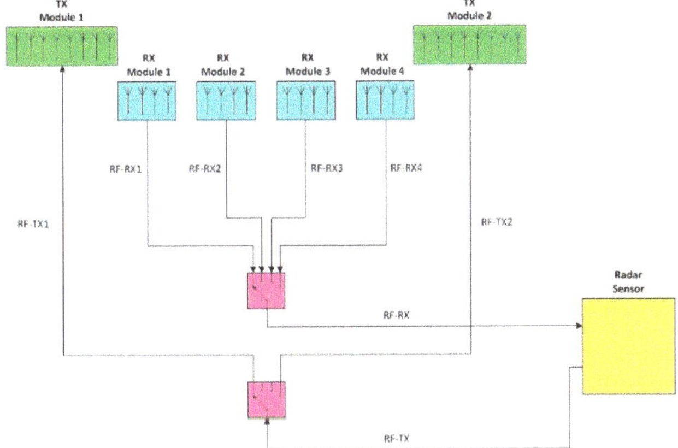

Figure 4. SPARX array block diagram.

The two transmitting antenna arrays have a uniform spacing of 18 mm and the receiving antenna array has a uniform spacing of 144 mm; as discussed in the previous section, this configuration allow to generate 256 virtual elements uniformly spaced by 9 mm, that correspond to a uniform linear array with a $\lambda/3.44$ spacing.

In Figure 5 it is shown a SPARX prototype with a reduced number of modules: two transmitting modules located in the upper external part and two receiving modules located in the inner lower part

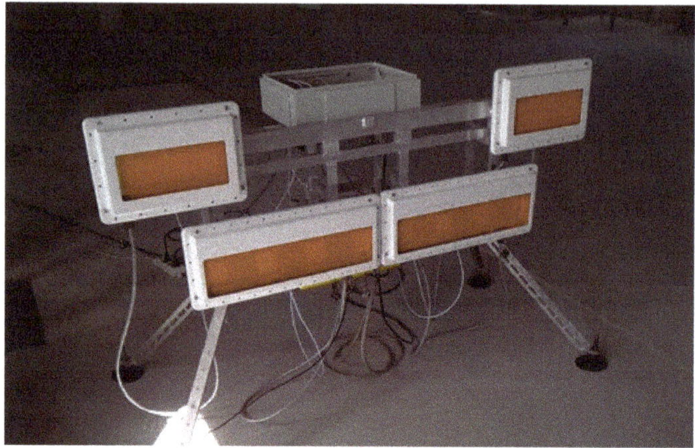

Figure 5. SPARX array prototype.

3.1. Transmitting Antenna Module

The transmitting antenna module consists in a microstrip switch matrix that route the RF signal from a single input to eight stacked patch antennas. The microstrip transmission line and the patch antenna technology allow to fully integrate the module in a single PCB (Figure 6). In Figure 7 is shown the Transmitting Antenna Module block diagram: the incoming RF signal is transmitted by a microstrip and pass through a SP4T switch followed by four SPDT switches, this matrix allow to switch the signal between the eight antennas feed lines; immediately before every antenna a power amplifier compensate the losses of the microstrip transmission line and the switch stages

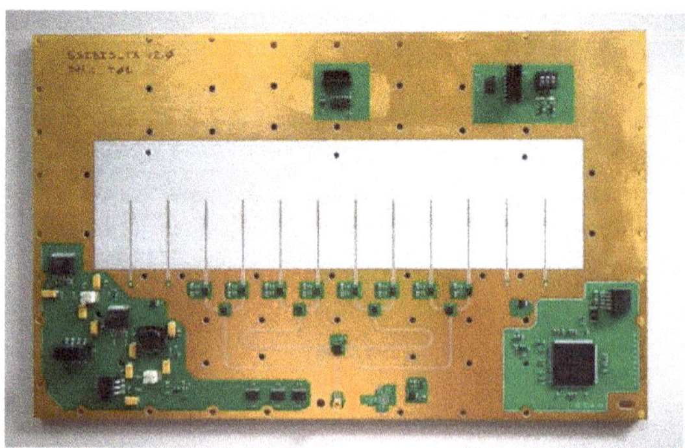

Figure 6. Transmitting antenna module.

The radiating elements are stacked microstrip patch antennas with vertical polarization, properly designed by IDS' laboratories in order to have 1 GHz bandwidth with a VSWR < 1.5 and a 3 dB beamwidth of 80° in the azimuth plane and 60° in the elevation plane. The total gain of the module including the power amplifier gain and transmission line losses is 15.7 dB.

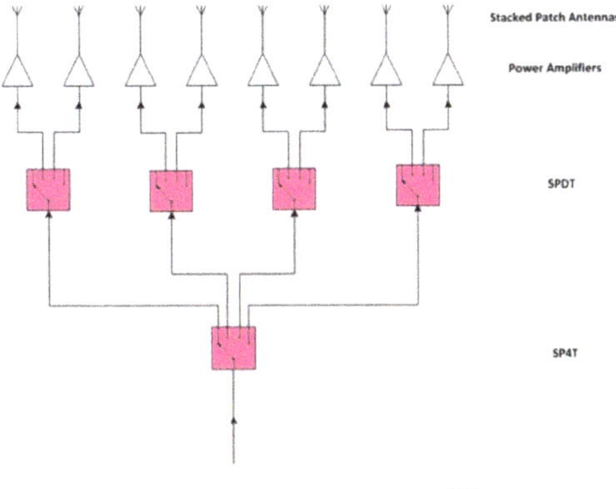

Figure 7. Transmitting antenna module block diagram.

3.2. Receiving Antenna Module

The receiving antenna module consists in a microstrip switch matrix that route the RF signal from four stacked patch antennas to a single output. The miscrostrip transmission line and the patch antenna technology allow to fully integrate the module in a single PCB (Figure 8). In Figure 9 is shown the receiving antenna module block diagram: immediately after every antenna a Low Noise Amplifier stage allow to keep low the noise figure of the system; the received RF signal pass through two SPDT switches followed by another SPDT switch, this matrix allow to switch the signal between the four antenna feed lines.

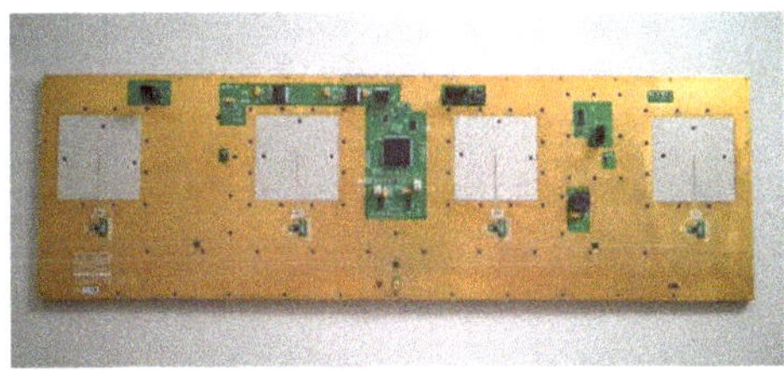

Figure 8. Receiving antenna module.

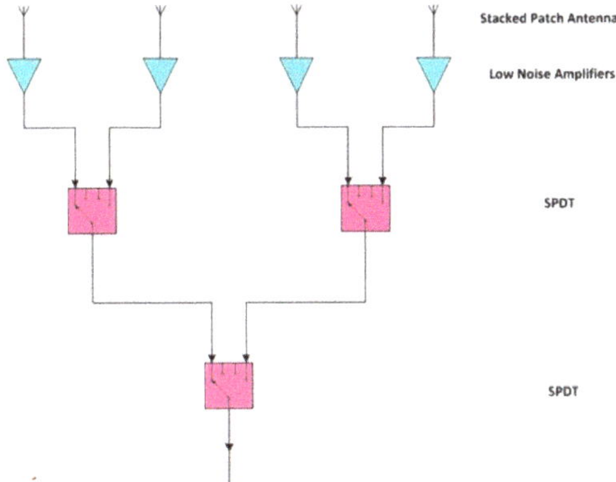

Figure 9. Receiving antenna module block diagram.

The radiating elements are stacked microstrip patch antennas with the same design of the transmitting one. The total gain of the module including the LNA gain and transmission line losses is 13.6 dB.

4. Field Test

The SPARX array operating principle has been tested with a reduced MIMO configuration, composed of two transmitting module and just one receiving module, for a total of 64 virtual channels. The system has been deployed in an external environment where it was possible to recognize various reflecting targets at different ranges and azimuth angles. The purpose of this preliminary test was to achieve the correct MIMO imaging in order to detect and identify all the relevant targets.

The acquisition scenario from the SPARX point of view is shown in Figure 10, while in Figure 11 the same scenario from the top view is shown. In both images the reflecting targets were highlighted in various colors to better distinguish them; in particular it is possible to recognize a paved road R (in yellow), various metallic poles P1, P2, P3, P4 and P5 (in purple) and some structures S1, S2 and S3 (in light blue). In range, all the relevant targets are located between 140 m (P1) and 450 m (S1), while in azimuth they are included between $-35°$ (P2) and $+20°$ (S3).

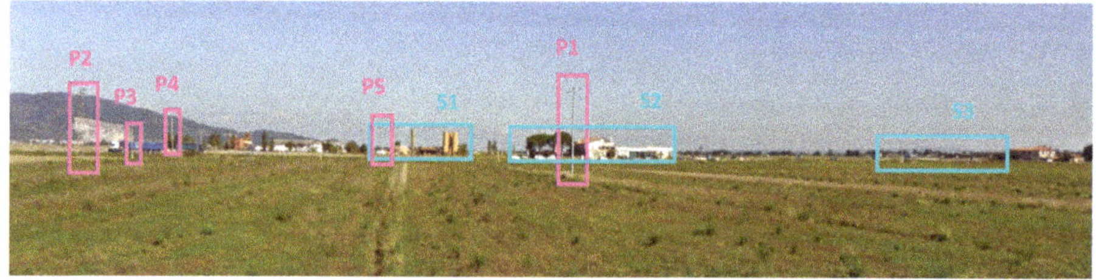

Figure 10. Acquisition scenario, SPARX point of view.

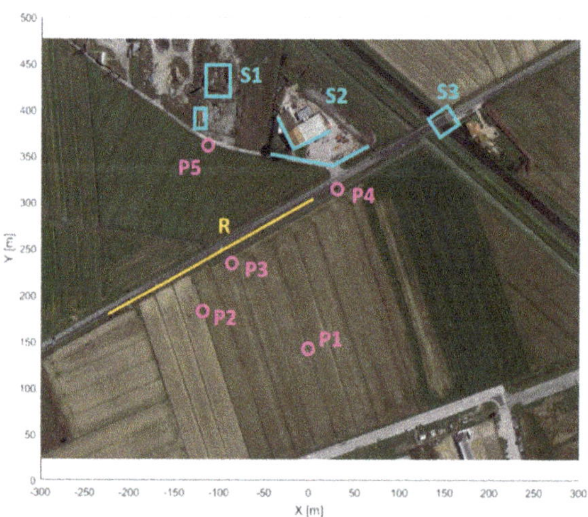

Figure 11. Acquisition scenario, top view.

5. Results

After a standard range-azimuth data focusing, it was possible to extract the scenario power maps; in Figure 12 the resulting SNR map obtained from the SPARX acquisitions is shown. In this map target SNR levels are estimated comparing their powers with respect to the background thermal noise level.

From this map it is possible to notice that all the relevant targets have been detected with a SNR greater than 35 dB, allowing an interferometric displacement measurement with precision greater than one tenth of a millimeter [5].

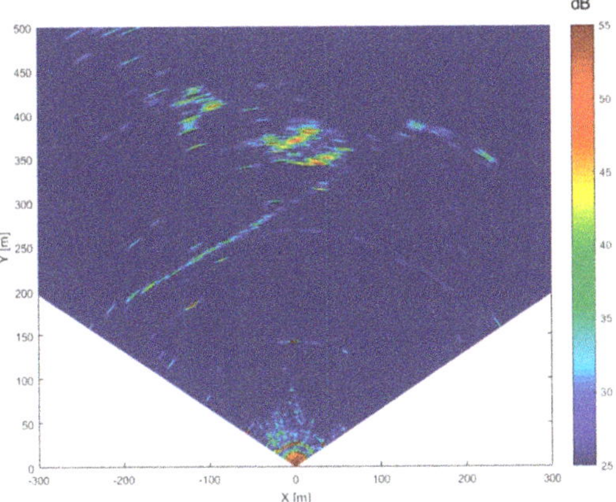

Figure 12. SPARX SNR estimated map (in dB).

Overlapping the acquired SNR map (Figure 12) with the scenario top view (Figure 11) it is possible to identify every strong measured signal with a specific reflecting target inside the acquisition scenario (Figure 13); although it should be noted that, due to prototype's low azimuth resolution, imaging of the farthest structures becomes rather coarse.

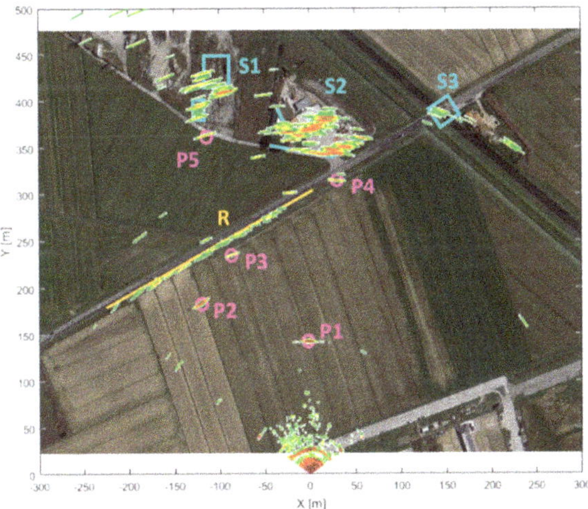

Figure 13. SPARX SNR estimated map superimposed on the acquisition scenario top view.

6. Conclusions

In this paper SPARX system have been introduced and described. It is a new MIMO system developed by IDS GeoRadar, in order to overcome some limitations of the current GB-InSAR systems. In particular, by decomposing the system into independent modules and using integrated technologies such as patch antennas and microstrip transmission line, SPARX development aims to reduce production costs and facilitate installation procedures compared to the current GB-InSAR MIMO prototypes. The field test conducted with the SPARX prototype showed that the MIMO imaging works effectively in detecting and identifying various target distributed inside the scenario.

However, it should be remarked that to obtain an azimuth resolution comparable to standard mechanical systems, a large number of modules is needed, thus greatly increasing the cost and complexity of the system. A possible solution to this difficulty is to exploit shorter wavelengths, in order to obtain high azimuth resolution compact systems [15], on the other hand, by reducing the transmitted wavelength, the operating range also decreases accordingly. From these considerations, it emerges that to reach a manufacturable GB-InSAR MIMO, the future developments will require a careful trade-off analysis between complexity and performances of the available technologies.

Author Contributions: Conceptualization, A.M., F.C. and G.A.; Formal Analysis, A.M.; Investigation A.M.; Supervision, F.C. and G.A.; Project Administration, A.B.; Resources A.B.; Writing—Original Draft Preparation, A.M.; Writing—Review & Editing, A.M.

Funding: This research was co-funded by Italian Ministry of Education, University and Research through the PON REC 2007-2013 Programme in the project "Sistemi radar per la sorveglianza e la Protezione delle InfrastRutture dI Trasporto" (SPIRIT - PON01_02408).

Conflicts of Interest: The authors declare no conflict of interest.

References

1. Monserrat, O.; Crosetto, M.; Luzi, G. A review of ground-based SAR interferometry for deformation measurement. *ISPRS J. Photogramm. Remote Sens.* **2014**, *93*, 40–48. [CrossRef]
2. Caduff, R.; Schlunegger, F.; Kos, A.; Wiesmann, A. A review of terrestrial radar interferometry for measuring surface change in the geosciences. *Earth Surf. Processes Landforms* **2015**, *40*, 208–228. [CrossRef]
3. Luzi, G. Ground based SAR interferometry: A novel tool for Geoscience. In Geoscience and Remote Sensing New Achievements. *InTech* **2010**. [CrossRef]
4. Farina, P.; Leoni, L.; Babboni, F.; Coppi, F.; Mayer, L.; Ricci, P. IBIS-M, an innovative radar for monitoring slopes in open-pit mines. In Proceedings of the International Symposium on Rock Slope Stability in Open Pit Mining and Civil Engineering, Vancouver, BC, Canada, 18–21 September 2011.
5. Rödelsperger, S. *Real-Time Processing of Ground Based Synthetic Aperture Radar (GB-SAR) Measurements*; (No. 33); Technische Universität Darmstadt, Fachbereich Bauingenieurwesen und Geodäsie: Darmstadt, Germany, 2010.
6. Atzeni, C.; Barla, M.; Pieraccini, M.; Antolini, F. Early warning monitoring of natural and engineered slopes with ground-based synthetic-aperture radar. *Rock Mech. Rock Eng.* **2015**, *48*, 235–246. [CrossRef]
7. Antonello, G.; Casagli, N.; Farina, P.; Leva, D.; Nico, G.; Sieber, A.J.; Tarchi, D. Ground-based SAR interferometry for monitoring mass movements. *Landslides* **2004**, *1*, 21–28. [CrossRef]
8. Viviani, F.; Michelini, A.; Mayer, L.; Coppi, F. IBIS-ArcSAR: An Innovative Ground-Based SAR System for Slope Monitoring. In Proceedings of the IGARSS 2018-2018 IEEE International Geoscience and Remote Sensing Symposium, Valencia, Spain, 22–27 July 2018; pp. 1348–1351.
9. Rödelsperger, S.; Coccia, A.; Vicente, D.; Meta, A. Introduction to the new metasensing ground-based SAR: Technical description and data analysis. In Proceedings of the 2012 IEEE International Geoscience and Remote Sensing Symposium (IGARSS), Munich, Germany, 22–27 July 2012; pp. 4790–4792.
10. Sammartino, P.F.; Tarchi, D.; Oliveri, F. GB-SAR and MIMO radars: Alternative ways of forming a synthetic aperture. In Proceedings of the International Conference on Synthetic Aperture Sonar and Synthetic Aperture Radar, Lerici, Italy, 13–14 September 2010; pp. 174–180.
11. Klare, J.; Saalmann, O. MIRA-CLE X: A new imaging MIMO-radar for multi-purpose applications. In Proceedings of the 7th European Radar Conference (EuRAD), Paris, France, 30 September–1 October 2010; pp. 129–132.
12. Li, J.; Stoica, P. *MIMO Radar Signal Processing*; John Wiley & Sons: Hoboken, NJ, USA, 2008.
13. Ender, J.H.G.; Klare, J. System architectures and algorithms for radar imaging by MIMO-SAR. In Proceedings of the IEEE Radar Conference, Pasadena, CA, USA, 4–8 May 2009; pp. 1–6.
14. Tarchi, D.; Oliveri, F.; Sammartino, P.F. MIMO radar and ground-based SAR imaging systems: Equivalent approaches for remote sensing. *IEEE Trans. Geosci. Remote Sens.* **2013**, *51*, 425–435. [CrossRef]

15. Biallawons, O.; Klare, J.; Saalmann, O. Technical realization of the MIMO radar MIRA-CLE Ka. In Proceedings of the 10th European Radar Conference (EuRAD), Nuremberg, Germany, 9–11 October 2013; pp. 21–24.

16. Sammartino, P.F.; Tarchi, D.; Baker, C.J. MIMO radar topology: A systematic approach to the placement of the antennas. In Proceedings of the 2011 International Conference on IEEE Electromagnetics in Advanced Applications (ICEAA), Torino, Italy, 12–16 September 2011; pp. 114–117.

Article

Interferometric DEM-Assisted High Precision Imaging Method for ArcSAR

Yanping Wang [1], Yang Song [1], Yun Lin [1,*], Yang Li [1], Yuan Zhang [1] and Wen Hong [2]

[1] School of Information Science and Technology, North China University of Technology (NCUT), Beijing 100144, China

[2] Institute of Electronics, Chinese Academy of Science (IECAS), Beijing 100190, China

* Correspondence: ylin@ncut.edu.cn

Received: 4 June 2019; Accepted: 28 June 2019; Published: 1 July 2019

Abstract: Ground-based arc-scanning synthetic aperture radar (ArcSAR) is the novel ground-based synthetic aperture radar (GBSAR). It scans 360-degree surrounding scenes by the antenna attached to rotating boom. Therefore, compared with linear scanning GBSAR, ArcSAR has larger field of view. Although the feasibility of ArcSAR has been verified in recent years, its imaging algorithm still presents difficulties. The imaging accuracy of ArcSAR is affected by terrain fluctuation. For rotating scanning ArcSAR, even if targets in scenes have the same range and Doppler with antenna, if the heights of targets are different, their range migration will be different. Traditional ArcSAR imaging algorithms achieve imaging on reference plane. The height difference between reference plane and target in scenes will cause the decrease of imaging quality or even image defocusing because the range migration cannot be compensated correctly. For obtaining high-precision ArcSAR image, we propose interferometric DEM (digital elevation model)-assisted high precision imaging method for ArcSAR. The interferometric ArcSAR is utilized to acquire DEM. With the assist of DEM, target in scenes can be imaged on its actual height. In this paper, we analyze the error caused by ArcSAR imaging on reference plane. The method of extracting DEM on ground range for assisted ArcSAR imaging is also given. Besides, DEM accuracy and deformation monitoring accuracy of proposed method are analyzed. The effectiveness of the proposed method was verified by experiments.

Keywords: synthetic aperture radar (SAR); ground-based synthetic aperture radar (GBSAR); arc-scanning synthetic aperture radar (ArcSAR); interferometric ArcSAR; DEM assisted SAR imaging

---

1. Introduction

Synthetic aperture radar (SAR) is capable of high-resolution imaging all-day and all-weather conditions [1,2]. As a complex image, the SAR image contains amplitude and phase information. We can get the deformation of the scenes by differential interferometric SAR (D-InSAR). Therefore, SAR is widely used in the field of ground deformation monitoring [3–5]. Especially the spaceborne SAR can achieve ground deformation monitoring in a wide range of scenes [6]. However, the spaceborne SAR systems and airborne SAR system require a long revisit cycle. They cannot realize continuous and repeated monitoring of a region. As an alternative, the GBSAR can achieve continuous and repeated monitoring of a region and feedback monitoring information in real time. In view of the above advantages, GBSAR has become one of the important means for deformation monitoring of dams' walls, buildings and slope [7–10].

The conventional GBSAR system scans the scenes along the linear rail. Its synthetic aperture is generated by moving the radar on the rail [11]. This type of working mode limits its field of view. A new mode GBSAR called ArcSAR can solve this problem. The ArcSAR system scans the surrounding scenes by the antenna attached to the rotating boom which extending from the center of the rotating platform. Its synthetic aperture is generated by the rotation of the antenna [12]. Therefore, under the premise of ensuring the resolution of the system, ArcSAR can cover the 360-degree scenes in once scanning, which effectively improve the field of view. Currently, several teams are working on the ArcSAR system [12–20], and the imaging algorithms for ArcSAR are also proposed.

Lee et al. deduced the geometric model and signal model of the ArcSAR system and proposed the imaging algorithm for ArcSAR [17,18]. Luo Yunhua et al. proposed the fast imaging algorithm for ArcSAR and used differential interferometric ArcSAR for ground deformation monitoring [19,20]. However, the above ArcSAR imaging algorithms perform the imaging on the reference plane. The imaging accuracy of the ArcSAR system is affected by the terrain fluctuation. For rotating scanning ArcSAR system, even if the targets in the scenes have the same range and Doppler with the antenna, the targets with different height have different range migration. Therefore, if the height of reference plane is inconsistent with the actual height of target in the scenes, the height difference between the reference plane and the target will cause the decrease of imaging quality or even image defocusing because the range migration cannot be compensated correctly.

To acquire the high precision ArcSAR image, the DEM of the scenes can be used to assist ArcSAR imaging. In this paper, we propose an interferometric DEM-assisted high precision imaging method for ArcSAR. Firstly, the DEM image of the scenes is acquired by interference with the ArcSAR slant range images. The acquired DEM image is on the slant range. However, we require the DEM image on the ground range to assist ArcSAR imaging. Thus, we next transform the DEM image on the slant range to the ground range. Finally, with the assist of the DEM image on the ground range, the target in the scenes can be imaged on its actual height. The proposed method does not rely on external DEM data. It can effectively avoid the decrease of ArcSAR imaging quality.

This paper proceeds as follows. The ArcSAR geometric model and signal model that consider the terrain of the scenes is introduced in Section 2. The error caused by ArcSAR imaging on the reference plane is analyzed in Section 3. The principle of the interferometric DEM-assisted high precision imaging method is given in Section 4. The DEM accuracy and deformation monitoring accuracy of proposed method are analyzed in Section 5. The effectiveness of the proposed high precision imaging method is verified by experiment in Section 6. In Section 7, we discuss the concluding remarks.

2. The Geometric Model and Signal Model of ArcSAR

Figure 1 shows the geometric model of ArcSAR system. The antenna of ArcSAR mounted by the boom rotates counterclockwise. It transmits and receives signals at equal intervals. The coordinate z-axis is the rotation axis of the ArcSAR system. We select *xoy* plane as the rotation plane. Point o is the rotation center. Point P represents the target in the scenes. Its coordinates can be expressed as:

$$\left(\sqrt{R_0^2 - h^2} \cos\varphi, \ \sqrt{R_0^2 - h^2} \sin\varphi, h \right) \tag{1}$$

where R_0 is the distance between the target P and the point o. h is the height of the target P. φ stands for the azimuth angle of the target position. Point S_0 is the position of the antenna phase center (APC). Its coordinates are

$$(r \cos\theta, r \sin\theta, h) \tag{2}$$

where r represents the length of the boom and θ is the rotation angle of the boom.

Taking example for linear frequency modulation (LFM) signal, the echo signal of the target P in ArcSAR system is shown as:

$$S(\theta, t_r) = \delta_p rect\left(\left(t_r - 2R_p/c \right)/T_p \right) \cdot rect((\theta - \varphi)/\theta_{bw})$$
$$exp\left(j\pi K_r \left(t_r - 2R_p/c \right)^2 \right) \cdot exp\left(-j4\pi f_c R_p/c \right) \tag{3}$$

where δ_p is the backscattering coefficient, t_r is the fast time of the ArcSAR system, c represents the speed of light, T_p stands for the signal pulse width, θ_{bw} is the antenna beam width, K_r represents the linear frequency modulation (LFM) rate, and f_c is the center frequency. R_p represents the distance between target P and the APC, which can be expressed as:

$$R_p = \sqrt{R_0^2 + r^2 - 2r\sqrt{R_0^2 - h^2}\cos(\theta - \varphi)} \tag{4}$$

We perform pulse compression of $S(\theta,t_r)$ in the frequency domain on the range direction. The result can be expressed as:

$$S(\theta, f) = \delta_p \cdot \text{rect}((\theta - \varphi)/\theta_{bw}) \cdot \text{rect}(f/B_r) \cdot \exp[-j4\pi(f + f_c)R_p/c] \tag{5}$$

where f is the current frequency in frequency domain after the spectral transformation of signal $S(\theta,t_r)$, B_r is the bandwidth. Since the echo signal has a shift-invariant characteristic in the azimuth direction, targets with different φ have the same form of range migration if they are the same distance from the rotation center. Therefore, for the convenience of derivation, in this paper, we assume $\varphi = 0$. The backscattering coefficient is also neglected. R_p and $S(\theta, f)$ can be rewritten as a current frequency in a frequency domain:

$$R_p = \sqrt{R_0^2 + r^2 - 2r\sqrt{R_0^2 - h^2}\cos\theta} \tag{6}$$

$$S(\theta, f) = \text{rect}(\theta/\theta_{bw}) \cdot \text{rect}(f/B_r) \cdot \exp[-j4\pi(f + f_c)R_p/c] \tag{7}$$

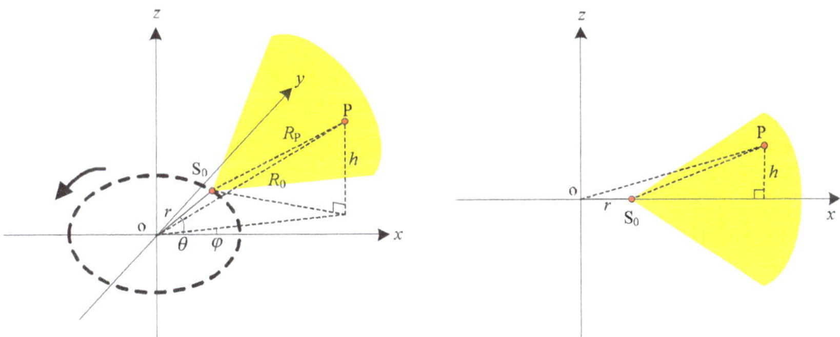

(**a**) Geometric model of ArcSAR system in 3-D space (**b**) Side view of ArcSAR geometric model

Figure 1. *Cont.*

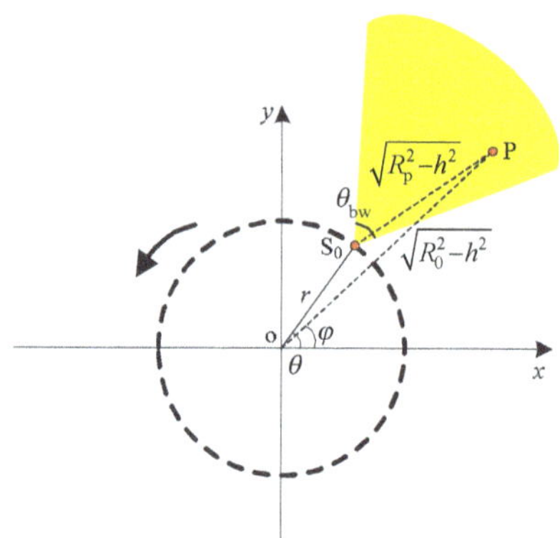

(c) Top view of the ArcSAR geometric model

Figure 1. The geometric model of ArcSAR system.

3. Analyzing the Error Caused by ArcSAR Imaging on the Reference Plane

In this section, the error model of ArcSAR imaging on the reference plane is first built. Then, the height difference threshold for determining whether the ArcSAR image will severely defocus is calculated. We also explore the relationship between height difference threshold and system parameters.

3.1. The Error Model of ArcSAR Imaging on the Reference Plane

As shown in Figure 2, the antenna rotates counterclockwise from S_1 to S_2. When the antenna rotates to S, it is closest to the target P. The xoy plane is defined as the reference plane (the xoy plane is defined as the reference plane in subsequent parts of this paper). We use the backward projection algorithm (BP algorithm) to achieve the imaging of the target P on the reference plane. The P_0 is the imaging result on the reference plane. It has the same range and Doppler from the antenna with target P. However, referring to the geometric model given in Figure 2, target P and P_0 has different range migration because the height of the reference plane is inconsistent with the actual height of the target P.

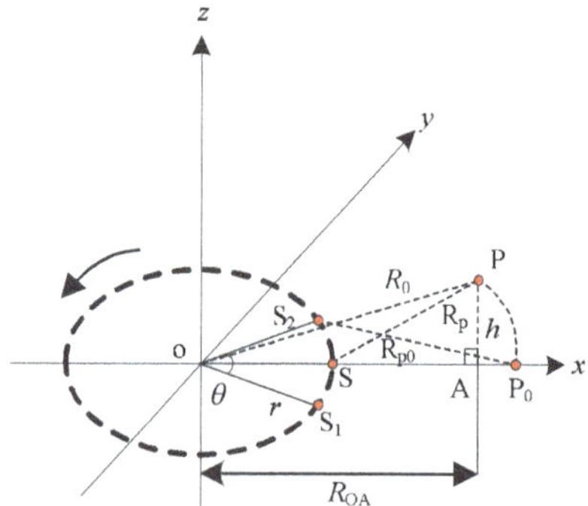

Figure 2. Target P and its imaging result on the reference plane.

The inverse Fourier transform is performed on $S(\theta, f)$ at the range direction to obtain its time domain form, which can be expressed as:

$$S_c(\theta, t_r) = \sin c\left(B_r\left(t_r - 2R_P/c\right)\right) \exp\left(-j\frac{4\pi f_c}{c}R_P\right) \tag{8}$$

The exponential term of $S_c(\theta,t_r)$ represents the range migration, which must be compensated during the process of focusing. Therefore, the matched filter is used to compensate for the range migration of $S_c(\theta,t_r)$. The ideal matched filter is expressed as follows:

$$H(\theta,t_r) = \exp\left\{ j\frac{4\pi f_c}{c} \cdot R_p \right\} \tag{9}$$

However, the actual matched filter used to compensate for the range migration of target P is different from the ideal matched filter because target P and P_0 have different range migrations. The actual matched filter contains the phase error, which means the range migration of target P cannot be compensated correctly. The actual matched filter can be expressed as:

$$H(\theta,t_r) = \exp\left\{ j\left(\frac{4\pi f_c}{c} \cdot R_p + \Delta p\right) \right\} \tag{10}$$

where Δp represents the phase error:

$$\Delta p = \frac{4\pi f_c}{c} \cdot (R_{p0} - R_p) \tag{11}$$

where R_{p0} represents the distance between P_0 and the APC, which can be expressed as:

$$R_{p0} = \sqrt{(R_{OA} - r)^2 + r^2 + h^2 - 2r\sqrt{(R_{OA} - r)^2 + h^2}\cos\theta} \tag{12}$$

$$R_{OA} = \sqrt{R_0^2 - h^2} \tag{13}$$

where R_{OA} is the ground range from the target to the rotation center. Therefore, the imaging result of the target P on the reference plane is

$$G_{p0} = \int_\theta S_c(\theta,t_r) \cdot H(\theta,t_r)\mathrm{d}\theta = \int_\theta B_r \sin c\left(B_r\left(t_r - 2R_p/c\right)\right)\exp(j\Delta p)\mathrm{d}\theta \tag{14}$$

The matched filter in Equation (10) cannot correctly compensate for the range migration. Thus, the imaging quality of the target P on the reference plane will decrease.

Then, we derive the condition for the defocusing of the ArcSAR imaging result on the reference plane. According to Equation (11), the occurrence of Δp is due to the slant range error caused by R_p and R_{p0}. We define the slant range error as the ΔR, which can be expressed as:

$$\Delta R = R_{p0} - R_p \tag{15}$$

Equations (6) and (12) indicate that the ΔR is the function of θ. The relationship curve of θ and ΔR is shown in Figure 3. S_1, S and S_2 are marked in the curve. As can be seen from the curve, when the antenna is at point S, the value of slant range error is zero. As the antenna deviates from position S, the value of slant range error gradually increases. When the antenna is at S_1 and S_2, the value of slant range error is the largest. We define the maximum slant range error as ΔR_{max}.

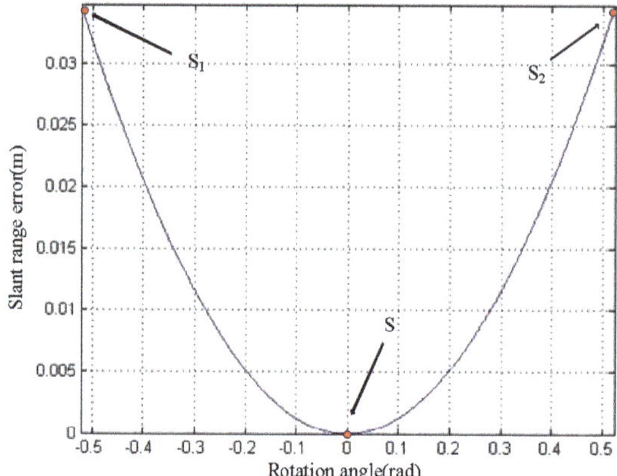

Figure 3. The θ–ΔR relationship curve.

The maximum phase error acceptable for SAR image is $\pi/4$ [21]. If the phase error exceeds $\pi/4$, the SAR image will severely defocus (when the phase error is less than $\pi/4$, this error is too small for SAR imaging to be ignored). Therefore, the condition for severe defocusing of the ArcSAR imaging result on the reference plane is:

$$\Delta p_{max} \geq \pi/4$$
$$\frac{4\pi f_c}{c} \cdot \Delta R_{max} \geq \pi/4 \tag{16}$$
$$\Delta R_{max} \geq \frac{\lambda}{16}$$

where λ represents the wavelength.

3.2. The Relationship between Height Difference Threshold and System Parameters

In this part, we calculate the height difference threshold for determining whether the ArcSAR imaging result on the reference plane will severely defocus, and analyze the relationship between the height difference threshold and the system parameters. We define the height of the reference plane as h_{ref}, which the value is 0 m. The geometric relationship is shown in Figure 2. The necessary parameters are given in Table 1.

Table 1. The ArcSAR system parameters.

r (m)	θ_{bw} (rad)	θ (rad)	λ (mm)	R_{sp} (m)	Br (MHz)
1	$\pi/3$	$\pi/3$	17.50	300	150

R_{sp} is the minimum slant range between the antenna and the target during the rotation. The relationship between height of the target and ΔR_{max} is as follows:

$$R_{p0max} - R_{pmax} = \Delta R_{max} \tag{17}$$

$$R_{p0max} = \sqrt{\left(r\sin\frac{\theta}{2}\right)^2 + \left(R_{sp} + r - r\cos\frac{\theta}{2}\right)^2} \tag{18}$$

$$R_{pmax} = \sqrt{\left(r\sin\frac{\theta}{2}\right)^2 + \left(\sqrt{R_{sp} - h^2} + r - r\cos\frac{\theta}{2}\right)^2 + h^2} \tag{19}$$

where R_{pmax} is the maximum slant range between the antenna and the target P during system rotation, R_{p0max} represents the maximum slant range between the antenna and P_0 during system rotation.

According to Equations (17)–(19) and the parameters given in Table 1, we can get the h–$\Delta R_{\max}$ relationship curve (Figure 4).

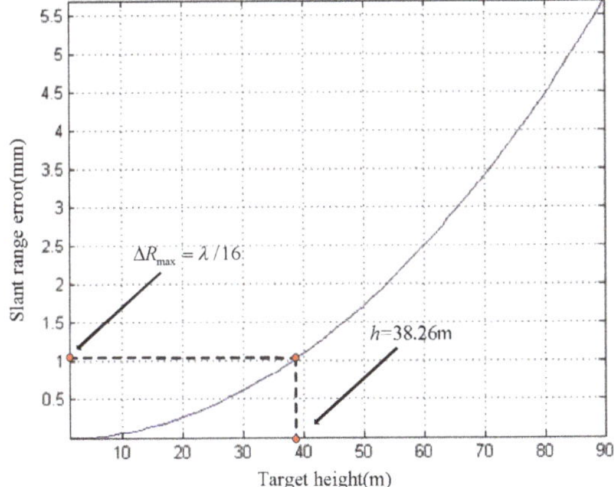

Figure 4. The h–$\Delta R_{\max}$ relationship curve.

With reference to Figure 4, it can be seen that the $\Delta R_{\max}$ increases with h. Applying the parameters of Table 1 to Equations (17)–(19), we calculate that the height difference threshold is 38.26 m. When h exceeds 38.26 m, the ArcSAR image of target P on the reference plane will severely defocus. We give the simulated imaging results of the target P on the reference plane in Figure 5. When h is 0 m, the imaging result of target P on the reference plane has no defocusing because the height of the reference plane is consistent with the actual height of the target at this time. When h is 80 m, the height of the target P exceeds height difference threshold, thus the imaging result of target P on the reference plane appears severe defocusing.

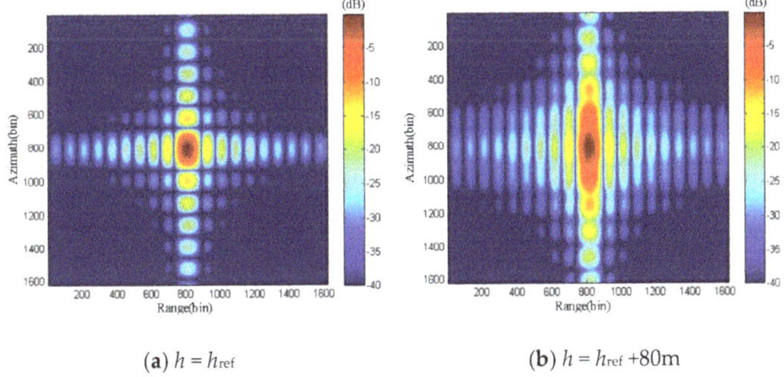

(**a**) $h = h_{\text{ref}}$ (**b**) $h = h_{\text{ref}} + 80$m

Figure 5. Imaging simulation results of target P on the reference plane.

The height difference threshold is determined by the system parameters given in Table 1. Therefore, the change of system parameters has an impact on height difference threshold: under the premise of changing only single system parameter, the increase of θ_{bw}, θ and R_{sp} will cause the height difference threshold to rise, while the increase of the f_c, r will cause the height difference threshold to decrease.

4. The Principle of Interferometric DEM-Assisted High Precision Imaging Method for ArcSAR

From the analysis of previous section, the ArcSAR imaging result on the reference plane is affected by the terrain fluctuation. For acquiring high precision ArcSAR image, an interferometric DEM-assisted high precision imaging method for ArcSAR is proposed in this paper. The interferometric ArcSAR is utilized to acquire the scenes DEM on the slant range. Since DEM-assisted ArcSAR imaging requires the DEM image on the ground range, we propose a polar coordinate transformation method for transforming DEM image from slant range to ground range. With the assist of DEM image on the ground range, the target in the scenes can be imaged on its actual height.

The steps of the above method are as follows: (1) The ArcSAR system is utilized to scan the same scenes at two different heights for getting ArcSAR Image 1 and ArcSAR Image 2. It should be noted that the acquired two ArcSAR images are the imaging results on the reference plane. Although the imaging accuracy of them is affected by the terrain fluctuation, their phase information can be used to obtain the interferometric phase. (2) ArcSAR Image 1 and ArcSAR Image 2 are used for interference to obtain interferometric phase. (3) We perform the operations of flat phase removing and phase unwrapping for the interferometric phase. (4) The unwrapped interferometric phase is used to inverse the DEM of the scenes. (5) The DEM image in slant range is transformed to the ground range by proposed polar coordinate transformation method. (6) The DEM of the scenes is used to assist ArcSAR imaging. A flow chart of the above method is shown in Figure 6.

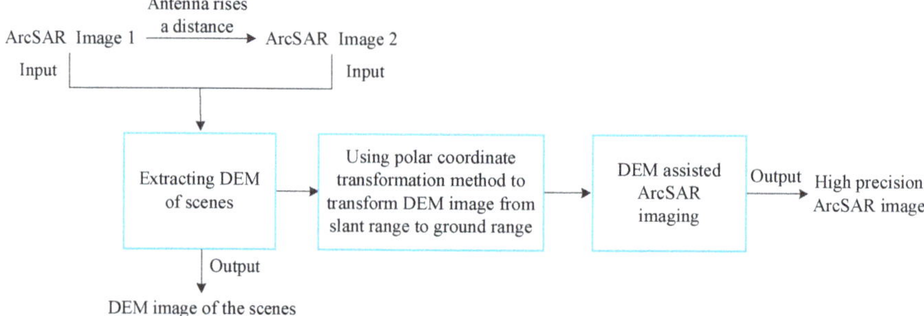

Figure 6. Flow chart of the proposed high precision ArcSAR imaging method.

4.1. Interferometric ArcSAR Extraction DEM of Scenes

Since the method of extracting DEM of the scenes using InSAR is relatively mature [22,23], we only briefly describe the main principle of interferometric ArcSAR in this section.

The model of interferometric ArcSAR is shown in Figure 7. Firstly, the antenna scans the target P at two different heights to obtain the imaging results on the reference plane, which can be shown as:

$$G_1 = E_s \exp\left(-j\frac{4\pi}{\lambda}R_{sp}\right) \tag{20}$$

$$G_2 = E_s \exp\left(-j\frac{4\pi}{\lambda}R_{sp2}\right) \tag{21}$$

where G_1 is the ArcSAR image taken when antenna is rotating on the xoy plane and G_2 is the ArcSAR image taken after raising the antenna to leave the xoy plane. The height baseline is Δz. E_s stands for the amplitude information of the SAR image. R_{sp2} represents the distance from S_3 to the target P.

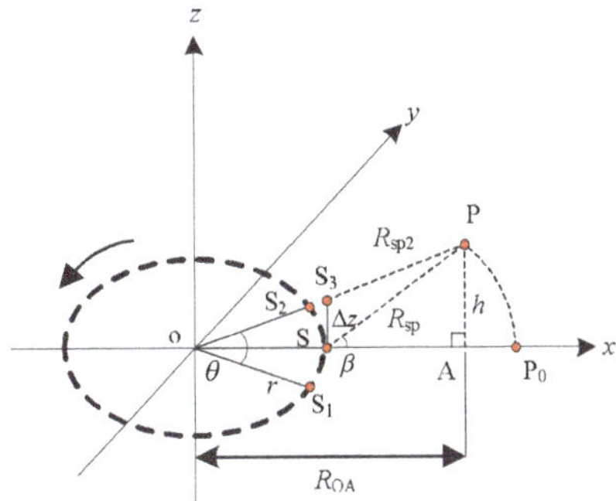

Figure 7. The model of interferometric ArcSAR.

Then, we can acquire the complex conjugate result of G_1 and G_2:

$$G_2 G_1^* = E_s^2 \exp\left(-j\frac{4\pi}{\lambda}R_{sp2}\right) \cdot \exp\left(j\frac{4\pi}{\lambda}R_{sp}\right) = E_s^2 \exp\left(j\frac{4\pi}{\lambda}\left(R_{sp} - R_{sp2}\right)\right) \tag{22}$$

According to Equation (22), the interferometric phase p_1 can be expressed as:

$$p_1 = \frac{4\pi}{\lambda}\left(R_{sp} - R_{sp2}\right) \tag{23}$$

Based on the work in [22], plane wave approximation is used to simplify the derivation process. Thus, R_{sp2} can be approximated as:

$$R_{sp2} = R_{sp} - \delta \tag{24}$$

where δ is the difference of R_{sp} and R_{sp2}, which can be expressed as

$$\delta = \Delta z \frac{h}{R_{sp}} = \Delta z \sin\beta \tag{25}$$

Thus, the p_1 is rewritten as:

$$p_1 = \frac{4\pi}{\lambda}\Delta z \sin\beta \tag{26}$$

β is the angle between R_{sp} and the reference plane, which can be expressed as:

$$\beta = \sin^{-1}\left(\frac{\lambda}{4\pi\Delta z}p_1\right) \tag{27}$$

The height of the target P can be expressed as:

$$h = R_{sp}\sin\beta = R_{sp}\left(\frac{\lambda}{4\pi\Delta z}p_1\right) \tag{28}$$

4.2. DEM Image Transforming from Slant Range to Ground Range

In this paper, the ArcSAR images used for extracting DEM are the slant range images. Therefore, on the range direction, the pixel units of the ArcSAR image are equally spaced according to the slant range. The DEM image obtained by the interference with the ArcSAR images is also the slant range image (each pixel unit on the DEM image represents DEM data). However, assisted ArcSAR imaging requires the use of DEM image on the ground range. Thus, it is necessary to transform the DEM image from slant range to ground range.

The process of DEM image from slant range to ground range in linear scanning GBSAR is generally achieved in Cartesian coordinate system. This is because the range direction and azimuth direction of the linear scanning GBSAR are along the coordinate axes of the Cartesian coordinate. Thus, in the Cartesian coordinate system, the linear scanning of GBSAR only requires operating in one dimension to complete the transformation of the DEM image.

However, if we realize the transformation of ArcSAR DEM image in Cartesian coordinate system, we have to operate in two dimensions. Because the azimuth direction of the ArcSAR system is the rotation direction of the antenna, and the range distance of the ArcSAR is the radial direction of the antenna motion track. To simplify the transformation process of ArcSAR DEM image, we propose a transformation method in polar coordinate system.

The DEM image in polar coordinate system is shown in Figure 8. The θ-axis of the coordinate system represents the azimuth direction. The pixel units in azimuth direction are equally spaced according to the rotation angle (θ) of the ArcSAR system, and there are N pixel units in this direction. The Rs-axis of the coordinate system represents the range direction. The pixel units in this direction are equally spaced according to the slant range (Rs), and there are M pixel units in this direction. The size of the DEM image is M×N. The coordinates of the pixel units are also indicated in Figure 8. Taking the pixel unit with coordinate (θ_i, Rs_j) as an example, we discuss the process of transforming the pixel unit from slant range to ground range. The specific operation steps of the proposed transformation method are as follows,

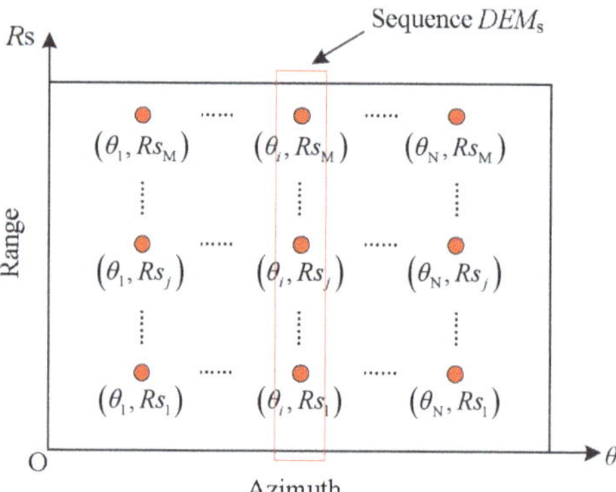

Figure 8. The DEM image in the polar coordinate system.

1. When $\theta = \theta_i$ ($i = 1, 2, ..., N$), the sequence *DEMs* are taken from the DEM image. This sequence stores the DEM data in the DEM image at $\theta = \theta_i$. Its size is M×1. At the same time, we define a slant range sequence R_n. It stores the slant range corresponding to each element in the sequence *DEMs*, which can be expressed as:

$$R_n = \left[Rs_1, Rs_2, ..., Rs_j, ..., Rs_M\right](j = 1, 2, ..., M) \tag{29}$$

2. We calculate the ground range sequence R_g using the geometric relationship between the sequence *DEMs* and the sequence R_n. The flow chart of the calculation process is shown in Figure 9. Sequence R_g stores the ground range corresponding to each DEM data in the sequence *DEMs*.

3. We define a ground range sequence R_{ge}, which is an increasing sequence. Its size is M × 1. Furthermore, the largest element of this sequence is R_{ge}(M), which is equal to the R_g(M). The difference of adjacent elements in R_{ge} is fixed.

4. We acquire the element in the sequence R_g that is numerically closest to the element R_{ge} (j), which we define as R_{ne}.

5. According to the position of R_{ne} in the sequence R_g, the DEM data corresponding to the element R_{ne} in the sequence *DEMs* can be found. We named these DEM data as H_{ne}.

6. Since $R_{ge}(j)$ is numerically close to R_{ne}, the position of their corresponding DEM data on the DEM image will be very close. Therefore, we assume that the DEM data of $R_{ge}(j)$ are also H_{ne}.

7. According to $R_{ge}(j)$ and H_{ne}, the slant range corresponding to $R_{ge}(j)$ can be calculated:

$$R_{new} = \sqrt{R_{ge}^2(j) + H_{ne}^2} \tag{30}$$

8. Let R_{new} interpolate on the sequence R_n (the purpose of this operation is to find the position of R_{new} on the sequence R_n). According to the position of R_{new} on the sequence R_n, we can find the DEM data corresponding to R_{new} on the sequence $DEMs$, which we define as H_{new}.

9. Since $R_{ge}(j)$ is the ground range corresponding to R_{new}, the DEM data of $R_{ge}(j)$ are also H_{new}.

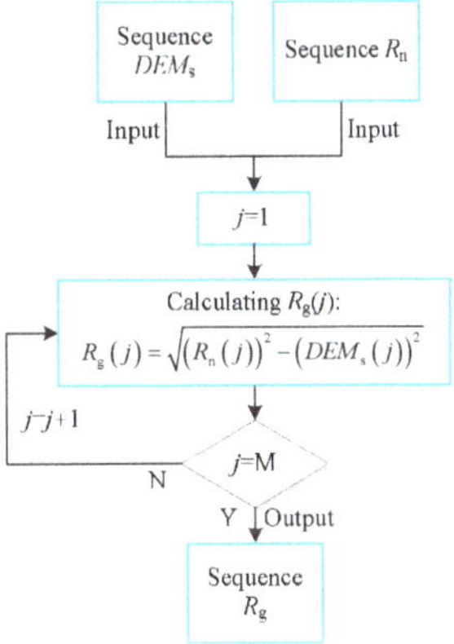

Figure 9. Flow chart of calculating the sequence R_g.

Through the above steps, the DEM data with coordinate (θ_i, Rs_j) on the slant range are transformed to the ground range. The coordinate of these DEM data on the ground range is $(\theta_i, R_{ge}(j))$. The above steps only describe the process of transforming one DEM datum in the DEM image from the slant range to ground range. The complete transformation process is shown in Figure 10.

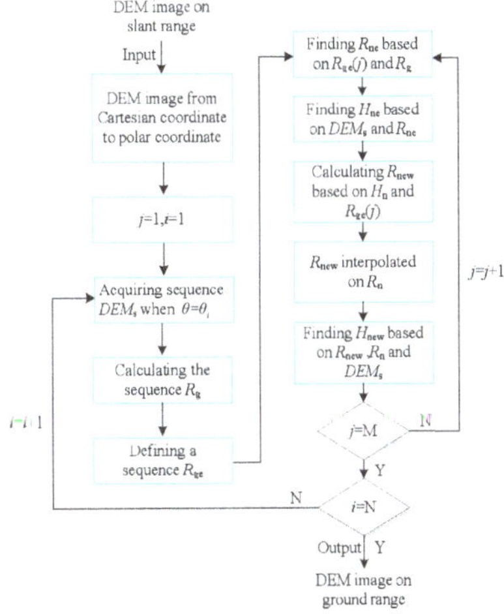

Figure 10. The polar coordinate transformation method proposed in this paper.

5. Accuracy Analysis

We extract the DEM using the SAR images imaged on the reference plane. The defocusing of SAR image causes the larger phase error, which has the impact on the accuracy of the DEM and further affects the imaging quality. The defocusing can cause two types of phase error. The first is the random phase error due to the decrease of the SAR image's signal-to-noise ratio (SNR). The second is the phase error due to the difference in phase distortion of the SAR image pair used for the interference. In this section, we analyze the impact of the above two phase errors on DEM accuracy. Since the application background of the interferometric ArcSAR in this paper is deformation monitoring, the deformation monitoring accuracy is also analyzed.

In addition, it should be noted that the purpose of analyzing DEM accuracy and deformation monitoring accuracy is to demonstrate the effect of DEM assist on improving the accuracy of the results. In practice, the analysis of DEM accuracy and deformation monitoring accuracy also needs to consider more complex factors such as baseline decoherence and time decoherence. Since these factors are not relevant to our topic, they are not considered in this paper.

5.1. The DEM Accuracy Analysis

5.1.1. Phase Error Analysis as the Decrease of the SNR Caused by Image Defocusing

Defocusing causes the decrease of the SAR image's SNR, which results in larger phase error. We analyze this type of phase error in this section. The SNR loss of the imaging result is defined as S_{loss}. According to the geometric model in Figure 7, we apply the numerical analysis method to obtain the h–S_{loss} curve of the imaging result P_0. The relationship between h and S_{loss} can be expressed as:

$$S_{\text{loss}} = 10\lg\left(\left|\frac{\left|\sum_{i=-\frac{\theta}{2}}^{i=\frac{\theta}{2}} \exp\left(j\frac{4\pi f}{c}\Delta R(i,h)\right)\right|}{\theta}\right|\right)^2 \tag{31}$$

The necessary parameters are given in Table 1. The h–S_{loss} curve is shown in Figure 11.

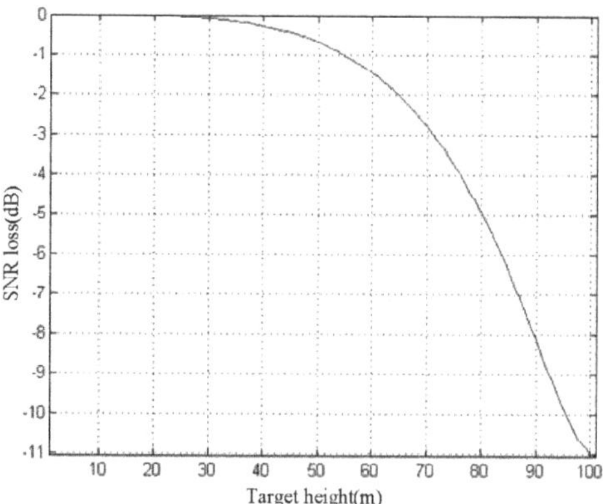

Figure 11. h–S_{loss} *curve.*

It can be seen in Figure 11 that, as h increases, the SNR loss gradually becomes severe. We assume that the SNR of the imaging result is 20 dB when $h = 0$ m. The SNR will decrease to 9.03 dB when $h = 100$ m, which means that the imaging result is seriously defocused.

According to the authors of [24], the root mean square error σ_{ps} of the phase error due to the decrease of the SNR can be expressed as:

$$\sigma_{ps} = \frac{1}{\sqrt{2NL}} \cdot \frac{\sqrt{1-\rho^2}}{\rho} \tag{32}$$

where NL stands for the number of looks. ρ represents the coherence of the SAR image pair. It is related to the SNR, which is [24]:

$$\rho = \frac{1}{1 + SNR^{-1}} \tag{33}$$

With reference to Equations (31) and (33), the $h-\rho$ curve can be obtained, as shown in Figure 12.

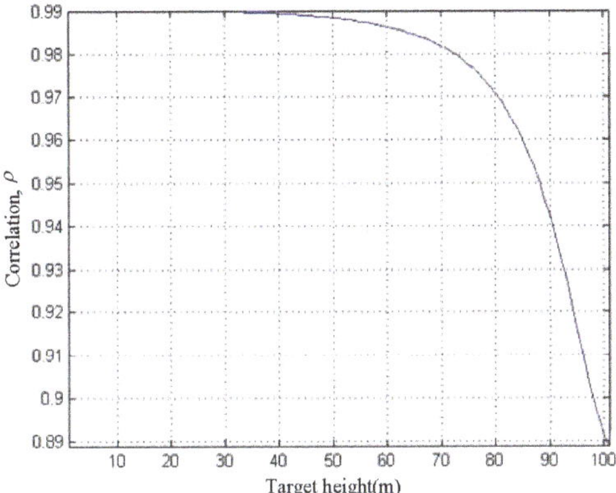

Figure 12. $h-\rho$ curve.

It can be seen in Figure 12 that as h increases, ρ gradually decreases. When $h = 100$ m, ρ decreases to 0.89. We use the 3×3 filter window for phase filtering, thus $NL = 9$. According to Equation (32) and Figure 12, the $h-\sigma_{ps}$ curve can be acquired, as shown in Figure 13.

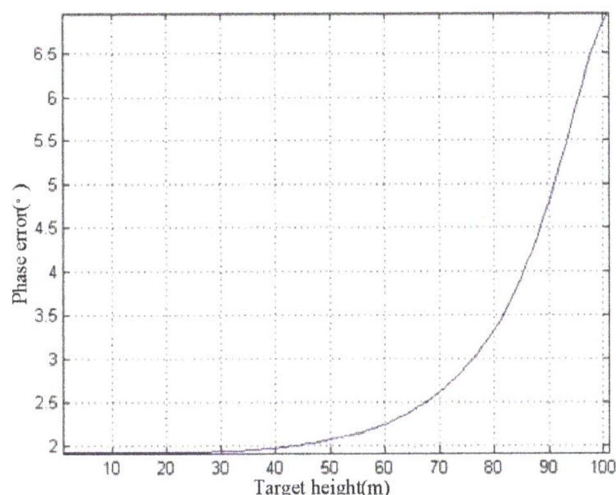

Figure 13. $h-\sigma_{ps}$ curve.

As shown in Figure 13, as h increases, σ_{ps} gradually rises. When $h = 100$ m, the σ_{ps} rises to 6.96.

5.1.2. Phase Error due to the Difference in Phase Distortion of the SAR Images

The defocusing can cause phase distortion in SAR images. If there is the difference in phase distortion of the SAR image pair used for interference, the phase error will be introduced during the interference process. In this section, we analyze this kind of phase error. We define the distortion phase as p_d. The h–p_d curves of the SAR image pair are shown in Figure 14.

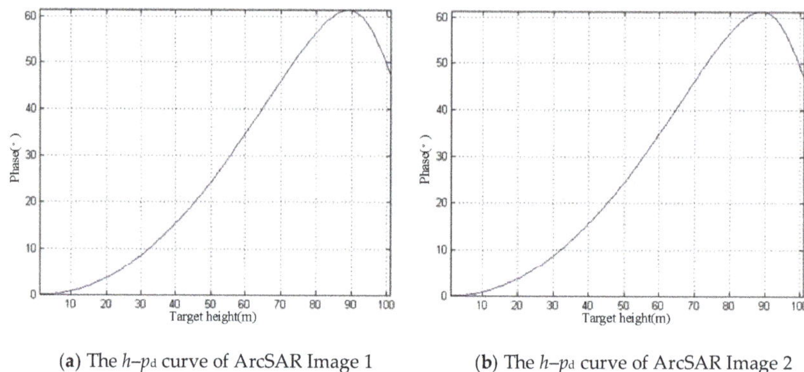

(**a**) The h–p_d curve of ArcSAR Image 1   (**b**) The h–p_d curve of ArcSAR Image 2

Figure 14. The h–p_d curves.

It can be found that the phase distortions of the above SAR image pair are similar, which means that most of the distortion phase will be offset during the interference process. Therefore, the phase error introduced by the process of interference is small, as shown in Figure 15.

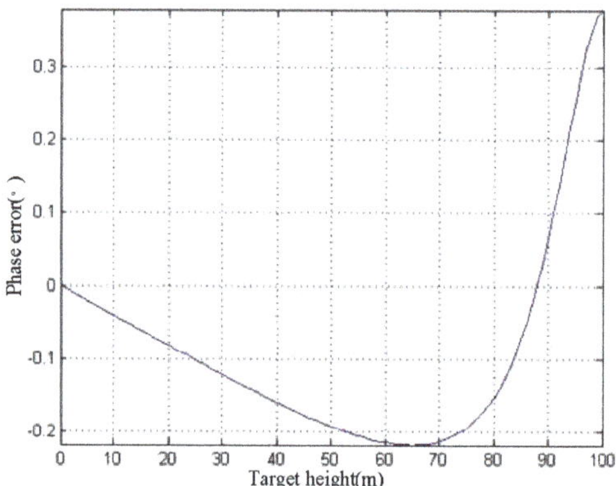

Figure 15. h–σ_{pd} curve.

σ_{pd} represents the phase error due to the difference in phase distortion of the SAR image pair.

5.1.3. The Effects of Phase Errors on DEM Accuracy

With reference to Equations (25), (26) and (28), we can derive the relationship between DEM accuracy σ_h and the above two types of phase error:

$$\sigma_h = \frac{\lambda R_{sp}}{4\pi \Delta z}\left(\left|\sigma_{ps}\right| + \left|\sigma_{pd}\right|\right) \tag{34}$$

According to Equation (34), Figures 13 and 15, the h–σ_h curve can be obtained, as shown in Figure 16.

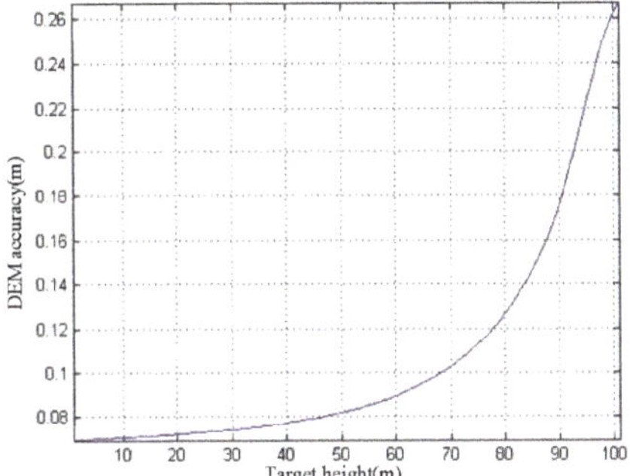

Figure 16. h–σ_h curve.

Referring to the conclusion of Section 3.2, when the R_{sp} = 300 m, the height difference threshold is 38.26 m. Therefore, if the accuracy of the DEM for assisting imaging is better than 38.26 m, the proposed imaging method can obtain high-precision ArcSAR imaging results. The curve in Figure 16 shows that the DEM accuracy does not exceed 0.5 m, which is much smaller than the height difference threshold. Thus, we can draw the following conclusion: Even if the SAR images used for interference appears defocused, the obtained DEM can still be used to assist ArcSAR imaging and acquire high precision image.

5.2. Deformation Monitoring Accuracy Analysis

We utilize the DEM obtained by interferometric ArcSAR to assist imaging, which can significantly improve the imaging quality. The improvement of image quality also helps to improve the deformation monitoring accuracy. In this part, the contribution of the proposed method to the improvement of deformation monitoring accuracy is analyzed. As a comparison, the impact of traditional ArcSAR imaging algorithm (the imaging method with the image on the reference plane) on the deformation monitoring accuracy is also discussed. With reference to Section 5.1, defocusing causes two types of phase error. They will affect the accuracy of deformation monitoring. We define the deformation monitoring accuracy as σ_d, which can be expressed as [7]:

$$\sigma_d = \frac{\lambda}{4\pi}\left(|\sigma_{ps}| + |\sigma_{pd}|\right) \tag{35}$$

According to Equation (35), Figures 13 and 15, the h–σ_d curve can be acquired, as shown in Figure 17.

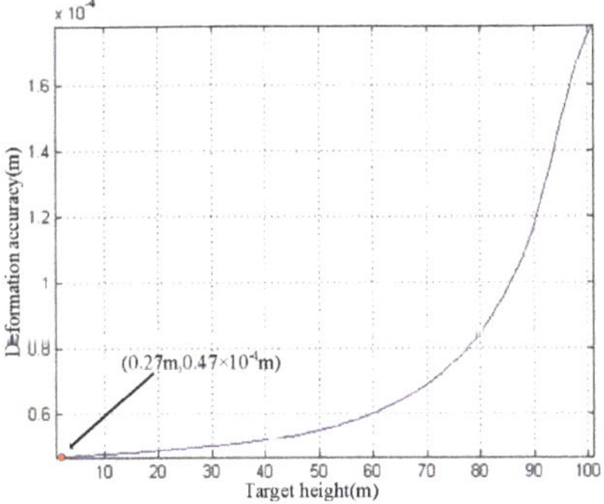

Figure 17. h–σ_d curve.

For the convenience of analysis, we take the deformation monitoring accuracy at $h = 100$ m as the example. Referring to the curve in Figure 17, the deformation monitoring accuracy of the traditional ArcSAR imaging algorithm is 1.78×10^{-4} m. If we use the proposed imaging method, the target in the scenes will be imaged on its real position. According to Figure 16, the DEM accuracy at h = 100 m is $\sigma_h(100$ m$) = 0.27$ m. Therefore, the deformation monitoring accuracy of the proposed method can reach to $\sigma_d(0.27$ m$) = 0.47 \times 10^{-4}$ m, as marked in Figure 17. Through the above analysis, we can conclude that, compared with the traditional ArcSAR imaging algorithm, the imaging method proposed in this paper can effectively improve the deformation monitoring accuracy.

6. Experiment

We used a distributed scenes imaging experiment of ArcSAR to verify the effectiveness of the proposed high-precision imaging method. The experiment contained two parts. First, we used the ArcSAR images of the distributed scenes for interference to extract the DEM image of the scenes. Then, the extracted DEM image was transformed from the slant range to the ground range using the polar coordinate transformation method proposed in Section 4.2. Second, we performed the distributed scenes ArcSAR imaging simulation experiment. The DEM image obtained in Experiment 1 was used for assisted imaging.

6.1. The Simulation Experiment of Interferometric ArcSAR Extraction DEM

Based on the geometric model and working mode of the ArcSAR system, we uses the existing RCS scenes and DEM data to simulate the ArcSAR image on the slant range. To facilitate subsequent experiment, the simulated ArcSAR image is shown in the polar coordinate system in Figure 18. The necessary parameters are given in Table 2.

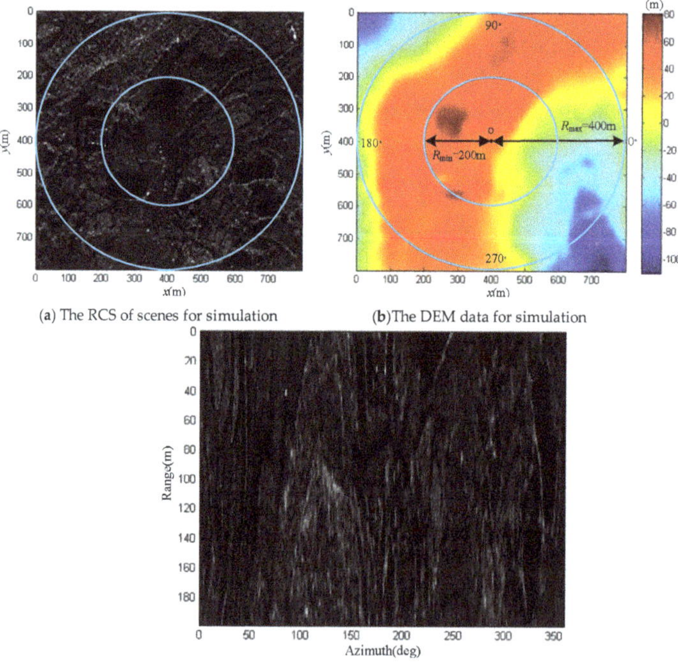

(a) The RCS of scenes for simulation   (b)The DEM data for simulation

(c) The simulated ArcSAR image in polar coordinate system

Figure 18. The RCS scenes for simulation, the DEM data for simulation and the simulated ArcSAR image in polar coordinate system.

Table 2. Experiment parameters.

r (m)	θ_{bw} (rad)	θ (rad)	λ (mm)	Br (MHz)	R_{max} (m)	R_{min} (m)
1	$\pi/3$	2π	17.50	150	400	200

R_{max} represents the maximum slant range from the imaging scenes to the ArcSAR system and R_{min} represents the minimum slant range from the imaging scenes to the ArcSAR system. They are marked in Figure 18b. The point o in Figure 18b is the rotation center. We defined the height of rotation plane to be 0 m. In Figure 18c, the horizontal axis represents the rotation angle of the ArcSAR system, and the vertical axis is the interval from R_{min} to R_{max}.

The ArcSAR imaging result in Figure 18c was considered to be the main complex image for interference. The height baseline Δz was 0.2 m. We simulated the sub SAR complex image for interference. The interferometric phase can be obtained by interference with the main SAR complex image and the sub SAR complex image. The interferometric phase image is shown in Figure 19.

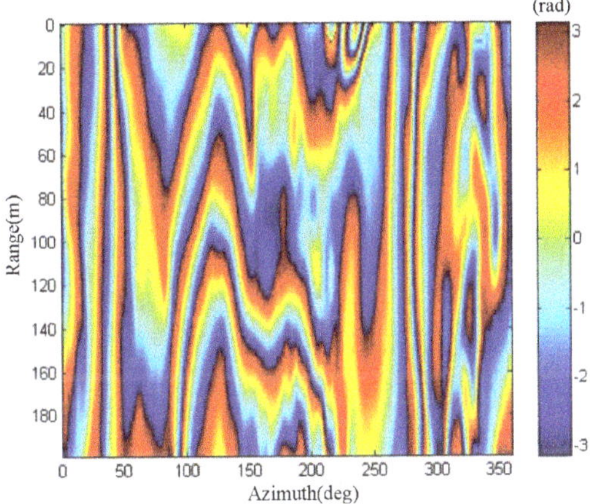

Figure 19. Interferometric phase image of phase wrapping.

The measured interferometric phase value shown in Figure 19 was modulated by 2π, ranging from $-\pi$ to π, and there was an ambiguity of many cycles in the interferometric phase value. Thus, it was necessary to perform phase unwrapping on the interferometric phase image. The branch cut method was applied to interferometric phase unwrapping [25,26]. In addition, to reduce the complexity of phase unwrapping, we also removed the flat phase of the interferometric phase. After the flat phase removal and phase unwrapping operations, the interferometric phase image could be used for DEM inversion. The DEM image obtained by the interferometric phase inversion is the slant range image, as shown in Figure 20a. We utilized the proposed polar coordinate transformation method to transform it to the ground range. The DEM image on the ground range is shown in Figure 20b.

We also calculated the mean square error (MSE) of the DEM image in Figure 20b with the existing DEM data. The result of the MSE was only 1.69. It can be seen that the interferometric ArcSAR could acquire high-precision DEM image of the scenes.

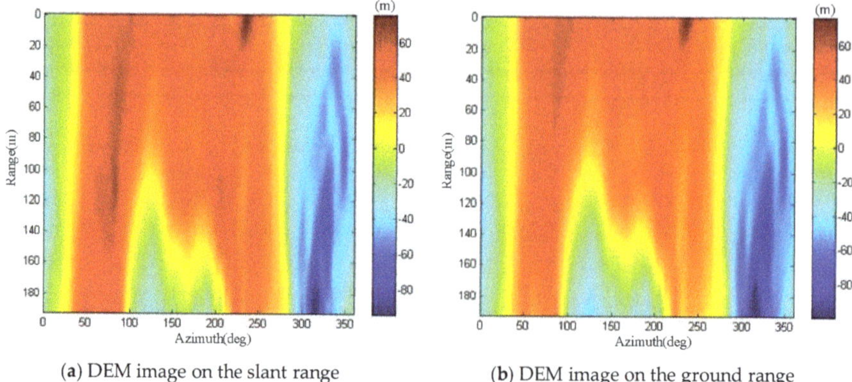

(**a**) DEM image on the slant range   (**b**) DEM image on the ground range

Figure 20. Interferometric ArcSAR extraction the DEM image of scenes.

6.2. Distributed Scenes Imaging Simulation Experiment Verification the DEM-Assisted High Precision Imaging Method for ArcSAR

We used the distributed scenes imaging simulation experiment to image the scenes given in Experiment 1. The necessary parameters are shown in Table 2. The DEM image acquired in Experiment 1 was applied to assist ArcSAR imaging. In addition, we also used the traditional imaging method of ArcSAR to image the distributed scenes on the reference plane as the comparison. The height of the reference plane was 0 m. Based on Equations (17)–(19) in Section 3.2, we found the defocused area and focused area of the distributed scenes imaging results on the reference plane, as shown in Figure 21.

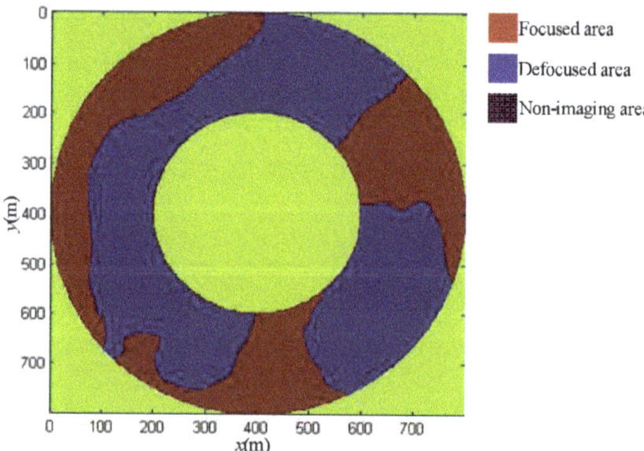

Figure 21. The defocused area and focused area of imaging result on the reference plane.

The imaging results by the two imaging methods in Cartesian coordinate system are shown in Figure 22.

It can be seen that the ArcSAR imaging result of the scenes on the reference plane showed severe defocusing, and the ArcSAR image obtained by the proposed imaging method was not defocused. To further analyze the imaging accuracy of the proposed imaging method, we set a strong scattering target in the scenes and analyzed the quality of its imaging result. Its imaging result is marked in Figure 22. We sliced the imaging result of the target and up sampled it 20 times, as shown in Figures 23 and 24. The imaging quality parameters of the strong scattering are shown in Table 3.

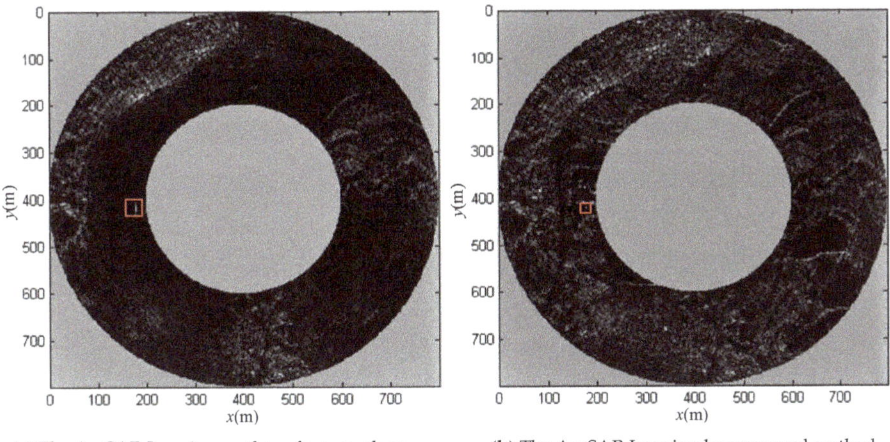

(**a**) The ArcSAR Imaging on the reference plane　　　(**b**) The ArcSAR Imaging by proposed method

Figure 22. The distributed scenes imaging simulation results of ArcSAR in Cartesian coordinate system by traditional method and proposed method.

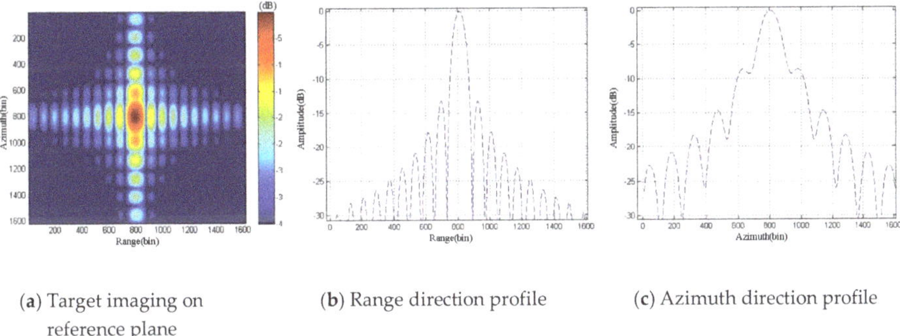

(**a**) Target imaging on　　　(**b**) Range direction profile　　　(**c**) Azimuth direction profile
reference plane

Figure 23. Imaging simulation results of the strong scattering target on the reference plane.

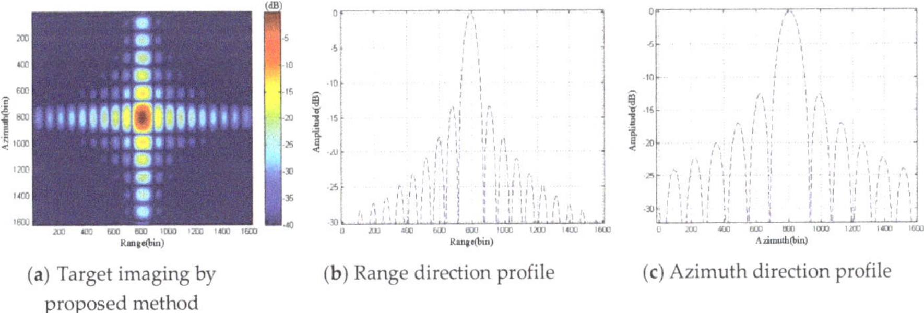

(**a**) Target imaging by　　　(**b**) Range direction profile　　　(**c**) Azimuth direction profile
proposed method

Figure 24. Imaging simulation results of the strong scattering target by proposed method.

According to Equations (17)–(19) in Section 3.2, we calculated that the height difference threshold of the strong scattering target was 28.28 m. The height of the strong scattering target in Figure 22 was 36.82 m, which exceeded height difference threshold. Therefore, its imaging result on the reference plane showed severe defocusing, as shown in Figure 23. The imaging result of the strong scattering target obtained by the proposed high-precision imaging method was not defocused, as shown in

Figure 24. Besides, it can be seen from the analysis results of the imaging quality in Table 3 that the proposed imaging method could effectively improve the quality of ArcSAR image and achieve the high-precision imaging.

Table 3. The quality parameters of the strong scattering target imaging result.

Parameters	Imaging on Reference Plane	Imaging by Proposed Method
Coordinates (m,deg)	(220.91,185.19)	(220.91,185.19)
Height (m)	36.82	36.82
Range direction resolution (m)	1.00	1.00
Range direction PSLR (dB)	−13.21	−13.22
Azimuth direction resolution (m)	1.92	1.91
Azimuth direction PSLR (dB)	−8.63	−12.57

7. Conclusions

In this paper, an interferometric DEM-assisted high precision imaging method for ArcSAR is proposed. The proposed method applies the interferometric ArcSAR to extract the DEM of scenes. The extracted DEM is utilized to assist ArcSAR imaging. This operation enables the target in the scenes image on its actual height. The proposed imaging method does not rely on external DEM data. Compared with the traditional ArcSAR algorithm imaged on the reference plane, the proposed method can effectively improve the accuracy of ArcSAR imaging.

Author Contributions: Y.L. (Yun Lin) and Y.S. performed the experiments and analysis. Y.W. and Y.S. wrote the manuscript. Y.L. (Yun Lin) contributed materials. Y.L. (Yang Li) and Y.Z. gave valuable advices on manuscript writing.

Funding: This research was funded by National Key R&D Program of China (grant number: 2018YFC1505103); Key international cooperation projects of the National Natural Science Foundation of China (grant number: No.61860206013); Natural Science Foundation of Beijing, China (grant number: No. 4192019); and Project No. 1921/008 Supported by "The Fundamental Research Funds for Beijing Universities".

Conflicts of Interest: The authors declare no conflict of interest.

References

1. Guaragnella, C.; D'Orazio, T. A Data-Driven Approach to SAR Data-Focusing. *Sensors* **2019**, *19*, 1649. [CrossRef] [PubMed]
2. Xin, L.; Tingting, L.; Kaizhi, W.; Xingzhao, L. A novel concept of plane grid resolution for high-resolution SAR imaging systems. In Proceedings of the 2014 IEEE Geoscience and Remote Sensing Symposium, Quebec City, QC, Canada, 13–18 July 2014; pp. 358–361.
3. Colesanti, C.; Locatelli, R.; Novali, F. Ground Deformation Monitoring Exploiting SAR Permanent Scatterers. In Proceedings of the IEEE International Geoscience and Remote Sensing Symposium, Toronto, ON, Canada, 24–28 June 2002; pp. 1219–1221.
4. Xie, P.; Zhang, M.; Zhang, L.; Wang, G. Residual Motion Error Correction with Backprojection Multisquint Algorithm for Airborne Synthetic Aperture Radar Interferometry. *Sensors* **2019**, *19*, 2342. [CrossRef] [PubMed]
5. Cao, N.; Lee, H.; Zaugg, E.; Shrestha, R.; Carter, W.E.; Glennie, C.; Lu, Z.; Yu, H. Estimation of Residual Motion Errors in Airborne SAR Interferometry Based on Time-Domain Backprojection and Multisquint Techniques. *IEEE Trans. Geosci. Remote. Sens.* **2018**, *56*, 2397–2407. [CrossRef]
6. Ruiz-Armenteros, A.M.; Manuel Delgado, J.; Ballesteros-Navarro, B.J.; Lazecky, M.; Bakon, M.; Sousa, J.J. Deformation monitoring of the northern sector of the Valencia Basin (E Spain) using Ps-InSAR (1993–2010). In Proceedings of the IGARSS 2018-2018 IEEE International Geoscience and Remote Sensing Symposium, Valencia, Spain, 23–27 July 2018; pp. 2244–2247.
7. Qi, Y.; Wang, Y.; Yang, X.; Li, H. Application of Microwave Imaging in Regional Deformation Monitoring using Ground Based SAR. In Proceedings of the Asian-Pacific Conference on Synthetic Aperture Radar (APSAR 2015), Singapore, 1–4 September 2015.

8.   Pieraccini, M.; Miccinesi, L.; Rojhani, N. A GBSAR Operating in Monostatic and Bistatic Modalities for Retrieving the Displacement Vector. *IEEE Geosci. Remote. Sens. Lett.* **2017**, *14*, 1–5. [CrossRef]

9.   Chan, Y.K.; Chu, C.Y. Ground based synthetic aperture radar for land deformation monitoring: Preliminary result. In Proceedings of the 2016 Progress in Electromagnetic Research Symposium (PIERS), Shanghai, China, 8–11 August 2016; pp. 2540–2542.

10.   Liu, Y.; Lee, C.; Yong, H.; Jia, L.; Youshi, W.; Placidi, S.; Roedelsperger, S. FastGBSAR case studies in China: Monitoring of a dam and instable slope. In Proceedings of the 2015 IEEE 5th Asia-Pacific Conference on Synthetic Aperture Radar (APSAR), Marina Bay Sands, Singapore, 1–4 September 2015; pp. 849–852.

11.   Yang, X.; Wang, Y.; Qi, Y.; Tan, W.; Hong, W. Experiment Study on Deformation Monitoring Using Ground-Based SAR. In Proceedings of the Asian-Pacific Conference on Synthetic Aperture Radar (APSAR 2013), Tsukuba, Japan, 23–27 September 2013.

12.   Garmatyuk, D.; Narayanan, R. Ultra-wideband continuous-wave random noise arc-SAR. *IEEE Trans. Geosci. Remote. Sens.* **2002**, *40*, 2543–2552. [CrossRef]

13.   Zhang, J. The static small object detection based on ground-based arc SAR. In Proceedings of the 2012 International Conference on Microwave and Millimeter Wave Technology (ICMMT), Shenzhen, China, 5–8 May 2012.

14.   Huang, Z.; Tan, W.; Huang, P.; Sun, J.; Qi, Y.; Wang, Y. Imaging algorithm study on ARC antenna array ground-based SAR. In Proceedings of the 2017 IEEE International Geoscience and Remote Sensing Symposium (IGARSS), Fort Worth, TX, USA, 23–28 July 2017; pp. 1634–1637.

15.   Rodriguez, A.C.F.; Fraidenraich, G.; Soares, T.A.P.; Filho, J.C.S.S.; Miranda, M.A.M.; Yacoub, M.D. Optimal and Suboptimal Velocity Estimators for ArcSAR with Distributed Target. *IEEE Geosci. Remote. Sens. Lett.* **2018**, *15*, 252–256. [CrossRef]

16.   Viviani, F.; Michelini, A.; Mayer, L.; Conni, F. IBIS-ArcSAR: an Innovative Ground-Based SAR System for Slope Monitoring. In Proceedings of the IGARSS 2018 IEEE International Geoscience and Remote Sensing Symposium, Valencia, Spain, 23–27 July 2018; pp. 1348–1351.

17.   Luo, Y.; Song, H.; Wang, R.; Xu, Z.; Li, Y. Signal processing of Arc FMCW SAR. Proceedings of 2013 Asia-Pacific Conference on Synthetic Aperture Radar (APSAR), Tsukuba, Japan, 23–27 September 2013.

18.   Luo, Y.; Song, H.; Wang, R.; Deng, Y.; Zhao, F.; Xu, Z. Arc FMCW SAR and Applications in Ground Monitoring. *IEEE Trans. Geosci. Remote. Sens.* **2014**, *52*, 5989–5998. [CrossRef]

19.   Lee, H.; Cho, S.-J.; Kim, K.-E. A ground-based Arc-scanning synthetic aperture radar (ArcSAR) system and focusing algorithms. In Proceedings of the 2010 IEEE International Geoscience and Remote Sensing Symposium, Honolulu, HI, USA, 25–30 July 2010; pp. 3490–3493.

20.   Lee, H.; Lee, J.-H.; Kim, K.-E.; Sung, N.-H.; Cho, S.-J. Development of a Truck-Mounted Arc-Scanning Synthetic Aperture Radar. *IEEE Trans. Geosci. Remote Sens.* **2014**, *52*, 2773–2779. [CrossRef]

21.   Cumming, I.G.; Wong, F.H. Digital processing of synthetic aperture radar data: Algorithms and implementation. ARTECH HOUSE: Norwood, MA, USA, 2004; ISBN 978-1-580-53058-3.

22.   Pieraccini, M.; Miccinesi, L. ArcSAR for detecting target elevation. *Electron. Lett.* **2016**, *52*, 1559–1561. [CrossRef]

23.   Pieraccini, M.; Miccinesi, L. ArcSAR: Theory, Simulations, and Experimental Verification. *IEEE Trans. Microw. Theory Tech.* **2017**, *65*, 293–301. [CrossRef]

24.   Zebker, H.; Werner, C.; Rosen, P.; Hensley, S. Accuracy of topographic maps derived from ERS-1 interferometric radar. *IEEE Trans. Geosci. Remote. Sens.* **1994**, *32*, 823–836. [CrossRef]

25.   Li, F.; Han, B.; Lin, X.; Hu, D.; Ding, C. A method of airborne InSAR DEM reconstruction in layover areas. In Proceedings of the 2012 IEEE International Geoscience and Remote Sensing Symposium, Munich, Germany, 22–27 July 2012.

26.   Nico, G.; Leva, D.; Antonello, G.; Tarchi, D. Ground-based SAR interferometry for terrain mapping: Theory and sensitivity analysis. *IEEE Trans. Geosci. Remote. Sens.* **2001**, *42*, 1344–1350. [CrossRef]

Article

Joint Sparsity Constraint Interferometric ISAR Imaging for 3-D Geometry of Near-Field Targets with Sub-Apertures

Yang Fang [1],*, Baoping Wang [2],*, Chao Sun [1], Shuzhen Wang [3],*, Jiansheng Hu [4] and Zuxun Song [1]

[1] School of Electronics and Information, Northwestern Polytechnical University, Xi'an 710072, China; sunchao2013@mail.nwpu.edu.cn (C.S.); zxsong@nwpu.edu.cn (Z.S.)

[2] National Key Laboratory of Science and Technology on UAV, Northwestern Polytechnical University, Xi'an 710065, China

[3] School of Computer Science and Technology, Xidian University, Xi'an 710071, China

[4] Department of Information Engineering, PAP of Engineering University, Xi'an 710068, China; hujiansheng121@163.com

* Correspondence: fangyang@mail.nwpu.edu.cn (Y.F.); baoping-wang@nwpu.edu.cn (B.W.); shuzhenwang@xidian.edu.cn (S.W.); Tel.: +86-29-8845-1041 (ext. 802) (B.W.)

Received: 9 October 2018; Accepted: 31 October 2018; Published: 2 November 2018

Abstract: This paper proposes a new interferometric near-field 3-D imaging approach based on multi-channel joint sparse reconstruction to solve the problems of conventional methods, i.e., the irrespective correlation of different channels in single-channel independent imaging which may lead to deviated positions of scattering points, and the low accuracy of imaging azimuth angle for real anisotropic targets. Firstly, two full-apertures are divided into several sub-apertures by the same standard; secondly, the joint sparse metric function is constructed based on scattering characteristics of the target in multi-channel status, and the improved Orthogonal Matching Pursuit (OMP) method is used for imaging solving, so as to obtain high-precision 3-D image of each sub-aperture; thirdly, comprehensive sub-aperture processing is performed using all sub-aperture 3-D images to obtain the final 3-D images; finally, validity of the proposed approach is verified by using simulation electromagnetic data and data measured in the anechoic chamber. Experimental results show that, compared with traditional interferometric ISAR imaging approaches, the algorithm proposed in this paper is able to provide a higher accuracy in scattering center reconstruction, and can effectively maintain relative phase information of channels.

Keywords: joint sparse reconstruction; interferometric inverse synthetic aperture radar; compressed sensing; near-field 3-D imaging; wide angle

---

1. Introduction

Near-field 3-D imaging is a microwave imaging technique that is developed on the basis of two-dimensional (2-D) synthetic aperture imaging. As it has higher spatial resolution capability, and has easy availability for engineering realization, near-field 3-D imaging is widely applied in Radar Cross Section (RCS) [1], non-destructive testing and evaluation (NDTE) [2,3], security check [4], concealed weapon detection [5–7], through-wall and inner wall imaging [8,9], breast cancer detection [10,11], etc. So far, a variety of techniques have been applied in near-field 3-D imaging to improve its performance, such as imaging based on range migration algorithm (RMA) [12–14] and polar format algorithm (PFA) [14], tomography imaging method [15], microwave holography method [16], confocal radar-based imaging [17], and NUFFT-based imaging [18,19].

Imaging approaches as mentioned above are all based on the traditional Nyquist sampling principle and matched filtering. In general, high range resolution is obtained by transmitting wideband signals, and high azimuth resolution is obtained in longer synthetic aperture time. However, with higher requirements for imaging resolution, traditional imaging approaches are encountering problems such as the sampling rate is too high, the data volume is too large, and the fast processing is difficult to carry out. According to the Compressed Sensing (CS) [20,21], a sparse signal or a sparse signal in a particular transform domain is sampled in a way that is lower than or far below than requirements of the Nyquist sampling theorem. The high-dimensional target signal can be accurately reconstructed by applying low-dimensional observation data by solving a minimum L-norm constrained optimization problem. At present, many scholars are combining CS with high-resolution near-field imaging and have achieved a number of research results, which fully demonstrate great potential of CS in reducing data sampling rate and improving imaging resolution [3,8,22,23].

At present, research on near-field 3-D imaging mainly focuses on planar scanning 3-D imaging. Although high-resolution target 3-D images can be obtained by applying the planar scanning 3-D imaging, the size of sampling data is huge, and conducting the measurement is time-consuming and the imaging efficiency is low. For the interferometric inverse synthetic aperture radar (InISAR) [24,25] method, 3-D views of the target can be obtained by applying the multi-antenna phase interference method, and data acquisition and signal processing are relatively simple, which make it easy for the system to perform functions. Thus, InISAR can be widely used in near-field 3-D imaging. InISAR 3-D imaging based on CS technology has the following advantages: (1) high-resolution ISAR images can be obtained only by short-time observation data. At the same time, rotation of the target can be approximately considered to be uniform in short-phase processing interval and thus the occurrence probability of range cell migration is reduced; (2) imaging results are not affected by sidelobe, and image resolution can be improved by increasing imaging grids, so that it is helpful to suppress angular glint phenomenon in InISAR imaging and (3) through the CS technology, the ISAR images can be further reconstructed by adopting sparse sampling data, thus reducing the pressure of data acquisition.

In interferometric imaging, compressed sampling and sparse reconstruction can be separately performed for each channel, so as to reduce the system sampling rate and improve the quality of radar imaging. For example, 3-D InISAR imaging method based on sparse constraint model is proposed using the sparsity of ISAR images in reference [24]. However, traditional InISAR imaging methods have the following problems: (1) because the observation objectives are consistent, multi-channel echoes in the InISAR system have a strong correlation, i.e., images of each channel have the same target support set. However, in single-channel independent processing, such prior information is not considered, and consistent location and number of scattering points among channel images cannot be ensured, which means it cannot ensure that all scattering points on the target are located in positions with the same pixel in two interferometric images, thus reducing the estimation accuracy of interferometric phase information and (2) in InISAR imaging, scattering characteristics of the target vary with the observation angle, and the imaging azimuth accuracy is limited by scattering anisotropy of the target.

Motived by the above problems in traditional InISAR imaging methods, this paper proposes the interferometric near-field 3-D imaging based on multi-channel joint sparse reconstruction. Firstly, a more universal multi-channel interferometric near-field echo signal model is set up; secondly, the two observed full apertures are divided into several sub-apertures according to the same criteria. By analyzing sparse characteristics of the target echo in each channel, a joint sparse constrained optimization model is set up and the problem of multi-channel high-resolution imaging is transformed into an optimization problem based on multi-channel joint sparse reconstruction. The improved orthogonal matching pursuit (OMP) is applied for high-resolution imaging solving to obtain 3-D images of each sub-aperture target; thirdly, the 3-D images of each sub-aperture are synthesized to obtain final 3-D imaging results of the target under full aperture; finally, the effectiveness of the proposed approach is verified by processing point target simulation data and Backhoe electromagnetic simulation data, and the InISAR system is set up in the microwave anechoic chamber to verify the practical applications of the proposed approach by processing the measured data obtained. Compared with traditional InISAR imaging methods, the proposed method in this paper has the following advantages: (1) reconstruction accuracy of strong scattering centers in ISAR images is improved owing to utilization of correlation among cross channels, and relative phase information of cross channels is kept effectively, so as to obtain interferometric phase information with higher accuracy; (2) owing to sub-aperture synthesis method applied, the target with scattering anisotropy in all directions can be accurately described, and the problem that accuracy of imaging azimuth angle is limited can be overcome; (3) because of information complementation and redundancy among multi-channel signals applied, given relatively large compression sampling ratio, generation of false scattering points can be effectively suppressed, thereby improving the imaging quality.

2. Signal Model of Inisar Near-Field Imaging

The InISAR system includes multiple antennas. This paper proposes a dual-antenna ISAR imaging system. Figure 1 shows geometric relationship between antennas and the target, where antenna TR_2 is located at origin O'. Antennae TR_1 and TR_2 form the vertical baseline along axis Z'. Define two coordinate systems, where $T'(x',y',z')$ is the radar coordinate system, axis y' is the line of sight of radar, x' and y' represent the horizontal and vertical directions, respectively. $T(x,y,z)$ is the target coordinate system, in which axis y is coincident with axis y', and axes x and y represent the azimuth direction and range direction of ISAR, respectively, and distance between origins of the two coordinate systems is R_0 ($R_0 < 4D^2/\lambda$, D is the maximum size of the target and λ is the wavelength of the incident wave). The target is moving at a constant speed in plane (x,y) at an angular velocity ω, and plane (x',y') is parallel to plane (x,y). Assuming that coordinate of any point P on the target is (x,y,z), and the coordinate in the cylindrical coordinate system is (r_0,θ_0,z). Then, at the moment t, the distance from the antenna I ($i \in TR_1, TR_2$) to the point P is:

$$R_i(t) = \sqrt{(R_0 + y)^2 + x^2 + (R_0 \tan\alpha_i - z)^2}$$
$$= \sqrt{(R_0 + r_0 \cos(\theta_0 + \omega t))^2 + (r_0 \sin(\theta_0 + \omega t))^2 + (R_0 \tan\alpha_i - z)^2} \tag{1}$$

where, α_i refers to the pitch angle from the antenna i to the origin of target coordinate system. Assuming that the antenna transmits a step frequency wideband signal [26]:

$$s_{it}(t) = \sum_{m=0}^{M-1} rect\left(\frac{t - t_{mp}}{T}\right) \exp[j2\pi f_m t] \tag{2}$$

where:

$$rect\left(\frac{t - t_{mp}}{T}\right) = \begin{cases} 1, & 0 < \frac{t-t_{mp}}{T} < 1; \\ 0, & \frac{t-t_{mp}}{T} < 0 \text{ and } \frac{t-t_{mp}}{T} > 1; \end{cases} \tag{3}$$

$f_m = f_0 + \Delta f$ is the frequency of the pulse centered at time t_{mp} and for pulses spaced equally in frequency and time; $t_{mp} = (m + pM)T$; Δf is the frequency difference for each step in the pulse burst; T is the time interval between pulses (pulse repetition period of pulses in the burst); M the number of pulses in each burst; $p = \overline{0, N-1}$-the index of emitted burst. The echo of the target received by antenna i after the coherent demodulation is (to understand easily, rectangular coordinates are used):

$$s_{ir}(t) = \iint_D g_i(x,y) \sum_{m=0}^{M-1} rect\left(\frac{t - t_{mp}}{T}\right) \exp[j2\pi f_m \frac{R_i(t,x,y)}{c}]dxdy, \tag{4}$$

where, D refers to the imaging scene area, (x,y) refers to the position coordinate of the target, and $g_i(x,y)$ indicates the backscatter coefficient of the target at position (x,y) received by antenna i. Since the imaging process is a linear system, and in the high frequency region, the total scattering of the target can be seen as a linear superposition of multiple strong scattering points, i.e., $s_{ir}(t)$ represents superposition of echo signals of all scattering points at (x, y) in the entire imaging region $g_i(x, y)$. The signal is a complex signal, amplitude $g_i(x, y)$ represents scattering intensity of scattering point (x, y), and phase $\exp(-j4\pi f_m R_i(t, x, y)/c)$ contains position information (x, y) of the scattering point. $\iint_D [\,]dxdy$ represents the summation of all the scattered echoes in the imaging scene. With signal processing technology, the target image can be obtained by separating position (x, y) and amplitude $g_i(x, y)$ from complex signal $s_{ir}(t)$.

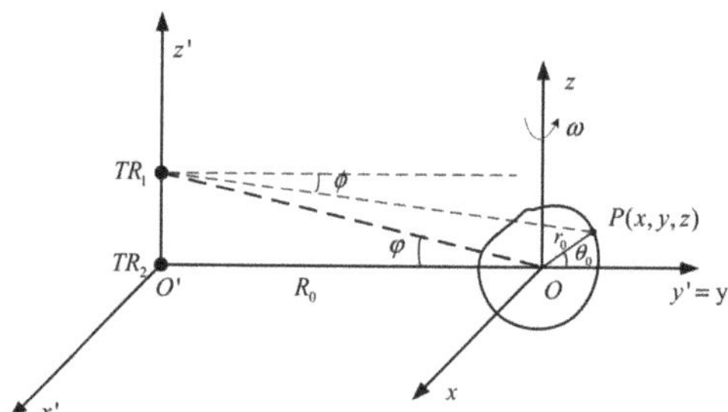

Figure 1. Geometric sketch of dual-antenna InISAR imaging.

The imaging scene is discretized. In order to represent the radar signal as a matrix, the corresponding 2-D backscatter coefficient matrix is concatenated into a one-dimensional column vector by row or line:

$$g_i = [g_i(1,1), \cdots, g_i(P,1), g_i(1,2), \cdots, g_i(P,2), \cdots, g_i(1,Q), \cdots, g_i(P,Q)]^T, \tag{5}$$

where, g_i refers to the vector of $PQ \times 1$, and P is the discrete grid number of axis x. Q is the discrete grid number of axis y.

According to the Formula (4), the discrete echo data can be indicated as follows:

$$s_{ir}(t) = \sum_{l=1}^{PQ} \sum_{m=0}^{M-1} g_i(l) \exp\left[-j\frac{4\pi}{\lambda_m} R_i(t,l)\right], \tag{6}$$

where, λ_m refers to wave length corresponding to the frequency f_m. $R_i(t,l)$ refers to the distance between the ith antenna and the lth target at time t.

The range (or frequency) and azimuth angle are also discrete in actual situations. Assuming that the range sampling point is N, the azimuth sampling point is M. The frequency of nth (frequency) pulse in the burst is defined by $f_n = f_0 + (n-1)\Delta f, n = 1, 2, \cdots, N$, where f_0 is the initial frequency, Δf is the step interval, and the azimuth is discretized as $\theta_m = (m-1)\Delta\theta, m = 1, 2, \cdots, M$, ($\Delta\theta$ is the angular interval). Formula (6) can be discretized as:

$$s_{ir}(f_n, \theta_m) = \sum_{l=1}^{PQ} g_i(l) \exp\left[-j\frac{4\pi f_n}{c} R_i(\theta_m, l)\right] \tag{7}$$

Equation (7) is represented as a matrix considering the effect of noise in actual situations:

$$s_{ir} = A_i g_i + e_i, \tag{8}$$

where s_i is a vector with the size of $MN \times 1$, which is formed through signal sampling; A_i is a dictionary matrix with the size of $MN \times PQ$, which is formed through mapping relationship between the target and the signal; g_i is a vector with the size of $PQ \times 1$, which is composed of scene backscatter coefficients; e_i is the additive complex noise in the channel. Specified composition of each vector and matrix in formula (8) is as follows:

$$s_{ir} = [s_{ir}(f_1, \theta_1), \cdots, s_{ir}(f_1, \theta_M), s_{ir}(f_2, \theta_1), \cdots,$$
$$s_{ir}(f_2, \theta_M), \cdots s_{ir}(f_N, \theta_1), \cdots, s_{ir}(f_N, \theta_M)] \tag{9}$$

Make:

$$a(f_n, \theta_m, l) = \exp\left[-j\frac{4\pi f_n}{c} R_i(\theta_m, l)\right] \tag{10}$$

Following vectors are defined as:

$$a(f_n, \theta_m) = [a(f_n, \theta_m, 1), a(f_n, \theta_m, 2), \cdots, a(f_n, \theta_m, PQ)]^T. \tag{11}$$

Then the dictionary matrix can be obtained as:

$$A_i = [\alpha(f_1, \theta_1), \cdots, \alpha(f_1, \theta_M), \alpha(f_2, \theta_1), \cdots, \alpha(f_2, \theta_M), \cdots, \alpha(f_N, \theta_1), \cdots, \alpha(f_N, \theta_M)] \tag{12}$$

Therefore, the projection relationship between the scene and signal in i channel is obtained. The radar signal model in the interferometric channel is represented as follows:

$$s = \begin{bmatrix} s_{1r} \\ s_{2r} \end{bmatrix} = Ag + e = \begin{bmatrix} A_1 & 0 \\ 0 & A_2 \end{bmatrix} \begin{bmatrix} g_1 \\ g_2 \end{bmatrix} + \begin{bmatrix} e_1 \\ e_2 \end{bmatrix}, \tag{13}$$

where, s corresponds with the echo signal in the interferometric channel, A corresponds with the dictionary matrix in the interferometric channel, g corresponds with the backscatter coefficient in each interferometric channel, and e corresponds with the additive noise in each interferometric channel. It should be noted that the dictionary matrix corresponding with each interferometric channel may be the same. However, in order to keep their generality, dictionary matrixes of the two interferometric channels should be represented separately.

In order to reduce the sampling data of each interferometric channel effectively, the $M'(M' \leq M)$ angular position is randomly selected to transmit signal in the azimuth direction, and then $N'(N' \leq N)$ frequency point is randomly selected in the distance direction. After compressed sampling, the interferometric echo signal model can be represented as follows:

$$s' = \Phi s = \begin{bmatrix} \Phi_1 0 \\ 0 \Phi_2 \end{bmatrix} s = \Phi Ag + \Phi e = A'g + e', \tag{14}$$

where, s' refers to CS echo signal with the size of $2N'M' \times 1$, $\Phi_i = \Phi_i^a \otimes \Phi_i^r$ refers to the measurement matrix corresponding with each channel, $\otimes$ refers to the Kronecker product, A' refers to the sensing matrix with the size of $2N'M' \times PQ$, g refers to the scene backscatter coefficient with the size of $2PQ \times 1$, and e' refers to noise with the size of $2PQ \times 1$.

Since $M' \leq M$, $N' \leq N$, recovery of the signal s from the measurements s' is ill-posed in general. However, according to the CS theory, when the matrix $\Phi A = A'$ has the Restricted Isometry Property (RIP) [27], it is indeed possible to recover the K largest g_i's from a similarly sized set of $M'N' = O(K\log(MN/K))$ measurements s'. The RIP is closely related to an incoherency property between Φ and A, where the rows of Φ do not provide a sparse representation of the columns of A, and vice versa.

In order to obtain the target images of each channel, the problem of interferometric near-field imaging can be converted into an optimization and reconstruction problem of two independent channels according to CS theory and sparse space distribution characteristics of the target scene:

$$\begin{cases} \min\|g_1\|_0 & s.t. \ \|s_1' - A_1'g_1\|_F \leq \xi_1 \\ \min\|g_2\|_0 & s.t. \ \|s_2' - A_2'g_2\|_F \leq \xi_2 \end{cases} \tag{15}$$

where, $\|\bullet\|_0$ refers to the zero norm of vector, namely the number of non-zero elements in the vector, $\|\bullet\|_F$ refers to the Frobenius norm of matrix, and ξ_i is a positive number which depends on the noise level. In Formula (15), it is quite important to select an appropriate noise level for the final optimization result: too high noise level will lead to the loss of some weak scattering points, while too low noise level will make it difficult to suppress strong noises. In Formula (15), optimization solving can be performed by applying the OMP method which not only guarantees reconstruction accuracy, but also has high computational efficiency.

3. Joint Sparsity Constraint Near-Field 3-D Imaging of Inisar Based on CS

3.1. Algorithm Flow Description

We first give the flow of the proposed approach in this paper, and describe the solution precisely in the following subsections. The process of the proposed approach is presented as follows:

Step 1: Full apertures of the two channels are divided into several sub-apertures by the same criteria, and each sub-aperture has a very small azimuth angle range.

Step 2: Global sparsity constraint and improved OMP algorithm are applied to obtain the 2-D complex images I_1 and I_2 of each sub-aperture in the two channels.

Step 3: Two images of each sub-aperture are performed with interferometric processing to obtain projection coordinates of the scattering points along the baseline.

Step 4: 3-D images of each sub-aperture target are constructed by synthesizing the 2-D ISAR images of the interferometric processing results.

Step 5: 3-D images of all sub-apertures are processed synthetically to obtain the final 3-D images.

Step 6: The imaging flow is shown in Figure 2.

3.2. Joint Sparse Constrained Optimization Model

As mentioned in Section 2, a small amount of compressed sampled data is used to achieve high-resolution reconstruction of the scene in the single-channel independent CS approach. However, it cannot guarantee consistency of positions and numbers of all scattering points in each channel, and integrity of the cross-information of channels is destroyed, which are not favorable for target scattering information extraction. In addition, complementarity and redundancy among multi-channel data does not fully exploit in the single-channel independent CS approach, so it cannot further improve the SNR gain and reduce data volume of the system.

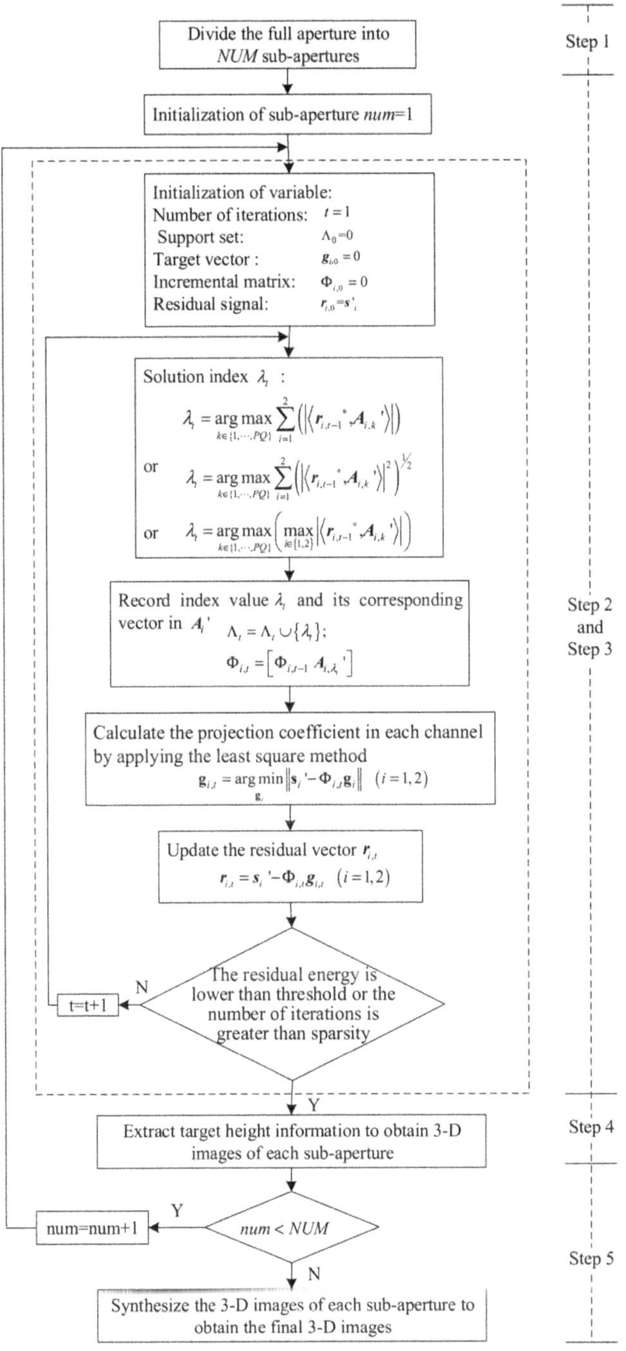

Figure 2. Imaging flow of proposed approach.

Two ISAR images used in the InISAR imaging are usually highly correlated, so they are more obvious in joint sparsity. Based on such prior information, the following joint sparse metric functions with global sparsity can be obtained as:

$$\|g\|_{p,0} = \left\| \left(|g_1|^p + |g_2|^p \right)^{1/p} \right\|_0, \ p \geq 1, \tag{16}$$

where, $\|\bullet\|_0$ is zero norm, which represents the number of non-zero elements in the vector. Different p values correspond with different mixed norm forms. This paper defines the following three global sparsity constraints:

(1)   when $p = 1$, $\|g\|_{1,0} = \|(|g_1| + |g_2|)\|_0$, is termed as mixed sum norm, i.e., sparsity of amplitude sum of the two ISAR images is taken as the global sparsity constraint;

(2)   when $p = 2$, $\|g\|_{2,0} = \left\| \left(|g_1|^2 + |g_2|^2 \right)^{1/2} \right\|_0$ is termed as mixed Euclidean norm;

(3)   when $p = \infty$, $\|g\|_{\infty,0} = \|\max(|g_1|, |g_2|)\|_0$ is termed as mixed infinite norm, i.e., the sparsity of one of the two ISAR images (with larger amplitude) is taken as the overall sparsity constraint.

The mixed sum norm and mixed Euclidean norm are both measured by taking the number of non-zero elements of amplitude sum of images in all channels as the global sparsity. Since the scattering points are aligned at different angles, the additivity is reasonable. The mixed infinite norm takes the number of non-zero elements of the pixel point with the largest amplitude in each image as the joint sparsity, and can also ensure that scattering points in the reconstructed sub-aperture images are aligned.

3.3. Optimal Solution Algorithm Constrained by Joint Sparse

According to the constructed global sparsity constraint function, the problem of InISAR imaging solution is transformed into the problem of multi-channel joint sparse optimal reconstruction problem:

$$\min_{g_1,g_2} \|g\|_{p,0} \quad s.t. \begin{cases} \|s'_1 - A'_1 g_1\|_2^2 \leq \xi \\ \|s'_2 - A'_2 g_2\|_2^2 \leq \xi \end{cases}, \tag{17}$$

where, ξ is determined by the minimum noise level of each channel to ensure that each channel can generate the target image. In Formula (17), data cannot be processed directly by applying the traditional CS optimal reconstruction algorithm. In [28], an improved convex optimization approach was proposed to solve the joint sparse imaging problem of interferometric channels. As sparse reconstruction is only performed for azimuth, the computational complexity is not high. While it is applied to the 2-D sparse reconstruction concerning range and azimuth as studied in this paper, the computational complexity becomes intensive, especially in high-resolution imaging. The target coefficient vector to be reconstructed usually has a large size, and thus the memory shortage of the computational platform will be inevitably encountered for practical applications. OMP algorithm is a commonly used greedy algorithm, which has high computational efficiency and can guarantee excellent reconstruction results. Therefore, this paper proposes an effective and improved OMP algorithm to solve the multi-channel joint sparse reconstruction problem based on OMP algorithm. Regarding the three mixed norm solutions proposed in this paper, step (2) is different in the way of index finding:

(1)   Initialization: number of iterations $t = 1$, and support set $\Lambda_0 = 0$. For the ith channel, its initialized target vector $g_{i,0} = 0$, and incremental matrix $\Phi_{i,0} = 0$, which is composed of column vectors in the support set. Make $r_{i,t}$ the residual signal after t iterations, and initialize $r_{i,0} = s'_i$.

(2)   Obtain index λ_t by solving the following formulas:

$$\text{Mixed sum norm: } \lambda_t = \arg\max_{k \in \{1, \cdots, PQ\}} \sum_{i=1}^{2} \left(\left| \left\langle r_{i,t-1}^*, A_{i,k}' \right\rangle \right| \right), \tag{18}$$

$$\text{Mixed Euclidean norm: } \lambda_t = \underset{k\in\{1,\cdots,PQ\}}{\arg\max} \sum_{i=1}^{2}\left(|\langle r_{i,t-1}{}^{*}, A_{i,k}{}'\rangle|^{2}\right)^{1/2}, \tag{19}$$

$$\text{Mixed infinite norm: } \lambda_t = \underset{k\in\{1,\cdots,PQ\}}{\arg\max}\left(\underset{i\in\{1,2\}}{\max}|\langle r_{i,t-1}{}^{*}, A_{i,k}{}'\rangle|\right), \tag{20}$$

where, $A_{i,k}{}'$ is the k column vector in perception matrix.

(3) Record the obtained index λ_t to the support set and its corresponding vector in A_i' to the incremental matrix:

$$\begin{aligned} \Lambda_t &= \Lambda_t \cup \{\lambda_t\}; \\ \Phi_{i,t} &= [\Phi_{i,t-1}\ A_{i,\lambda_t}{}'] \end{aligned} \tag{21}$$

(4) Adopt the least square method to calculate the projection coefficient of each channel:

$$g_{i,t} = \underset{g_i}{\arg\min}\|s_i' - \Phi_{i,t}g_i\| \quad (i = 1,2) \tag{22}$$

(5) Update residual signal $r_{i,t}$:

$$r_{i,t} = s_i{}' - \Phi_{i,t}g_{i,t} \quad (i = 1,2) \tag{23}$$

(6) For the number of iteration $t = t + 1$, repeat step (2) to (4) until the energy of the residual signal is lower than the preset threshold *Thres* or the number of iterations reaches the preset sparsity K.

Compared with the standard OMP algorithm, the proposed algorithm is mainly improved in step (2). In the standard OMP algorithm, the support set for different channels may be different, because the index $\lambda_{t,i}$ is determined only for the single-channel signal itself:

$$\lambda_{t,l} = \underset{k\in\{1,\cdots,PQ\}}{\arg\max}\left\langle r_{i,t-1}{}^{*}, A_{i,k}\right\rangle(i = 1,2). \tag{24}$$

Such independent processing makes inconsistent positions and number of non-zero coefficients in the target vector finally reconstructed in each channel, which is not favorable for extraction of target scattering information. For the improved OMP algorithm, the multi-channel target scattering information is used to determine candidate vectors and solve the projection coefficients of each channel in the same support set, so as to ensure the consistency of the position and number of non-zero coefficients in target vector reconstructed in each channel.

Setting of the preset threshold *Thres* is related to the noise level of the echo signal. Considering the high SNR ratio, the threshold can generally be set as about 0.05 of the energy of the echo signal. At this time, most of the scattering centers on the target can be accurately reconstructed; whereas, with the increase of noise in echo signal, the set threshold value also increases, so as to avoid more false scattering points caused by noise in the generated image. If the sparsity K is known, the reconstruction results obtained by applying the CS matching pursuit reconstruction algorithm are quite excellent but it is difficult to obtain accurate sparsity in actual engineering. In such case it always requires a large number of SAR images for statistical analysis to determine approximate sparsity K range of different kinds of observation scenes, and then obtain the optimal sparsity K within the determined range by the optimization criteria.

3.4. Extraction of Target Scattering Information

When the distance between the target and the antenna satisfies the near-field condition and the baseline of the antennas is much smaller than R_0, according to plane spectrum theory [29], the distance in Formula (1) can be represented as:

$$R_i(t) = (R_0 + r_0\cos(\theta_0 + \omega t))\cos\alpha_i\cos\phi + r_0\sin(\theta_0 + \omega t)\cos\alpha_i\sin\phi + (R_0\tan\alpha_i - z)\sin\alpha_i, \quad (25)$$

where, $\phi = \arctan\left(\frac{r_0\sin(\theta_0+\omega t)}{R_0+r_0\cos(\theta_0+\omega t)}\right)$ is the angle between the target and antenna in the plane Oxy. To simplify the expression, the Cartesian coordinates of the target are represented in the cylindrical coordinate system, so Equation (25) can be expressed as:

$$R_i(t) = \sqrt{(R_0 + r_0\cos(\theta_0 + \omega t))^2 + (r_0\sin(\theta_0 + \omega t))^2}\cos\alpha_i + (R_0\tan\alpha_i - z)\sin\alpha_i. \quad (26)$$

Then, the echo signal of the scattering point P can be represented as:

$$s_i(t) = g_i\exp\left(-j4\pi f\frac{\Re\cos\alpha_i + (R_0\tan\alpha_i - z)\sin\alpha_i}{c}\right), \quad (27)$$

where, $\Re = \sqrt{(R_0 + r_0\cos(\theta_0 + \omega t))^2 + (r_0\sin(\theta_0 + \omega t))^2}$. This item is the same for the two antennas. According to Formula (26), it is found that phase information of the scattering point in target ISAR image contains height information of the scattering point, interferometric processing for two ISAR images is conducted, and the interferometric phase difference of P images is:

$$\begin{aligned}\Delta\varphi &= \frac{4\pi f}{c}(\Re\cos\alpha_2 + (R_0\tan\alpha_2 - z)\sin\alpha_2 - (\Re\cos\alpha_1 + (R_0\tan\alpha_1 - z)\sin\alpha_1))\\ &= \frac{4\pi f}{c}(\Re(\cos\alpha_2 - \cos\alpha_1) + R_0(\tan\alpha_2\sin\alpha_2 - \tan\alpha_1\sin\alpha_1) - z(\sin\alpha_2 - \sin\alpha_1))\end{aligned} \quad (28)$$

In the InISAR imaging system, the baseline length is much less than the distance between the antennas and target, so pitch angle difference between the antenna TR_1 and TR_2 is very small, namely $\alpha_2 = \alpha_1 + \Delta\alpha$, $\Delta\alpha \ll 1$. In this paper, the antenna TR_2 is at the origin, so $\alpha_1 = 0$, $\alpha_2 = \Delta\alpha$, and the height of the scattering point can be estimated as:

$$z = \frac{\frac{\lambda}{4\pi}\Delta\varphi - \Re\cos\alpha_2 - R_0\tan\alpha_2\sin\alpha_2}{-\sin\alpha_2} \quad (29)$$

When the antenna TR_1 and TR_2 are symmetrically distributed in the coordinate origin, namely $\alpha_1 = \alpha_2$, the height of the scattering point is estimated as:

$$z = \frac{\Delta\varphi\lambda}{8\pi\sin\alpha_1} \quad (30)$$

For target containing K scattering points, the height of each scattering point can be estimated by interferometric processing of the corresponding pixel points in the two ISAR images.

In actual processing, in order to avoid possible ambiguity of interferometric phase difference, the following judgements on the interferometric phase difference are required:

$$\begin{cases} if\,\Delta\varphi > \pi, \Delta\varphi - 2\pi \\ if\,\Delta\varphi < -\pi, \Delta\varphi + 2\pi \end{cases}. \quad (31)$$

4. Experiments and Analysis

In order to verify the effectiveness of the algorithm in this paper, point target simulation data, electromagnetic software simulation data, and measured data in anechoic chamber is adopted to carry out imaging verification and performance analysis, respectively.

4.1. Numerical Simulations

The simulation target is composed of 46 scattering centers withe shape of plane model, and its distribution is shown in Figure 3. Stepped frequency signal of radar transmission and system parameters setting are as shown in Table 1.

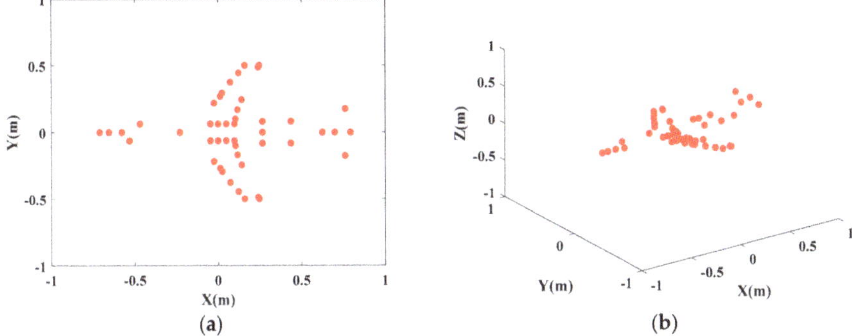

Figure 3. Geometric distribution for scattering model of plane point. (**a**) 2D distribution diagram of scattering point; (**b**) 3-D distribution diagram of scattering point.

Table 1. Simulation parameter.

Parameter	Parameter Value
Carrier frequency	10 GHz
Bandwidth	4 GHz
Frequency step interval	40 MHz
Azimuth accumulation angle	20°
Azimuth sampling interval	0.2°
Distance between antenna and target	2 m
Baseline length	0.02 m

Independent CS processing and joint CS processing based on global sparsity are separately used for InISAR imaging. The 3-D distribution of scattering points is shown in Figure 4. The above experiments are carried out with full data and signal without noise. Figure 4a is the single-channel independent processing reconstruction result, Figure 4b–d are the global sparse joint sparse reconstruction results. It can be seen from the figures that traditional single-channel independent processing can basically reflect the 3-D distribution of scattering points of the target, but cannot guarantee location consistency of all scattering points in different channels, which results in deviated location estimation. While with the proposed method, consistency of location and number of scattering points and more accurate reconstruction results can be ensured.

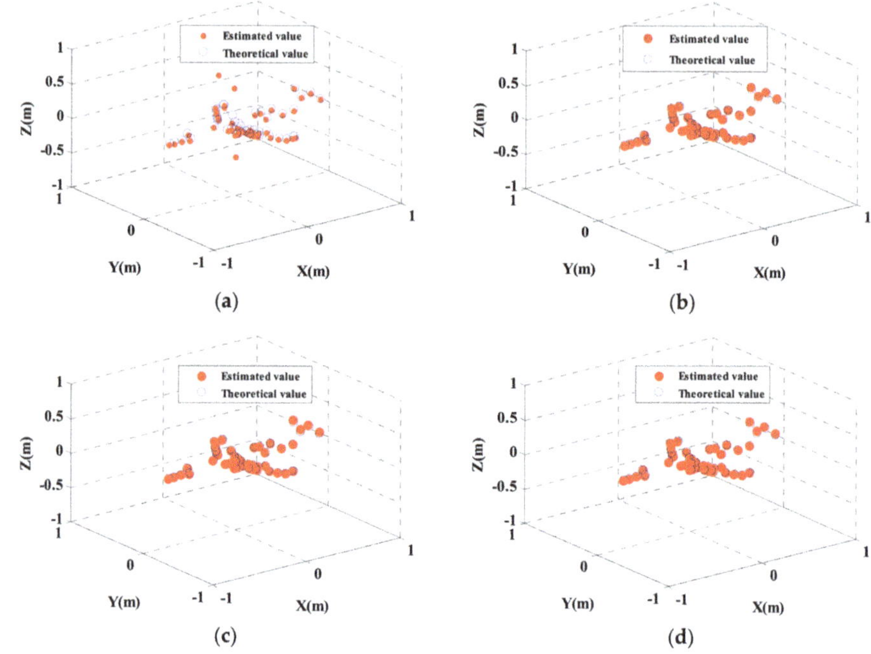

Figure 4. 3-D Imaging results of near-field InISAR. (**a**) Traditional imaging approach, (**b**) Global sparsity mixed sum norm processing; (**c**) Global sparsity mixed Euclidean norm processing (**d**) Global sparsity mixed infinite norm processing.

4.1.1. Precision Analysis of Scattering Point Coordinate Estimation

Based on the above experiments, three typical scattering points are selected for statistical comparison. It can be seen from the coordinate values of scattering points in the Table 2 that the conventional imaging approach has obvious deviation in estimating the height coordinate values of scattering points, while the proposed approach is more accurate in estimating the location of scattering points. For example, when estimating the height information, the maximum deviation of the traditional imaging approach is 0.6438, while the maximum deviation of the approach in this paper is 0.046. Compared with the traditional method, the accuracy of the proposed approach is improved by roughly 90% as indicated by statistical analysis of the location errors of all scattering points.

Table 2. Comparison of coordinates of typical scattering points.

	Scattering Point 1 (x, y, z)	Scattering Points 2 (x, y, z)	Scattering Points 3 (x, y, z)
Theoretical coordinate	(0.0125, 0.2688, 0)	(0.1438, 0.2438, 0)	(0.1188, 0.1688, 0.0467)
Traditional imaging	(0, 0.24, 0.6438)	(0.12, 0.23, 0.14)	(0.1, 0.1, 0.13)
Mixed sum norm	(0.009, 0.26, 0)	(0.1387, 0.24, 0.046)	(0.11, 0.1, 0.046)
Mixed infinite norm	(0.008, 0.26, 0)	(0.1387, 0.24, 0.046)	(0.1063, 0.1, 0.046)
Mixed Euclidean	(0.01, 0.265, 0)	(0.139, 0.241, 0.03)	(0.1163, 0.15, 0.046)

4.1.2. Precision Analysis of Interferometric Phase

For near-field InISAR imaging system, to estimate the height information of target scattering point through the phase difference of corresponding pixel among complex images of various channels, it must ensure the consistency of strong scattering position among various channels images during imaging of various channels. The accuracy of interferometric phase has a direct influence on imaging quality. Figure 5 is the interferometric phase distribution of the complex image of two channels after processing. According to the comparative study on Figure 5a–d, the accuracy of scattering point in obtaining interferometric phase is higher and broader for the proposed approach comparing with that of the independent single-channel processing imaging approach.

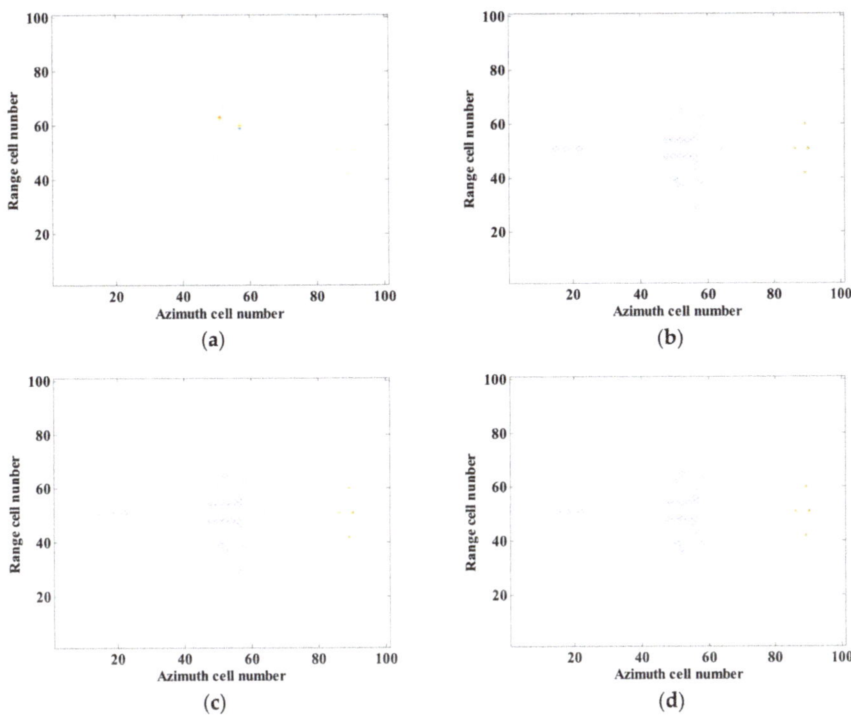

Figure 5. Interferometric phase image of near-field InISAR 3-D imaging. (**a**) Traditional imaging approach, (**b**) Global sparsity mixed sum norm processing; (**c**) Global sparsity mixed Euclidean norm processing (**d**) Global sparsity mixed infinite norm processing.

4.1.3. Noise Suppression Performance

Figure 6 shows the results obtained by applying the traditional imaging approach and the proposed approach when there is noise in echo signal, where the noise is 5 dB complex white Gaussian noise. Add the complex white Gaussian noise with SNR ratio of -15 dB to 25 dB in the echo data, repeat Monte-Carlo simulation test for 100 times under every noise level, and calculate MSE estimated on the basis of height.

In Figure 6, imaging results show that there is a considerable deviation in estimation of traditional processing approach on height of target scattering point. While by applying the imaging approach (global sparsity Euclidean norm) proposed in this paper, the height of target scattering point can be estimated more accurate.

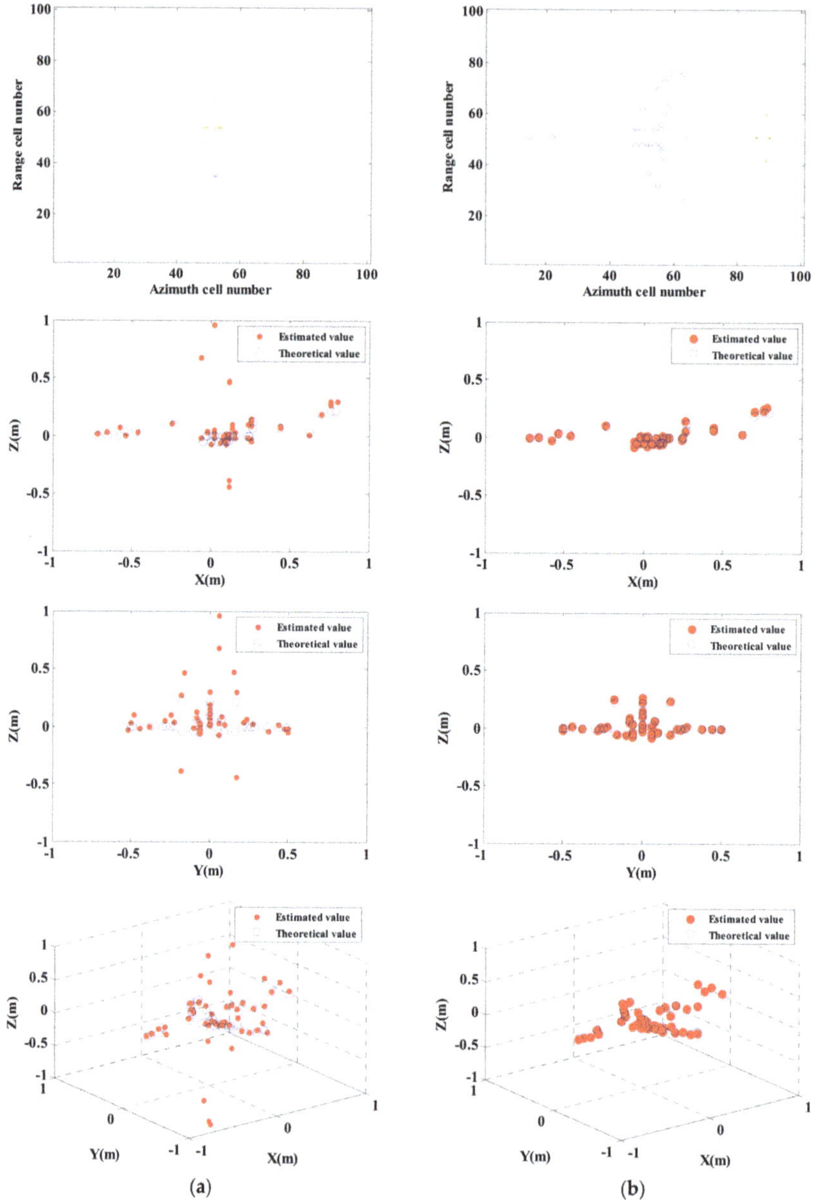

Figure 6. Imaging results obtained by adopting traditional approach and proposed approach respectively under 5 dB Noise. (**a**) Traditional imaging processing; (**b**) imaging processing of proposed approach (mixed Euclidean norm); interferometric phase images (top layer); complex images of channels 1 and 2 (second and third layers); final 3-D imaging results (last layer).

Figure 7 describes MSE estimated by heights of four compressed sensing approaches under different SNRs. Under low SNR, the performance estimated by the heights of the four approaches is low. With the increase of SNR, height estimation performance increases. However, the estimation performance based on global sparsity in this paper is obviously superior to that obtained by the traditional processing approach.

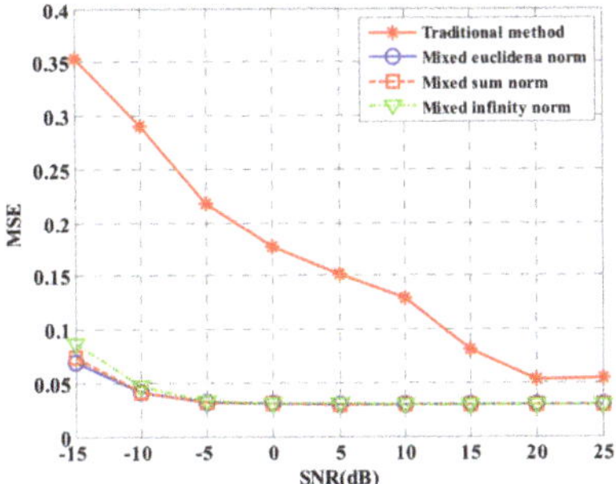

Figure 7. Comparison for imaging performance of four approaches under different SNRs.

4.1.4. Performance of Sparsity Sampling Imaging

Randomly select a certain quantity of data from echo data to carry out InISAR imaging for further investigating the influence of sparsity sampling on height estimation. Then, estimate the height information of the scattering point through interference processing. The SNR is fixed as 10 dB, and the pitch angle of antenna TR1 is fixed as 0.05°. Carry out Monte-Carlo simulation for 100 times to every sparsity sampling scheme, and calculate the MES estimated on the basis of height.

Figure 8 shows that application of the traditional approach fails to perform effective imaging of target given 80% under-sampling rate. However, with the proposed approach, accurate imaging of target can be performed. Figure 9 shows that the height estimation performance of processing based on global sparsity is still superior to that of results obtained on the basis of independent processing under sparsity sampling condition. Even when the measurement quantity is low (10% of full data), the overall sparsity constraint can still ensure a better interferometric imaging performance.

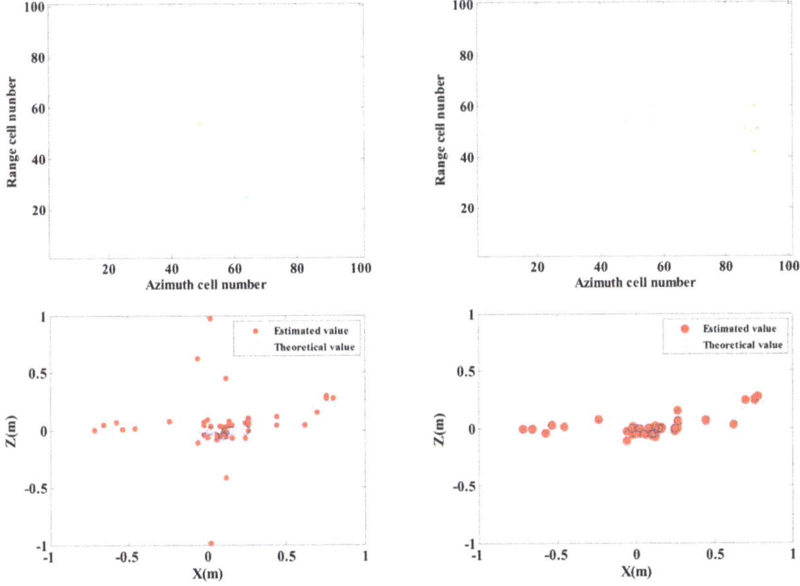

Figure 8. *Cont.*

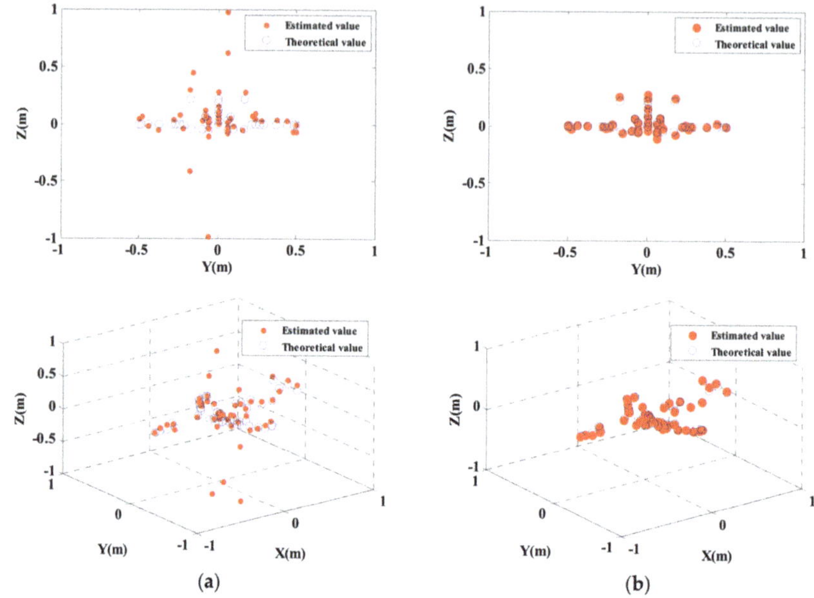

Figure 8. Imaging results of traditional approach and proposed approach under 20% of effective data. (**a**) Traditional imaging processing; (**b**) imaging processing of proposed approach (mixed Euclidean norm); interferometric phase images (top layer); complex images of channels 1 and 2 (second and third layers); final 3-D imaging results (last layer).

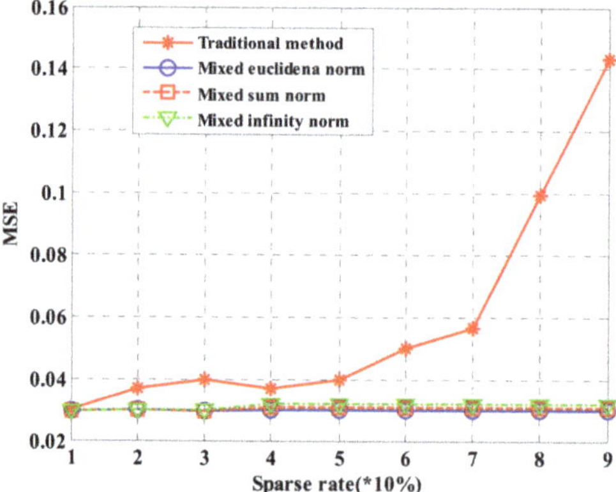

Figure 9. Comparison for imaging performance of four approaches under different sparsity samplings.

4.1.5. Computational Complexity

The running time of independent CS approach of traditional single-channel depends on step (2). Its computing cost is $O(Lt_{end}N_sPQ)$, wherein, the t_{end} is the times of algorithm iterative circulation, and the N_s is the signal sampling times. The proposed approach has higher calculation efficiency and is only added with $O(Lt_{end}PQ)$ times of addition calculation compared to independent CS approach of single-channel. The increased calculation times by applying the proposed approach can be nearly ignored in practical application. On the basis of the space storage efficiency, the approach proposed in the paper needs to occupy more memory space compared to that needed by applying the independent CS approach of single-channel, but it can be effectively released through parallelization.

4.2. Experiments and Analysis of Backhoe

The electromagnetic scattering echo data which is more closed to actual measurement is obtained by using high-frequency electromagnetic software and 3-D model of target. In this paper, Backhoe electromagnetic simulation data is adopted to verify the effectiveness of the proposed approach. In the experiment, select two groups of data with the adjacent pitch angle of 42° and 42.07° to divide the whole aperture into 17 sub-apertures. Specific parameters are shown in Table 3, and Figure 10 shows the CAD model of Backhoe.

Table 3. Parameters of electromagnetic simulation system.

Parameter	Parameter Value
Carrier frequency	10 GHz
Bandwidth	6 GHz
Sampling point number of frequency	512
Azimuth accumulation angle	51°
Sampling point number of direction	71 * 17
Pitch angle	0.07°

Figure 10. 3-D CAD model of Backhoe.

Adopt traditional imaging approach and the proposed approach respectively to carry out InISAR imaging, and the 3-D distribution of target scattering points is shown in Figure 11.

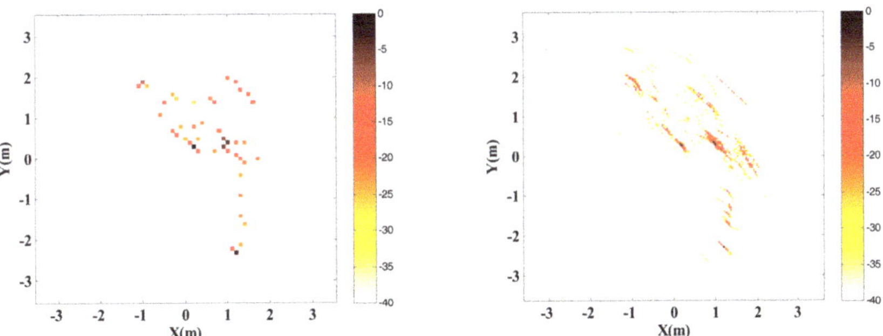

Figure 11. *Cont.*

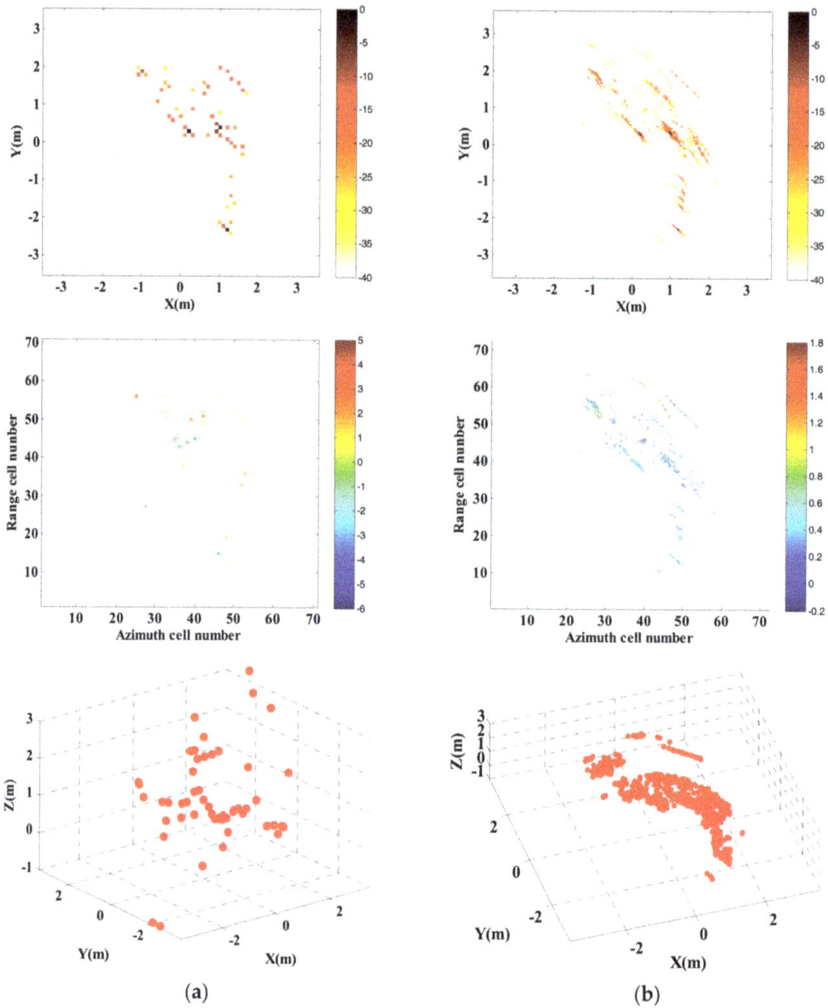

Figure 11. InISAR imaging results of Backhoe with complete data. (**a**) Traditional imaging processing; (**b**) imaging processing of proposed approach (mixed Euclidean norm); complex images of channels 1 and 2 (first and second layers); interferometric phase images (third layer); final 3-D imaging results (last layer).

Randomly select 25% observed data from full data to generate the sparsity sampling data, and adopt traditional imaging approach and the proposed approach respectively to carry out InISAR imaging. The 3-D distribution of target scattering points is shown in Figure 12. Figures 11 and 12 show that the reconstructed target information of traditional approach has a larger deviation. Regardless of complete data or insufficient data provided, the target can be imaged effectively. What's more, with the decrease of data quantity, the imaging performance gets worse. On the contrary, by applying the proposed approach, 3-D target images with higher quality can be obtained. When the effective data is 25%, it can still maintain accuracy of height information in estimation.

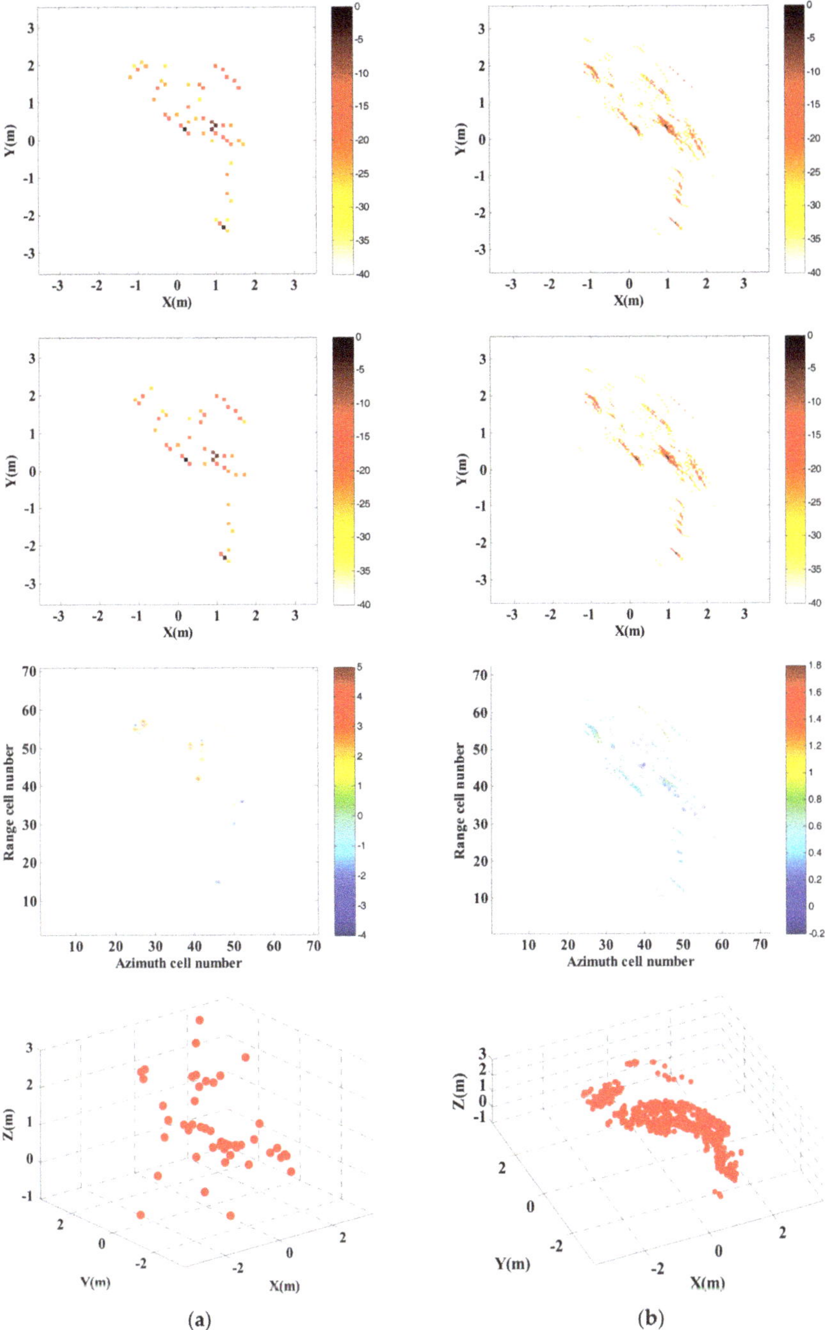

Figure 12. InISAR imaging results of Backhoe with 25% data. (**a**) Traditional imaging processing; (**b**) imaging processing of proposed approach (mixed Euclidean norm); complex images of channels 1 and 2 (first and second layers); interferometric phase images (third layer); final 3-D imaging results (last layer).

4.3. Actual Measurement Experiment in Anechoic Chamber

In order to further verify the effectiveness of the proposed approach in terms of practical applications perspective, a near-field InISAR test platform in an anechoic chamber is established, and the test system is shown as Figure 13. The target is put on the low scattering foam bracket, under which there is a turntable. The two antennas are fixed with interval of 0.2 m, and achieve azimuth accumulation through the rotation of turntable. Test parameters are shown in Table 4.

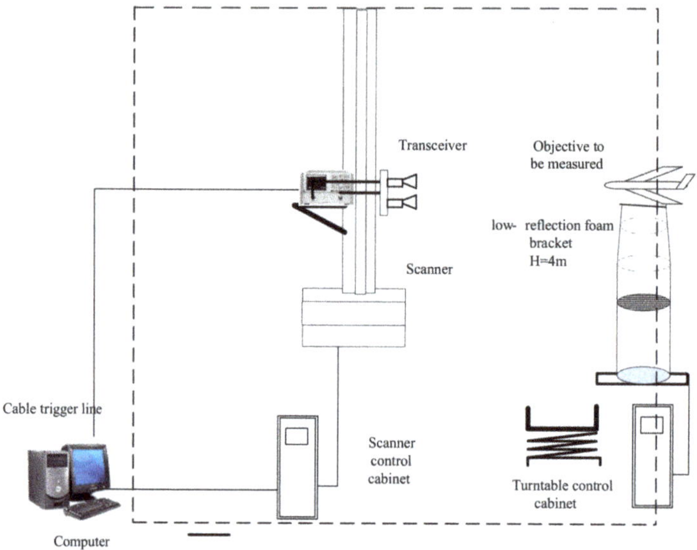

Figure 13. Frame diagram for near-field InISAR imaging system in anechoic chamber.

Table 4. Parameters of near-field InISAR test system.

Parameter	Parameter Value
Carrier frequency	10 GHz
Bandwidth	4 GHz
Frequency step interval	40 MHz
Azimuth accumulation angle	20°
Azimuth sampling interval	0.2°
Distance between antenna and target	0.02 m
Baseline length	2 m

The parameter calculation rules are as follows:

(1) Antenna baseline

Height information of the target is mainly calculated by phase difference of dual-antenna propagation path. In general, the interferometric phase difference is a periodic function for the period with 2π. In order to avoid fuzzy height, the interferometric phase difference shall meet the requirement of $\Delta\varphi \leq 2\pi$, so that the baseline length meet the requirement of $d \leq \frac{\lambda R_0}{2H}$, wherein λ represents transmitting frequency, R_0 represents distance from the receiving/transmitting antenna to the target, and $\delta_y = \frac{c}{2B}$ represents maximum height of the target.

(2) Sampling principle

Range resolution in ISAR imaging is $\delta_y = \frac{c}{2B}$, and azimuth resolution is $\delta_x = \frac{\lambda}{2\theta}$, wherein B represents signal bandwidth and θ represents azimuth accumulation angle. In actual imaging, the range resolution is generally equal to the azimuth resolution. Concerning resolution requirements, the signal bandwidth and azimuth accumulation angle can be determined through $\delta_y = \frac{c}{2B}$ and $\delta_x = \frac{\lambda}{2\theta}$.

Concerning range resolution requirements, the frequency sampling interval is $\Delta f \leq \frac{c}{2R_0}$, also taken as step frequency interval.

Concerning azimuth resolution requirements, the azimuth sampling interval is $\Delta\theta \leq \frac{\lambda}{2D}$, wherein D represents maximum size of the target.

(3) Distance from antenna and target

This paper focuses on near-field imaging, in principle, distance from the antenna to the target is represented as: $R_0 < \frac{4D^2}{\lambda}$, wherein D represents maximum size of the target and λ represents length of incident electromagnetic wave.

In addition, scanning in vertical direction does not exist because the two antennas are located fixedly, the beam center is fixed and only the target rotates in InISAR. As mentioned in Section 4.3, In the experiment, the range sampling interval and azimuth sampling interval are $\Delta f' = n\Delta f(n = 1, 2, \cdots, N)$ and $\Delta\theta' = m\Delta\theta(m = 1, 2, \cdots, M)$ respectively.

The test is adopted with stepped frequency signal, which features easy achievement of wideband and low requirements for hardware system. In order to make easier application of CS in the test, this paper adopts the deterministic sparsity observation approach based on Cat sequence for distance-oriented compressed sampling. The following shows the steps of Cat mapping to produce random sequence and construct observation matrix:

(1) Produce chaos sequence according to Cat mapping equation, and the mapping is defined as:

$$\begin{bmatrix} x_{n+1} \\ y_{n+1} \end{bmatrix} = \begin{bmatrix} 1 & a \\ b & ab+1 \end{bmatrix} \begin{bmatrix} x_n \\ y_n \end{bmatrix} (\text{mod } 1), \tag{32}$$

where, $(\text{mod } 1)$ represents the integer whose real number is casted out, namely $x \bmod 1 = x - \lfloor x \rfloor$. x_n sequence is selected to construct the needed deterministic random sequence.

(2) For using the chaos sequence construction Φ_r in stable area, cast out the g value in front of the sequence. It means to select x_{g+1} as the starting point of the sampling. Meanwhile, sample the produced sequence with the interval of d for ensuring the mutual independence of elements in chaos sequence:

$$z_k = x_{g+kd}, \quad k = 0, 1, 2, \cdots, N - 1 \tag{33}$$

After obtaining the output sequence z_k of Formula (33), directly divide the z_k into $N' = N/U$ with equal interval. Select the corresponding position of maximum value in various intervals, and assign 1 to corresponding position of Φ_r, and others a zero:

$$\Phi_r = \begin{bmatrix} \underbrace{1, \cdots, 0, \cdots, 0}_{u} & 0 & \cdots & & 0 \\ \vdots & \ddots & \cdots & & \cdots \\ 0 & 0 & \cdots & \underbrace{0, \cdots, 0, \cdots, 1}_{u} \end{bmatrix}, \tag{34}$$

where, each row has a $(0, 1)$ random sequence with length of U, and value 1 is at the position of the maximum value in the corresponding interval of the chaos sequence produced in $\{1, 2, \cdots, U\}$, and others are zero.

In practical applications, adopt parallelization to transmit N' random single frequency signal, and its data rate is N'/T. Its achievement process is shown in Figure 14.

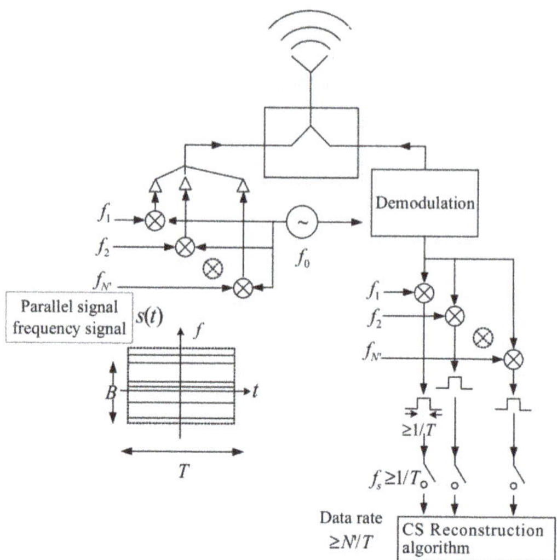

Figure 14. Range compressed sampling process of stepped frequency signal.

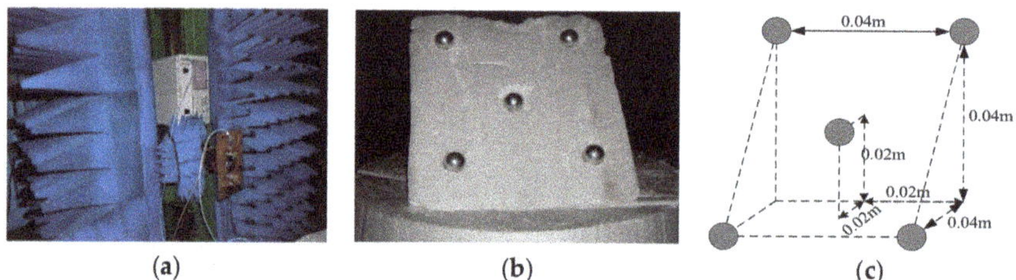

(a)   (b)   (c)

Figure 15. Target model of five metal balls. (**a**) Scanning frame and probe; (**b**) optical picture of five balls; (**c**) distribution of target spatial position.

Figure 15 is the optical pictures of the measurement system and target distribution. Figures 16 and 17 show imaging results by applying the two approaches provided with complete data and compressed sampling data. It can be found that by applying the traditional approach, target height information cannot be estimated completely because of incomplete correspondence of scattering point position, especially that the effective height information of target cannot be extracted if the compressed sampling proportion is large. Comparatively, by adopting the proposed approach, the height information of target can be estimated accurately given full data and compressed sampling data.

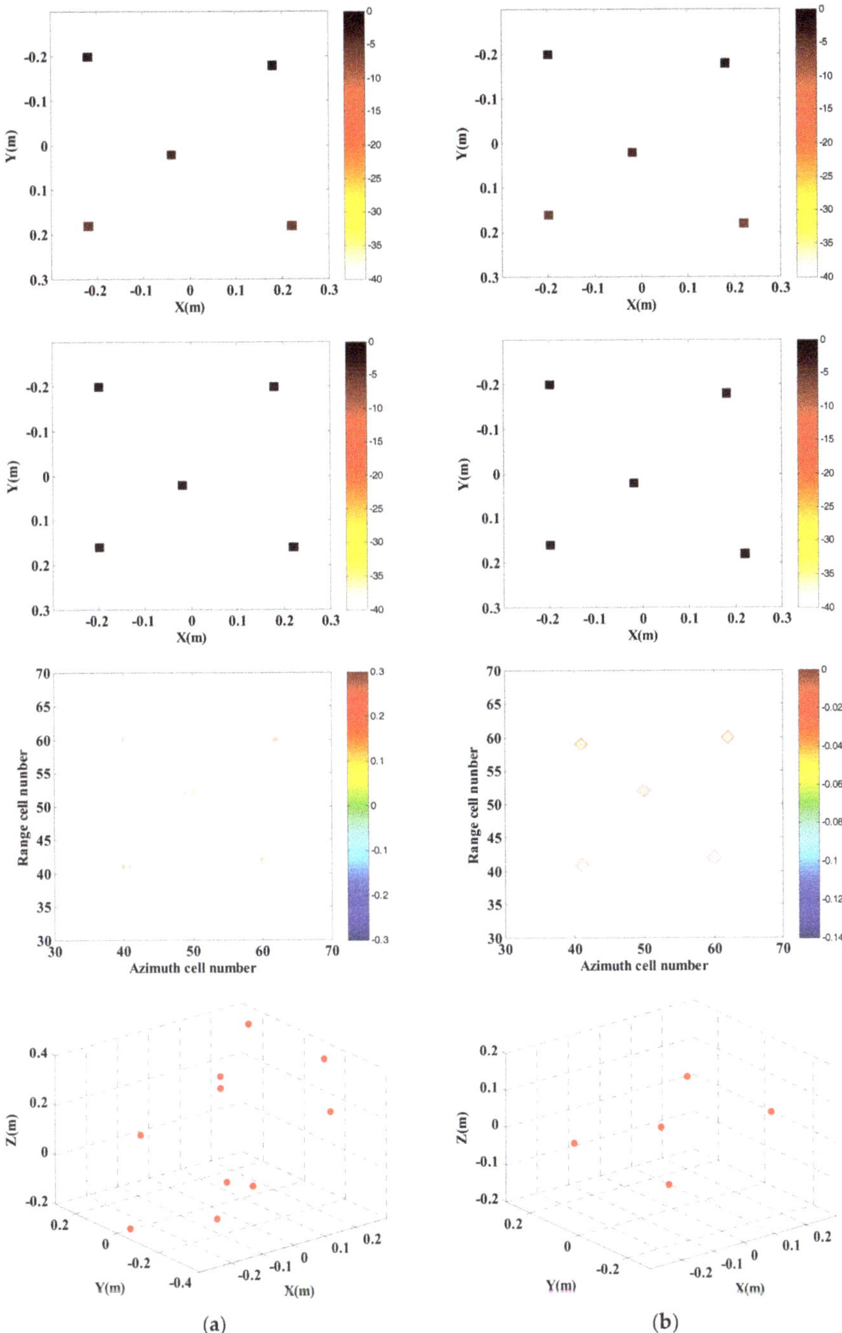

Figure 16. InISAR imaging results of five metal balls with complete data. (**a**) Traditional imaging processing; (**b**) imaging process of proposed approach (mixed Euclidean norm); complex images of channels 1 and 2 (first and second layers); interferometric phase images (third layer); final 3-D imaging results (last layer).

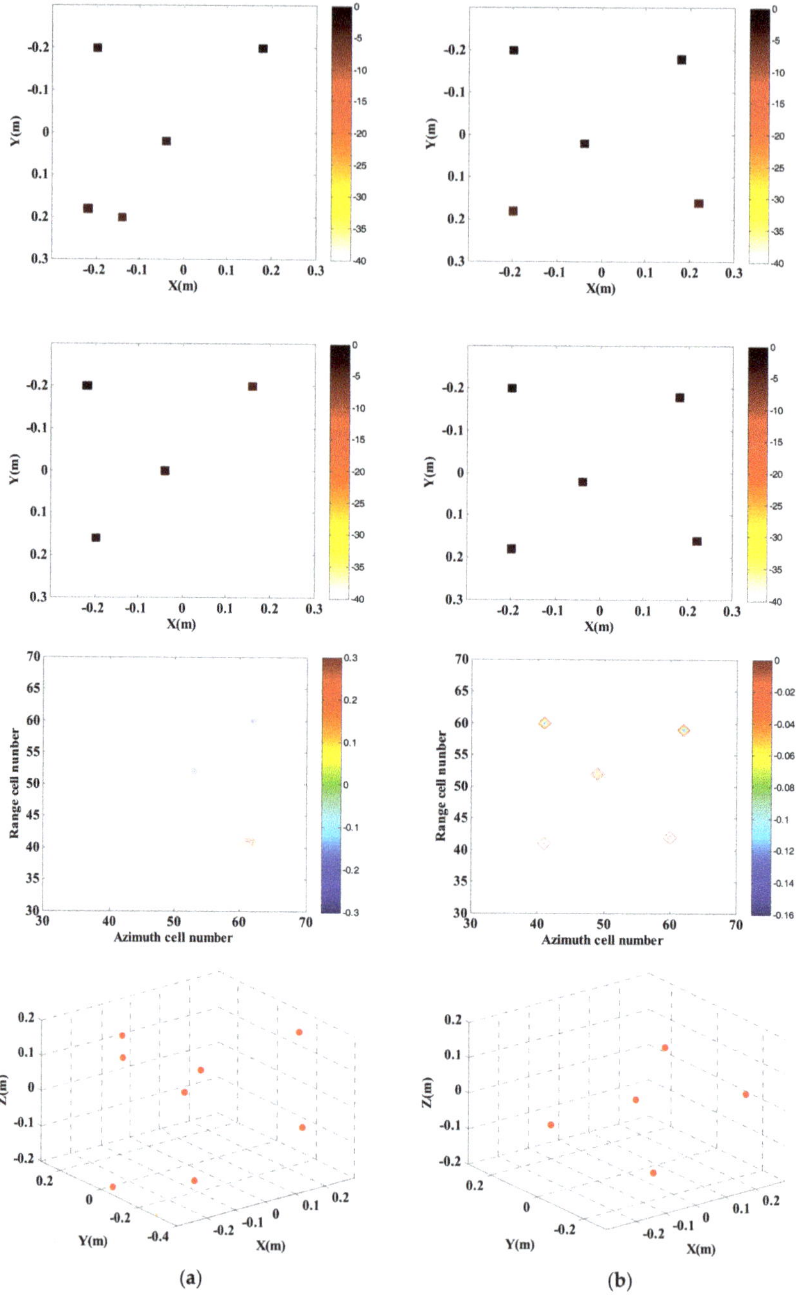

Figure 17. InISAR imaging results of five metal balls with 20% data. (**a**) Traditional imaging processing; (**b**) imaging processing of proposed approach (mixed Euclidean norm); complex images of channels 1 and 2 (first and second layers); interferometric phase images (third layer); final 3-D imaging results (last layer).

From Tables 5 and 6, it can be found that by applying the traditional approach, position consistency of all scattering points in images of different channels cannot be ensured. If the compression rate is high, the consistency is more obvious. However, by applying the proposed approach, reconstruction is performed through multi-channel joint sparsity, so as to ensure that the scattering points are at the same pixel of the images from different channels, which is more favorable for extraction of target height information.

Table 5. Target position information provided with complete data.

Approach	Channel	Ball 1	Ball 2	Ball 3	Ball 4	Ball 5
Initial	–	(−0.20, −0.20)	(0.20, −0.20)	(0.00, 0.00)	(−0.20, 0.20)	(0.20, 0.20)
Traditional	1	(−0.22, −0.19)	(0.18, −0.18)	(−0.05, 0.02)	(−0.22, 0.19)	(0.22, 0.22)
approach	2	(−0.21, −0.18)	(0.18, −0.19)	(−0.03, 0.02)	(−0.22, 0.18)	(0.23, 0.23)
Proposed	1	(−0.20, −0.20)	(0.19, −0.19)	(−0.01, 0.01)	(−0.20, 0.19)	(0.21, 0.21)
approach	2	(−0.20, −0.20)	(0.19, −0.19)	(−0.01, 0.01)	(−0.20, 0.19)	(0.21, 0.21)

Table 6. Target position information provided with compressed sampling data.

Approach	Channel	Ball 1	Ball 2	Ball 3	Ball 4	Ball 5
Initial	–	(−0.20, −0.20)	(0.20, −0.20)	(0.00, 0.00)	(−0.20, 0.20)	(0.20, 0.20)
Traditional	1	(−0.20, −0.20)	(0.19, −0.19)	(−0.04, 0.01)	(−0.22, 0.19)	(−0.14, 0.20)
approach	2	(−0.21, −0.20)	(0.18, −0.19)	(−0.04, 0.00)	(−0.20, 0.17)	(Null, Null)
Proposed	1	(−0.20, −0.20)	(0.19, −0.19)	(−0.01, 0.01)	(−0.20, 0.19)	(0.20, 0.21)
approach	2	(−0.20, −0.20)	(0.19, −0.19)	(−0.01, 0.01)	(−0.20, 0.19)	(0.20, 0.21)

One of the practical applications of the proposed approach is security check. Here we conduct detecting and imaging on closed chamber in the anechoic chamber. The chamber is used to simulate luggage carrier, which is placed with one knife, two bottles of water (one is full and the other is half-full) and two bottles of Coco-cola. The test parameters are consistent with parameters as shown in Table 4. With 10% echo data adopted, Figure 18 shows the test system and target scene and distribution of targets within the box. The imaging results are as shown in Figure 19.

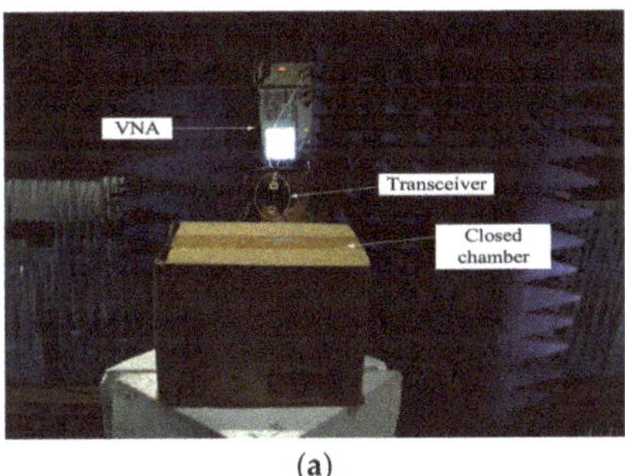

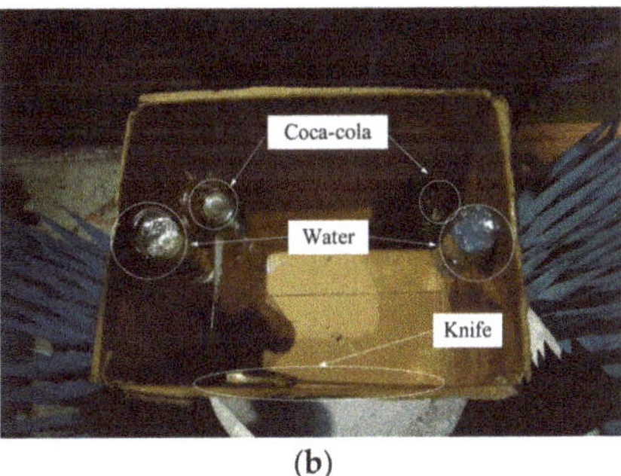

(a)   (b)

Figure 18. Imaging test for closed chamber. (a) Imaging system and target scene; (b) distribution of targets.

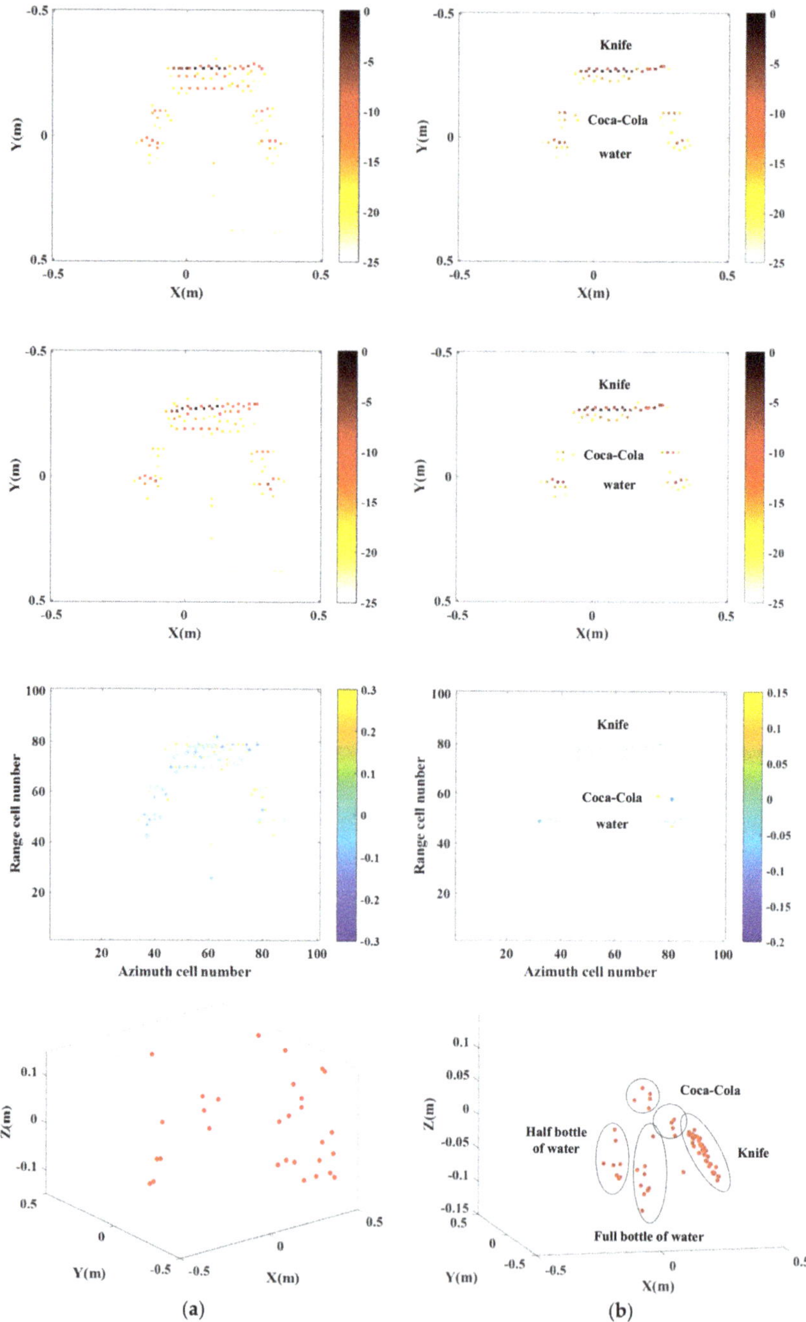

Figure 19. InISAR imaging results of anechoic chamber target with complete data. (**a**) Traditional imaging processing; (**b**) imaging processing of proposed approach (mixed Euclidean norm); complex images of channels 1 and 2 (first and second layers); interferometric phase images (third layer); final 3-D imaging results (last layer).

It can be seen from Figure 20 that in the case of only 20% echo data applied, effective imaging on the targets is unable to be achieved with the traditional imaging approach. Through adoption of the proposed imaging approach, clear target images are available, provided with the shapes and location information of the knife, full bottle of water, half bottle of water and bottles of Coco-cola in the chamber.

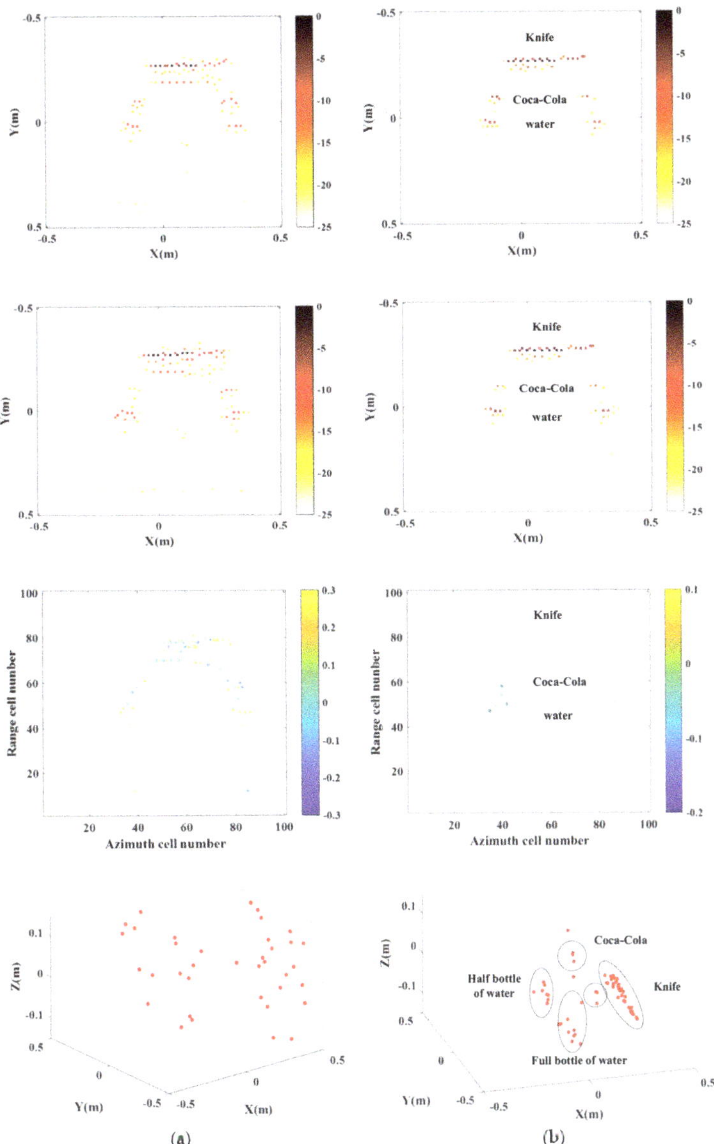

Figure 20. InISAR imaging results of anechoic chamber target with 20% data. (**a**) Traditional imaging processing; (**b**) imaging processing of proposed approach (mixed Euclidean norm); complex images of channels 1 and 2 (first and second layers); interferometric phase images (third layer); final 3-D imaging results (last layer).

5. Conclusions

Focusing on the near-field ISAR imaging, this paper puts forward an interferometric near-field 3-D imaging approach for joint sparsity reconstruction. Since scattering characteristics of targets in different channels are effectively made use of in joint sparsity, the imaging results feature a combination of interferometric processing and sparsity optimization. In addition to acquisition of near-field high-resolution 3-D images with less observation echoes applied, it can also accurately reflect the position information of scattering points. Moreover, it can effectively solve the problem that the accuracy of target scattering azimuth is not high in different directions by adopting sub-aperture synthesis. As verified by tests, target 3-D views with higher quality can be obtained by applying the imaging approach as proposed in this paper, so as to provide reliable judgment basis for target identification and other applications. Since this paper adopts an OMP-based reconstruction approach, the calculation complexity is not high. Also, it requires more research on rapid InISAR near-field 3-D imaging approach in combination with the traditional near-field imaging approach for future study.

Author Contributions: Y.F. was responsible for all of the theoretical work, performed the simulation and analyzed experimental data; B.W., Z.S. and C.S. conceived and designed the experiments; S.W. and J.H. revised the paper.

Funding: This research was funded by [National Natural Science Foundation of China] grant number [61472324, 61771369] And The APC was funded by [61472324].

Acknowledgments: The authors would like to thank the journal manager, the handling editor, and the anonymous reviewers for their valuable and helpful comments.

Conflicts of Interest: The authors declare no conflicts of interest.

References

1. Yu, D.; Liu, W.L.; Zhang, Z.H. Near field scattering measurement based on ISAR imaging technique. In Proceedings of the IEEE International Symposium on Antennas, Propagation & EM Theory, Xi'an, China, 22–26 October 2012; pp. 725–728.
2. Kharkovsky, S.; Zoughi, R. Microwave and millimeter wave nondestructive testing and evaluation—Overview and recent advances. *IEEE Instrum. Meas. Mag.* **2007**, *10*, 26–38. [CrossRef]
3. Yang, X.H.; Zheng, Y.R.; Ghasr, T.; Donnell Kn, M. Microwave imaging from sparse measurements for near-field synthetic aperture radar. *IEEE Trans. Instrum. Meas.* **2017**, *66*, 2680–2692. [CrossRef]
4. Sheen, D.M.; McMarkin, D.L.; Hall, T.E. Near-field three-dimensional radar imaging techniques and applications. *Appl. Opt.* **2010**, *49*, E83–E93. [CrossRef] [PubMed]
5. Sheen, D.M.; McMarkin, D.L.; Hall, T.E. Three-dimensional millimeter-wave imaging for concealed weapon detection. *IEEE Trans. Microw. Theory Tech.* **2001**, *49*, 1581–1592. [CrossRef]
6. Sheen, D.M.; McMarkin, D.L.; Hall, T.E. Combined illumination cylindrical millimeter-wave imaging technique for concealed weapon detection. In *Passive Millimeter-Wave Imaging Technology IV, Proceedings of the AeroSense, Orlando, FL, USA, 24–28 April 2000*; SPIE: Washington, DC, USA, 2000; pp. 52–60.
7. Sheen, D.M.; McMarkin, D.L. Three-dimensional radar imaging techniques and systems for near-field applications. In Proceedings of the SPIE Defense + Security, Baltimore, MD, USA, 12 May 2016.
8. Fang, Y.; Wang, B.P.; Sun, C.; Song, Z.X.; Wang, S.Z. Near field 3-D imaging approach for joint high-resolutionimaging and phase error correction. *J. Syst. Eng. Electron.* **2017**, *28*, 199–211. [CrossRef]
9. Fallahpour, M.; Zoughi, R. Fast 3-D qualitative method for through-wall imaging and structural health monitoring. *IEEE Geosci. Remote Sens. Lett.* **2015**, *12*, 2463–2467. [CrossRef]
10. Jalilvand, M.; Li, X.Y.; Zwirello, L.; Zwick, T. Ultra wideband compact near-field imaging system for breast cancer detection. *IET Microw. Antennas Propag.* **2015**, *9*, 1009–1014. [CrossRef]
11. Fear, E.C.; Hagness, S.C.; Meaney, P.M.; Okoniewski, M.; Stuchly, M.A. Enhancing breast tumor detection with near-field imaging. *IEEE Microw. Mag.* **2002**, *3*, 48–56. [CrossRef]
12. Zhu, R.Q.; Zhou, J.X.; Jiang, G.; Fu, Q. Range migration algorithm for near-field MIMO-SAR imaging. *IEEE Geosci. Remote Sens. Lett.* **2017**, *14*, 2280–2284. [CrossRef]
13. Zhuge, X.D.; Yarovoy, A.G. Three-dimensional near-field MIMO array imaging using range migration techniques. *IEEE Trans. Image Process.* **2012**, *21*, 3026–3033. [CrossRef] [PubMed]

14. Fortuny, J. Efficient Algorithms for Three-Dimensional Near-Field Synthetic Aperture Radar Imaging. Ph.D. Thesis, University of Karslruhe, Karslruhe, Germany, 2001; pp. 15–43.

15. Demirci, S.; Cetinkaya, H.; Tekbas, M.; Yigit, E.; Ozdemir, C.; Vertiy, A. Back-projection algorithm for ISAR imaging of near-field concealed objects. In Proceedings of the 2011 XXXth URSI General Assembly and Scientific Symposium, Istanbul, Turkey, 13–20 August 2011; pp. 1–4.

16. Zhang, Y.; Deng, B.; Yang, Q.; Gao, J.K.; Qin, Y.L.; Wang, H.Q. Near-field three-dimensional planar millimeter-wave holographic imaging by using frequency scaling algorithm. *Sensors* **2017**, *17*, 2438. [CrossRef] [PubMed]

17. Li, X.; Bond, E.J.; van Veen, B.D. An overview of ultra-band microwave imaging via space-time beamforming for early-stage breast-cancer detection. *IEEE Antennas Propag. Mag.* **2005**, *47*, 19–34.

18. Kan, Y.Z.; Zhu, Y.F.; Tang, L.; Fu, Q.; Pei, H.C. FGG-NUFFT-based method for near-field 3-D imaging using millimeter waves. *Sensors* **2016**, *16*, 1–15. [CrossRef] [PubMed]

19. Li, S.Y.; Zhu, B.C.; Sun, H.J. NUFFT-Based Near-Field Imaging Technique for Far-Field Radar Cross Section Calculation. *IEEE Antennas Wirel. Propag. Lett.* **2010**, *9*, 550–553. [CrossRef]

20. Donoho, D.L. Compressed sensing. *IEEE Trans. Inf. Theory* **2006**, *52*, 1289–1306. [CrossRef]

21. Candes, E.J.; Romberg, J.; Tao, T. Robust uncertainty principles: Exact signal reconstruction from highly incomplete frequency information. *IEEE Trans. Inf. Theory* **2006**, *52*, 489–509. [CrossRef]

22. Li, S.Y.; Zhao, G.Q.; Li, H.M.; Ren, B.L. Near-field radar imaging via compressive sensing. *IEEE Trans. Antenna Propag.* **2015**, *63*, 828–833. [CrossRef]

23. Bi, D.J.; Xie, Y.L.; Ma, L.; Li, X.F.; Yang, X.H.; Zheng, Y.R. Multifrequency compressed sensing for 2-D near-field synthetic aperture radar image reconstruction. *IEEE Trans. Instrum. Meas.* **2017**, *66*, 777–791. [CrossRef]

24. Liu, Y.B.; Li, N.; Wang, R.; Deng, Y.K. Achieving high-quality three-dimensional InISAR imaging of maneuvering via super-resolution ISAR imaging by exploiting sparseness. *IEEE Geosci. Remote Sens. Lett.* **2014**, *11*, 828–832.

25. Zhang, Q.; Yeo, T.S.; Du, G.; Zhang, S.H. Estimation of three-dimensional motion parameters in interferometric ISAR imaging. *IEEE Trans. Geosci. Remote Sens.* **2004**, *42*, 292–300. [CrossRef]

26. Andon, D.L.; Chavdar, N.M. ISAR imaging reconstruction technique with stepped frequency modulation and multiple receivers. In Proceedings of the 24th Digital Avionics Systems Conference (DASC'05), Washington, DC, USA, 30 October–3 November 2005. 14E2-11.

27. Morabito, A.F.; Palmeri, R.; Isernia, T. A Compressive-Sensing-inspired procedure for array antenna diagnostics by a small number of phaseless measurements. *IEEE Trans. Antennas Propag.* **2016**, *64*, 3260–3265. [CrossRef]

28. Tropp, J.A.; Gilbert, T.C. Signal recovery from random measurements via orthogonal matching pursuit. *IEEE Trans. Inf. Theory* **2007**, *53*, 4655–4666. [CrossRef]

29. Antoin, B.; Josep, P.; Luis, J.; Angel, C. Spherical wave near-field imaging and radar cross-section measurement. *IEEE Trans. Antennas Propag.* **1998**, *46*, 730–735.

Article

Performance Analysis of Ionospheric Scintillation Effect on P-Band Sliding Spotlight SAR System

Lei Yu, Yongsheng Zhang *, Qilei Zhang, Yifei Ji and Zhen Dong

College of Electronic Science and Technology, National University of Defense Technology,
Changsha 410073, China; yulei17@nudt.edu.cn (L.Y.); zhangqilei@nudt.edu.cn (Q.Z.); jyfnudt@163.com (Y.J.);
dongzhen@nudt.edu.cn (Z.D.)
* Correspondence: zyscn@163.com; Tel.: +86-135-7483-8648

Received: 24 March 2019; Accepted:7 May 2019; Published: 9 May 2019

Abstract: The space-borne P-band synthetic aperture radar (SAR) maintains excellent penetration capability. However, the low carrier frequency restricts its imaging resolution. The sliding spotlight mode provides an operational solution to meet the requirement of high imaging resolution in P-band SAR design. Unfortunately, the space-borne P-band SAR will be inevitably deteriorated by the ionospheric scintillation. Compared with the stripmap mode, the sliding spotlight SAR will suffer more degradation when operating in the scintillation active regions due to its long integration time and complex imaging geometry. In this paper, both the imaging performance and scintillation effect for P-band sliding spotlight mode are studied. The theoretical analysis of scintillation effect is performed based on a refined model of the two-frequency and two-position coherence function (TFTPCF). A novel scintillation simulator based on the reverse back-projection (ReBP) algorithm is proposed to generate the SAR raw data for sliding spotlight mode. The proposed scintillation simulator can also be applied to predict the scintillation effect for other multi-mode SAR systems such as terrain observation by progressive scans (TOPS) and ScanSAR. Finally, a group of simulations are carried out to validate the theoretical analysis.

Keywords: ionosphere; P-band; reverse back-projection (ReBP); synthetic aperture radar (SAR); sliding spotlight; scintillation

1. Introduction

It is widely known that synthetic aperture radar (SAR) system working at P-band shows its superiority in penetrating the forest foliage and the ground surface, which will have an extensive application prospect in biomass measurement and geological observation [1–3]. Therefore, there has been an upward trend of developing P-band SAR, for example the BIOMASS mission [4,5]. Despite the remarkable advantages, there are two main drawbacks existing in the space-borne P-band SAR systems. One is the severe susceptivity of the ionospheric impact [6–9], especially for the equatorial scintillation effect. The other is the limitation of azimuth resolution which is restricted by the low central frequency.

To elevate the azimuth resolution and maintain adequate imaging swath, the sliding spotlight mode has been used in many SAR systems such as TerraSAR-X and PAMIR [10–13]. The sliding spotlight mode controls the scanning velocity of beam footprint by steering the antenna, thus obtaining longer integration time than stripmap mode and larger scene than spotlight mode. Consequently, the sliding spotlight mode is a practical way for P-band high-resolution SAR system. However, little literature has been proposed to evaluate the ionospheric effect for P-band sliding spotlight SAR system.

The intensive solar radiation results in the ionization of ionospheric molecule. Varying from the scale of spatial distribution, the ionosphere is typically categorized into the background ionosphere (larger than 10 km) and ionospheric irregularities (less than 10 km) [8,14]. The background ionospheric effect can be mitigated using the split-spectrum method or using the ionospheric prior knowledge acquired from the global navigation satellite system (GNSS)/BeiDou system [15–18]. In recent papers, the multi-squint (MS) interferometry methodology is proposed [5], which provides a new ionospheric mitigation approach for SAR system with limited bandwidth. The scintillation effect is caused by small scale ionospheric turbulent irregularities, which typically occurs after the sunset in the equatorial and polar regions [14]. The strong scintillation effect, which usually shown as streaks in SAR images, has been extensively reported by the Advanced Land Observing Satellite (ALOS)/the Phased Array-type L-band Synthetic Aperture Radar (PALSAR) system. The scintillation turbulence on both amplitude and phase will introduce serious spatial and frequency decorrelation within the SAR integration time and further distort the imaging performance. The former research usually focuses on the analysis of the scintillation effect for stripmap SAR systems [19–26]. The generalized ambiguity function (GAF) proposed by Ishimaru [19] in 1999, provides a comprehensive model to evaluate the degradation of signal coherence introduced by the ionospheric effect. Based on the GAF model, Li et al. [20] introduced the two-frequency and two-position coherence function (TFTPCF) into the traditional GAF model to evaluate the scintillation-induced signal decorrelation. The analysis of anisotropic irregularity is performed by C. Wang [23] and a statistical evaluation of L-band equatorial scintillation is carried out by Meyer [24]. The SAR scintillation simulator (SAR-SS) is studied by Carrano [27] based on the phase screen theory for predicting the scintillation effect on the L-band SAR. In Carrano's work, the inverse range-Doppler algorithm (RDA) is applied to generate the unaffected SAR signal. However, it cannot actually simulate the observation geometry of SAR system, thus it is not suitable to reconstruct the SAR raw data for the sliding spotlight SAR system.

The former research builds the foundation of our work. However, these achievements only take the stripmap mode into consideration and further research still needs to be accomplished by considering the sliding spotlight observation geometry. Compared with the stripmap mode, the P-band sliding spotlight SAR system has an ultra-long integration time and more complicated observation geometry which means a longer exposure time and a longer ionospheric penetration length (IPL) in scintillation active regions. Due to the beam scanning, the incident angle of beam center, which is an important parameter in scintillation simulations, varies within the acquisition time. All these characteristics will make the scintillation effect on sliding spotlight mode show different patterns.

In this paper, we firstly introduce the observation geometry of sliding spotlight SAR system in Section 2. Then, in Section 3, the theoretical analysis of scintillation effect is performed based on the refined GAF and TFTPCF model. The comparisons between sliding spotlight mode and stripmap mode are presented. In Section 4, a novel SAR-SS is proposed by considering the beam scanning of sliding spotlight mode. The reverse back-projection (ReBP) algorithm [28] is applied to generate the sliding spotlight mode SAR raw data. Finally, the simulations are performed on both point target and extended target to demonstrate the scintillation-induced imaging distortion. A group of 500-time Monte-Carlo simulations are carried out to validate the theoretical analysis.

2. The Observation Geometry of Sliding Spotlight SAR System

The sliding spotlight mode SAR can make a good balance between the azimuth resolution and imaging scene by controlling the velocity of beam footprint [10,11]. The observation geometry of space-borne sliding spotlight SAR is shown in Figure 1. The beam center points at the steering point position O' within the entire acquisition time. Technically, the turbulent ionosphere can be considered to be a very thin phase screen at an equivalent altitude and the radar beam scans over the phase screen within the integration time. H_{sat} is the orbit height of radar platform and H_{iono} is the equivalent ionospheric height at 350 km. I_P represents the ionospheric penetration point (IPP) and X_{iono} is the IPL within acquisition time. V_g and V_{sat} represent the ground velocity and the platform velocity, respectively. θ_{i0} is the ionospheric incident angle of beam center in zeros Doppler plane. Due to the steering of antenna, the beam central incident angle θ_i varies within the acquisition time and is determined by the instant squint angle $\theta_{sq}(\eta)$. The relationship is given as follow

$$\theta_i = \arccos\left(\cos\left(\theta_{i0}\right) \cdot \cos\left(\theta_{sq}(\eta)\right)\right) \tag{1}$$

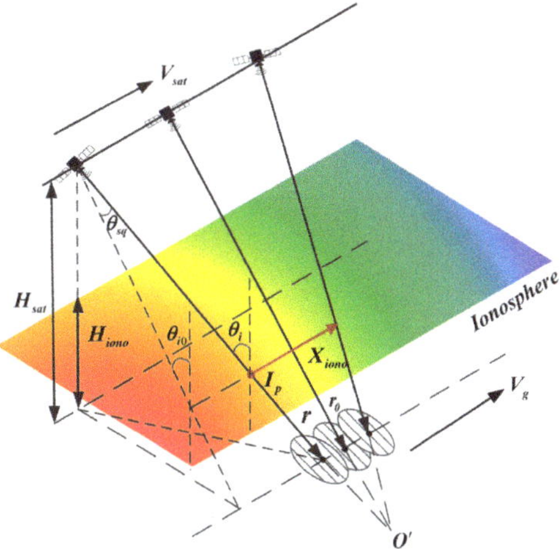

Figure 1. The observation geometry of sliding spotlight SAR with ionosphere.

We define that R_c is the closest range between the radar platform and the scene center and R_{rot} is the closest range between the radar platform and the steering point. Based on the imaging geometry, the relationship between R_c and R_{rot} is expressed as

$$\frac{R_{rot} - R_c}{R_{rot}} = \frac{V_r - |k_\omega| R_c}{V_r} \tag{2}$$

where V_r denotes the effective radar velocity and k_ω is the angular velocity of antenna steering. In particular, when $R_{rot} = +\infty$, the system is equivalent to the stripmap mode and when $R_{rot} = R_c$ the system is equal to the spotlight mode. The beam scanning prolongs the integration time of sliding spotlight mode which can be calculated as $T_a = 2R_c \tan\left(\lambda/2D\right)/V_g$, where λ is the wavelength, D is the size of azimuth antenna and V_g is the ground velocity which is given as follow

$$V_g = V_r - \frac{H_{sat}}{\cos\theta_i'} \cdot |k_\omega| \cdot \sec^2\left(\theta_{sq}(\eta)\right) \tag{3}$$

where θ_i' is the ground incident angle and $\theta_{sq}(\eta) = \theta_{sq0} + k_\omega \eta$ is the instant squint angle. It can be seen that k_ω is the key factor which determines the ground velocity and the integration time. The integration time and theoretical azimuth resolution as a function of k_ω are shown in Figure 2. It is obvious that both the integration time and azimuth resolution significantly increase with k_ω. When $|k_\omega| = 0.007$ rad/s, the integration time reaches 86.49 s compared to 10.67 s for stripmap ($|k_\omega| = 0$ rad/s), meanwhile the theoretical resolution will increase to less than 0.75 m.

As a consequence of beam scanning, the ionospheric incident angle of beam center varies within the acquisition time, whereas the traditional scintillation simulator cannot accurately simulate the beam scanning of sliding spotlight mode. Thus, in this paper, a novel scintillation simulator is proposed to accommodate the sliding spotlight geometry and exactly reconstruct the SAR raw data of sliding spotlight mode. Furthermore, a refined TFTPCF model is applied to perform the theoretical analysis for sliding spotlight mode.

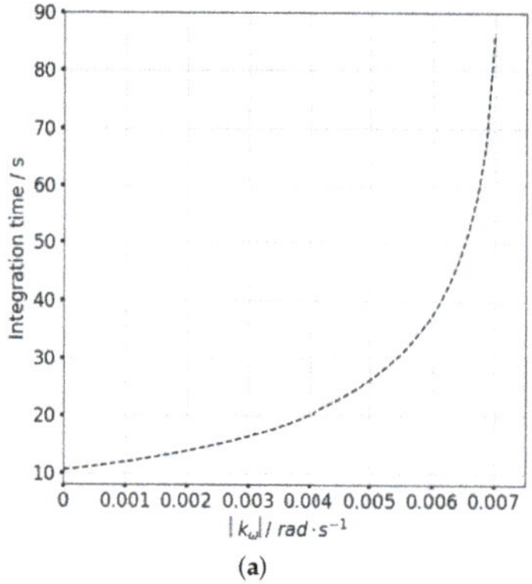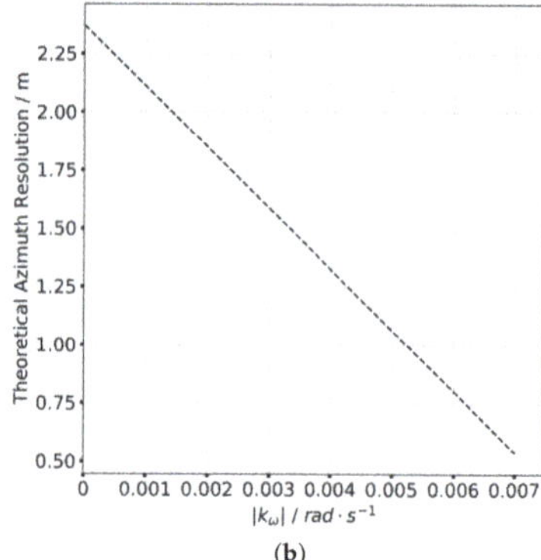

Figure 2. The integration time and azimuth resolution of P-band sliding spotlight mode as function of k_ω. (**a**) Integration time. (**b**) Theoretical azimuth resolution.

3. Theoretical Analysis Based on TFTPCF Model

The GAF model which is first proposed by Ishimaru [19], provides a succinct model to analyze the ionospheric effect. As is shown in Figure 1, the GAF can be expressed as the coherent accumulation of the SAR signal received from the target at position **r** and the reference signal focus at the position $\mathbf{r}_0$, which is expressed as

$$\chi\left(\mathbf{r},\mathbf{r}_0\right) = \sum_n 2\pi \int g_n\left(\omega, r_n\right) \cdot f_n^*\left(\omega, r_{0n}\right) d\omega \tag{4}$$

$$\begin{cases} f_n\left(\omega, r_{0n}\right) = u_i\left(\omega\right) \exp\left(j\dfrac{\omega}{c} 2 r_{0n}\right) \\[2mm] g_n\left(\omega, r_n\right) = u_i\left(\omega\right) \dfrac{\exp\left(j2 \int \beta\left(\omega\right) dl + 2j\phi\left(\omega, \boldsymbol{\rho}_n\right)\right)}{\left(4\pi r_n\right)^2} \end{cases} \tag{5}$$

where $f_n\left(\omega, r_{0n}\right)$ and $g_n\left(\omega, r_n\right)$ represents the reference signal and received signal at n_{th} sampling point, $\psi_{iono} = 2 \int \beta\left(\omega\right) dl$ is the dispersive phase introduced by background ionosphere and $\phi\left(\omega, \boldsymbol{\rho}_n\right)$ is the scintillation phase corresponding to the IPP position $\boldsymbol{\rho}_n$ on the scintillation phase screen and the signal frequency ω.

Based on the stop-go assumption, the SAR signal penetrates the ionospheric phase screen twice at one sampling point with different instant frequency (the SAR transmit signal is linear modulated). Thus, the random phase induced by ionospheric irregularities can be properly analyzed by the two-position two-frequency function. Based on the phase screen theory, Li et al. [20] proposed the proper TFTPCF model to study the scintillation effect from the second moment of the GAF, which is expressed as

$$\left\langle\left|\chi\left(\mathbf{r},\mathbf{r}_0\right)\right|^2\right\rangle = (2\pi)^2 \sum_m \sum_n \int_{+\infty}^{-\infty} \int_{+\infty}^{-\infty} \Gamma_{1,1} \cdot \exp\left(j2\left(\int_{r_m} k\left(\omega_1\right)dl \int_{r_n} k\left(\omega_2\right)dl\right)\right) \cdot$$
$$\exp\left(-j2\left(\frac{\omega_1 r_{0m} - \omega_2 r_{0n}}{c}\right)\right) d\omega_1 d\omega_2 \tag{6}$$

$$\Gamma_{1,1} = \left\langle\exp\left\{j2\left[\phi_1\left(\omega_1, \boldsymbol{\rho}_n\right) - \phi_2\left(\omega_2, \boldsymbol{\rho}_m\right)\right]\right\}\right\rangle \tag{7}$$

where ρ_m and ρ_n represent the IPP position at different azimuth position and $\langle\cdot\rangle$ is the mathematical expectation. $\Gamma_{1,1}$ is the TFTPCF which serves as a window function in the accumulation process of Equation (6). The signal decorrelation will be introduced when $\Gamma_{1,1} < 0.707$ (-3 dB threshold). Based on the law of large number, the scintillation phase error tends to follow the Gaussian distribution. Thus, the Gaussian approximation can be used to simplify the TFTPCF, which is expressed as follows.

$$
\begin{aligned}
\Gamma_{1,1}\left(\omega_1,\omega_2;\boldsymbol{\rho}_n-\boldsymbol{\rho}_m\right) &= \left\langle \exp\left\{j2\left[\phi_1\left(\omega_1,\boldsymbol{\rho}_n\right)-\phi_2\left(\omega_2,\boldsymbol{\rho}_m\right)\right]\right\}\right\rangle \\
&\approx \exp\left(-2\cdot\left\langle\left[\phi_1\left(\omega_1,\boldsymbol{\rho}_n\right)-\phi_2\left(\omega_2,\boldsymbol{\rho}_m\right)\right]^2\right\rangle\right) = R_\phi\left(\omega_1,\omega_1;0\right)+R_\phi\left(\omega_2,\omega_2;0\right)-2R_\phi\left(\omega_1.\omega_2;\Delta x\right)
\end{aligned}
\tag{8}
$$

where $R_\phi\left(\omega_n,\omega_m;\Delta x\right)$ is the auto-correlation function (ACF) of the scintillation phase which is determined by the frequency separation (ω_1 and ω_2) and the 1-D IPP spatial separation ($\Delta x = \|\boldsymbol{\rho}_m-\boldsymbol{\rho}_n\|$). The ACF in Equation (8) is derived from the inverse Fourier transform of the power spectral density (PSD) function of the scintillation phase. In this paper, the Rino power law spectrum [29] is applied to simulate the scintillation phase screen and the ACF based on Rino's spectrum can be expressed as

$$
R_\phi\left(\omega_1,\omega_2;\Delta x\right) = r_e^2\lambda_1\lambda_2 C_s L\sec\theta_n\sec\theta_m\cos\theta_i\cdot G\left|\frac{\Delta x}{2q_0}\right|^{v-1/2}\frac{K_{v-1/2}\left(q_0\Delta x\right)}{2\pi\cdot\Gamma_0\left(v+1/2\right)}
\tag{9}
$$

$$
C_s L = C_k L\left(\frac{2\pi}{1000}\right)^{p+1}
\tag{10}
$$

where G is the gain factor, $K_\varepsilon\left(\cdot\right)$ is the modified Bessel function and $\Gamma_0\left(\cdot\right)$ is the gamma function, both $C_s L$ and $C_k L$ are the symbols of scintillation strength and $p = 2v$ is the phase spectral index, $q_0 = 2\pi/L_0$ is the wavenumber corresponding to the outer scales, θ_n and θ_m are the ionospheric incident angle of beam center at different sampling points.

In previous work, the variation of beam central incident angle is never considered due to the observation geometry of the stripmap mode. However, for sliding spotlight mode, the beam scanning leads to the increase of incident angle which will prolong the signal propagation path in irregularity layer and further aggravates the signal decorrelation. The instant ionospheric incident angle is applied as a modification of ACF for sliding spotlight mode which is presented as

$$
R_\phi\left(\omega_1,\omega_2;\Delta x\right) = r_e^2\lambda_1\lambda_2 C_s L\sec\theta_n\sec\theta_{sq}\left(\eta_m\right)\cdot G\left|\frac{\Delta x}{2q_L}\right|^{v-1/2}\frac{K_{v-1/2}\left(q_0\Delta x\right)}{2\pi\cdot\Gamma_0\left(v+1/2\right)}
\tag{11}
$$

where $\theta_{sq}\left(\eta_m\right)$ is the squint angle at the m_{th} sampling point. The influence of different scintillation parameters is analyzed from the TFTPCF curves. Based on the refined ACF in (11), the TFTPCF curves with different scintillation parameters are given in Figures 3–5, and Table 1 presents the default value of irregularity parameters.

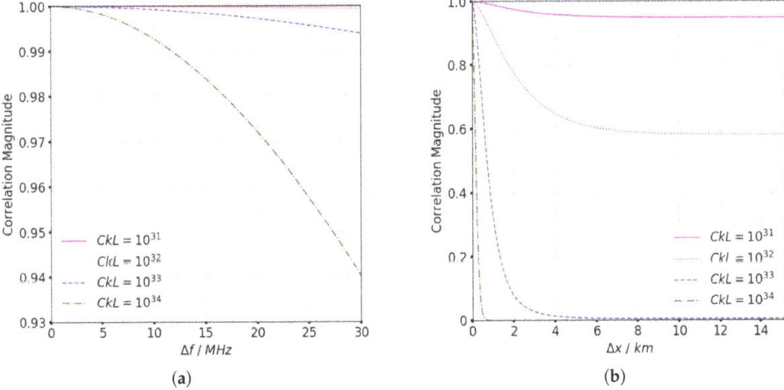

Figure 3. TFTPCF curves with different $C_k L$. (**a**) TFTPCF versus frequency separation. (**b**) TFTPCF versus spatial separation.

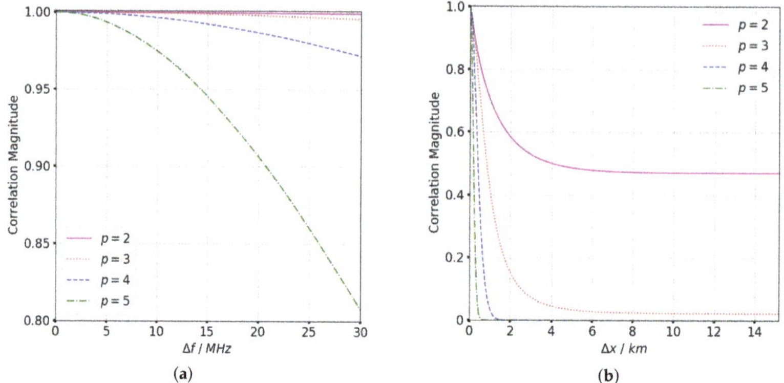

Figure 4. TFTPCF curves with different p. (**a**) TFTPCF versus frequency separation. (**b**) TFTPCF versus spatial separation.

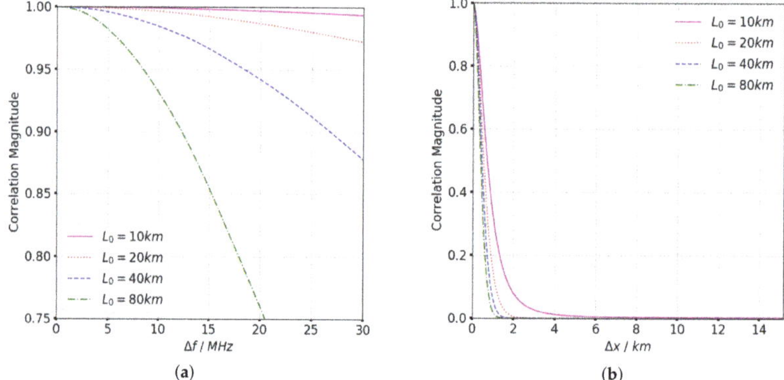

Figure 5. TFTPCF with different L_0. (**a**) TFTPCF versus frequency separation. (**b**) TFTPCF versus spatial separation.

Table 1. The default value of ionospheric irregularity parameters

Parameters	Symbol	Value
Scintillation strength	$C_k L$	10^{33}
Spectral index	p	3
Outer scale	L_0	10 km
Irregularity structure scale	a/b	10/1

According to Figure 3, the frequency correlation shows a significant declination with frequency separation when $C_k L > 10^{34}$. The signal decorrelation becomes more serious with the increase of spectral index as is shown in Figure 4. In Figure 5 it is clear that the signal frequency coherence decays dramatically for $L_0 \geq 40$ km, whereas the spatial coherence shows little difference with the increase of outer scales. For a general comparison, the signal spatial coherence is more sensitive to the scintillation strength and spectral index than outer scales. Furthermore, the signal correlation declines more significant with the increasing of spatial separation which means the spatial variation of beam central incident angle is considerable. The comparison of TFTPCF curves between the stripmap mode and sliding spotlight mode with different $C_k L$ is shown in Figure 6. The red curves in Figure 6 represent the TFTPCF value of stripmap mode ($k_\omega = 0$ rad/s) and the blue curves represent the modified TFTPCF of sliding spotlight mode. It is obvious that the sliding spotlight mode is more susceptible to the scintillation effect, which also consists with the aforementioned analysis. Furthermore, the ionospheric coherent length is applied to analyze the decorrelation of P-band sliding spotlight mode, which is defined as the spatial separation Δx when $\Gamma\left(\omega_0; \Delta x\right) \leq 0.707$. The imaging degeneration need to be considered when the IPL is longer than the ionospheric coherent length. The coherent length with different spectral index is illustrated in Figure 7, and the scintillation strength is set as $C_k L = 10^{32}$ refers to the mildly scintillation condition.

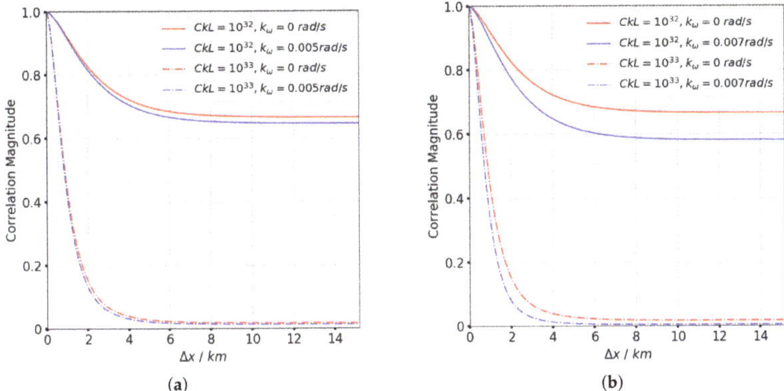

Figure 6. The comparison of TFTPCF curves between stripmap mode (red curves) and sliding spotlight mode (blue curves) with different $C_k L$ (real line: $C_k L = 10^{32}$, dashed line: $C_k L = 10^{33}$). (a) $k_\omega = -0.005$ rad/s (b) $k_\omega = -0.007$ rad/s.

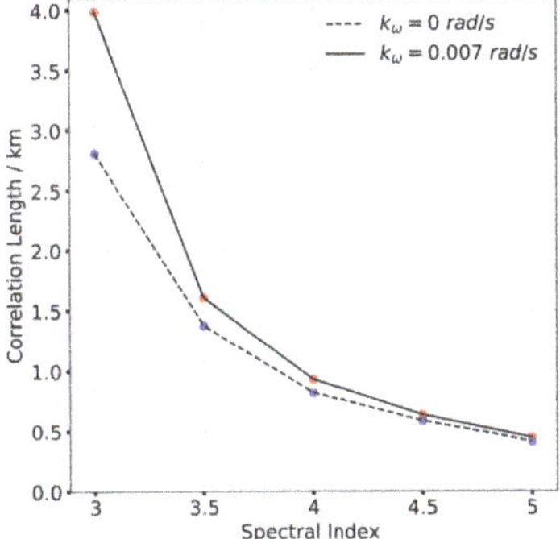

Figure 7. Coherent length with different spectral index.

According to Figure 7, the ionospheric correlation length dramatically declines with the increase of spectral index. The correlation length of sliding spotlight mode is less than stripmap mode, which means the sliding spotlight mode is more sensitive to the scintillation effect. The IPL of the space-borne P-band sliding spotlight SAR system is 142.03 km ($IPL = V_{sat} T_a$, V_{sat} is the platform velocity obtained from the orbit roots). Since the IPL is significantly longer than the ionospheric coherent length, the P-band sliding spotlight mode will definitely be influenced by ionospheric irregularities. In Li's work, the SAR resolution is defined as the absolute range separation $\delta_r = \|\mathbf{r} - \mathbf{r}_0\|$ by using the criterion of ambiguity function, which is expressed as

$$\left\langle |\chi\left(\mathbf{r}, \mathbf{r}_0\right)|^2 \right\rangle \Big/ \left\langle |\chi\left(\mathbf{r}_0, \mathbf{r}_0\right)|^2 \right\rangle = \exp\left(-2\right) \tag{12}$$

However, the redefined SAR resolution in GAF model dose not conform to the general concept of the SAR resolution based on the -3 dB criterion. Thus, the simulation of real scene is required to evaluate the scintillation effect for sliding spotlight mode quantitatively.

4. The ReBP-Based Scintillation Simulator for Sliding Spotlight Mode

4.1. Basic of Scintillation Simulator

The scintillation simulator proposed by Carrano [27] is based on the phase screen theory and has been widely acknowledged. The phase screen theory assumes that the turbulent irregularities are constrained within a very thin layer. Therefore, the ray-bending and multi-scattering effect can be neglected within the layer.

The complete SAR–SS consists of two essential steps: the phase screen simulator and propagation simulator which corresponds to the wave propagation history. When the radio wave penetrates through the ionosphere, the scintillation phase is introduced into the signal by the phase screen simulator. The 2-D scintillation phase screen is generated by multiplying the irregularity's phase spectrum by complex white noise with unit power. After that when the radio wave transmits into the free space from the ionosphere down to the ground, the diffraction effect is simulated by the propagation simulator. It is calculated by solving the parabolic wave equation (PWE). Finally, the ionospheric transfer function (ITF) is obtained by incorporating the phase screen and propagation simulator. In this paper, a novel ReBP–based SAR–SS is proposed based on the observation geometry of sliding spotlight mode.

4.2. The Modified Propagation Simulator for Sliding Spotlight Mode

The propagation of transionospheric radio waves from the free space down to the ground follows the scalar Helmholtz equation which is expressed as

$$\nabla^2 E\left(\boldsymbol{\rho}\right) + k_0^2 \left[1 + \Delta\varepsilon_r\right] E\left(\boldsymbol{\rho}\right) = 0 \tag{13}$$

$$E\left(\boldsymbol{\rho}\right) = U\left(\boldsymbol{\rho}\right) \cdot e^{j\mathbf{k}\cdot\boldsymbol{\rho}} \tag{14}$$

where $k_0 = 2\pi/\lambda$ is the wavenumber corresponds to the signal frequency, $\Delta\varepsilon_r$ is the fluctuation term of the dielectric permittivity mainly induced by the dispersive background ionosphere, $E\left(\boldsymbol{\rho}\right)$ is the electronic field, $U\left(\boldsymbol{\rho}\right)$ is the complex amplitude and $\boldsymbol{\rho} = \left(x, y, z\right)$ is the space vector of the electromagnetic waves defined in the geomagnetic coordinate as is shown in Figure 8. The coordinate center is chosen at the IPP position and the x-axis, y-axis, and z-axis are defined as the magnetic north, magnetic east, and vertical down to the earth. θ and φ are the ionospheric incident angle and magnetic heading of radar beam center. To make an explicit description, we neglect the inclined angle between the magnetic heading of radar platform and the magnetic east. Thus, φ is considered to be the squint angle of beam center. In Figure 8, the squint angle rotates with a constant angular velocity in acquisition time. $\mathbf{k} = k_0 \left(\sin\theta\cos\varphi, \sin\theta\sin\varphi, \cos\theta\right)$ is the transmit vector of the radio waves as well as $\mathbf{k}_\perp = k_0 \left(\cos\varphi, \sin\varphi\right)$ is the projection of the transmit vector in horizontal plane.

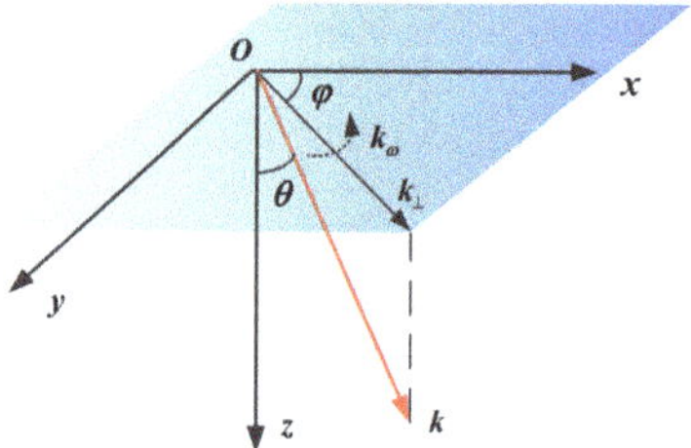

Figure 8. The propagation coordinate system of sliding spotlight SAR signal.

By substituting the equations above and considering the Fresnel assumption (which means $\partial^2 U/\partial z^2 \approx 0$), the PWE in geomagnetic coordinate is expressed as

$$\nabla_\perp^2 U = -k_0^2 \Delta \varepsilon_r U + 2j \frac{\partial U}{\partial x} k_0 \sin \theta \cos \varphi + 2j \frac{\partial U}{\partial y} k_0 \sin \theta \sin \varphi + 2j \frac{\partial U}{\partial z} k_0 \cos \theta \tag{15}$$

Please note that the dispersion induced by background ionosphere is not considered in the simulator, so the fluctuating part $\Delta \varepsilon_r$ in Equation (15) is neglected in the following derivations. Since the ionosphere irregularities are considered to be a very thin phase screen, the diffraction effect is neglected within the irregularity layer and the propagation path is considered to be a straight line for SAR signals. Based on the aforementioned assumptions, the complex amplitude of transmitted waves which has penetrated the phase screen is expressed as $U\left(\boldsymbol{\rho}_\perp, 0^+\right) = U\left(\boldsymbol{\rho}_\perp, 0\right) \cdot e^{j\phi(\boldsymbol{\rho}_\perp)}$, where $\boldsymbol{\rho}_\perp$ is the distance vector in the x-y plane and the $\phi\left(\boldsymbol{\rho}_\perp\right)$ is the scintillation phase corresponds to the penetration point on phase screen. The Fourier split-step method is used to the PWE to solve the second-order derivative terms in Equation (16). Then, we derive the complex amplitude for the SAR signal as follow

$$U\left(\boldsymbol{\rho}_\perp, z\right) = U\left(\boldsymbol{\rho}_\perp, 0\right) \cdot T\left(\boldsymbol{\rho}_\perp\right) \tag{16}$$

$$T\left(\boldsymbol{\rho}_\perp\right) = F^{-1} \left\{ \exp \left[j \left(\frac{\kappa^2 \cdot z}{2k_0} \sec \theta \right) \right] \cdot F \left\{ e^{j\phi(\boldsymbol{\rho}_\perp)} \right\} \right\} \tag{17}$$

where $\kappa = \left(\kappa_x, \kappa_y\right)$ is the transverse wavenumber. The spherical wave propagation is considered in the simulator by scaling the horizontal coordinate and the propagation distance with the factor $z = z_1 z_2/(z_1 + z_2)$, where z_1 is the distance between the radar platform and the ionospheric height, z_2 is the ionospheric height, $\sec \theta$ is applied to convert the vertical distance to the oblique distance. $T\left(\boldsymbol{\rho}_\perp\right)$ is the ITF which includes both the phase and amplitude fluctuations. The upward and downward ITF are the same since the symmetric propagation history. Therefore, the two-way ITF is calculated by squaring $T\left(\boldsymbol{\rho}_\perp\right)$.

For sliding spotlight SAR system, the azimuthal temporal variation of beam central incident and squint angle are considered to be a modification into the original model. Here we use the penetration point at the edge of the phase screen as a reference, then the squint angle φ and incident angle θ are expressed as

$$\begin{cases} \varphi(m) = \theta_{sq0} + k_\omega \dfrac{\Delta x_a \cdot m}{V_{IPP}} \\[2ex] \theta(m) = \arccos \left\{ \cos\left(\theta_{i0}\right) \cdot \cos\left(\theta_{sq0} + k_\omega \dfrac{\Delta x_a \cdot m}{V_{IPP}}\right) \right\} \end{cases} \tag{18}$$

where $\Delta x_a = V_{IPP}/PRF$ is the sampling distance of ionospheric phase screen at the azimuth direction and the $V_{IPP} = R_e \cdot V_g/(R_e + H_{iono}) \cdot \sin\left(\theta_i\right)$ is the velocity of the IPP, where R_e is the radius of Earth.

4.3. The Modified Phase Screen Simulator for Sliding Spotlight Mode

The 2-D scintillation phase screen is typically generated by applying the Gaussian noise with unit power passes through a linear filter with a specified PSD. Some research has been accomplished to study the ionospheric spectrum including the Shkarofsky spectrum, the modified Kolmogorov spectrum, and Rino power law spectrum. The Rino's spectrum has been proved by real measured data and widely used in global ionospheric scintillation model (GISM) and wide band model (WBMOD) [24,27]. The PSD function of Rino spectrum is expressed as

$$P_\phi(\kappa) = \frac{r_e^2 \lambda^2 \sec^2\left(\theta\left(\kappa\right)\right) \cdot C_s L \cdot a \cdot b}{\left[q_0 + \left(A \kappa_x^2 + B \kappa_x \kappa_y + C \kappa_y^2 \right) \right]^{(p+1)/2}} \tag{19}$$

where r_e is the classical electron radius, λ is the signal wavelength. Both a and b are structural scaling factors of irregularities along and across the magnetic field. A, B and C are the coefficients determined by the transmit direction and geomagnetic field whose expression has been discussed in Carrano's work [27]. $\theta(\kappa)$ is the incident angle of beam center correlates to the spatial wavenumber. In our work, the spatial variant incident angle is applied as a modification for the original Rino spectrum and the scintillation phase screen is then derived based on the modified phase spectrum in Equation (19).

4.4. The Structure of ReBP-Based Scintillation Simulator

Due to the shortage of space-borne P-band SAR data, the scintillation-contaminated SAR echo is required to be reconstructed from the SAR images. However, the existing method such as the inverse RDA cannot exactly accommodate the sliding spotlight observation geometry. The ReBP algorithm [28] provides an efficient and flexible method to simulate the SAR raw data for arbitrary imaging geometry which has been validated by the real data of Sential-1 mission. The ReBP algorithm takes the advantages of the accuracy and the expandability for analyzing the atmospheric propagation. Furthermore, the parallelization can be used in ReBP process to improve the computational efficiency. In this paper, the modified two-steps scintillation simulator is merged into the ReBP process to exactly accommodate the observation geometry of sliding spotlight mode and derive the SAR raw echo. The block diagram of the SAR-SS proposed in this paper is shown in Figure 9. According to Figure 9, the single look complex (SLC) image is given as an input and the outer loop runs for each image range line. After the up-sampling process, a projection of the azimuth beam is used to limit the illumination time of each target in the scene (shown as the SAR image pixels). In this procedure the beam scanning is considered for sliding spotlight mode. Then the interpolation is performed for the whole range line followed by the remodulation process where the ITF is introduced into the range-compressed raw data and finally after the range decompression the scintillation-contaminated raw data is acquired. By adjusting the beam projection procedure, the ReBP–based SAR–SS can also be applied to simulate the scintillation effect for TOPS and ScanSAR modes, the modifications of incident and squint angle follow the discussions in this section.

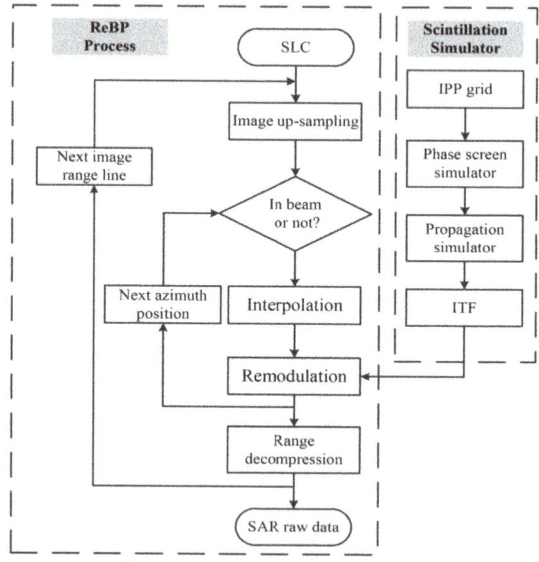

Figure 9. The block diagram of the ReBP–based SAR–SS.

5. Simulation

In this section, the point target and extended target simulation are performed to present the scintillation effect on P-band sliding spotlight system, and a group of 500-time Monte-Carlo simulations are carried out to validate the theoretical analysis. The typical P-band LEO SAR system parameters are applied to carry out the simulation. The radar system and orbit parameters are shown in Table 2. The contrast simulation on stripmap mode SAR system also follows the parameters in Table 2. The simulations are performed by using the ReBP–based SAR–SS which is shown in Figure 9 and the detailed process is described as follow: The SLC image is used as the input and the image scene is defined in the earth-centered earth-fixed (ECEF) coordinate. Then the IPP grids are calculated by the positions of radar platform and scene targets and the corresponding ITFs are derived from the SAR–SS. The ReBP algorithm is used to generate the scintillation-contaminated SAR echo. Finally, the BP algorithm is used to reconstruct the SAR image from the raw data. Besides the imaging resolution, the peak power loss, peak to sidelobe ratio (PSLR) and integrated sidelobe ratio (ISLR) are considered to evaluate the imaging performance.

Table 2. Radar System and Orbit Parameters.

Parameters	Value	Unit
Carrier frequency	0.6	GHz
Bandwidth	60	MHz
Altitude of radar	700	km
Scanning angular velocity(k_ω)	−0.0055	rad/s
Semi-major Axis	7071	km
Inclination	98.6	deg
The Argument of Latitude	40	deg

5.1. Point Target Simulation

The simulation on point targets are shown in Figures 10 and 11. A 6 km × 6 km point array is applied to carry out the simulation. The origin point array is shown in Figure 10. The contour map, range and azimuth slices are shown in Figure 11. The central point target in red square is selected to perform a detail analysis. The ideal imaging result in Figure 11a has sub-meter level azimuth resolution with 0.713 m in azimuth and 2.26 m in range by using the default system parameters. The ideal image demonstrates the excellent performance of sliding spotlight mode in high-resolution SAR imaging. The scintillation strength in Figure 11b,c are $C_k L = 10^{33}$ and $C_k L = 10^{34}$ which refers to the moderate and strong strength of scintillation. Other ionospheric parameters are shown in Table 1 as the default value. Compared with the ideal imaging result in Figure 11a, the resolution in azimuth degenerates from 0.713 m to 4.894 m and 7.625 m in the case of $C_k L = 10^{33}$ and $C_k L = 10^{34}$, respectively. The more significant deteriorations are shown as the degeneration of PSLR and ISLR. According to Figure 11b, both the PSLR and ISLR decay to −3.19 dB and −3.26 dB and in Figure 11c the PSLR and ISLR drop to −1.45 dB and −0.17 dB, respectively. The extremely high PSLR and ISLR indicate that the scintillation effect will lead to serious expand of the azimuth mainlobe and further not only degenerate the azimuth resolution but also induce the peak loss. Compared with the azimuth imaging result, the distortion in range is not as serious as that in azimuth. The asymmetric sidelobe can be seen in Figure 11b mainly due to the power leakage of azimuth mainlobe.

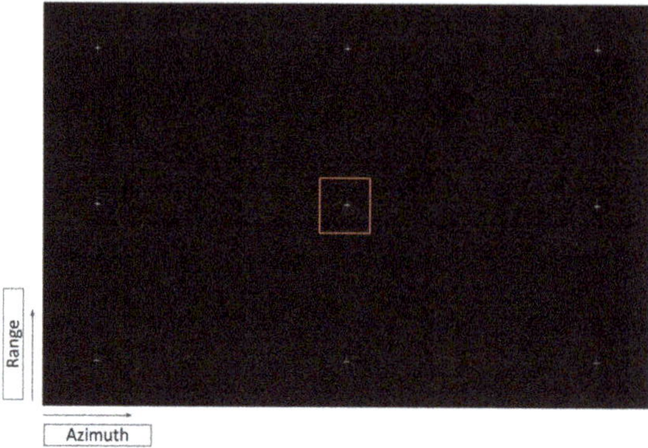

Figure 10. The point array target used in simulation.

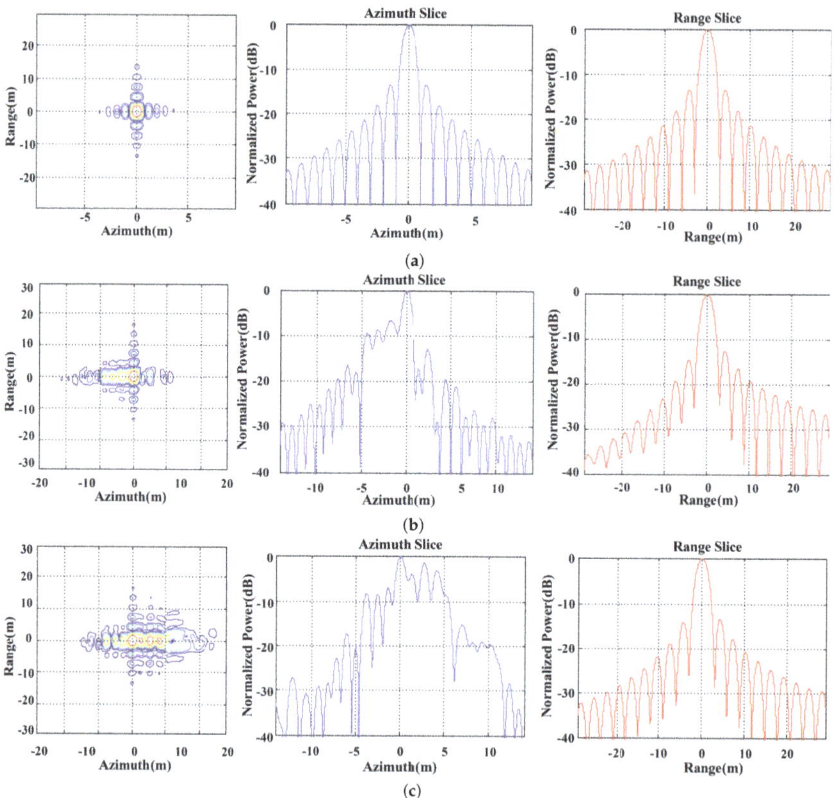

Figure 11. The simulation results of point targets. (**a**) Ideal imaging results. (**b**) Scintillation imaging result for $C_k L = 10^{33}$. (**c**) Scintillation imaging result for $C_k L = 10^{34}$.

The scintillation mitigation on point target is performed in Figure 12. The peak loss induced by ionosphere scintillation will weaken the SAR image contrast which makes the dominant scatters hard to select. Therefore, in this paper, the minimum-entropy autofocusing is applied to mitigate the scintillation effect instead of phase gradient autofocusing method. Since the spatial variation of scintillation phase screen, the autofocusing performance is limited in strong scintillation conditions. The scintillation parameters are set as $C_k L = 10^{33}$ and $p = 3$. The PSLR/ISLR before the autofocusing are -6.26 dB and -2.33 dB in Figure 12a. After the minimum-entropy autofocusing the PSLR/ISLR become -10.76 dB and -5.65 dB, respectively. The autofocusing result indicates that the existing scintillation mitigation method does not work perfectly even in moderate scintillation condition. As is mentioned before, the MS interferometric method [5] shed some new light on the mitigation of ionospheric scintillation, especially for sliding spotlight mode with large squint angle variations.

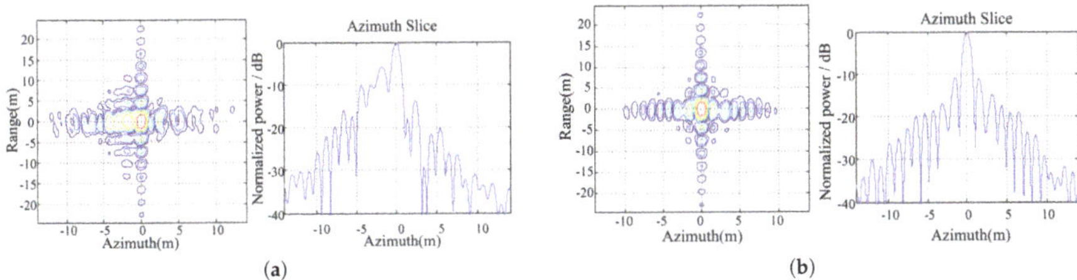

Figure 12. The scintillation mitigation on point target. (**a**) Scintillation imaging result for $C_k L = 10^{33}$. (**b**) The autofocusing result.

5.2. Extended Target Simulation

The extended target simulations are carried out by using a 2000×2000 pixels real SAR image acquired from a P-band air-borne SAR system working at 600 MHz as is shown in Figure 13a. Since the observation geometry is redefined in the beam projection and interpolation process of the ReBP algorithm, the geometry difference between two systems can be neglected in the simulation. The simulation is performed by considering the influence of different spectral index from 3 to 5, and $C_k L$ is set as 10^{34} to present a significant demonstration. Based on the theoretical analysis in Section 3, the TFTPCF serious degenerates with the increasing of spectral index. The decrease of TFTPCF will lead to the signal decorrelation and the azimuth defocusing.

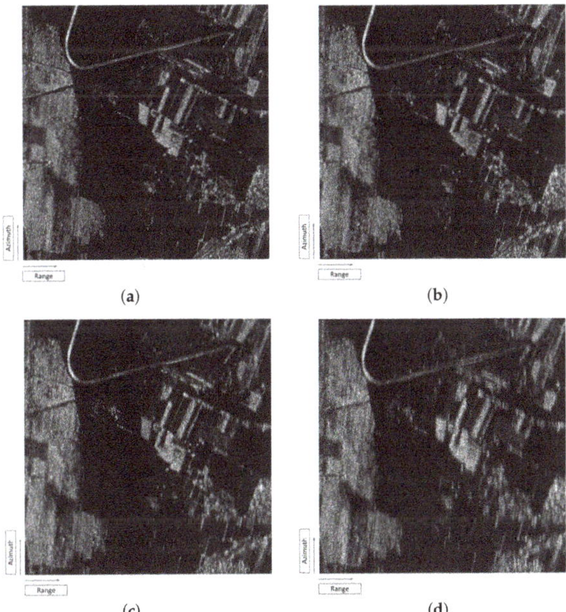

Figure 13. The simulation results of extended target. (**a**) Original SAR image. (**b**) Scintillation imaging result for $p = 3$. (**c**) Scintillation imaging result for $p = 4$. (**d**) Scintillation imaging result for $p = 5$.

Based on the quantitative analysis from the Monte-Carlo simulation listed in Figure 14, both PSLR and ISLR in azimuth increase with the rising of spectral index, which will distort the azimuth imaging performance and weaken the SAR image contrast. It can be seen from the extended target result that the image blur become more serious with the increase of spectral index. The image is still cognizable in the case of $p = 3$. However, in Figure 13d for $p = 5$, the scintillation-contaminated image is nearly unable to recognize. The image blur can be seen from the houses and trees in the middle of the scene. The extended target simulation corroborates the experiment result of point target that the degeneration of PSLR and ISLR induced by scintillation will seriously distort the imaging performance.

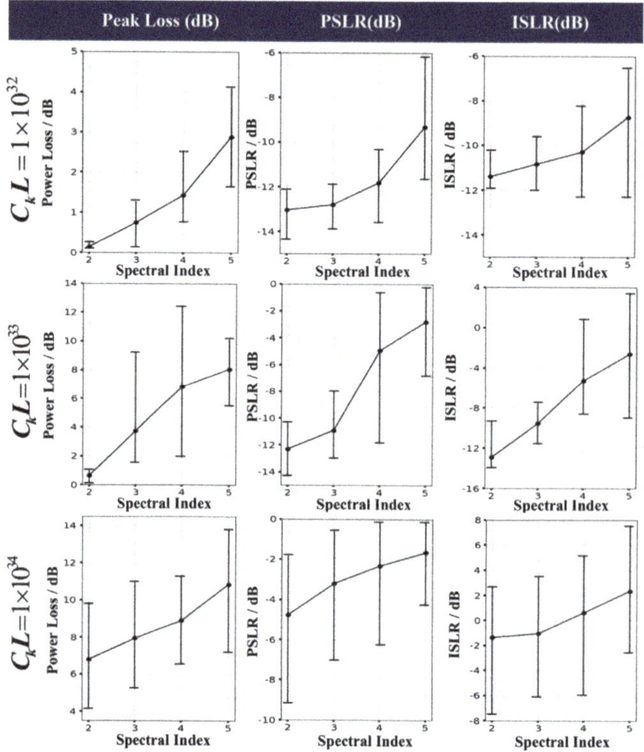

Figure 14. The Monte-Carlo simulation results of the scintillation effect on point targets.

5.3. Monte-Carlo Simulation

As is mentioned before, the phase and amplitude scintillation is a random process. Therefore, the Monte-Carlo simulation is required to perform a statistical analysis. In our work, the simulation is iteratively performed on the point array target as is shown in Figure 10. For each group of scintillation parameters, the iterations are performed for 500 times and the statistical results are shown in Figure 14. As is discussed in Section 3, the signal decorrelation is not sensitive to the outer scale. Thus, the simulation focus on the imaging performance with different scintillation strength (from 10^{32} to 10^{34} as is shown in different rows) and different spectral index (from 2 to 5 as is shown in the x-axis of each line graph). The peak loss, PSLR and ISLR are counted and plotted as the line graph in different rows. The black spot represents the mean value of the statistical data and the vertical short lines represent the variation scope of the variables. The Monte-Carlo simulation has a good agreement with the theoretical analysis that the imaging quality degenerates with the increase of scintillation strength and spectral index. The positive ISLR happens in the case of $C_k L \geq 10^{33}$ and $p \geq 4$ due to the serious expand of the azimuth mainlobe. The peak loss is considerable under the scintillation, which will reduce the visibility of weak scatters and the contrast of SAR images. The experiment results also demonstrate that the scintillation effect is less serious in the case of $C_k L \leq 10^{32}$ and $p \leq 2$ and this can be considered to be a threshold to evaluate the influence of scintillation effect on space-borne P-band SAR system.

Another group of Monte-Carlo simulations are carried out to make a comparison between stripmap mode and sliding spotlight mode as is shown in Table 3. The simulation is performed in the case of $C_k L = 10^{34}$, the mean value of peak loss, PSLR and ISLR are illustrated in Table 3. It is shown that all the indicators of sliding spotlight mode are lower than stripmap mode which means the scintillation effect will bring more serious distortion in sliding spotlight mode in the same ionospheric condition which validates the theoretical analysis in Section 3.

Table 3. The comparison of scintillation effect on point targets between stripmap mode and sliding spotlight mode from Monte-Carlo simulation.

	Peak Loss/dB				PSLR/dB				ISLR/dB			
Spectral Index	2	3	4	5	2	3	4	5	2	3	4	5
Sliding spotlight	6.81	7.95	8.88	10.84	−4.77	−3.21	−2.33	−1.67	−1.35	−1.04	0.61	2.33
Stripmap	2.08	4.41	6.20	7.87	−6.57	−4.39	−2.76	−1.99	−3.96	−2.73	−0.45	0.36

6. Conclusions

The space-borne P-band SAR system has a splendid prospect for its advantage in penetration ability. However, the P-band SAR imaging resolution is limited for its low central frequency and sensitivity of the ionospheric effect. In this paper, an in-depth analysis of scintillation effect is performed on P-band sliding spotlight SAR. Based on the refined TFTPCF model, the theoretical analysis indicate that the beam scanning and longer IPL will aggravate the signal decorrelation and make the sliding spotlight mode more sensitive to the ionospheric scintillation than stripmap mode. To accommodate the sliding spotlight geometry, a novel ReBP-based SAR-SS is proposed to generate the scintillation-contaminated SAR echo. The simulations on both point and extended target indicate that the scintillation-induced azimuth degeneration becomes more serious with the increasing of scintillation strength and spectral index. The Monte-Carlo simulation shows that the scintillation effect will be insignificant in the case of $C_k L \leq 10^{32}$ and $p \leq 2$ which can be considered to be a threshold. Since the ReBP algorithm also accommodates to TOPS and ScanSAR modes, the SAR-SS proposed in this paper can also be used to analyze the scintillation effect for these multi-mode SAR systems working in L-band or P-band. The mitigation of scintillation distortion will be further researched in the future work.

Author Contributions: L.Y. and Y.J. conceived and designed the experiments; L.Y. and Y.Z. performed the experiments; L.Y. and Q.Z. analyzed the data; Z.D. contributed reagents/materials/analysis tools; L.Y. wrote the paper.

Funding: This work was supported in part by the National Natural Science Foundation of China (NSFC) under Grant Nos. 61501477 and 61171123 and in part by the International Science and Technology Cooperation Program of China (ISTCP) under Grand No. 2015DFA10270.

Conflicts of Interest: The authors declare no conflict of interest.

References

1. Rignot, E.; Zimmermann, J.R.; Vanzyl, J.J. Spaceborne applications of P-band imaging radars for measuring forest biomass. *IEEE Trans. Geosci. Remote Sens.* **1995**, *33*, 1162–1169. [CrossRef]
2. Toan, T.L.; Beaudoin, A.; Riom, J.; Guyon, D. Relating forest biomass to SAR data. *IEEE Trans. Geosci. Remote Sens.* **1992**, *30*, 403–411. [CrossRef]

3. Dobson, M.C.; Ulaby, F.T.; Toan, T.L.; Beaudoin, A. Dependence of radar backscatter on coniferous forest biomass. *IEEE Trans. Geosci. Remote Sens.* **1992**, *30*, 412–415. [CrossRef]

4. Arcioni, M.; Bensi, P.; Davidson, M.W.; Drinkwater, M.; Fois, F.; Lin, C.C.; Meynart, R.; Scipal, K.; Silvestrin, P. ESA'S BIOMASS mission candidate system and payload overview. In Proceedings of the IEEE International Geoscience and Remote Sensing Symposium, Munich, Germany, 22–27 July 2012; pp. 5530–5533.

5. Mancon, S.; Giudici, D.; Tebaldini, S. The ionospheric effects mitigation in the BIOMASS mission exploiting multi-squint coherence. In Proceedings of the 12th European Conference on Synthetic Aperture Radar, Aachen, Germany, 4–7 June 2018; pp. 1–6.

6. Meyer, F.; Bamler, R.; Jakowski, N.; Fritz, T. The Potential of Low-Frequency SAR Systems for Mapping Ionospheric TEC Distributions. *IEEE Geosci. Remote Sens. Lett.* **2006**, *3*, 560–564. [CrossRef]

7. Rogers, N.C.; Quegan, S.; Kim, J.S.; Papathanassiou, K.P. Impacts of ionospheric scintillation on the BIOMASS P-band satellite SAR. *IEEE Trans. Geosci. Remote Sens.* **2014**, *52*, 1856–1868. [CrossRef]

8. Liu, J.; Kuga, Y.; Ishimaru, A.; Pi, X.; Freeman, A. Ionospheric effects on SAR imaging: A numerical study. *IEEE Trans. Geosci. Remote Sens.* **2003**, *41*, 939–947.

9. Xu, Z.W.; Wu, J.; Wu, Z.S. Potential effects of the ionosphere on space-based SAR imaging. *IEEE Trans. Geosci. Remote Sens.* **2008**, *56*, 1968–1975.

10. Mittermayer, J.; Moreira, A.; Loffeld, O. Spotlight SAR data processing using the frequency scaling algorithm. *IEEE Trans. Geosci. Remote Sens.* **1999**, *37*, 2198–2214. [CrossRef]

11. Eldhuset, K. Ultra high resolution spaceborne SAR processing. *IEEE Trans. Aerosp. Electron. Syst.* **2004**, *40*, 370–378. [CrossRef]

12. Wang, P.; Liu, W.; Chen, J.; Niu, M.; Yang, W. A high-order imaging algorithm for high-resolution spaceborne SAR based on a modified equivalent squint range model. *IEEE Trans. Geosci. Remote Sens.* **2015**, *53*, 1225–1235. [CrossRef]

13. He, F.; Chen, Q.; Dong, Z.; Sun, Z. Processing of Ultrahigh-Resolution Spaceborne Sliding Spotlight SAR Data on Curved Orbit. *IEEE Trans. Aerosp. Electron. Syst.* **2013**, *49*, 819–839. [CrossRef]

14. Xu, Z.W.; Wu, J.; Wu, Z.S. A survey of ionospheric effects on space-based radar. *Waves Random Media* **2004**, *14*, S189–S273. [CrossRef]

15. Meyer, F.J. Performance Requirements for Ionospheric Correction of Low-Frequency SAR Data. *IEEE Trans. Geosci. Remote Sens.* **2011**, *49*, 3694–3702. [CrossRef]

16. Mannix, C.R.; Belcher, D.P.; Cannon, P.S.; Angling, M.J. Using GNSS signals as a proxy for SAR signals: Correcting ionospheric defocusing. *Radio Sci.* **2016**, *51*, 60–70. [CrossRef]

17. Xiao, W.; Liu, W.; Sun, G. Modernization milestone: BeiDou M2-S initial signal analysis. *GPS Solut.* **2015**, *20*, 125–133. [CrossRef]

18. Azcueta, M.; Tebaldini, S. Non-Cooperative Bistatic SAR Clock Drift Compensation for Tomographic Acquisitions. *Remote Sens.* **2017**, *9*, 1087. [CrossRef]

19. Ishimaru, A.; Kuga, Y.; Liu, J.; Kim, Y.; Freeman, T. Ionospheric effects on synthetic aperture radar imaging at 100 MHz to 2 GHz. *Radio Sci.* **1999**, *34*, 257–268. [CrossRef]

20. Li, L.L.; Li, F. SAR imaging degradation by ionospheric irregularities based on TFTPCF analysis. *IEEE Trans. Geosci. Remote Sens.* **2007**, *45*, 1123–1130. [CrossRef]

21. Belcher, D.; Cannon, P. Amplitude scintillation effects on SAR. *IET Radar Sonar Navig.* **2014**, *8*, 658–666. [CrossRef]

22. Meyer, F.J. A review of ionospheric effects in low-frequency SAR–Signals, correction methods, and performance requirements. In Proceedings of the 2010 IEEE International Geoscience and Remote Sensing Symposium, Honolulu, HI, USA, 25–30 July 2010; pp. 29–32.

23. Wang, C.; Zhang, M.; Xu, Z.W.; Chen, C.; Sheng, D.S. Effects of anisotropic ionospheric irregularities on space-Borne SAR imaging. *IEEE Trans. Geosci. Remote Sens.* **2014**, *62*, 4664–4673. [CrossRef]

24. Meyer, F.; Chotoo, K.; Chotoo, S.; Huxtable, B.; Carrano, C. The influence of equatorial scintillation on L-band SAR image quality and phase. *IEEE Trans. Geosci. Remote Sens.* **2016**, *54*, 869–880. [CrossRef]

25. Hu, C.; Li, Y.; Dong, X.; Wang, R.; Ao, D. Performance Analysis of L-Band Geosynchronous SAR Imaging in the Presence of Ionospheric Scintillation. *IEEE Trans. Geosci. Remote Sens.* **2017**, *55*, 159–172. [CrossRef]

26. Ji, Y.F.; Zhang, Q.L.; Zhang, Y.S.; Dong, Z. L-band geosynchronous SAR imaging degradations imposed by ionospheric irregularities. *China Sci. Inf. Sci.* **2017**, *60*, 060308. [CrossRef]

27. Carrano, C.S.; Groves, K.M.; Caton, R.G. Simulating the impacts of ionospheric scintillation on L band SAR image formation. *Radio Sci.* **2012**, *47*, RS0L20. [CrossRef]
28. Li, D.; Rodriguez-Cassola, M.; Prats-Iraola, P.; Wu, M.; Moreira, A. Reverse Backprojection Algorithm for the Accurate Generation of SAR Raw Data of Natural Scenes. *IEEE Geosci. Remote Sens. Lett.* **2017**, *14*, 2072–2076. [CrossRef]
29. Rino, C.L. A power law phase screen model for ionospheric scintillation 2. Strong scatter. *Radio Sci.* **1979**, *14*, 1147–1155. [CrossRef]

Article

Improving the Topside Profile of Ionosonde with TEC Retrieved from Spaceborne Polarimetric SAR

Cheng Wang [1],*, Wulong Guo [1], Haisheng Zhao [2], Liang Chen [1], Yiwen Wei [3] and Yuanyuan Zhang [3]

[1] Qian Xuesen Laboratory of Space Technology, China Academy of Space Technology, Haidian district, Beijing 100094, China; guo.wulong@163.com (W.G.); chenliang@qxslab.cn (L.C.)
[2] National Key Laboratory of Electromagnetic Environment, China Research Institute of Radiowave Propagation, Qingdao 266107, China; zhaohaisheng213@163.com
[3] School of Physics and Optoelectronic Engineering, Xidian University, Xi'an 710071, China; ywwei@xidian.edu.cn (Y.W.); yyzhang1@xidian.edu.cn (Y.Z.)
* Correspondence: solskjaer2006@126.com; Tel.: +86-186-1815-4639

Received: 14 December 2018; Accepted: 23 January 2019; Published: 26 January 2019

Abstract: Signals from spaceborne polarimetric synthetic aperture radar will suffer from Faraday rotations when they propagate through the ionosphere, especially those at L-band or lower frequencies, such as signals from the Phased Array type L-band Synthetic Aperture Radar (PALSAR). For this reason, Faraday rotation compensation should be considered. On the other hand, Faraday rotation could also be retrieved from distorted echoes. Moreover, combining Faraday rotation with the radar parameters and the model of magnetic field, we could derive the total electron content (TEC) along the signal path. Benefiting from the high spatial resolution of the SAR system, TEC obtained from PALSAR could be orders of magnitude higher in spatial resolution than that from GPS. Besides, we demonstrated that the precision of TEC from PALSAR is also much higher than that from GPS. With the precise TEC available, we could fuse it with data from other ionosphere detection devices to improve their performances. In this paper, we adopted it to help modify the empirically modeled topside profile of ionosonde. The results show that the divergence between the modified profile and the referenced incoherent scattering radar profile reduced by about 30 percent when compared to the original ionosonde topside profile.

Keywords: polarimetric synthetic aperture radar; total electron content; ionospheric electron density distribution

---

1. Introduction

Ionospheric variations have been studied for earthquake prediction, solar activity analysis, radar image modification and geomagnetic storm research [1–5]. Among all kinds of ionospheric characteristics, the total electron content (TEC) is one of the most used parameters. For TEC evaluation, widely used instruments include Incoherent Scatter Radar (ISR), ionosonde, and Global Navigation Satellite Systems (GNSS). Beyond their popularity, some inconveniences still exist for each instrument. The ISR is the most powerful piece of equipment for TEC detection. However, it is poorly distributed on the Earth due to its expensive cost for developing and operation. The ionosonde station is easy to build and has a reasonable cost, so it has been set up worldwide [6]. Nevertheless, an inherent issue of the ionosonde is that it can only directly detect the electron density under the peak height. Though many methods have been proposed for modeling the topside profile of ionosphere from the ionosonde measurement, it is still a subject for ongoing investigation [7–9]. The GNSS can map the TEC of the ionosphere globally and in real-time, but the spatial resolution of the observation is too low for the fine analysis of ionosphere above a certain area [10]. Recently, full polarimetric spaceborne synthetic aperture radar (PolSAR) was demonstrated to be qualified for ionospheric inhomogeneities imaging [11]. Specifically, Faraday rotation (FR) and TEC images were derived from PolSAR data, and ionospheric perturbations observed from variations of these images were verified using ground-based GPS receivers and network. Instead of analyzing relative variations in TEC images, we would like to quantitatively determine the precision of TEC retrieved from PolSAR, i.e., how precise the TEC could be when compared to that obtained from the powerful ISR. Furthermore, we would like to present a method for improving the topside electron density profile of ionosonde using the precise TEC retrieved from PolSAR.

This work relates to two parts of ionosphere detection. First, we evaluated the precision of TEC measured by the spaceborne synthetic aperture radar. Experiments showed impressive results whereby the precision was within 1 TECU when compared with the TEC obtained from ISR. Specifically, as examples, we used the L band PolSAR data from the Phased Array type L-band Synthetic Aperture Radar (PALSAR) onboard the Advanced Land Observation Satellite. The ISR station that observes the same space as the PolSAR was found. Then, TECs derived from these two facilities were compared. The results verified their consistency.

The second part of this paper presents a method to improve the topside profile of the ionosonde with TEC retrieved from PolSAR data. The ionosonde station located in the scene of PolSAR was found. Since the ionosonde could directly observe the electron densities under the peak height, we were then able to derive the TEC of the topside profile by subtracting the TEC of the bottomside profile from the PolSAR TEC. This topside TEC could then be used to help model the electron density topside profile. To evaluate our method, we managed to find an ISR which was also located in the same area as the ionosonde station. A comparison of the electron density profile of the ISR and the modified topside profile of ionosonde validated our algorithm.

The outline of this paper is given as follows: Section 2 presents the method of TEC retrieval from PolSAR and the method for improving the topside profile model of ionosonde with the known TEC. Then, details of data used in this paper are introduced in Section 3. Section 4 illustrates our experimental results to validate our proposed methods. Finally, Section 5 concludes the paper.

2. Methods for TEC Retrieval and Topside Profile Modification

We first present the methods used in this paper for retrieving TEC from PolSAR data, and then introduce the algorithm for improving the topside profile of ionosonde with the retrieved TEC.

2.1. TEC Retrieval from PolSAR

Radar signals at the L band and lower frequencies will suffer strong Faraday rotation (FR) effects when they pass the ionosphere. However, these FR effects could serve as useful information for deriving the total electron density along the signal propagation path. According to [11,12], the TEC along the radio path of PolSAR is related to FR by

$$\Omega = \frac{2.365 \times 10^4}{f^2} \int n_e B \cos\theta \, ds = \frac{2.365 \times 10^4}{f^2} \langle B \cos\theta \rangle TEC, \tag{1}$$

where Ω stands for FR, f is the radar operating frequency, B denotes the magnitude of the ambient magnetic field, and θ is the angle between the radio wave and ambient magnetic field vectors. The second equality in Equation (1) is deduced from the first mean value theorem for definite integrals [13], where $TEC = \int n_e ds$ is the TEC along the radar signal path, and $\langle B \cos\theta \rangle$ corresponds to the mid-value of $B \cos\theta$ at a specific altitude of the path. Therefore, once FR is estimated from distorted radar signals, we can easily derive TEC under a specified $\langle B \cos\theta \rangle$.

FR retrieval has been well studied in the last two decades [14–21]. In this paper we chose the B&B method [16] as the FR estimator due to its robustness and reliability:

$$\begin{bmatrix} Z_{11} & Z_{12} \\ Z_{21} & Z_{22} \end{bmatrix} = \begin{bmatrix} 1 & j \\ j & 1 \end{bmatrix} \begin{bmatrix} M_{hh} & M_{vh} \\ M_{vh} & M_{vv} \end{bmatrix} \begin{bmatrix} 1 & j \\ j & 1 \end{bmatrix}$$
$$\Omega = \tfrac{1}{4}\arg\left(\langle Z_{12} Z_{21}^* \rangle\right). \tag{2}$$

In Equation (2), the M matrix represents the measured full polarimetric scattering matrix. Therefore, when the full polarimetric data is obtained, TEC can be calculated from Equations (1) and (2).

2.2. Improving the Topside Profile Model of Ionosonde with Known TEC

Many models are available for the topside profile of ionosonde. In this paper, we aim to improve the most commonly used model, the α-Chapman model, with a constant scale height H_T [7]:

$$Ne = N_m F_2 \times e^{\frac{1-z-e^{-z}}{2}},$$
$$z = \frac{h - h_m F_2}{H_T} \tag{3}$$

where $N_m F_2$ represents the electron density at the peak height $h_m F_2$ of the F_2 layer. Both $N_m F_2$ and $h_m F_2$ can be directly obtained from the ionosonde station. Therefore, H_T is the only parameter that we need to estimate to determine a topside profile.

Traditional method estimates H_T from the bottomside profile of each measurement [9]. Though empirically feasible, this method does not take into account any truly measured characteristic about the topside profile. In this paper, we propose a method to evaluate H_T with the TEC of the topside density profile. Integrating Equation (3) with respect to height gives

$$\frac{1}{(N_m F_2)^2} \int Ne^2 dh = \int e^{1-z-e^{-z}} dh. \tag{4}$$

According to [22–24], if the electron density profile is Chapman-model based, we have

$$\int Ne^2 dh = 0.66 Ne_{\max} TEC, \tag{5}$$

where $Ne_{\max}$ is the maximal electron density along the path. Referring to the topside profile only, we can obtain from Equation (5)

$$\int_{h_m F_2}^{H_s} Ne^2 dh = 0.66 Ne_{\max} TEC_{Top}. \tag{6}$$

TEC_{Top} is the TEC of the topside profile of the ionosonde. When we can measure the TEC through other equipment precisely, TEC_{Top} is simply obtained by subtracting $TEC_{Bott} = \int_0^{h_m F_2} Ne dh$ from TEC. Substituting Equation (6) into the left side of Equation (4) gives

$$\frac{0.66 Ne_{\max} TEC_{Top}}{(N_m F_2)^2} = \int_{h_m F_2}^{H_s} e^{1-z-e^{-z}} dh. \tag{7}$$

In Equation (7), H_s is the maximum altitude to which the TEC is measured, and in this paper, it is the altitude of PALSAR. Integrating the right side of Equation (7) while recalling Equation (3), we can obtain

$$\int_{h_m F_2}^{H_s} e^{1-z-e^{-z}} dh = H_T \int_0^{\frac{H_s - h_m F_2}{H_T}} e^{1-z-e^{-z}} dz = H_T \left(\exp\left(1 - e^{-\frac{H_s - h_m F_2}{H_T}} \right) - 1 \right). \tag{8}$$

Joining Equations (7)–(8) leads to

$$\frac{0.66 N e_{\max} TEC_{Top}}{(N_m F_2)^2} = H_T\left(\exp\left(1 - e^{-\frac{H_s - h_m F_2}{H_T}}\right) - 1\right) \tag{9}$$

Solving the nonlinear Equation (9), we can obtain H_T. Thus, the topside profile can be obtained from Equation (3) with the TEC-related constant scale height H_T.

3. Data Details

3.1. Data for Estimating TEC Precision of PolSAR

The spaceborne L band PolSAR data for TEC retrieval were obtained from the Phased Array type L-band Synthetic Aperture Radar (PALSAR) of the Japanese Advanced Land Observation Satellite (ALOS) which orbits Sun-synchronously at about 692 km of altitude (data access https://vertex. daac.asf.alaska.edu/). Each scene of the radar data covers a rectangle area of the Earth. The TEC of each scene is estimated as the average TEC of the whole scene. For a quantitative evaluation of the reliability of TEC derived from PolSAR, we compared it to the TEC derived from ISR, because ISR is believed to be the most powerful device for monitoring the ionosphere. However, since ISRs are sparsely distributed on the earth, despite trying our best, we only managed to find three groups of corresponding data. The ISR data were collected from Poker Flat ISR station [25], and corresponding PolSAR scenes were chosen to make sure the distance from the Poker Flat ISR to the scene was less than 40 km. Three selected scenes are shown in Figure 1.

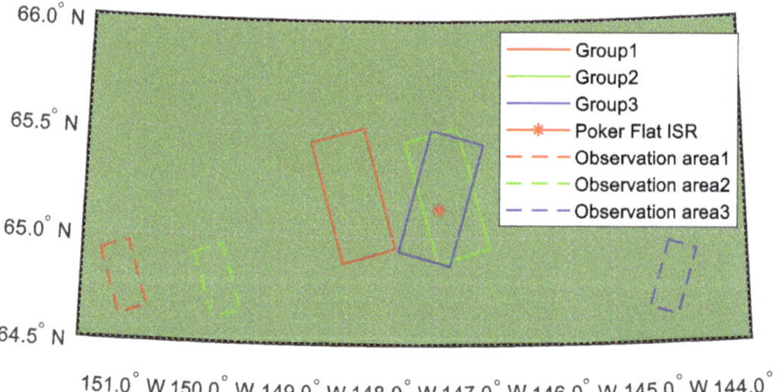

Figure 1. Position relationship of three ground tracks of PALSAR scenes and Poker Flat ISR. The red rectangle corresponds to the PALSAR scene of Group 1, while the green and the blue ones correspond to the measurement of Group 2 and Group 3, respectively. The position of Poker Flat ISR is shown as a red asterisk, while the observation area of each group is shown as a dashed rectangle in its corresponding color.

From Figure 1, we can find that Poker Flat ISR station is fully covered by PALSAR scenes of Group 2 and 3, while the range between the ISR station and PALSAR scene 1 is within 40 km. The PALSAR illumination modes of three scenes can also be observed in Figure 1. Specifically, the dashed rectangles represent the actual observation areas of PALSAR on ionosphere at 300 km, i.e., the coverage of PALSAR beam on the ionosphere at 300 km. Therefore, we know that the distance from each observation area to its ground scene center is around 100 km and that the TEC obtained from PALSAR is slant TEC rather than vertical TEC (VTEC). The VETC should be estimated as

$$VTEC = TEC \times \cos\eta, \tag{10}$$

where η is the off-nadir angle of the PALSAR. To ensure that the ISR and PALSAR observe the same space, the best choice is to find an observation of ISR that just to illuminate the area of PALSAR. However, this kind of observation is not available, so we picked the observation closest to that of PALSAR as a representation and assumed that the ionosphere is stable within the observation area. Details about the observations are given in Table 1. The Piercing Lat. (Lon.) in the table is the center latitude (longitude) of the observation area of the corresponding instrument. From the Table 1, we can see that the two devices observed the same ionosphere at the same time.

Table 1. Corresponding observing times and positions of three groups of data. The off-nadir angle of PALSAR (Phased Array type L-band Synthetic Aperture Radar) in this table is 21.5 degrees. ISR: Incoherent Scatter Radar.

Data Group	Instrument	Center Latitude	Center Longitude	Center Observation Time (UTC:Y/M/D HH/MM)	Piercing Latitude	Piercing Lontitude
1	PALSAR	65.193	−148.439	2011/03/19 07/32	64.786	−151.022
	Poker Flat ISR	65.130	−147.471	2011/03/19 07/30	64.650	−148.000
2	PALSAR	65.194	−147.369	2011/03/31 07/28	64.782	−149.949
	Poker Flat ISR	65.130	−147.471	2011/03/31 07/28	65.130	−147.471
3	PALSAR	65.183	−147.450	2010/08/06 21/06	64.800	−144.832
	Poker Flat ISR	65.130	−147.471	2010/08/06 21/08	65.370	−145.070

3.2. Data for Modeling the Topside Profile of Ionosonde

Once the TEC of an area was calculated from PolSAR data, it was possible to utilize it to improve the topside profile of the ionosonde that observes the same area with PolSAR. In this paper, the ionosonde data was accessed from Digital Ionogram DataBase (DIDB) [26]. The topside profile of the data was based on what we discussed previously, i.e., the α-Chapman model with a constant scale height H_T. Therefore, a comparison of the original profile and the modified profile is available. As a reference of the true measured electron profile, ISR data was again adopted in the experiment to validate our proposed method.

Despite doing our best, we only found one group of corresponding data, as shown in Table 2. The ISR data was still obtained from Poker Flat ISR station, the ionosonde data was from EIELSON station (ID: EI764), and the PolSAR data was from PALSAR-2 on board Advanced Land Observing Satellite-2 (ALOS-2). Their geographic relationship is shown in Figure 2 where the PolSAR scene fully covers the Poker ISR station and EIELSON station is adjacent to the right bottom of the scene. From the "Piercing Lat." and "Piercing Lon." in Table 2, we could know that both ISR and ionosonde observe the ionosphere vertically. Therefore, we only need to pay attention to the off-nadir angel of PALSAR while estimating the VTEC.

Table 2. Observing time and positions of three instruments. The off-nadir angle of PALSAR-2 in this table is 30.8 degree.

Instrument	Center Latitude	Center Longitude	Center Observation Time (UTC: Y/M/D HH/MM)	Piercing Latitude	Piercing Longitude
PALSAR	65.193	−148.439	2014/08/29 22/24	64.603	−144.362
Poker Flat ISR	65.130	−147.471	2014/08/29 22/20	65.130	−147.471
EIELSON station	64.660	−147.070	2014/08/29 22/15	64.660	−147.070

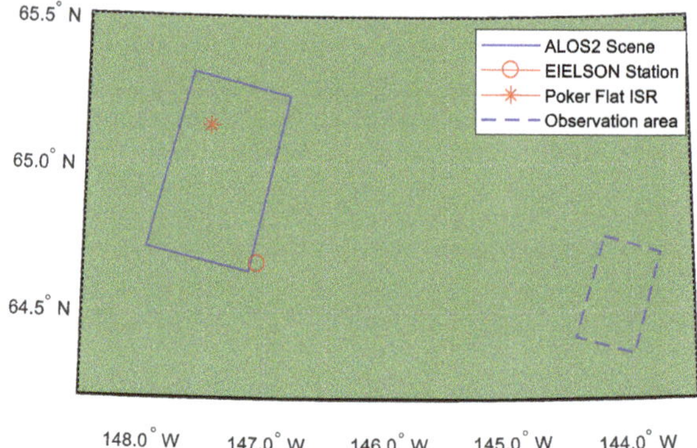

Figure 2. Geographic relationship of ALOS-2 scene, EIELSON station, and Poker ISR station.

4. Results and Discussions

4.1. Validation of the Precision of PolSAR in Estimating TEC

Integrating the electron density profile of ISR up to the altitude of PALSAR gave the TEC of ISR along the observing direction. The VTEC of ISR was obtained by using Equation (10), but the η of ISR is the angle between the beam direction and the vertical direction.

Equation (1) tells us that the value of PolSAR TEC is affected by the mid-value $\langle B \cos \theta \rangle$ [13]. However, it is not easy to theoretically determine such a value. In our experience, the value of $B \cos \theta$ linearly decreases as the altitude increases, so the mid-value is determined mostly by the electron density profile. Here, we give a brief discussion about how to determine the $\langle B \cos \theta \rangle$. Recall Equation (1) and explicitly represent it into the form of the first mean value theorem for definite integrals:

$$\Omega = \frac{2.365 \times 10^4}{f^2} \int n_e B \cos \theta ds = \frac{2.365 \times 10^4}{f^2} \int_0^{H_s} B \cos \theta dTEC = \frac{2.365 \times 10^4}{f^2} \langle B \cos \theta \rangle TEC. \quad (11)$$

Comparing Equation (11) to the first mean value theorem for definite integrals, we get

$$\int_a^b f(x)dx = f(\xi)(b-a), \xi \in [a,b] \quad (12)$$

It can be found that $B \cos \theta$ corresponds to $f(x)$ and TEC corresponds to x. Therefore, to determine the mid-value $\langle B \cos \theta \rangle$, we should know the functional relationship between $B \cos \theta$ and TEC. Though the function cannot be explicitly represented, we can show it numerically in a figure and obtain some useful information.

Take the data from ionosonde and PolSAR in Table 2 as an example. The $B \cos \theta$ as a function of altitude could be obtained from the ambient magnetic field of PALSAR, as shown in Figure 3a. Clearly in Figure 3a, $B \cos \theta$ decreases nearly linearly with the increase of altitude. TEC as a function of altitude could be obtained by integrating the electron density profile of the ionosonde in Figure 3b. Taking advantage of the same variable, i.e., the altitude in the above two functions, we can plot $B \cos \theta$ as a function of TEC, as shown in Figure 3c. For clarity, the relationship between altitude and TEC is plotted in Figure 3c, so we can get a direct idea about the altitude where the mid-value $\langle B \cos \theta \rangle$ lies. $B \cos \theta$ still monotonically decreases in Figure 3c, but the curve changes from convex to concave at the peak height.

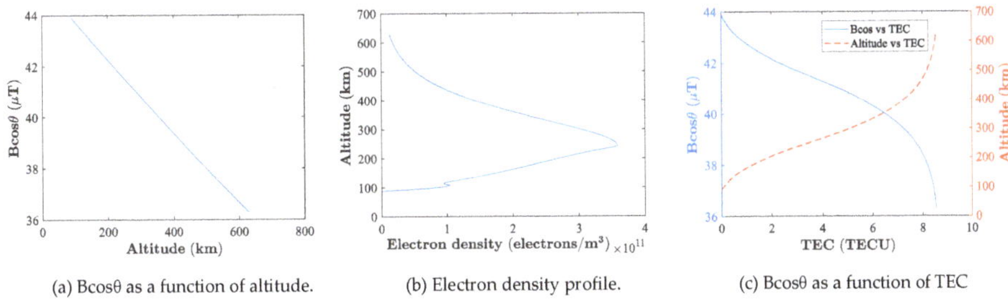

(a) Bcosθ as a function of altitude. (b) Electron density profile. (c) Bcosθ as a function of TEC

Figure 3. Relationship between altitude and total electron content (TEC) as well as $B\cos\theta$.

The first mean value theorem for definite integrals shows that $\langle B\cos\theta\rangle TEC$ is actually the area of the region under the "Bcosθ vs. TEC" curve of Figure 3c. Intuitively, the mid-value $\langle B\cos\theta\rangle$ of Figure 3c should be located at around 5 TECU which corresponds to an altitude of about 400 km. Though we only take one group of data as an example, most cases satisfy this mode where the altitude for the mid-value $\langle B\cos\theta\rangle$ is a little higher than the peak height of the ionosphere as long as the TEC of the topside profile is larger than the TEC of the bottom side. However, it was still not possible for us to determine a specific mid-value for each scene of PolSAR, so we experimented on three different altitudes to see how precisely the PolSAR could estimate TEC and how the magnetic field influences the TEC value. The magnetic field at 300 km, 400 km, and 500 km was picked as the mid-value along each radio path, respectively. For each scene of PolSAR, FR estimation from Equation (2) may be biased by residual calibration errors, though PALSAR has been reported to be well calibrated [27,28]. Therefore, following [29], only pixels of SNR >10 dB were selected to estimate the final TEC. This corresponds to select pixels of a signal amplitude higher than −17 dB, since the noise equivalent sigma zero (NESZ) is estimated to be about −27 dB for PALSAR. Here, FR estimates as a function of the signal amplitude are given in Figure 4 for each PolSAR scene, where the signal amplitude is defined as the "circular cross-pol product" $abs\left(Z_{12}Z_{12}^{*}\right)$ from Equation (2). From Figure 4, we can see that the signal amplitude of each PolSAR scene is higher than −17 dB. Therefore, the TEC derived from averaging the whole PolSAR TEC image is countable. One should note that the average window used to reduce the speckle noise in Equation (2) was set to be 21×41 pixels. Actually, we tested window sizes from 10×10 to 200×200, and the changes between the resulting TECs were within 0.01 TECU.

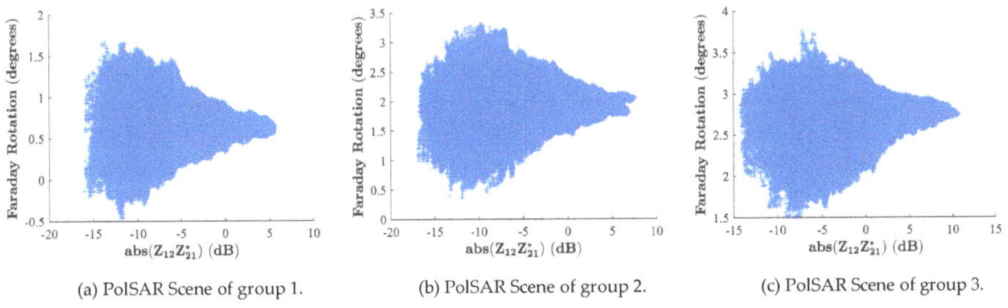

(a) PolSAR Scene of group 1. (b) PolSAR Scene of group 2. (c) PolSAR Scene of group 3.

Figure 4. Faraday rotation (FR) estimates as a function of signal amplitude. PolSAR: polarimetric spaceborne synthetic aperture radar.

Results of TECs retrieved from different instruments are given in Table 3. The TEC deviation at 400 km is defined as the absolute value of the difference between PALSAR VTEC and ISR VTEC. Since TEC varies in a PolSAR scene, we calculated the standard deviation of TEC for each PolSAR scene and present it following each term in parentheses. All figures in Table 3 are given in units of TECU ($1\ TECU = 10^{16}$ electrons/m^2).

Table 3. TEC measurements of Poker Flat ISR and PolSAR. GNSS: Global Navigation Satellite System.

Instrument.	Group 1	Group 2	Group 3
Poker Flat ISR	1.484	5.700	7.365
PolSAR (300 km)	1.514 (0.324)	4.646 (0.444)	6.812 (0.351)
PolSAR (400 km)	1.586 (0.340)	4.864 (0.465)	7.112 (0.367)
PolSAR (500 km)	1.660 (0.355)	5.092 (0.487)	7.434 (0.384)
Deviation (400 km)	0.102	0.836	0.253
Poker Flat ISR (average)	1.46	5.71	7.86
GNSS	6	12.5	10.7

From Table 3, some phenomena are observed. First, the standard deviation of TEC in each PolSAR scene is small. This feature demonstrates that no strong variation occurred in the ionosphere during the observations. Therefore, the average TEC of the scene can be represented by the PolSAR TEC. Second, the derived PolSAR TEC increases along with the altitude of the magnetic field, which can be easily explained from Equation (1). Since $\langle B \cos \theta \rangle$ decreases with an increase in altitude, the derived TEC will get larger. Third, the TEC deviations between ISR and PolSAR are within 1 TECU. This illustrates the feasibility of measuring TEC with PolSAR. For a better understanding of the performance of PolSAR, we also present here the TECs obtained from GNSS. Clearly, the TEC differences between GNSS and ISR range from 3 to 5 TECU. Since GNSS measures TEC up to about 20,000 km, the differences are acceptable. However, for group 1 and group 2, the TECs from GNSS are way too large to be believed as precise. For a further illustration that the ionosphere is stable during the measuring time, we also give the average TEC of ISR in different directions. The ISR data belonging to different beam directions were first collected together, and a polynomial fitting algorithm was adopted to form a smooth electron density profile. Then average TEC of ISR was obtained by integrating the electron densities with respect to the altitude. The small differences between "Poker Flat ISR" and "Poker Flat ISR (average)" further verify the stable circumstance of the ionosphere.

4.2. Modeling the Topside Profile of Ionosonde with New H_T

We have demonstrated that PolSAR data can be used to estimate the TEC under 700 km with a considerable precision within 1 TECU. Therefore, after the bottomside electron density profile of ionosphere has been precisely measured by an ionosonde, the corresponding PolSAR TEC can be used to calculate the TEC_{Top} for the ionosonde. Resorting to Equation (9), we can derive the parameter H_T. Then, the topside profile could be easily calculated from Equation (3). In this section, PolSAR TEC derived with magnetic field at 400 km is used, which is 11.4 TECU. Still, to account for potential residual calibration errors, only pixels of SNR > 10 dB are employed. For clarity, the TECs of "Poker Flat ISR" and "Poker Flat ISR (average)" are also given here, which are 12.37 TECU and 12.23 TECU, respectively. Clearly, the deviation between PolSAR TEC and ISR TEC is still within 1 TECU, which again validates the precision of PolSAR for TEC estimation. The small difference between "Poker Flat ISR" and "Poker Flat ISR (average)" indicates a stable ionosphere. For comparison, the TEC obtained from GNSS is 15.7 TECU, which is acceptable but not as precise as that of PolSAR.

Figure 5 shows the result of the proposed method where "ISR" represents the ISR data after polynomial fitting. Note that we adopted the ISR data from all beams for polynomial fitting, rather than the vertical beam only, because the TEC deviations between "Poker Flat ISR" and "Poker Flat ISR (average)" are small, and vertical data is too sparse to form a fine polynomial fitting. In Figure 5, "Ionosonde" stands for the original data obtained from DIDB, and "Improved" is our result with the topside profile calculated from the known TEC_{Top}. It is clear that the bottomside profile of ionosonde matches well with that of ISR, while the two topside parts are divergent. The reason for this was discussed previously, i.e., that no topside information is considered in the original method. After modifying the topside profile with the known TEC, we can easily see that the new topside profile is more consistent with that of ISR. The divergence between the ionosonde topside profile and the

ISR topside profile reduced by 30.41 percent after the presented algorithm was adopted. Here, the divergence is defined as the average absolute difference between the ISR profile and the ionosonde profile, and the percentage of reduction was calculated from Equation (13). This result proves the validity of our method.

$$percentage = 1 - mean\left(\frac{|ISR - Improved|_{Top}}{|ISR - Ionosonde|_{Top}}\right). \tag{13}$$

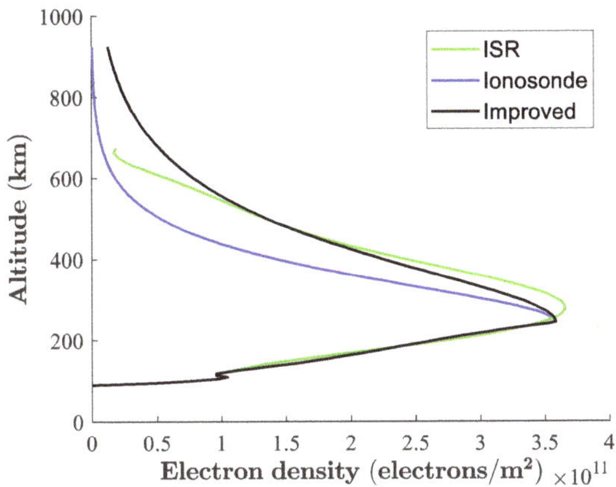

Figure 5. Electron density profiles for ISR and ionosonde.

5. Conclusions

This paper first validated the ability of PolSAR as an effective device in measuring the TEC of the ionosphere, and then demonstrated the feasibility of using known TEC to help improve the topside profile of ionosonde. The results show that PolSAR is able to measure the TEC with higher precision compared to GNSS. Furthermore, the improved topside profile proved to be much more consistent with the profile of ISR than the original profile.

In this paper, the α-Chapman model was used for the topside profile because of its popularity. Actually, if any other model is developed for the topside profile, the proposed TEC modification strategy could also be taken as an auxiliary process. One should note that the proposed method could only detect the TEC under the altitude of the satellite. However, it is still reliable to expand the modified ionosonde profile beyond this altitude, since TECs under 700 km cover the majority of TECs under 10,000 km.

Author Contributions: C.W., W.G. and H.Z. initiated the research. Under the supervision of L.C., C.W. performed the analysis and wrote the manuscript. Y.W. and Y.Z. revised the manuscript. All authors read and approved the final version of the manuscript.

Funding: This research was funded by the National Natur1al Science Foundation of China (NSFC) under Grants 41604157, 41601483, and 61871352, and by the National Key Laboratory of Electromagnetic Environment.

Acknowledgments: Special thanks goes to Bodo Reinisch for his help in offering the account of DIDB. We also thank the Japanese Aerospace and Exploration Agency and the Alaska Satellite Facility for making the PALSAR data publicly available. Madrigal Database made the incoherent scattering radar data available. We appreciate their works.

Conflicts of Interest: The authors declare no conflict of interest. The funders had no role in the design of the study; in the collection, analyses, or interpretation of data; in the writing of the manuscript, or in the decision to publish the results.

References

1. Yizengaw, E.; Dyson, P.L.; Essex, E.A.; Moldwin, M.B. Ionosphere dynamics over the southern hemisphere during the 31 March 2001 severe magnetic storm using multi-instrument measurement data. *Ann. Geophys. ANGEO* **2005**, *23*, 707–721. [CrossRef]
2. Wen, D.; Yuan, Y.; Ou, J.; Zhang, K. Ionospheric response to the geomagnetic storm on August 21, 2003 over China using GNSS-based tomographic technique. *IEEE Trans. Geosci. Remote. Sens.* **2010**, *48*, 3212–3217. [CrossRef]
3. Zhao, H.-S.; Xu, Z.-W.; Wu, J.; Wang, Z.-G. Ionospheric tomography by combining vertical and oblique sounding data with TEC retrieved from a tri-band beacon. *J. Geophys. Res. Space Phys.* **2010**, *115*. [CrossRef]
4. Zhao, H.-S.; Xu, Z.-W.; Wu, J.; Quegan, S. Ionospheric tomography of small-scale disturbances with a triband beacon: A numerical study. *Radio Sci.* **2010**, *45*. [CrossRef]
5. Liu, Y.; Fu, L.; Wang, J.; Zhang, C. Studying ionosphere responses to a geomagnetic storm in June 2015 with multi-constellation observations. *Remote. Sens.* **2018**, *10*, 666. [CrossRef]
6. Reinisch, B.W.; Galkin, I.A. Global ionospheric radio observatory (GIRO). *Earth Planets Space* **2011**, *63*, 377–381. [CrossRef]
7. Reinisch, B.W.; Huang, X. Deducing topside profiles and total electron content from bottomside ionograms. *Adv. Space Res.* **2001**, *27*, 23–30. [CrossRef]
8. Reinisch, B.W.; Huang, X.; Galkin, I.A.; Paznukhov, V.; Kozlov, A. Recent advances in real-time analysis of ionograms and ionospheric drift measurements with digisondes. *J. Atmos. Sol. Terr. Phys.* **2005**, *67*, 1054–1062. [CrossRef]
9. Huang, X.; Reinisch, B.W. Vertical electron content from ionograms in real time. *Radio Sci.* **2001**, *36*, 335–342. [CrossRef]
10. Wang, C.; Chen, L.; Liu, L.; Yang, J.; Lu, Z.; Feng, J.; Zhao, H.-S. Robust computerized ionospheric tomography based on spaceborne polarimetric SAR data. *IEEE J. Sel. Top. Appl. Earth Obs. Remote. Sens.* **2017**, *10*, 4022–4031. [CrossRef]
11. Pi, X.; Freeman, A.; Chapman, B.; Rosen, P.; Li, Z. Imaging ionospheric inhomogeneities using spaceborne synthetic aperture radar. *J. Geophys. Res. Space Phys.* **2011**, *116*, A04303. [CrossRef]
12. Wang, C.; Zhang, M.; Xu, Z.-W.; Zhao, H.-S. TEC retrieval from spaceborne SAR data and its applications. *J. Geophys. Res. Space Phys.* **2014**, *119*, 8648–8659. [CrossRef]
13. Kim, J.S.; Papathanassiou, K.P.; Scheiber, R.; Quegan, S. Correcting distortion of polarimetric SAR data induced by ionospheric scintillation. *IEEE Trans. Geosci. Remote. Sens.* **2015**, *53*, 6319–6335. [CrossRef]
14. Qi, R.; Jin, Y. Analysis of the effects of Faraday rotation on spaceborne polarimetric SAR observations at P-band. *IEEE Trans. Geosci. Remote Sens.* **2007**, *45*, 1115–1122. [CrossRef]
15. Freeman, A. Calibration of linearly polarized polarimetric SAR data subject to Faraday rotation. *IEEE Trans. Geosci. Remote Sens.* **2004**, *42*, 1617–1624. [CrossRef]
16. Bickel, S.H.; Bates, R.H.T. Effects of magneto-ionic propagation on the polarization scattering matrix. *Proc. IEEE* **1965**, *53*, 1089–1091. [CrossRef]
17. Rogers, N.C.; Quegan, S.; Kim, J.S.; Papathanassiou, K.P. Impacts of ionospheric scintillation on the BIOMASS P-band satellite SAR. *IEEE Trans. Geosci. Remote Sens.* **2014**, *52*, 1856–1868. [CrossRef]
18. Chen, J.; Quegan, S. Improved estimators of Faraday rotation in spaceborne polarimetric SAR data. *IEEE Geosci. Remote Sens. Lett.* **2010**, *7*, 846–850. [CrossRef]
19. Wang, C.; Liu, L.; Chen, L.; Feng, J.; Zhao, H.-S. Improved TEC retrieval based on spaceborne PolSAR data. *Radio Sci.* **2017**, *52*, 2016RS006116. [CrossRef]
20. Li, L.; Zhang, Y.; Dong, Z.; Liang, D. New Faraday rotation estimators based on polarimetric covariance matrix. *IEEE Geosci. Remote Sens. Lett.* **2014**, *11*, 133–137. [CrossRef]
21. Meyer, F.J. Performance requirements for ionospheric correction of low-frequency SAR data. *IEEE Trans. Geosci. Remote Sens.* **2011**, *49*, 3694–3702. [CrossRef]
22. Hartmann, G.K.; Leitinger, R. Range errors due to ionospheric and tropospheric effects for signal frequencies above 100 MHz. *Bull. Géod.* **1984**, *58*, 109–136. [CrossRef]
23. Datta-Barua, S.; Walter, T.; Blanch, J.; Enge, P. Bounding higher-order ionosphere errors for the dual-frequency GPS user. *Radio Sci.* **2008**, *43*. [CrossRef]

24. Wang, C.; Chen, L.; Liu, L. A new analytical model to study the ionospheric effects on VHF/UHF wideband SAR imaging. *IEEE Trans. Geosci. Remote Sens.* **2017**, *55*, 4545–4557. [CrossRef]
25. Poker Flat ISR station. Available online: http://isr.sri.com/madrigal/ (accessed on 8 February 2018).
26. Digital Ionogram DataBase. Available online: http://ulcar.uml.edu/DIDBase/ (accessed on 5 February 2018).
27. Borner, T.; Papathanassiou, K.P.; Marquart, N.; Zink, M.; Meadows, P.J.; Rye, A.J.; Wright, P.; Meininger, M.; Tell, B.R.; Traver, I.N. ALOS PALSAR products verification. In Proceedings of the 2007 IEEE International Geoscience and Remote Sensing Symposium, Barcelon, Spain, 23–28 July 2007; pp. 5214–5217.
28. Eriksson, L.E.B.; Sandberg, G.; Ulander, L.M.H.; Smith-Jonforsen, G.; Hallberg, B.; Folkesson, K.; Fransson, J.E.S.; Magnusson, M.; Olsson, H.; Gustavsson, A.; et al. ALOS PALSAR calibration and validation results from Sweden. In Proceedings of the 2007 IEEE International Geoscience and Remote Sensing Symposium, Barcelona, Spain, 23–28 July 2007; pp. 1589–1592.
29. Meyer, F.J.; Nicoll, J.B. Prediction, detection, and correction of Faraday rotation in full-polarimetric L-band SAR data. *IEEE Trans. Geosci. Remote Sens.* **2008**, *46*, 3076–3086. [CrossRef]

Article

Monitoring Land Subsidence in Wuhan City (China) using the SBAS-InSAR Method with Radarsat-2 Imagery Data

Yang Zhang [1], Yaolin Liu [1,2,3], Manqi Jin [1], Ying Jing [1], Yi Liu [1], Yanfang Liu [1,*], Wei Sun [4], Junqing Wei [1] and Yiyun Chen [1,*]

1. School of Resource and Environmental Sciences, Wuhan University, Wuhan 430079, China; zhangy1010@whu.edu.cn (Y.Z.); liuyaolin2010@163.com (Y.L.); kingerin@163.com (M.J.); y.crystal@whu.edu.cn (Y.J.); liuyi2010@whu.edu.cn (Y.L.); weijunqing@whu.edu.cn (J.W.)
2. Key Laboratory of Geographic Information System, Ministry of Education, Wuhan University, Wuhan 430079, China
3. Collaborative Innovation Center for Geospatial Information Technology, Wuhan 430079, China
4. Wuhan Geomatics Institute, Wuhan 430022, China; gnss.wei@gmail.com
* Correspondence: yfliu59@126.com (Y.L.); chenyy@whu.edu.cn (Y.C.)

Received: 21 December 2018; Accepted: 5 February 2019; Published: 12 February 2019

Abstract: Wuhan city is the biggest city in central China and has suffered subsidence problems in recent years because of its rapid urban construction. However, longtime and wide range monitoring of land subsidence is lacking. The causes of subsidence also require further study, such as natural conditions and human activities. We use small baseline subset (SBAS) interferometric synthetic aperture radar (InSAR) method and high-resolution RADARSAT-2 images acquired between 2015 and 2018 to derive subsidence. The SBAS-InSAR results are validated by 56 leveling benchmarks where two readings of elevation were recorded. Two natural factors (carbonate rock and soft soils) and three human factors (groundwater exploitation, subway excavation and urban construction) are investigated for their relationships with land subsidence. Results show that four major areas of subsidence are detected and the subsidence rate varies from −51.56 to 27.80 millimeters per year (mm/yr) with an average of −0.03 mm/yr. More than 83.81% of persistent scattered (PS) points obtain a standard deviation of less than −6 mm/yr, and the difference between SBAS-InSAR method and leveling data is less than 5 mm/yr. Thus, we conclude that SBAS-InSAR method with Radarsat-2 data is reliable for longtime monitoring of land subsidence covering a large area in Wuhan city. In addition, land subsidence is caused by a combination of natural conditions and human activities. Natural conditions provide a basis for subsidence and make subsidence possible. Human activities are driving factors and make subsidence happen. Moreover, subsidence information could be used in disaster prevention, urban planning, and hydrological modeling.

Keywords: land subsidence; Radarsat-2 images; small baseline subset (SBAS) method; interferometric synthetic aperture radar (InSAR)

---

1. Introduction

Land subsidence is defined as a gradual settling or sudden sinking of the ground surface [1–3], which results from natural processes or human activities [4–7]. Over the past decades, numerous land subsidence events have been reported in many cities around the world where the rapid urban construction and the extensive groundwater exploitation are taking place [8–13]. Land subsidence can lead to serious environmental problems and considerable economic losses, such as damage to infrastructures and increased risk of urban pluvial flooding [14–18]. Thus, the demand for monitoring the spatial and temporal distribution of land subsidence is increasing.

Traditional point-based monitoring approaches such as ground leveling and global positioning system (GPS) techniques could not provide sufficient samples required by land subsidence mapping [19]. In recent years, interferometric synthetic aperture radar (InSAR) technology has been rapidly developed to cover a large geographic area. InSAR method is low-cost and effective [20,21]. Nevertheless, the InSAR method suffers from temporal decorrelation and atmospheric disturbance [22–24]. Therefore, many advanced InSAR methods based on multi-interferograms such as persistent scatterer interferometry (PS-InSAR) and small baseline subset interferometry (SBAS-InSAR) have been proposed to overcome these limitations [25–29]. Furthermore, high-resolution SAR images are gradually applied such as ALOS-PALSAR and Radarsat-2 images [30,31].

The primary cause of land subsidence is human activities, such as groundwater withdrawal, coal mining, petroleum extraction, land creation, subway excavation, and building loading [4,21,29,32–36]. Besides, natural factors might also be critical, such as soft soil, karst geomorphologic [37,38]. Previous studies have examined the cross-correlations between these factors and land subsidence [39,40]. However, it remains unclear whether human factor works alone or with natural factor. Thus, the roles of natural and human factors in land subsidence require further study.

Wuhan city, which is the biggest city in central China, has various types of natural conditions and has experienced rapid urbanization in recent years. It is a typical city to study the problem of land subsidence in China. Previous studies have mapped land subsidence in Wuhan city using advanced InSAR methods [5,41,42]. However, longtime monitoring of land subsidence covering all urban areas of Wuhan city is lacking. In addition, Radarsat-2 images have not yet been applied to subsidence monitoring in Wuhan city.

This study explores the application of SBAS-InSAR method with high-resolution Radarsat-2 images to long-term monitoring of land subsidence in Wuhan city, and the cause of land subsidence. Specifically, (i) we investigate the potentials of 20 Radarsat-2 images acquired between 17 October 2015 and 3 June 2018 to derive land subsidence rates in Wuhan city. (ii) The InSAR results are validated by 56 leveling benchmarks. (iii) We study the influence of natural conditions and human activities on land subsidence and their interrelationships.

2. Study Area and Data Preparation

2.1. Study Area

Wuhan city (29°58′ N–31°22′ N, 113°41′ E–115°05′ E) is located in the east of an alluvial plain called Jianghan Plain, see Figure 1. The Yangtze River, the world's third longest river, flows through the heart of the city. The average elevation of the city is about 37 m. About 26% of total area (2205.06 km^2) is covered by water [43], such as rivers, lakes, ponds and ditches. The city has a subtropical monsoon climate characterized by four distinct seasons, abundant precipitation, and considerable sunshine. The average annual temperature is 16.6 °C and the precipitation averages 1269 mm. The rainfall concentrates in early summer (May to July) [44].

Carbonate rock and soft soils, which might contribute to land subsidence, are widespread in Wuhan city, see Figure 1. There are six carbonate rock belts aligned in an East-West orientation, and they cover an area of more than 1100 km^2 [45–47]. Soft soils have high water content, high compressibility, high porosity and low shear strength. Soft soils are mainly distributed along the banks of two rivers, the Yangtze River and the Han River, and the maximum thickness exceeds 10 m [48,49]. Wuhan city has experienced rapid economic growth since the China's reform and opening up policy in 1979. It has become a megacity with a population in excess of 10 million.

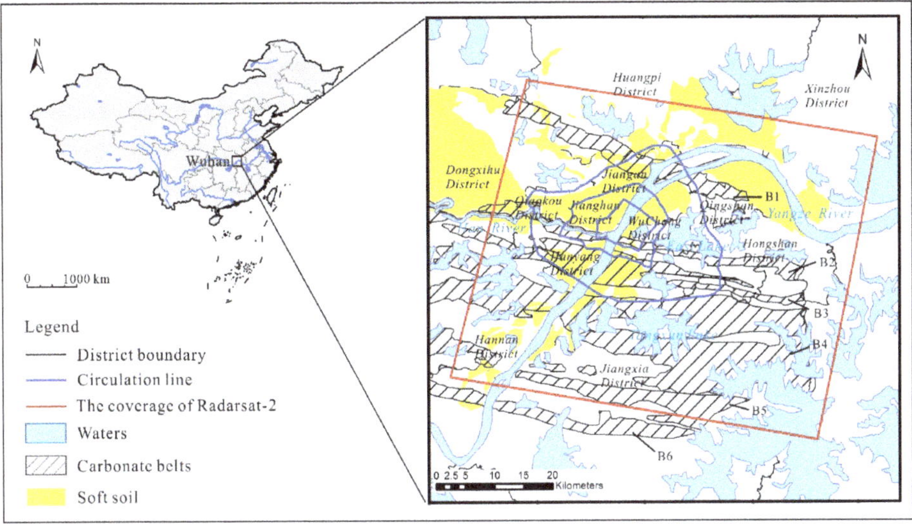

Figure 1. The location of Wuhan city in China and the study area. The red rectangle illustrates the coverage of Radarsat-2. B1–B6 represent six carbonate rock belts aligned in an East-West orientation, namely Tianxingzhou, Daqiao, Baishazhou, Zhuankou, Junshan, and Hannan.

2.2. Datasets

We employ 20 descending Radarsat-2 wide ultra-fine (WUF) single-look complex (SLC) images acquired from October 2015 to June 2018 at intervals of 24, 48, 72 or 96 days. These single horizontal-horizontal (HH) polarization images covered a 50 × 50 km area, see the red rectangle in Figure 1. Main parameters of Radarsat-2 WUF SLC data are detailed in Table 1. The Shuttle Radar Topography Mission (SRTM) 90 m DEM is used to simulate and remove topographic phases. To validate the InSAR results, we also employ 56 leveling benchmarks where two readings of elevation were recorded in September 2016 and March 2017, respectively.

Table 1. Parameters of Radarsat-2 WUF SLC images.

Parameters	Description
Product type	Radarsat-2 WUF SLC
Track no.	226
Band	C
Wavelength (cm)	5.5
Revisit frequency (day)	24
Incidence angle (degree)	30–50
Range resolution (m)	1.6
Azimuth resolution (m)	2.8
Orbit direction	Descending

We gathered data about natural and human factors that influence land subsidence. Two natural factors include soft soil and carbonate rock, see Figure 1. A map of soft soils distribution and a map of carbonate belts distribution are obtained from Wuhan municipal commission of urban-rural development and a geological study, respectively [47]. Three human factors are considered: groundwater exploitation, subway excavation and urban construction. The data of the three human factors include an official route map of the Wuhan subway system, the groundwater resources regionalization of Wuhan, two high resolution images of the year 2015 and 2017. In addition, impervious surface fraction is an index that measures the level of urban construction [50].

3. Methodology

The SBAS-InSAR method is used to process Radarsat-2 WUF SLC images in the ENVI SARScape module to obtain land subsidence information in Wuhan city [31]. The SBAS-InSAR method is an advanced InSAR technique that could improve the monitoring accuracy [51]. The SBAS-InSAR method relies on an appropriate combination of differential interferograms within the thresholds of temporal and spatial baselines, so the geometric decorrelation is minimal [26,31,36,52]. Figure 2 shows the main steps of SBAS-InSAR method to detect land subsidence.

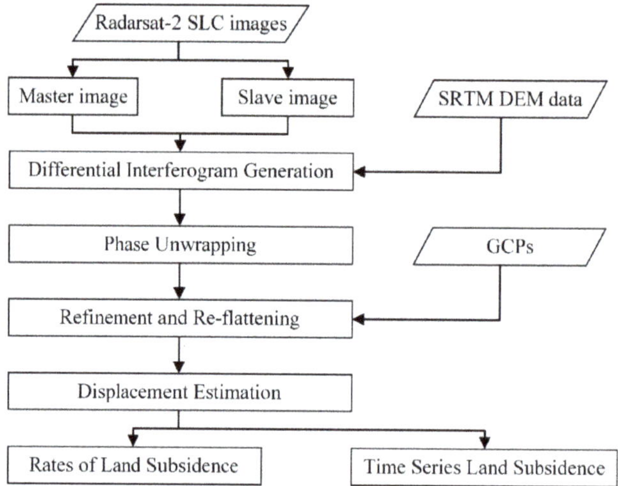

Figure 2. Flowchart of SBAS-InSAR data processing.

3.1. Differential Interferogram Generation

The image acquired on 17 September 2016 is selected as the super master image, and the remaining 19 images are slave images. The selection of interferograms is constrained by a maximum spatial baseline of 630 m (45% of the critical spatial baseline) and a maximum temporal baseline of 350 days. After topographic phase removal, 106 differential interferograms are generated, see Figure 3. The signal-to-noise ratio is improved by performing multi-looking factors of 4 × 4 in the range and azimuth directions, and Goldstein filtering method.

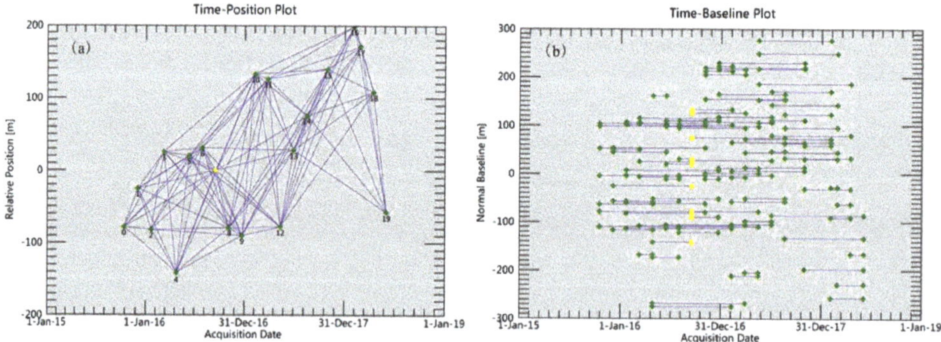

Figure 3. (a) Time–position of Radarsat-2 image interferometric pairs and (b) time–baseline of Radarsat-2 image interferometric pairs. The yellow diamond denotes the super master image. Blue lines represent interferometric pairs. Green diamonds denote slave images.

3.2. Phase Unwrapping

Both minimum cost flow (MCF) network and Delaunay 3D are employed for phase unwrapping, and a coherence threshold of 0.35 is chosen [35]. Then, 39 interferometric pairs with poor unwrapping and low coherence are eliminated.

3.3. Refinement and Re-flattening

After phase unwrapping, 46 Ground Control Points (GCPs) are selected to correct the unwrapped phase. The selection criteria are as follows: (1) the location has a high coherence value and good phase unwrapping, (2) land deformation is close to zero according to previous studies and leveling data, and (3) we should select as many GCPs as possible.

3.4. Displacement Estimation

Preliminary displacements are estimated by a linear model that is robust and commonly used [36]. Meanwhile, the residual topography is also removed. Then, atmospheric phase was removed by an atmospheric filtering. Subsequently, geocoding in the line of sight (LOS) direction with a resolution of 10 m is employed to calculate SBAS. Finally, subsidence rate and subsidence time series are obtained and mapped across the study area.

3.5. InSAR Data Validation by Using Leveling Benchmarks

The InSAR results are validated by 56 leveling benchmarks. Among these leveling benchmarks, a stable one located at East Lake Peony Garden (30°34′27″ N, 114°21′57″ E) is used as a reference point to measure land subsidence. Four parameters, namely, maximum discrepancy (MaxD), minimum discrepancy (MinD), mean absolute discrepancy (MD), and root mean square (RMS), are used to describe the reliability of SBAS-InSAR derived land subsidence rate map.

4. Results and Validation

4.1. Rates of Land Subsidence

Figure 4 shows the average subsidence velocity in the radar LOS from October 2015 to June 2018 across Wuhan city by using SBAS-InSAR technique. A negative value (in red color) indicates land subsidence, and a positive value (in blue color) indicates uplift. The total number of derived permanent scatter (PS) points was 8,680,765, and the average density was 3472 points/km^2. The subsidence rate varies from −51.56 to 27.80 millimeters per year (mm/yr) with an average of −0.03 mm/yr. Additionally, a pronounced subsidence area located in Hankou district, adjacent to the Xinrong Light Rail Transit station with a maximum velocity exceeding −50 mm/yr, is identified.

Land subsidence is widely found in most areas of the city, and land uplift in surrounding rural areas is also apparent (Figure 4). Four major areas of subsidence are detected: Hankou (HK), Qingshan Industrial Zone (QSIZ), Northern Shahu Lake (NSL), and Baishazhou (BSZ). HK covers the largest subsidence area, and is the main commercial district of the city. QSIZ is the city's oldest and biggest industrial area, and there are many large manufacturing plants, such as Wuhan Iron and Steel (Group) Corporation, Wuhan Petrochemical Complex, and Qingshan Thermal Power Plant. NSL has been undergoing rapid economic growth and high intensity of urban construction over the years. BSZ is located in the south of the city, and has speed up the construction of traffic facilities. Interestingly, all four major areas of subsidence are distributed along the banks of the Yangtze River. Other areas of subsidence are sinking slowly at a rate of less than −10 mm/yr.

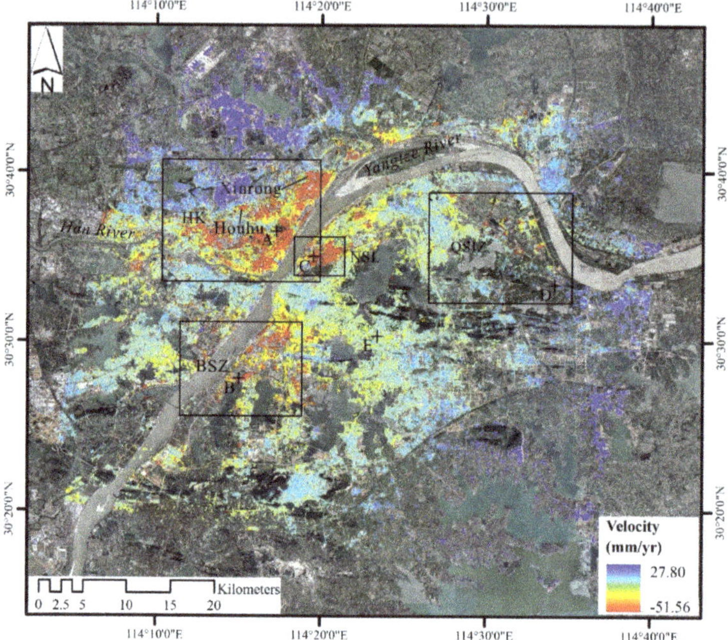

Figure 4. The average subsidence velocity in LOS from October 2015 to June 2018 across Wuhan city by using SBAS-InSAR technique. The four black rectangles are the four major areas of subsidence. A-E are five points of subsidence, detailed in Figure 6.

4.2. Evolution of Land Subsidence

Figure 5 illustrates the spatial distribution of subsidence and its changes over time. In most part of the city, the cumulative subsidence is stable in a range of −15 to 15 mm. But for the four major areas of subsidence, the cumulative subsidence gradually increases over time, and the area is constantly expanding. The maximum cumulative subsidence has reached up to −126.43 mm, is located in Xinrong of HK, see Figure 4.

The time series of subsidence at five typical PS points marked as A–E in Figure.4, is shown in Figure 6. Points A, B, C, and D are located in HK, BSZ, NSL, and QSIZ, respectively, which are the four major areas of subsidence. Point E is located in an urban area with minor subsidence of nearly zero mm. Points A, B, C, and D present nonlinear subsidence. One possible reason is that the seasonal variation of groundwater levels might influence the rate of subsidence. When in early summer (May, June, and July) rainfall concentrates, groundwater will be recharged and the rate of subsidence will slow down, see Figure 6. Points B, C, and D show similar trends of subsidence, and point B subsides more than points C and D. The subsidence at point A suddenly increases in 2017 probably due to the construction of Wuhan Metro Line No. 8.

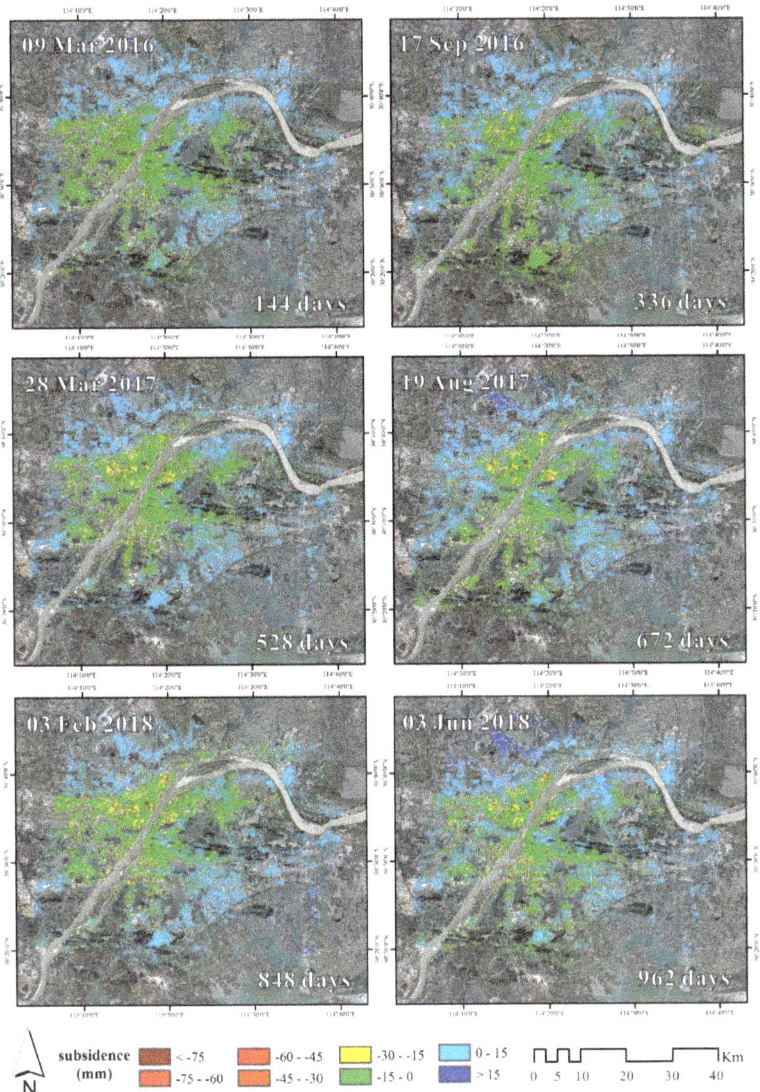

Figure 5. Spatio-temporal evolution of accumulated subsidence in Wuhan city derived from Radarsat-2 images. Only 6 of the 20 subsidence maps are shown.

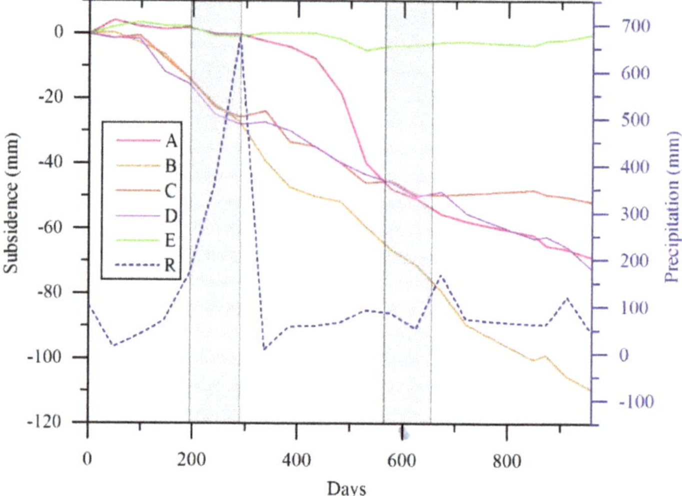

Figure 6. Time-series subsidence at the five typical points A–E. The gray rectangle denotes the early summer (May, June, and July).

4.3. InSAR Data Validation

Statistical analysis of the mean standard deviations is conducted to assess the internal precision of subsidence rates of subsidence rates. More than 83.81% of PS points obtain a standard deviation of less than -6 mm/yr, proving that applying SBAS-InSAR method to derive subsidence rates is reliable.

The land subsidence derived from Radarsat-2 images are compared to those derived from leveling data (Figure 7). 41 out of 56 leveling benchmarks are located within the generated grids, and are selected for validation. Figure 7 shows the results of leveling data against SBAS-InSAR method. For most validation points, the difference between the two methods is less than 5 mm/yr. MaxD, MinD, MD, and RMS are 9.22, 0.03, 1.38, and 4.03 mm/year, respectively. The result of SBAS-InSAR coincides with that of leveling data, which indicates that SBAS-InSAR method is able to monitor land subsidence with acceptable precision.

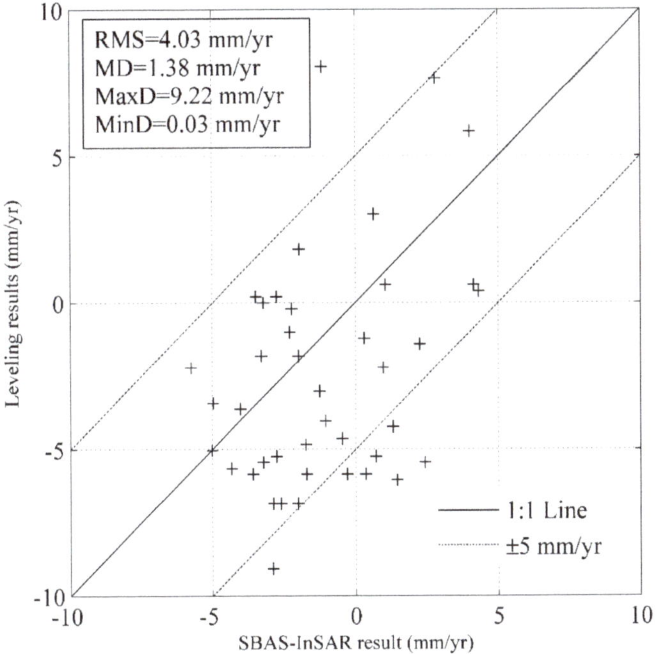

Figure 7. Leveling data versus SBAS-InSAR method plots of land subsidence.

5. Discussion

5.1. Comparison with Previous Studies

In this study, SBAS-InSAR method with Radarsat-2 data is reliable for longtime monitoring of land subsidence covering a large area in Wuhan city (October 2015 to June 2018). We also compare our results with those of the following studies (Table 2).

Table 2. Summary of the previous studies of land subsidence in Wuhan city.

Previous Studies	Data	Method	Subsidence Rate	Reference
Zhou et al.	15 C-band Sentinel-1A images, interferometric wide TOPS acquisition mode, VV polarization, ascending orbit, covering most of Wuhan city	SBAS-InSAR	−82–18 mm/yr	[5]
Bai et al.	12 X-band TerraSAR-X images, stripmap acquisition mode, HH polarization, ascending orbit, covering major urban areas of Wuhan city	PS-InSAR	−63.7–17.5 mm/yr	[41]
Costantini et al.	45 X-band COSMO-SkyMed images, stripmap acquisition mode, HH polarization, covering most of HK	PS Pair InSAR	−80–40 mm/yr	[42]
Benattou et al.	36 C-band Sentinel-1A images, interferometric wide TOPS acquisition mode, VV polarization, ascending orbit, covering major urban areas of Wuhan city	PS-InSAR	−127–23 mm/yr	[53]

Zhou et al. [5] obtained the rate of subsidence in Wuhan city by using SBAS-InSAR method with 15 Sentinel-1A images (April 2015 and April 2016) with 5 m × 20 m (range × azimuth) spatial resolution. Their results showed that subsidence rates varied from −82 mm/yr to 18 mm/yr, and the maximum rate of subsidence was detected in Houhu of HK. In addition, there are several centers of subsidence areas in Wuchang, Qingshan, Hanyang, and Hongshan district.

Bai et al. [41] investigated the rate and spatial patterns of subsidence in major urban areas in Wuhan city using PS-InSAR method with TerraSAR-X images (October 2009 and August 2010) with 2.0 m × 3.3 m (range × azimuth) spatial resolution. Subsidence rates varied from −63.7 mm/yr to 17.5 mm/yr, and HK is the largest subsidence area.

Costantini et al. [42] obtained subsidence information from high-resolution X-band COSMO-SkyMed data (June 2013 to June 2014) with 2.21 m × 1.63 m (range × azimuth) spatial resolution using PS pair InSAR method. Subsidence rates of most PS points in HK varied from −80 mm/yr to 40 mm/yr.

Benattou et al. [53] measured the rate of subsidence using 36 sentinel-1A images (June 2015 and April 2017) with 5 m × 20 m (range × azimuth) spatial resolution. The average deformation ranged from −127 mm/yr to 23 mm/yr and a new center of subsidence areas (Jiufengxiang) was found.

In our study, four major areas of subsidence are clearly identified, namely, HK, QSIZ, NSL, and BSZ, which are consistent with earlier research conducted by Zhou et al. However, the maximum rate of subsidence is −52 mm/yr, which is lower than the maximum rate of −82 mm/yr by Zhou et al. It is also lower than the rate of −67 mm/yr conducted by Bai et al. and −127 mm/yr conducted by Benattou et al. The reason behind this is that subsidence might occur over a short period of time and the rate of longtime monitoring would be relatively lower. Our longtime monitoring of land subsidence reflect a long term change of land subsidence relative to previous studies. The most severe ground settlement site of our study is located at Xinrong of HK, but in the study of Zhou et al. it is located at one other place named Houhu (Figure 4). Compared to the work of Bai et al. some places within major areas of subsidence exhibit a considerable increase in subsidence velocity. For example, the subsidence velocity in NSL is between −15 mm/yr and 5 mm/yr in the study of Bai et al. during 2009–2010, but it exceeds −15 mm/yr in our study during 2015–2018. By comparing and analyzing the results of subsidence monitoring at different times, the law of land subsidence over time in Wuhan city can be revealed.

5.2. Causes of Subsidence in Wuhan City

5.2.1. Natural Factors

In Wuhan city, carbonate rock and soft soils are widespread and might cause land subsidence (Figures 1 and 4). For the four major areas of subsidence, BSZ and QSIZ are located on the carbonate rock belts, and HK and NSL are located on the soft soils. Obviously, there exists a spatial correlation between land subsidence and the two natural factors. The rate of subsidence increases with the thickness of soft soils (Figure 8a). Taking Hongshan district and Jiangan district (Figure 1) as examples, we compare areas located on carbonate rock belts with the whole of the two urban areas (Figure 8b). The subsidence rate of areas on carbonate rock belts is higher than those of the whole of the two urban areas.

However, land subsidence is not significant in some other areas located on carbonate rocks or soft soil area. For example, the rate of land subsidence in Daqiao carbonate rock belt is lower than −5 mm/yr, indicating that the surface is relatively stable. Therefore, an area located on carbonate rock or soft soils is not sure to subside, but an area of subsidence requires natural conditions such as the carbonate rock or soft soils. In summary, natural factors are necessary but not sufficient conditions for land subsidence.

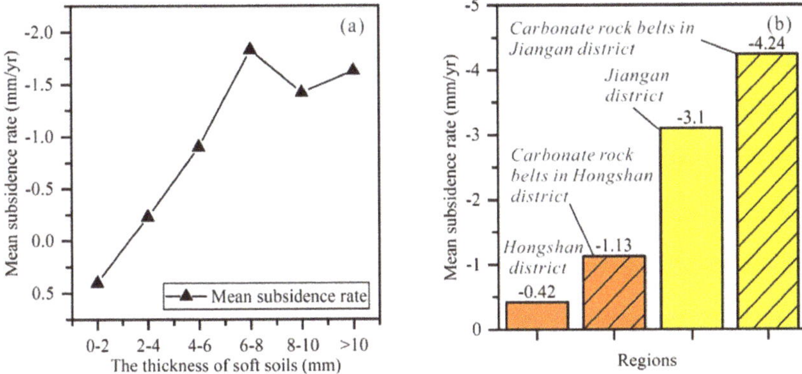

Figure 8. (**a**) Relationship between soft soil thickness and subsidence rate. (**b**) The subsidence rate of areas located on carbonate rock belts and those of the whole of the two urban areas.

5.2.2. Human Activities

According to the government's planning for utilization of the groundwater resource, all four major areas of subsidence are located in the groundwater exploitation regions (GERs) wherein large quantities of groundwater is continuously pumped (Figure 9). Groundwater extraction will increase the fluctuation of groundwater levels. That results in the compaction of highly compressible soft soils and the dissolution of carbonate rocks or suffusion processes. Therefore, land subsidence occurs.

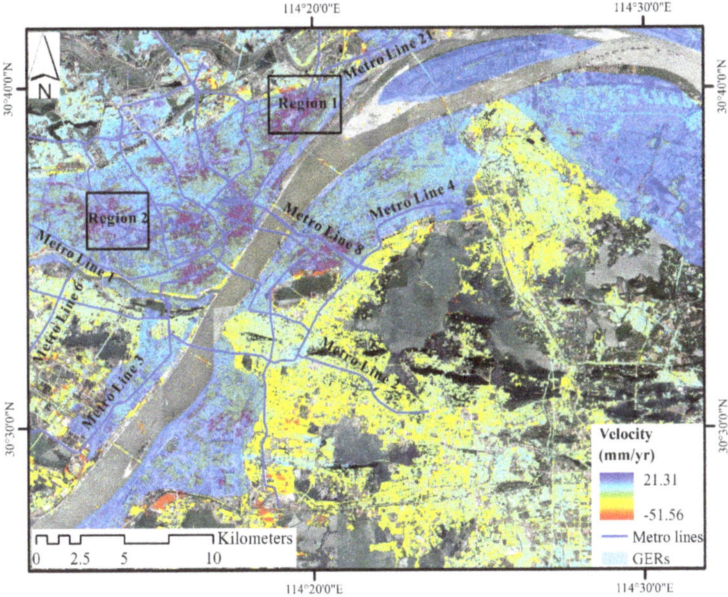

Figure 9. Map of the GERs and Metro Networks of Wuhan city.

Many subways have been built such as Metro Lines No. 3, 6, 8 and 21, or are under construction such as Metro Lines No. 5, 7 and 11, during our study period 2015−2018. Digging subway tunnels inevitably disturb the surrounding soil, and land subsidence is more likely to follow, especially in areas of soft soil and carbonate rock. As shown in Figure 9, several centers of severe subsidence areas are distributed along the metro lines such as Region 1.

In Region 1 (Figure 9), the subway lines have a high density and two metro lines intersect, namely Metro Lines No. 1 and 21, see Figure 10a. The intersection is near subway Station A and B that are situated at the center of subsidence area. The rate of subsidence reaches up to −44.30 mm/yr. A subsidence profile passing through stations A and B is shown in Figure 10b. The rate of subsidence decreases with the distance to subway stations. Therefore, subway construction can affect land subsidence.

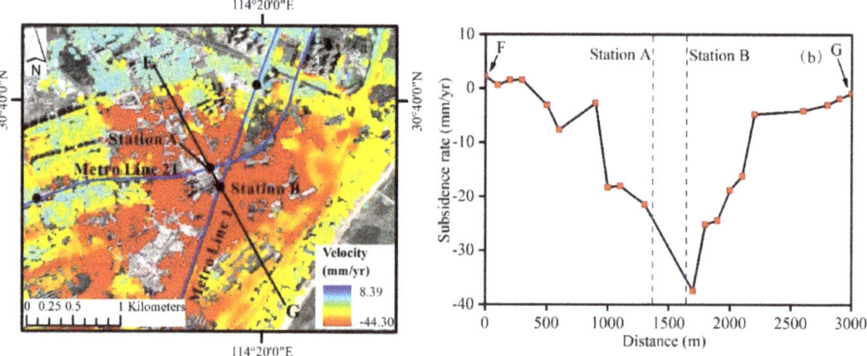

Figure 10. Maps show subsidence rate in Region 1 (**a**), and a subsidence profile passing through stations A and B (**b**).

Wuhan city's urban construction has entered into a stage of rapid growth during our study period 2015−2018. The annual investment in urban construction exceeds 20 billion dollars and many new buildings and transport facilities are constructed. Building a foundation often requires pumping groundwater during excavation, which could result in subsidence. In addition, when the soil underneath a building could no longer support the loading, the building will start to settle. Traffic loading also has much more influence on land subsidence because it can cause foundation deformation. Region 2 (Figure 9) is a new central business district (CBD) of the city where many high-rise buildings concentrated in, such as Wuhan Center Tower (438 m). Many new buildings and transport facilities have been constructed or being constructed. The rate of subsidence is shown in Figure 11 and severe subsidence are detected. Four typical PS points (i.e., H, I, J, and K) are selected to analyze the subsidence (Figure 11).

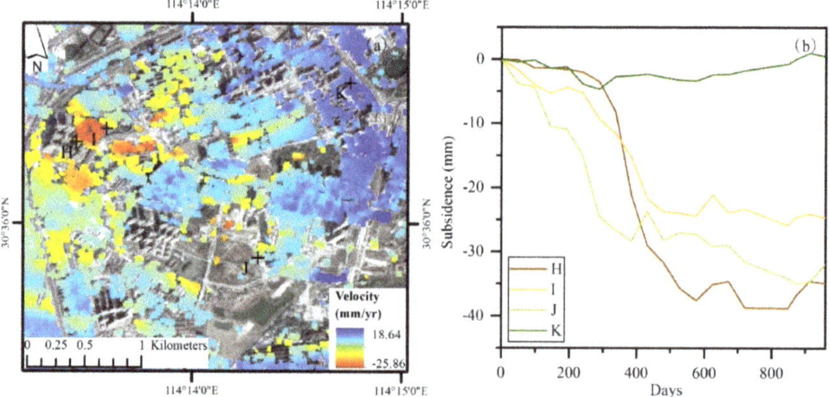

Figure 11. Maps show subsidence rate in Region 2 (**a**), and time-series subsidence at the four points H-K (**b**).

Points H, I and J are close to new buildings, new roads and a high-rise building, respectively (Figure 12). Point K is located on a stable surface. Points H, I and J subside greatly over time compare to point K. In addition, there is a correlation between subsidence and impervious surface fraction, see Figure 13. Thus, we can infer that urban construction such as buildings and transport facilities may drive subsidence.

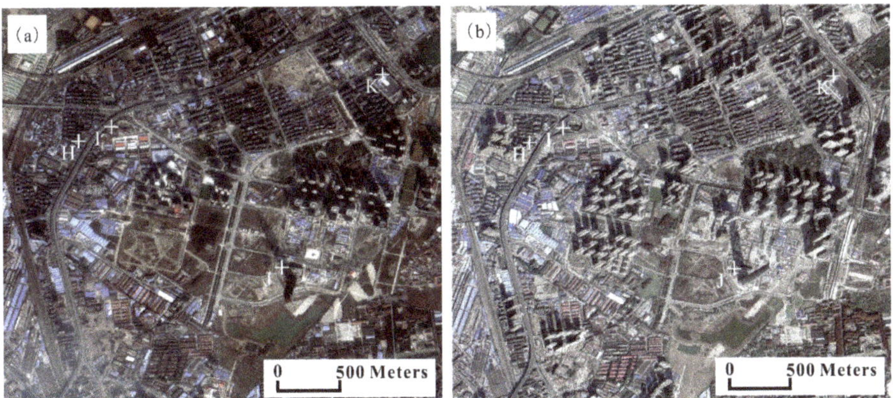

Figure 12. Maps show the satellite images of Region 2 on 21 January 2015 (**a**) and 9 December 2017 (**b**).

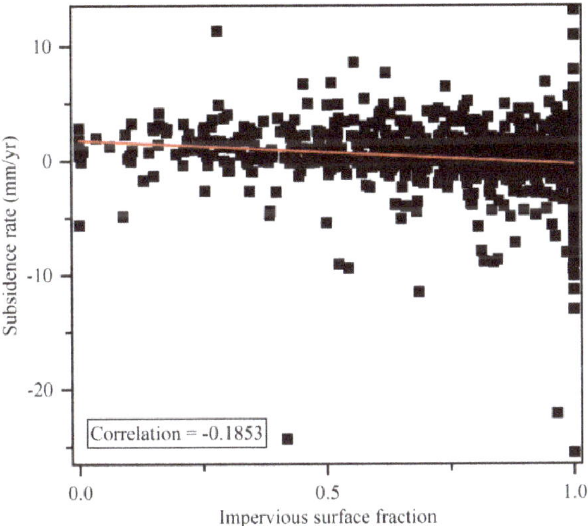

Figure 13. The correlation between subsidence rate and impervious surface fraction.

In this city, soft soils or carbonate rocks are widespread, but only these areas with intensive human activities show severe subsidence, so natural conditions provide a basis for subsidence and make subsidence possible. Human activities are driving factors and make subsidence happen. Therefore, land subsidence is caused by a combination of natural conditions and human activities.

6. Conclusions and Future Work

Our study employs SBAS-InSAR method with Radarsat-2 data for long-term monitoring of land subsidence in a megacity, Wuhan city. The InSAR results are validated by leveling data, and the causes of subsidence are investigated. The results allowed us to draw the following conclusions:

(i) SBAS-InSAR method with Radarsat-2 data could be used for longtime monitoring of land subsidence with acceptable accuracy in Wuhan city; (ii) natural conditions provide a basis for subsidence and make subsidence possible while human activities are driving factors and make subsidence happen.

Despite our success of longtime monitoring of subsidence in a megacity, Wuhan city, other advanced InSAR methods could also be investigated, such as PS-InSAR. Future study will be focused on the causes of subsidence and its spatial differences using spatial regression models. While much work has been conducted to derive land subsidence information in so many cities, the potential applications of subsidence information are rarely discussed. It is also important to explore the application of subsidence information to disaster prevention, urban planning and hydrological modeling.

Author Contributions: Conceptualization, Y.Z.; Methodology, Y.Z.; Validation, W.S.; Formal Analysis, Y.Z. and J.W.; Resources, Y.L. (Yanfang Liu); Data Curation, M.J., Y.L. (Yanfang Liu) and J.W.; Writing—Original Draft Preparation, Y.Z.; Writing—Review & Editing, Y.J., Y.L. (Yi Liu) and Y.C.; Visualization, Y.L. (Yi Liu); Supervision, Y.L. (Yaolin Liu); Funding Acquisition, Y.L. (Yaolin Liu).

Acknowledgments: This research was financially supported by the National Natural Science Foundation of China (No. 41771432). We greatly appreciated the editors and the reviews for their constructive suggestions and insightful comments which helped us greatly to improve this manuscript.

Conflicts of Interest: The authors declare no conflict of interest.

References

1. Bianchini, S.; Moretti, S. Analysis of recent ground subsidence in the sibari plain (Italy) by means of satellite sar interferometry-based methods. *Int. J. Remote Sens.* **2015**, *36*, 4550–4569. [CrossRef]

2. Du, Z.Y.; Ge, L.L.; Ng, A.H.M.; Zhu, Q.G.Z.; Yang, X.H.; Li, L.Y. Correlating the subsidence pattern and land use in bandung, indonesia with both sentinel-1/2 and alos-2 satellite images. *Int. J. Appl. Earth Obs. Géoinf.* **2018**, *67*, 54–68. [CrossRef]

3. Deng, Z.; Ke, Y.H.; Gong, H.L.; Li, X.J.; Li, Z.H. Land subsidence prediction in beijing based on ps-insar technique and improved grey-markov model. *Giscience Remote Sens.* **2017**, *54*, 797–818. [CrossRef]

4. Zhu, L.; Gong, H.L.; Li, X.J.; Wang, R.; Chen, B.B.; Dai, Z.X.; Teatini, P. Land subsidence due to groundwater withdrawal in the northern Beijing plain, China. *Eng. Geol.* **2015**, *193*, 243–255. [CrossRef]

5. Zhou, L.; Guo, J.M.; Hu, J.Y.; Li, J.W.; Xu, Y.F.; Pan, Y.J.; Shi, M. Wuhan surface subsidence analysis in 2015–2016 based on sentinel-1a data by sbas-insar. *Remote Sens.* **2017**, *9*, 982. [CrossRef]

6. Liu, Y.Y.; Zhao, C.Y.; Zhang, Q.; Yang, C.S. Complex surface deformation monitoring and mechanism inversion over Qingxu-Jiaocheng, China with multi-sensor sar images. *J. Geodyn.* **2018**, *114*, 41–52. [CrossRef]

7. Hwang, C.; Yang, Y.D.; Kao, R.; Han, J.C.; Shum, C.K.; Galloway, D.L.; Sneed, M.; Hung, W.C.; Cheng, Y.S.; Li, F. Time-varying land subsidence detected by radar altimetry: California, Taiwan and north China. *Sci. Rep.* **2016**, *6*, 28160. [CrossRef] [PubMed]

8. Gao, M.L.; Gong, H.L.; Chen, B.B.; Zhou, C.F.; Chen, W.F.; Liang, Y.; Shi, M.; Si, Y. Insar time-series investigation of long-term ground displacement at Beijing capital international airport, China. *Tectonophysics* **2016**, *691*, 271–281. [CrossRef]

9. Garcia, A.J.; Bakon, M.; Martinez, R.; Marchamalo, M. Evolution of urban monitoring with radar interferometry in madrid city: Performance of ers-1/ers-2, envisat, cosmo-skymed, and sentinel-1 products. *Int. J. Remote Sens.* **2018**, *39*, 2969–2990. [CrossRef]

10. Ghazifard, A.; Akbari, E.; Shirani, K.; Safaei, H. Evaluating land subsidence by field survey and d-insar technique in Damaneh city, Iran. *J. Arid Land* **2017**, *9*, 778–789. [CrossRef]

11. Pratesi, F.; Tapete, D.; Del Ventisette, C.; Moretti, S. Mapping interactions between geology, subsurface resource exploitation and urban development in transforming cities using insar persistent scatterers: Two decades of change in Florence, Italy. *Appl. Geogr.* **2016**, *77*, 20–37.

12. Qu, F.F.; Zhang, Q.; Lu, Z.; Zhao, C.Y.; Yang, C.S.; Zhang, J. Land subsidence and ground fissures in Xi'an, China 2005–2012 revealed by multi-band insar time-series analysis. *Remote Sens. Environ.* **2014**, *155*, 366–376. [CrossRef]

13. Sowter, A.; Amat, M.B.; Cigna, F.; Marsh, S.; Athab, A.; Alshammari, L. Mexico city land subsidence in 2014–2015 with sentinel-1 iw tops: Results using the intermittent sbas (isbas) technique. *Int. J. Appl. Earth Obs. Géoinf.* **2016**, *52*, 230–242. [CrossRef]

14. Gao, M.L.; Gong, H.L.; Chen, B.B.; Li, X.J.; Zhou, C.F.; Shi, M.; Si, Y.; Chen, Z.; Duan, G.Y. Regional land subsidence analysis in eastern beijing plain by insar time series and wavelet transforms. *Remote Sens.* **2018**, *10*, 365. [CrossRef]

15. Ge, L.L.; Ng, A.H.M.; Du, Z.Y.; Chen, H.Y.; Li, X.J. Integrated space geodesy for mapping land deformation over choushui river fluvial plain, Taiwan. *Int. J. Remote Sens.* **2017**, *38*, 6319–6345. [CrossRef]

16. Luo, X.G.; Wang, J.J.; Xu, Z.Y.; Zhu, S.; Meng, L.S.; Liu, J.K.; Cui, Y. Dynamic analysis of urban ground subsidence in Beijing based on the permanent scattering insar technology. *J. Appl. Remote Sens.* **2018**, *12*, 026001. [CrossRef]

17. Ng, A.H.M.; Ge, L.L.; Li, X.J.; Zhang, K. Monitoring ground deformation in Beijing, China with persistent scatterer sar interferometry. *J. Geod.* **2012**, *86*, 375–392. [CrossRef]

18. Yin, J.; Yu, D.P.; Wilby, R. Modelling the impact of land subsidence on urban pluvial flooding: A case study of downtown Shanghai, China. *Sci. Total Environ.* **2016**, *544*, 744–753. [CrossRef]

19. Luo, Q.L.; Perissin, D.; Lin, H.; Zhang, Y.Z.; Wang, W. Subsidence monitoring of tianjin suburbs by terrasar-x persistent scatterers interferometry. *IEEE J. Sel. Top. Appl. Earth Obs. Remote Sens.* **2014**, *7*, 1642–1650. [CrossRef]

20. Guo, J.M.; Zhou, L.; Yao, C.L.; Hu, J.Y. Surface subsidence analysis by multi-temporal insar and grace: A case study in Beijing. *Sensors* **2016**, *16*, 1495. [CrossRef]

21. Yan, S.Y.; Liu, G.; Deng, K.Z.; Wang, Y.J.; Zhang, S.B.; Zhao, F. Large deformation monitoring over a coal mining region using pixel-tracking method with high-resolution radarsat-2 imagery. *Remote Sens. Lett.* **2016**, *7*, 219–228. [CrossRef]

22. Ge, L.; Chang, H.-C.; Rizos, C. Mine subsidence monitoring using multi-source satellite sar images. *Photogramm. Eng. Remote Sens.* **2007**, *73*, 259–266. [CrossRef]

23. Jiang, L.M.; Lin, H.; Cheng, S.L. Monitoring and assessing reclamation settlement in coastal areas with advanced insar techniques: Macao city (China) case study. *Int. J. Remote Sens.* **2011**, *32*, 3565–3588. [CrossRef]

24. Samsonov, S.; d'Oreye, N.; Smets, B. Ground deformation associated with post-mining activity at the french-german border revealed by novel insar time series method. *Int. J. Appl. Earth Obs. Géoinf.* **2013**, *23*, 142–154. [CrossRef]

25. Le, T.S.; Chang, C.P.; Nguyen, X.T.; Yhokha, A. Terrasar-x data for high-precision land subsidence monitoring: A case study in the historical centre of Hanoi, Vietnam. *Remote Sens.* **2016**, *8*, 338. [CrossRef]

26. Nikos, S.; Ioannis, P.; Constantinos, L.; Paraskevas, T.; Anastasia, K.; Charalambos, K. Land subsidence rebound detected via multi-temporal insar and ground truth data in kalochori and sindos regions, northern Greece. *Eng. Geol.* **2016**, *209*, 175–186. [CrossRef]

27. Zhang, Z.; Wang, C.; Wang, M.; Wang, Z.; Zhang, H. Surface deformation monitoring in Zhengzhou city from 2014 to 2016 using time-series insar. *Remote Sens.* **2018**, *10*, 1731. [CrossRef]

28. Zhou, C.F.; Gong, H.L.; Chen, B.B.; Zhu, F.; Duan, G.Y.; Gao, M.L.; Lu, W. Land subsidence under different land use in the eastern Beijing plain, China 2005–2013 revealed by insar timeseries analysis. *Gisci. Remote Sens.* **2016**, *53*, 671–688.

29. Liu, Y.Y.; Zhao, C.Y.; Zhang, Q.; Yang, C.S.; Zhang, J. Land subsidence in Taiyuan, China, monitored by insar technique with multisensor sar datasets from 1992 to 2015. *IEEE J. Sel. Top. Appl. Earth Obs. Remote Sens.* **2018**, *11*, 1509–1519. [CrossRef]

30. Aimaiti, Y.; Yamazaki, F.; Liu, W. Multi-sensor insar analysis of progressive land subsidence over the coastal city of Urayasu, Japan. *Remote Sens.* **2018**, *10*, 1304. [CrossRef]

31. Castellazzi, P.; Arroyo-Dominguez, N.; Martel, R.; Calderhead, A.I.; Normand, J.C.L.; Garfias, J.; Rivera, A. Land subsidence in major cities of central Mexico: Interpreting insar-derived land subsidence mapping with hydrogeological data. *Int. J. Appl. Earth Obs. Géoinf.* **2016**, *47*, 102–111. [CrossRef]

32. Perissin, D.; Wang, Z.Y.; Lin, H. Shanghai subway tunnels and highways monitoring through cosmo-skymed persistent scatterers. *Isprs J. Photogramm. Remote Sens.* **2012**, *73*, 58–67. [CrossRef]

33. Liu, P.; Li, Q.Q.; Li, Z.H.; Hoey, T.; Liu, Y.X.; Wang, C.S. Land subsidence over oilfields in the yellow river delta. *Remote Sens.* **2015**, *7*, 1540–1564. [CrossRef]

34. Liu, X.T.; Cao, Q.X.; Xiong, Z.G.; Yin, H.T.; Xiao, G.R. Application of small baseline subsets d-insar technique to estimate time series land deformation of Jinan area, China. *J. Appl. Remote Sens.* **2016**, *10*. [CrossRef]

35. Aimaiti, Y.; Yamazaki, F.; Liu, W.; Kasimu, A. Monitoring of land-surface deformation in the karamay oilfield, Xinjiang, China, using sar interferometry. *Appl. Sci.* **2017**, *7*, 772. [CrossRef]

36. Chen, G.; Zhang, Y.; Zeng, R.Q.; Yang, Z.K.; Chen, X.; Zhao, F.M.; Meng, X.M. Detection of land subsidence associated with land creation and rapid urbanization in the chinese loess plateau using time series insar: A case study of Lanzhou new district. *Remote Sens.* **2018**, *10*, 270. [CrossRef]

37. Zhao, Q.; Lin, H.; Jiang, L.M.; Chen, F.L.; Cheng, S.L. A study of ground deformation in the guangzhou urban area with persistent scatterer interferometry. *Sensors* **2009**, *9*, 503–518. [CrossRef] [PubMed]

38. Gutierrez, F.; Parise, M.; De Waele, J.; Jourde, H. A review on natural and human-induced geohazards and impacts in karst. *Earth-Sci. Rev.* **2014**, *138*, 61–88. [CrossRef]

39. Erten, E.; Rossi, C. The worsening impacts of land reclamation assessed with sentinel-1: The rize (Turkey) test case. *Int. J. Appl. Earth Obs. Géoinf.* **2019**, *74*, 57–64. [CrossRef]

40. Chen, B.B.; Gong, H.L.; Lei, K.C.; Li, J.W.; Zhou, C.F.; Gao, M.L.; Guan, H.L.; Lv, W. Land subsidence lagging quantification in the main exploration aquifer layers in Beijing plain, China. *Int. J. Appl. Earth Obs. Géoinf.* **2019**, *75*, 54–67. [CrossRef]

41. Bai, L.; Jiang, L.M.; Wang, H.S.; Sun, Q.S. Spatiotemporal characterization of land subsidence and uplift (2009–2010) over wuhan in central China revealed by terrasar-x insar analysis. *Remote Sens.* **2016**, *8*, 350. [CrossRef]

42. Costantini, M.; Bai, J.; Malvarosa, F.; Minati, F.; Vecchioli, F.; Wang, R.L.; Hu, Q.; Xiao, J.H.; Li, J.P. Ground deformations and building stability monitoring by cosmo-skymed psp sar interferometry: Results and validation with field measurements and surveys. In Proceedings of the 2016 IEEE International Geoscience and Remote Sensing Symposium, Beijing, China, 10–15 July 2016.

43. Zhen, L.; Tan, Y.; Lin, L.; Yu, Z.; Lan, H. Study of land surface composition of Wuhan city based on linear spectral mixture analysis. *Remote Sens. Technol. Appl.* **2013**, *28*, 780–784.

44. Wang, Y.; Jun-Ling, W.U.; Wang, H.L.; Zhao, D.F. The statistical analysis of rainfall in wuhan in the past 50 years. *J. Hubei Univ. Technol.* **2006**, *21*, 98–100.

45. Luo, X.J.; Survey, C. Features of the shallow karst development and control of karst collapse in Wuhan. *Carsologica Sin.* **2013**, *32*, 419–432.

46. Guan, S.; Tao, L.; Xie, J.; Xia, D. The developmental characteristics of karst in Wuhan urban development area. *Urban Geotech. Investig. Surv.* **2017**, *41*, 157–162.

47. Luo, X.J.; Survey, C. Division of "six belts and five types" of carbonate region and control of karst geological disaster in Wuhan. *J. Hydraulic Eng.* **2014**, *45*, 171–179.

48. Zheng, X.C.; Tang, H.M.; Qin, Z.M. Study on the imperilments of soft foundation and land subsidence in Wuhan. *Geol. Sci. Technol. Inf.* **2003**, *22*, 95–99.

49. Chen, Z.; Chen, S.; Lisheng, W.U. Experimental analysis of soft soil characteristics in Wuhan. *Resour. Environ. Eng.* **2015**, *12*, 974–977.

50. Yang, Z; Yanfang, L.; Yi, L. Spatial and Temporal Patterns Analysis of Impervious Surface in Wuhan City. *Sci. Geogr. Sin.* **2017**, *37*, 1917–1924.

51. Liu, Y.L.; Huang, H.J.; Liu, Y.X.; Bi, H.B. Linking land subsidence over the yellow river delta, China, to hydrocarbon exploitation using multi-temporal insar. *Nat. Hazards* **2016**, *84*, 271–291. [CrossRef]

52. Berardino, P.; Fornaro, G.; Lanari, R.; Sansosti, E. A new algorithm for surface deformation monitoring based on small baseline differential sar interferograms. *IEEE Trans. Geosci. Remote Sens.* **2002**, *40*, 2375–2383. [CrossRef]

53. Benattou, M.M.; Balz, T.; Liao, M. Measuring surface subsidence in Wuhan, China with sentinel-1 data using psinsar. In Proceedings of the 2018 ISPRS TC III Mid-term Symposium Developments, Technologies and Applications in Remote Sensing, Beijing, China, 7–10 May 2018.

Article

An Improved Time-Series Model Considering Rheological Parameters for Surface Deformation Monitoring of Soft Clay Subgrade [†]

Xuemin Xing [1,2], Lifu Chen [1,3,4,*], Zhihui Yuan [1,3] and Zhenning Shi [2]

[1]   Laboratory of Radar Remote Sensing Applications, Changsha University of Science & Technology, Changsha 410014, China
[2]   School of Traffic and Transportation Engineering, Changsha University of Science & Technology, Changsha 410014, China
[3]   School of Electrical and Information Engineering, Changsha University of Science & Technology, Changsha 410014, China
[4]   School of Engineering, Newcastle University, Newcastle upon Tyne NE1 7RU, UK
*   Correspondence: Lifu.Chen@newcastle.ac.uk
†   This paper is an expanded version of "Investigation on INSAR Time Series Deformation Model Considering Rheological Parameters for Soft Clay Subgrade Monitoring" published in the Proceedings of the ISPRS TC III Mid-term Symposium 'Developments, Technologies and Applications in Remote Sensing', Beijing, China, 7–10 May 2018.

Received: 14 May 2019; Accepted: 9 July 2019; Published: 11 July 2019

Abstract: Building deformation models consistent with reality is a crucial step for time-series deformation monitoring. Most deformation models are empirical mathematical models, lacking consideration of the physical mechanisms of observed objects. In this study, we propose an improved time-series deformation model considering rheological parameters (viscosity and elasticity) based on the Kelvin model. The functional relationships between the rheological parameters and deformation along the Synthetic Aperture Radar (SAR) line of sight are constructed, and a method for rheological parameter estimation is provided. To assess the feasibility and accuracy of the presented model, both simulated and real deformation data over a stretch of the Lungui highway (built on soft clay subgrade in Guangdong province, China) are investigated with TerraSAR-X satellite imagery. With the proposed deformation model, the unknown rheological parameters over all the high coherence points are obtained and the deformation time-series are generated. The high-pass (HP) deformation component and external leveling ground measurements are utilized to assess the modeling accuracy. The results show that the root mean square of the residual deformation is ±1.6 mm, whereas that of the ground leveling measurements is ±5.0 mm, indicating an improvement in the proposed model by 53%, and 34% compared to the pure linear velocity model. The results indicate the reliability of the presented model for the application of deformation monitoring of soft clay highways. The estimated rheological parameters can be provided as a reference index for the interpretation of long-term highway deformation and the stability control of subgrade construction engineering.

Keywords: deformation model; time series deformation; rheological parameter; highway

1. Introduction

Stability control of highways built on a soft clay subgrade is one of the key technical problems for highway subgrade engineering. Due to the geological characteristics of large natural moisture content, high compressibility, low strength, and poor structure of soft clay, roads built on soft clay subgrade are more prone to displacement and instability, especially under large traffic loads.

Consequently, long-term surface deformation monitoring for infrastructure built on soft clay, after highway embankment settlement construction, is of considerable practical significance to the prevention of transportation safety accidents and the assurance of highway construction quality [1,2]. Although differential interferometric synthetic aperture radar (DInSAR) can cover a shortage of traditional ground measurement methods, its capacity in highway deformation monitoring is limited by its well-known spatial-temporal decorrelation and atmospheric delay effects [3,4]. Time-series technologies, such as permanent scatterer interferometry (PSI) [5], small baseline subset (SBAS) [6,7], temporally coherent point InSAR (TCP-InSAR) [8], and so on [9], have been proven to possess great capacities for large traffic infrastructure monitoring (i.e., railways, highways, and bridges) [10–21]. They can pick up ground displacement information with millimeter-level precision through high coherence points, maintaining long-term stable backscatter characteristics. Thus, they are insusceptible to spatial-temporal decorrelation [22].

Deformation modeling is a crucial step in time-series processing, determining the temporal and functional relationships between the phase component of displacement and the deformation parameters over highly coherent points. An accurate and reliable deformation model can not only improve the accuracy of deformation estimation, but also control the residual phase within a reasonable range of a whole phase cycle $[-\pi, \pi]$. Deformation modeling has a significant impact on the subsequent processing steps, including high coherence point identification, unknown parameter estimation, and phase unwrapping. It can also provide a reference for the interpretation of the final deformation results. Among traditional time-series models, the most commonly used is the linear velocity model, which simply assumes that temporal displacement follows linearly varying characteristics, and treats the deformation rate as a constant parameter over each time-adjacent interferometric period. This model was originally proposed as a PSI technique, and has been successfully applied in a large amount of cases. However, under the assumption of a pure linear varying characteristic among all temporal periods, the linear velocity model has significant limitations. When the real deformation of the monitored object is close to a linearly varying characteristic, the residual phase can be easily suppressed within the range of a whole phase cycle; however, when a strong non-linear component exists in the total displacement, the residual phase may possibly exceed the reasonable range of $[-\pi, \pi]$, thus inducing a non-unique solution of unknown parameters and large inaccuracy. Due to the deficiency of the linear model, some non-linear deformation models have been presented, such as the Seasonal [23,24], Polynomial [25], Hyperbola, and Spline function [26] models. Although these models have achieved fitting of the temporally varying process of deformation for different observed features with better experimental results, they are generally based on a combination of one or several empirical mathematical functions to fit the deformational variations, ignoring the physical mechanism of deformation of the monitored object. The parameters for those models are generally mathematical coefficients that lack physical significance. Soft clay has the properties of mellow soil, large natural water content, and high compressibility. Under the conditions of constant external load, deformation is related to natural compression and the extravasation of the inner water in the soft soil, combined with external environmental factors (such as rainfall and temperature), thus the deformation of soft clay is characterized as an obvious temporal non-linear variation. In particular, for highways built on soft clay, a single pure empirical mathematical function may not describe the actual dynamic evolution, due to its temporally complicated non-linear characteristics, and a negative impact could be imposed on the accuracy of the obtained measurements and the subsequent displacement prediction. This would be adverse to the corresponding long-term analysis and deformation interpretation following highway construction.

According to authoritative statistics, more than 70% of pavement structure damage is related to long-term rheological deformation of the subgrade [27]. The rheological property is one of the primary engineering properties of soft soil, representing the temporal effect of soil deformation. Rheology is a subject that studies the deformation laws of materials over time under certain conditions (e.g., stress and strain) [28]. Rheological parameters (elastic modulus and viscosity) are significant factors for characterizing the rheological properties of soft clay. During the operation step of highway post-construction, the external load can be considered as constant and the underground deformation increases with time, so the rheological deformation plays a dominant impact role. In the theory of rheology, the rheological model is a kind of mechanical model (composed of spring, dashpot, and slide rod) that represents the rheological characteristics of rocks and soil and describes the dynamic temporal evolution process. The most widely used rheological models can be divided into linear models and non-linear models. For linear rheological models, a qualitative analysis of the material is initially carried out, then the corresponding rheological state function is constructed, which quantitatively represents the functional relationship between the strain of soft soil material and physical variables (i.e., viscosity, elastic modulus, and time). These are mainly based on the series-parallel connection of basic mechanical components (i.e., Burgers model, Kelvin model, and Maxwell model, among others). This kind of model can easily and intuitively express complex mechanical properties, which is helpful for conceptually understanding the elastic and visco-elastic properties of soft soil deformation. Their mathematical expressions can directly describe the rheological deformation, and are applicable to the simulation of the initial or stable rheological deformation of rock and soil material [29,30]. However, theoretical models for time-series displacement that consider rheological parameters have been rarely mentioned in previous InSAR deformation studies.

Based on the background discussed above, we propose a time-series deformation model based on rheological theory. The Kelvin rheological model, a typical one-dimensional linear rheological model, is adopted to form a functional relationship between radar line-of-sight deformation and the rheological parameters (elastic modulus and viscosity). The method of rheological parameter estimation is also illustrated in this paper. The proposed model is tested by a simulated experiment and a real data experiment. In the real data scenario, the rheological parameters of a stretch of highway (namely, the Lungui Highway in Foshan, China) are obtained, and the time-series subsidence over the period of June 2014 to December 2015 is investigated using TerraSAR X imagery.

2. Time-Series Modeling Considering Rheological Parameters

2.1. Time-Series Deformation Model

Suppose $M + 1$ SAR images covering the same area are acquired in repeat orbits at different dates. Then, N interferometric pairs may be generated, according to certain temporal baseline and spatial baseline thresholds, where the N interferometric pairs are generated through two-orbit D-InSAR processing, while satisfying the inequality $\frac{M}{2} \leq N \leq \frac{M(M-1)}{2}$. In the processing, all images are registered and resampled to the same image first. Then, an external DEM is used to remove the topographic phase and, consequently, phase unwrapping is carried out for each interferometric pair. For each high coherence point P in the i-th interferogram, the wrapped interferometric phase can be expressed as [31]:

$$\Delta\varphi_i^p = \frac{4\pi}{\lambda}\Delta d_i + \Delta\varphi_i^{topo,p} + \Delta\varphi_i^{res,p} \tag{1}$$

where λ is the radar wavelength (the X band is used in our real data experiment, and the corresponding λ is 3.2 mm); Δd is the line-of-sight cumulative deformation over the time period of the i-th interferogram, indicating the low-pass (LP) component of the total deformation; $\Delta\varphi_i^{topo,p}$ represents the residual topographic phase component, which can be expressed a $\Delta\varphi_i^{topo,p} = \frac{4\pi B_i}{\lambda R_p \sin\theta}\Delta Z^p$, where B_i defines the vertical baseline, θ is the incident angle, R_p represents the distance between the sensor and the target P, and ΔZ^p defines the residual elevation, which is an unknown parameter; and $\Delta\varphi_i^{res,p}$ is the residual phase component, including phase noise, atmospheric delay, and the high-pass (HP) deformation component. Taking the pure linear model as an example, Δd_i in Equation (1) can be written as

$$\Delta d_i = v\, t, \tag{2}$$

where v is the linear deformation rate, which is regarded as a constant parameter over each time-adjacent interferometric period, and t defines the temporal baseline for the i-th interferogram.

2.2. Rheological Model Based on the Kelvin Model

As discussed above, a Linear rheological model can express the complex mechanical properties of the rheological deformation easily and directly. Consequently, a one-dimensional linear rheological model, the Kelvin model, was selected for our experiments. The Kelvin rheological model is a kind of commonly used delay model, based on mechanical composition elements. It is a parallel system with a spring (pure elastomer) and a glue pot (pure viscous body), illustrating the phenomenon that, under the action of stress, the strain of the material does not reach the final strain value immediately, but has a relative lag process. Figure 1 shows a schematic diagram of the combined elements in the Kelvin rheological model. The rheological state equation of the Kelvin model can be written as [32]:

$$\varepsilon = \frac{\sigma_c}{E}\left(1 - e^{-\frac{E}{\eta}t}\right) \tag{3}$$

where ε defines the strain related to the material and σ_c defines a constant external load. When the post-construction operation stage of a highway starts, the external load mainly includes the gravity of the surface layer and the load of the traffic vehicles. However, for practical analysis, the load of the vehicles can be ignored, due to its minor magnitude relative to the gravity of the highway layer. The gravity of the highway layer can be obtained through the investigation of the highway structure and soil mass sample testing in the upper part of the soft soil layer. E represents the elastic modulus of the material, which is also called the deformation modulus; whereas η defines viscosity, also known as the viscosity coefficient. E and η are significant rheological parameters, which are treated as unknown parameters in Equation (3). Finally, t represents the total time span of strain occurrence.

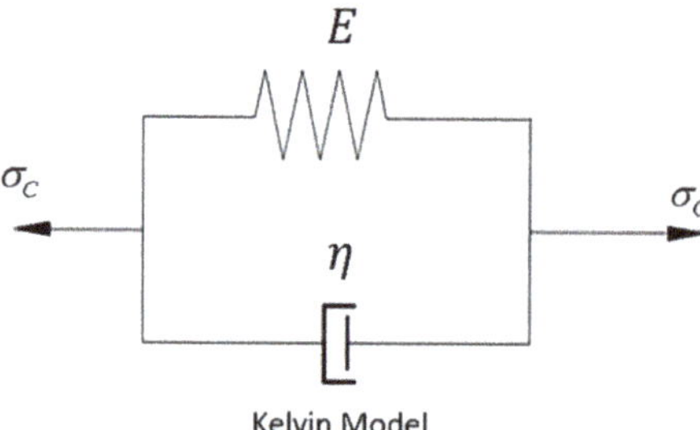

Figure 1. Kelvin rheological model (with a constant external load σ_c).

The functional relationship between the subsidence of the soft clay subgrade S_v and the strain ε can be expressed as [28]:

$$S_v = \int_{t_1}^{t_2} \int_0^H \varepsilon \cdot dh dt, \tag{4}$$

where H is the average thickness of the soft soil layer, which can be obtained by consulting the highway design materials; and h and t are integral variables, representing the soft clay thickness and the time span of subsidence, respectively. The subsidence S_v can be further written as

$$S_v = \frac{H\sigma_c}{E}(t_2 - t_1) - \frac{\eta H\sigma_c}{E^2}\left(e^{-\frac{E}{\eta}t_1} - e^{-\frac{E}{\eta}t_2}\right) \tag{5}$$

2.3. Improved Deformation Model Considering Rheological Parameters

When horizontal movement is ignored, the vertical settlement can be calculated according to the formula $S_{LOS} = S_v \cos\theta$. Combined with Equations (3)–(5), the deformation components related to rheology along the line-of-sight direction can be expressed as

$$S_{LOS\_rhe} = \frac{H\cos\theta\sigma_c}{E}(t_2 - t_1) - \frac{\eta\cos\theta H\sigma_c}{E^2}\left(e^{-\frac{E}{\eta}t_1} - e^{-\frac{E}{\eta}t_2}\right). \tag{6}$$

For each interferogram, t_1 and t_2 represent the acquisition date of master and slave images, respectively. Equation (6) is introduced into the original time-series deformation model, and the low-pass deformation component of the model can be rewritten as a combination of linear and rheological components:

$$S_{LOS} = v(t_2 - t_1) + S_{LOS\_rhe} \tag{7}$$

After substituting Equation (7) into Equation (1), the phase in Equation (1) can be expressed as

$$\Delta\varphi_i^p = \frac{4\pi\cos\theta}{\lambda}\left[\frac{H\sigma_c}{E}(t_2 - t_1) - \frac{\eta H\sigma_c}{E^2}\left(e^{-\frac{E}{\eta}t_1} - e^{-\frac{E}{\eta}t_2}\right) + v(t_2 - t_1)\right] + \frac{4\pi B_i}{\lambda R_p \sin\theta}\Delta Z^p + \Delta\varphi_i^{res,p}. \tag{8}$$

Suppose there are N interferometric pairs generated, and that the unknown parameters in the Equation are the rheological parameters E and η, linear rate v, and elevation correction ΔZ. Supposing that there are at least four interferometric pairs generated, the unknown parameters over all high coherent points of each image can be solved and, consequently, the corresponding rheological parameters can be estimated.

2.4. Unknown Parameter Estimation

The estimation of the unknown parameters in Equation (8) is a non-linear parameter estimation problem. The genetic algorithm for non-linear least-squares estimation is utilized here to estimate the unknown parameters. The genetic algorithm is a method based on global optimization searching that is insusceptible to both the number of unknown parameters and the specific form of the model. The basic idea of the genetic algorithm is to obtain the population individuals as the final solution of the parameters, which can satisfy the condition of minimizing the fitness function through the operations of selection, crossover, and mutation. The population size, iteration times, and individual gene magnitudes for each individual of the population need to be set preliminarily [33]. According to Equation (8), each individual gene of a population includes the rheological parameters (E and η), linear velocity v, and elevation correction ΔZ. The fitness function is mainly modeled following the residual minimum norm principle, which can be expressed as follows:

$$f = \|\Delta\varphi_i^{res,p}\| = \min, \tag{9}$$

where $\Delta\varphi_i^{res,p}$ represents the residual phase in Equation (8). The general search procedure includes the following steps. (1) The magnitude of each initial individual gene should be set, which means the initial value range of each parameter should be fixed, and the corresponding fitness function value of each individual population can be calculated. (2) Whether the iteration termination condition for the minimum fitness function is satisfied should be determined. If not, multiple steps of selection, crossover, and mutation should be carried out to generate a new population of individuals, after which the fitness function value will be calculated again. If it is satisfied, the generated individual genes will be selected as the final estimated parameter. (3) As mentioned in [34], the simplex method can improve the precision of the results generated by the genetic algorithm; thus, we introduce it into our experiment to optimize the searching results. The parameters obtained by the genetic algorithm are taken as the input initial value of the simplex method, and the output optimized searching results are determined as the final solutions.

3. Simulated Experiment

In order to verify the feasibility and accuracy for solving the aforementioned models, a simulated experiment was designed and implemented. The elastic modulus coefficient was set up by investigating the design materials and the structural morphology of the test highway. It was controlled within the interval $[0, 50]$ MPa. The v iscosity η was set within the interval $[0, 8] \times 10^6$ Mpa. The linear deformation velocity v was within the range of $[-0.2, 0.1]$ m/y. The real parameter fields of elastic modulus, viscosity, and linear velocity were simulated by a two-dimensional Gaussian function model. The elevation correction ΔZ was simulated through a Gaussian random simulator, with the value controlled within the range of $[-50, 50]$ m. Linear velocity was simulated using the Matlab peaks function, which can satisfy both positive and negative distribution characteristics of displacement [35]. There were 200 high coherence points and 10 interferograms generated in the simulation. With the known SAR sensor parameters (TerraSAR-X Stripmap data with descent orbital mode was used), including spatial and temporal baselines of each interferometric pair, values of all the parameters for over 200 high coherence pixels could be detected from the simulated field as true values in the following validation.

With the initial estimation of the unknown parameters obtained by the genetic algorithm, the simplex method was used to determine the final solutions. Compared with the real values detected from the simulated field, the accuracy of the model and algorithm were evaluated. Figure 2 shows the comparison between the estimated value of rheological parameters and the real values (the noise level here was 0.5 rad). From Figure 2, we can see that the red and blue broken lines show good consistency, indicating that the estimated parameters were in good agreement with the true values. Table 1 shows the quantitative comparisons of RMSE (root mean square error) for each unknown parameter in Figure 2. For the four unknown parameters, the magnitude of errors accounted for lower than 6% of the mean parameter estimations. The comparative results imply the feasibility and reliability of the aforementioned model and parameter estimation method.

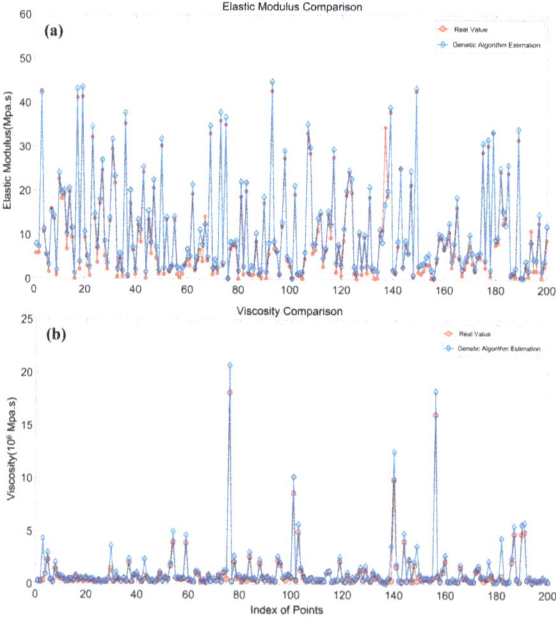

Figure 2. Estimated rheological parameters compared with real values in the simulation (the noise level is 0.5 rad).

Table 1. Comparison of root mean square error (RMSE) for each parameter in the simulated experiment.

Rheological Parameters	E (Mpa)	η (10^6 Mpa s.)	v (mm/y)	ΔZ (m)
RMSE	±1.5870 (3.5%)	±0.1741 (0.8%)	±3.5 (1.3%)	±0.29 (5.4%)

4. Real Data Analysis

4.1. Geological Background of Study Area

The test area selected in this paper was a stretch of a highway; namely, the Lungui Highway, located in Shunde district, Foshan city, Guangdong province, China. The construction of the Lungui road started in March 2011. It was opened to traffic segmentally during the construction process, and the whole route officially opened to traffic in January 2015. The Lungui Highway connects Longzhou Road, Nanguo Road, and Hengjiu Road northward, becoming one of the most important connection channels between the Shunde west district and the three main routes (from south to north) in Shunde Central City. Figure 3a shows the study area, featured at different scales. As Figure 3a shows, the red rectangle outlines the spatial coverage of the selected TerraSAR-X images, while the green rectangle shows the subset for generating the interferometric results. Figure 3b shows the location of the Lungui Highway with the average intensity map as background. As shown in Figure 3b, the Lungui Highway is located close to three hydrological systems: the Xi River and the Rongui and Shunde Branch Rivers. Plenty of ponds and a large amount of silt are distributed along the route.

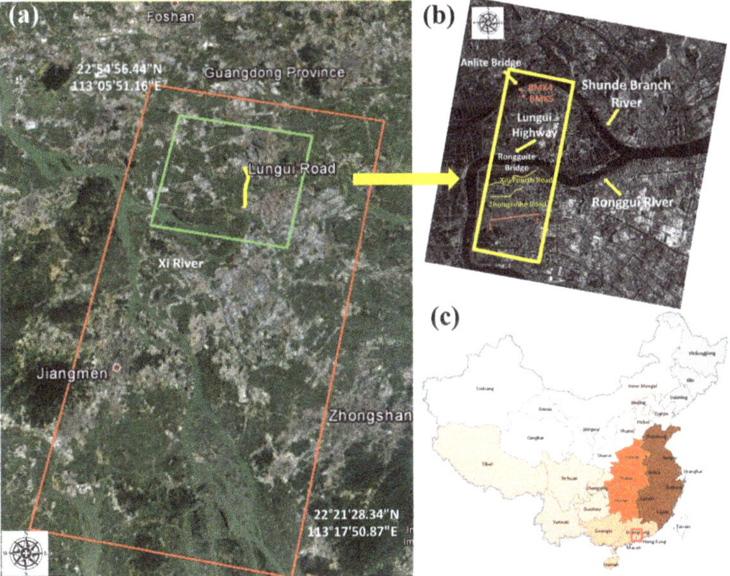

Figure 3. (**a**) Regional scale in China; (**b**) an amplified image of the area within the highway region of interest (outlined in the yellow rectangle); and (**c**) location of the study area in China.

According to the design criteria of the test highway, the permissible vertical post-construction settlement is 30 cm/y for regular road segments, 20 cm/y for culverts, and 10 cm/y for bridge connections. According to the statistics of the Fuoshan Transportation Bureau, the passenger flow volume in 2014 of the Shunde District, where the highways are located, was up to 2018.31 million people per kilometer, whereas the freight flow on the test highways was approximately 654.64 million tons per kilometer.

This huge traffic flow indicates the significant traffic situation of the Lungui Highway. According to our collected geological material, with a developed surface water system and extensive aquifers, the soft soil covering the upper layer is extremely soft and has high compressibility. Delta alluvial and silt plain dominates the topography of the area. Due to these geological characteristics, the subgrade of the highway is extremely prone to liquefaction and seismic subsidence. For this reason, long-term stability monitoring of this area is critically necessary. The yellow rectangle in Figure 3b defines the test stretch of Lungui road of interest in our experiment. Two major bridges, namely the Rongguite and Anlite Bridges, are contained in the test highway. Figure 3c shows the corresponding location of the test area on the China map.

Figure 4 shows the transversal profile of section LL′ in the test area (see the red solid line at the bottom of Figure 3b), where 2% and 0.82% represent the gradient, and the average thickness of the soft soil layer in this cross-section is 4.5 m. From the transversal distribution along the test highway, the main distribution characteristics of the geotechnical layer are as follows: ground layer distribution is pseudo-viscosity plain fill and a quaternary system of brand-new sea-land cross stratum. The surface quaternary is mainly composed of silty soil and mealy sand, deposited by sea and land, including mucky clay, silty soil, and mealy sand. The underground strata below the Rongguite and Anlite Bridges are mainly argillaceous siltstone and silty mudstone. From the longitudinal distribution characteristics of the route, the soft soil layer in the north of Rongguite Bridge is mainly composed of continuously distributed mucky clay, with a thickness of 12.93–19.50 m. In the section between Ronguite Bridge and Xiti Fouth Road, the main components of the soft soil are mucky clay and silty clay, with a thickness of 6.46–9.90 m. In the section between Xiti Fouth Road and Zhongxinhe Road, mucky clay and mealy sand dominate the soft clay layer, with a thickness of 5.57–11.62 m. In the last section, south of Zhongxinhe Road, the soft layer is mainly mealy sand, with a thickness of 3.57–7.62 m (as shown in Figure 4b).

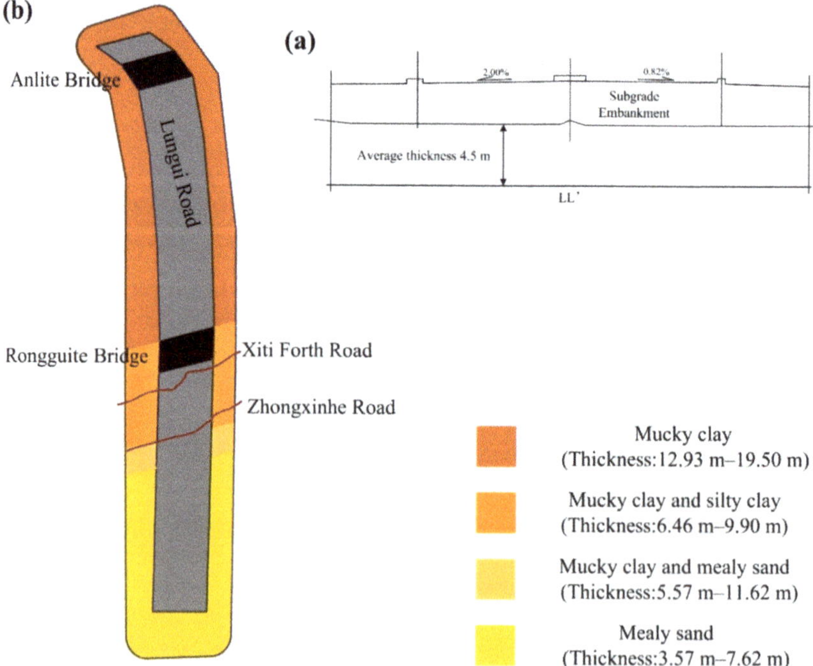

Figure 4. (a) Transversal profile at LL′ in Figure 3a and (b) geological distribution of the soft soil along the longitudinal direction of the Lungui Highway.

4.2. SAR Acquisition and Data Processing

A total of 17 repeat-pass TerraSAR X-band Stripmap images were collected (orbit no. 119, descending), with a spatial resolution of 3 m (3.29 m along azimuth, 2.64 m along range, average incidence angle of 26.4°). These acquisitions covered the period from 17 June 2014 to 27 November 2015. The parameters of these TerraSAR-X images are listed in Table 2. In the processing of the two-pass differential interferometry, a subset of 18 × 15 km was selected, covering about a quarter of the total area (see Figure 3a). SBAS processing was used to generate the unwrapped interferograms for the test area. Due to the narrow ribbon characteristics of our observed object, the multi-look ratio along range and azimuth directions was set as 1:1 to ensure the original resolution of the test highway. The thresholds for the temporal-spatial baseline of the interferometric combination were empirically set to 130 m and 300 days, respectively. SARScape 5.2 and Envi 5.3 were used in our experiment to generate a total of 57 small baseline interferometric pairs. Figure 5 shows the spatial and temporal baseline for all the interferometric combinations in our experiment. The numbers 0–16 in Figure 5 correspond to each SAR image, and number 7 represents the index of the selected super master image (acquired on 14 February 2015). In the two-pass D-InSAR processing, all the rest of the images were registered and resampled to the super master image. In order to remove the topographic phase, a 1-arc-second Shuttle Radar Topography Mission digital elevation model (SRTM DEM, ~30 m spacing) provided by NASA was utilized. In addition, a Gaussian filter was selected to suppress the phase noise. After the flat earth phase removal and phase filtering processing, a polynomial fitting method was used to remove the orbital error and, then, the commonly used minimum cost flow (MCF) method was utilized to unwrap the wrapped interferometric deformation phases [36]. Finally, a total of 57 unwrapped interferometric images were generated. Figure 6 shows the selected interferometric images and the average coherence map.

Table 2. List of the interferometric pairs and their parameters with image number 7 as the master (orbit no. 119, descending).

Image No.	Acquisition Date (yyyy/mm/dd)	Normal Baseline (m)	Temporal Baseline (Days)
1	2014/06/17	−71.50	198
2	2014/08/22	−137.97	132
3	2014/09/13	−286.33	110
4	2014/10/05	−110.85	88
5	2014/10/27	−249.06	66
6	2014/11/18	−74.56	44
7	2015/01/01	0	0
8	2015/02/14	−133.14	44
9	2015/03/08	−106.99	66
10	2015/05/13	−271.51	132
11	2015/06/26	−122.85	176
12	2015/08/09	−149.22	220
13	2015/08/31	−65.63	242
14	2015/09/22	−253.29	264
15	2015/10/14	−159.34	286
16	2015/11/05	−233.83	308
17	2015/11/27	−11.87	330

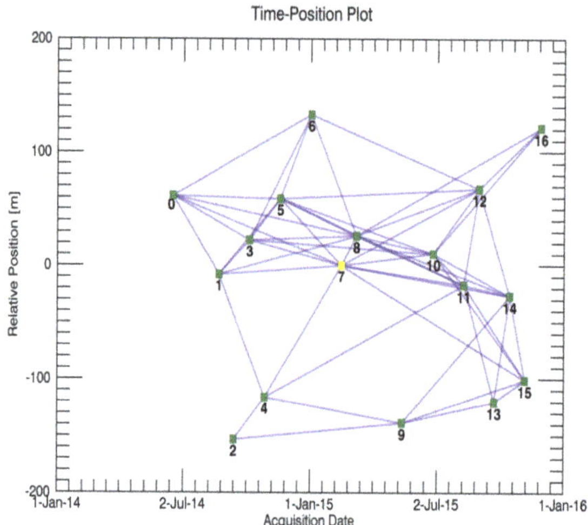

Figure 5. Temporal and perpendicular baselines of the available pairs.

Figure 6. Selected interferometric images and the average coherence map (the last picture, bottom right) of the area shown in Figure 3a.

During processing, high coherence candidates were selected, based on a coherence threshold of 0.6. In order to ensure that the most coherent points were distributed over the highway among the 57 total interferometric pairs, we selected the interferograms carefully and deleted those with bad coherence and less points along the route. Consequently, only 25 high-quality pairs with densely distributed coherence pixels over the highway region were selected. The subsequent experiments, including rheological parameter estimation and time-series deformation inversion, were carried out using MATLAB. Due to the large amount of densely distributed coherent points, the search operation of the genetic algorithm was extremely time consuming; thus, we masked the targets distributed along the route as our observed pixels. According to our in situ investigation and the design materials collected from the highway construction company, Fuoshan, China, we found a section of the test highway which was still under road surfacing from November 2014 to December 2014, and the whole route was opened to traffic in January 2015. We also downloaded the corresponding Google Earth maps covering the test area, which are shown as Figure 7. As shown in Figure 7, from November 2014 to December 2014 the area located in the yellow rectangle was without a road surface, whereas, in the map acquired on January 2015, the surfacing was finished; thus, the highway was opened to traffic entirely at that time. In order to ensure the accuracy of our deformation results, we deleted the coherent points located in the yellow rectangle (due to their low coherence, the number of coherent candidates in the highlighted area was actually significantly lower than those in the other stretches of the highway). Finally, 6657 highly coherent points were selected.

Figure 7. Study area on multi-temporal google maps. The area within the yellow rectangle was still under construction until December 2015.

The phase model of Equation (8) was established for each high coherence point, and the unknown parameters (v, E, η, and ΔZ) were obtained by the methods discussed in Section 2.4. Based on the investigation of geological data and rock structure characteristics in the test area, the initial individual gene range was set as follows: the elastic modulus coefficient E was set within the range of $[0, 50]$ MPa, the viscosity η within the range of $[0, 8] \times 10^6$ Mpa·s, the linear velocity v was in the interval $[-0.5, 0.2]$ m, and the elevation correction ΔZ was in the interval $[-50, 50]$ m. In the process of the genetic algorithm search, the upper threshold for the genetic population was set to 700 generations, with 1000 individuals in each population and a crossover probability of 0.7. The crossover mode was two-point crossover, and a Gaussian function was selected as the mutation function. The termination condition for the program iterations was minimization of the average fitness function value, which indicated a stable fitness value. Finally, the individual genes of population satisfying the fitness function value condition were selected as the estimated unknown parameters for the coherent point. Substituting the obtained final solutions of the unknowns into Equation (5), the low-pass (LP) component of time-series deformation (LP-deformation) could be acquired. Subsequently, in order to obtain the high-pass (HP) deformation component (HP-deformation), the residual phase in Equation (8) was processed with temporal high-pass filtering and spatial low-pass filtering [31]. The final time-series of the deformation were obtained through the sum of HP- and LP-deformation components on each pixel.

4.3. Experimental Results

Figure 8 shows the results of the four unknowns, for all coherent points, in Equation (8). All images were in the slant-range projection. It can be seen, from Figure 8a,b that the elastic modulus was generally distributed within the range of [1.5, 5] Mpa, whereas the viscosity was distributed in [2, 6] × 10⁶ Mpa. From the color distribution, both rheological parameters gradually varied near Xiti Fourth Road, whereas a clear color change boundary appeared next to Zhongxinhe Road. According to the field investigation of the area, from Xiti Fourth Road to Anlite Bridge along the route, breeding ponds (see the little black rectangles in Figure 8) and villages were densely distributed, and few typical urban buildings could be found near this stretch. The soil along this segment was mainly mucky clay and silty clay, as mentioned in Section 4.1. In contrast, the area below Zhongxinhe Road was generally urban districts, with densely distributed residential constructions including banks, office buildings, and other urban infrastructure. Correspondingly, the mucky content in the soil along this stretch was relatively low. Figure 8c shows the linear velocities in Equation (8), with an overall distribution of −50 to 20 mm/y. Similar color distribution characteristics can be found in the figure, with Zhongxinhe Road as an obvious boundary. The detected subsidence rate in the upper region was relatively obvious, with maximum value of 67 mm/y, whereas the area below Zhongxinhe Road was more stable, with the deformation velocity generally lower than 10 mm/y. Figure 8d shows the overall distribution of height corrections, and the results are generally within the interval of [−50, 40] m, with a maximum DEM error of 95 m.

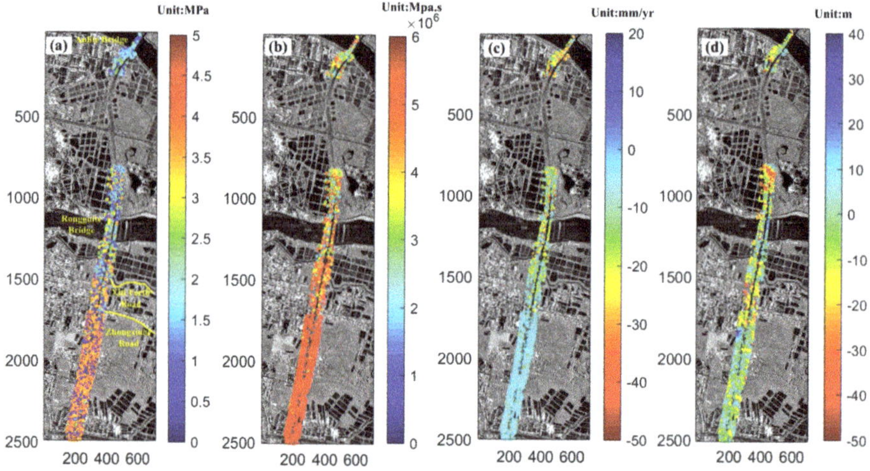

Figure 8. Estimated model parameters: (**a**) elasticity, E; (**b**) viscocity, η; (**c**): linear velocity, v; and (**d**) height correction, ΔZ.

Figure 9 shows the overall time-series deformation results obtained for the test highway. From the spatial characteristics of the color distribution, we can see that Zhongxinhe Road is still the obvious dividing line in the images (with a dark orange color in the upper area), and that the maximum subsidence was 124 mm (on 27 November 2015). The deformation was significantly weaker in the bottom part (generally blue-green), with a maximum subsidence of only 48 mm. It can also be seen, from the temporal color variation in Figure 9, that the deformation was rapid subsidence from the initial time to May 2015, with the subsidence velocity decreasing slowly. However, from June 2015, the deformation showed a relatively stable performance, and even a slight uplift.

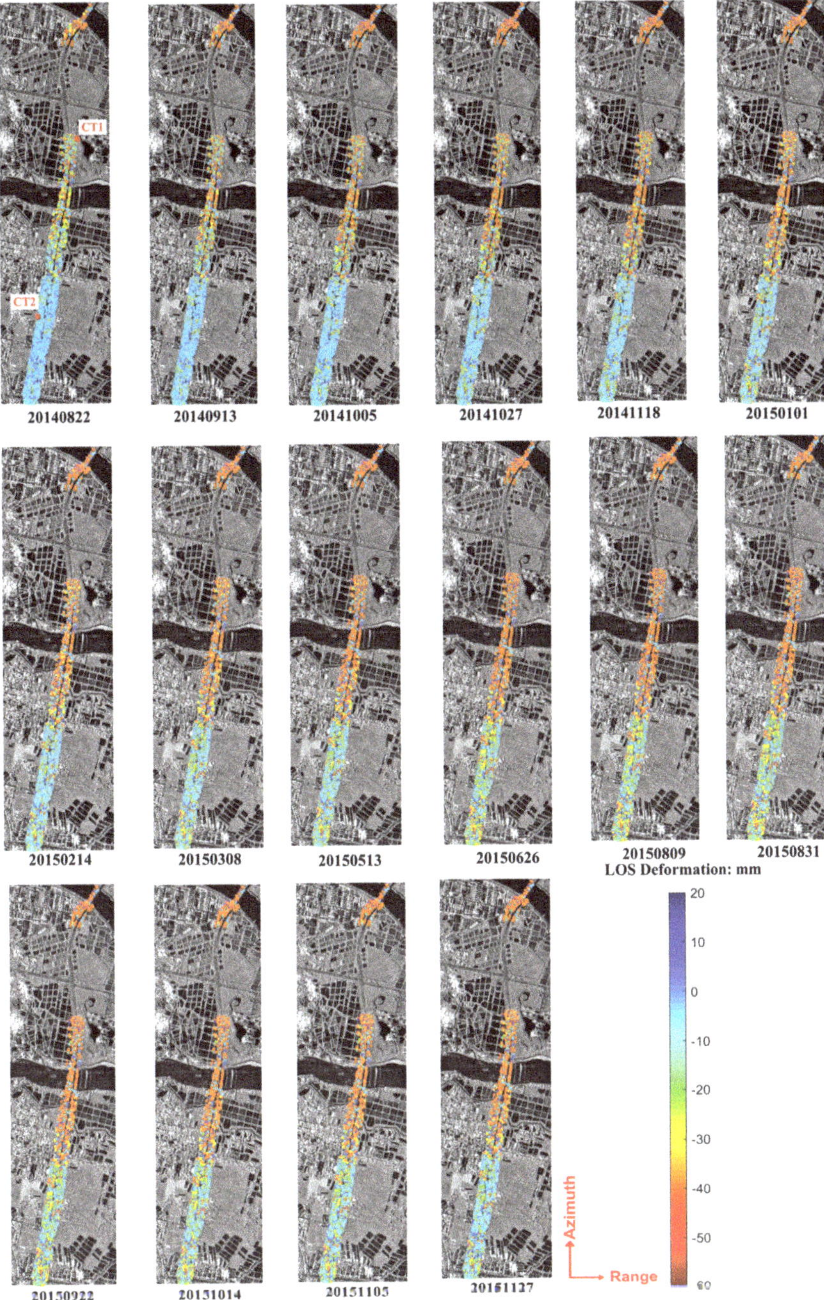

Figure 9. Time-series deformation over the tested area (with reference to 17 June 2014).

5. Discussions

5.1. Potential Reasons for the Deformation

According to the above analysis, the spatial characteristics of subsidence were related to the following:

(1). The magnitude of the obtained elastic modulus and viscosity parameters reflect the aforementioned deformation characteristics. It can be obviously seen, from Figures 8 and 9, that the bottom area was under a relatively stable deformation, with higher elastic modulus and viscosity values. Under the condition of unidirectional stress, the elastic modulus equals the stress divided by the strain along the direction [37]. As Equation (4) shows, deformation can be understood as a temporal integration of strain. Therefore, when the external load is constant, the stress can be considered a constant, and the higher the elastic modulus is, the lower the deformation performs. The physical parameter viscosity (also known as the viscosity coefficient) is a measure to describe the viscosity of a fluid, which is a demonstration of the fluid flow dynamics for its internal friction phenomenon [38]. Higher viscosity reflects greater friction in fluid. In this paper, viscosity is treated as the parameter that reflects the internal friction property of soil mass and its ability to resist deformation. The higher the value of the viscosity, the greater the friction resistance between the soil mass is and, thus, less strain and deformation. This is the key reason why the areas with low deformation showed a higher magnitude of elastic modulus and viscosity.

(2). As described in Section 4.1, with densely distributed ponds around the upper stretch of the highway, mucky clay and silt dominated the geological content of the clay, and the soft soil layer of the upper segment was relatively thicker (with a thickness of 12.93–19.50 m). In addition, the water system around this stretch was well-developed and, thus, the ground subsidence was more obvious. In contrast, the areas below Zhongxinhe Road in the image were mainly urban districts, where the silt content in the soil of the road foundation was lower, with mealy sand and silty clay as the dominant geotechnical content. Furthermore, compared to the upper stretch of the highway, the average thickness of the soft soil layer was only 3.57–7.62 m, with a lower water discharge flow under the surface and an advanced drainage system in the urban areas. As a result, the settlement was much weaker.

5.2. Temporal Deformation Characteristics over Feature Points

In order to further investigate the temporal variation characteristics of deformation, two feature points (CT1 and CT2) were selected for analysis (the locations are shown in the first image of Figure 9), with comparison to the results obtained through the pure linear model (see Figure 10). CT1 was located in the bottom area of the image, where the subsidence was quite obvious (with an accumulated subsidence up to 185 mm), whereas the maximum subsidence in the linear velocity model was only 47 mm. The deformation difference between the two models was mainly due to the large rheological component of the deformation obtained at point CT1. According to the estimation of Equation (8), the linear component at CT1 only accounted for 24% of the total deformation; in contrast, the rheological component accounted for 71%, and the residual deformation isolated from the residual phase accounted for 5%. As shown in the results of the pure linear velocity model, the linear deformation component accounted for 84% of the total deformation, and the non-linear part of the residual phase accounted for 16%. This indicates that a majority of the non-linear deformation may not be reflected in the residual phase of the linear velocity model when a substantial real non-linear deformation has occurred. Therefore, the obtained non-linear deformation result, isolated from the residual phase of the linear model, may have a large deviation from the real value; thus, it shows a significant difference from the rheological model. As shown in Figure 10, the overall deformation sequences obtained by the rheological model displayed an obvious non-linear trend, whereas temporally continuous linear subsidence characteristics were displayed by the linear model. As shown in Figure 9, CT2 was located in the bottom, relatively stable area, and the corresponding time-series deformation obtained through both models are shown in Figure 10b. Due to the relatively low residual phase component in the linear model and the low rheological deformation component estimated in the rheological model, the results for the two models at CT2 were consistent during the period of June 2014 to August 2015, with a maximum difference of only 7 mm, and a maximum subsidence of 32 mm for the rheological model.

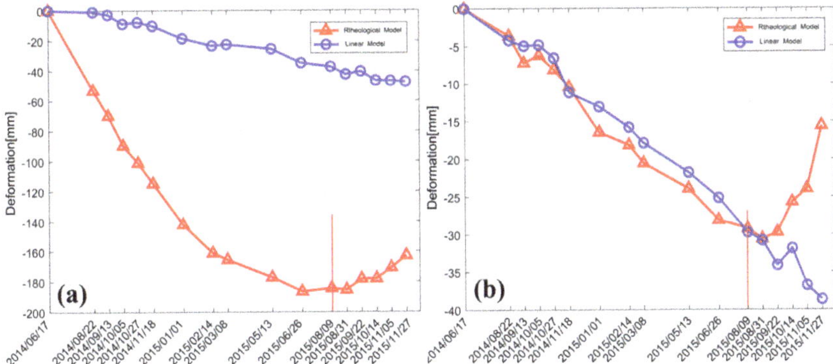

Figure 10. Time-series results on two feature points: (**a**) CT1 and (**b**) CT2 (with reference to 2014/6/17).

It can also be seen, from Figure 10b, that the rheological model results showed a slow subsidence recovery from August 2015, with a magnitude of 14 mm at CT1 and 15 mm at CT2. The reasons for this are proposed to be related to the following:

(1). Soft clay has the property of mellow soil, large natural water content, and high compressibility. During the period of June 2014 to June 2015, under the conditions of constant external load, the void between the soil mass was being compressed and the inner water was being released; thus, the deformation during this period was characterized as obvious subsidence with a decreasing velocity.

(2). Second, as discussed in Section 4.2, a stretch in the middle of the test highway was still undergoing road surfacing from June 2014 to November 2014, and compaction of the soft soil layer in the middle section may have accelerated the subsidence phenomena in nearby stretches.

(3). With the passage of time, when the natural compression of the soil reaches its limit and the porosity ratio drops to the minimum, the deformation caused by the early external load and extravasation of the inner water in the soft soil ceases. Consequently, the subsequent deformation was mainly affected by external environmental factors. According to precipitation data provided by the Fuoshan Meteorological Bureau, the annual precipitation in 2015 was 2055.2 mm, 20% higher than previous years. The spatial and temporal distributions of annual precipitation were extremely asymmetric, being three or four times higher in October and December. Extreme weather events, such as thunderstorms, wind, hail, and tornadoes occurred frequently, and disasters induced by rainstorms and typhoons were obvious. Under the combined influence of high trough and low vortex, continuous precipitation had occurred in the city from 27 August 2015. The average rainfall amount was 127.3 mm in Fuoshan city, with 100–250 mm recorded at 65% of the automatic stations, and 250 mm at Shunde automatic station [39]. Due to the increase of rainfall, both water content and discharge in the water system correspondingly increased. With accelerated flow speed, perched ground water in the upper layer of the subgrade increased due to the impact of rainfall and the supply from the surrounding water system. This is the key reason we suppose as the cause of the subsidence recovery that occurred from August 2015.

5.3. Comparative Analysis with other Non-Linear Time-Series Models

We also conducted an experiment based on a polynomial model to generate the time-series deformation over this route, according to [25]. The temporal displacement over the two feature points are shown in Figure 11 (we only showed the LP-deformation component). As Figure 11 shows, we can see that in the early stage (the period from June 2014 to February 2015) the temporal variation characteristic was a stable deformation velocity, whereas an obvious significant increase in deformation velocity was present in the later stage (the period from March 2015 to November 2015). As discussed above, the subsidence velocity should more reasonably follow a temporally gradual decrease for soft clay, which indicates that the polynomial model is not suitable here. Additionally, the accumulated displacements over both points were close to 300 mm over the test period, which exceeds the critical permissible maximum subsidence for a highway area (according to the design materials for the test road, the permissible subsidence is 20 cm). Our suggested reason for this incorrect result is related to the polynomial model itself. A polynomial model is a certain mathematical empirical modeling function with the significant advantage of spatial approximation. However, when the real temporal displacement variation does not follow the characteristics of a polynomial function, the estimated deformation results may be incorrect. Similar unexpected results may occur in other, similar time-series models (e.g., the logistic and hyperbolic models).

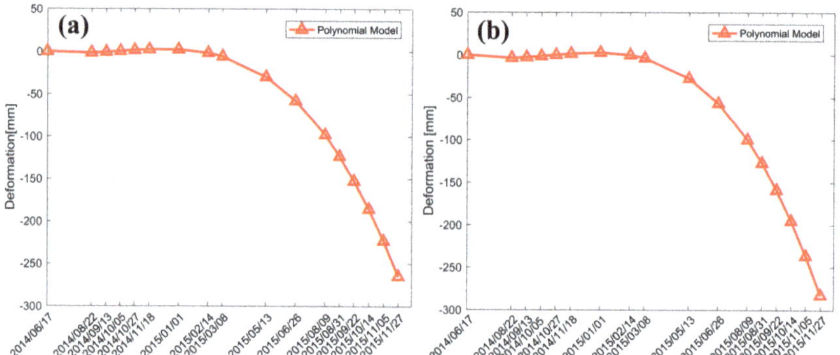

Figure 11. Low pass (LP)-deformation of the two feature points, derived from a polynomial model: (**a**) CT1 and (**b**) CT2 (with reference to 17 June 2014).

5.4. Accuracy Evaluation

According to [25], the fitting accuracy of a deformation model can be reflected by the HP-deformation component. The smaller the HP-deformation is, the higher the accuracy of the selected model. The HP-deformation of each interferogram obtained through the rheological model was compared with that of the linear velocity model. Figure 12 shows a comparison of the average residual deformation over all high coherence points for each interferogram. It can be seen from the figure that all deformation was within 8 mm, where the deformation in the 2nd, 4th, 12th, and 18th interferograms was obviously higher, indicating that the residual phases of these images were large. For those images, the HP-deformation of the rheological model was obviously smaller than that of the linear model. For the rheological model, the variance of HP-deformation over all the interferograms was 1.6 mm, whereas that of the linear model was 2.4 mm, indicating an accuracy increase of 33% in the rheological model.

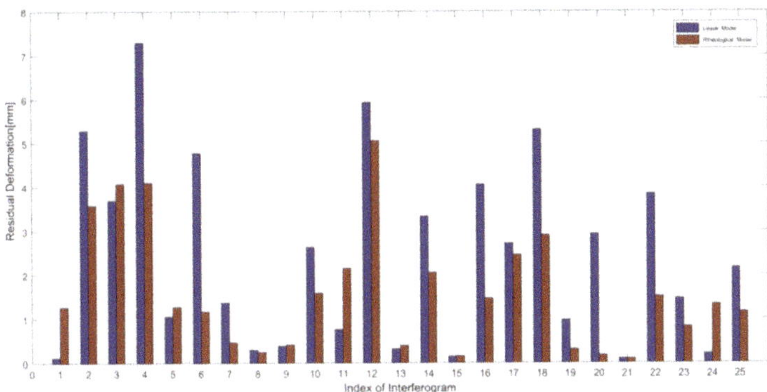

Figure 12. RMS of residual deformation of a 25-interferogram comparison for two models.

In addition, ground measurements of two leveling points in the test area were collected (the locations of the leveling points are shown in Figure 3a, close to Anlite Bridge). The temporal span of leveling measurement was from June 2014 to February 2015. In order to carry out an accurate comparison, we transferred the generated Line of Sight (LOS) deformation into vertical displacement according to the equation $S_{LOS} = S_v \cos\theta$, and extracted the eight dates of the measurement data that coincided temporally with our SAR acquisition dates. The total reference point of leveling and all the SBAS processing methods were the same pixel, which was selected according to our in situ investigation and registered deformation material collection. The comparison results are shown in Figure 13, where the blue solid squares represent deformation results obtained by the rheological model, and the purple solid triangles illustrate the linear model results. It can be clearly seen in Figure 13 that the rheological model results were closer to the leveling results. Table 3 shows the quantitative comparison results of the root mean square error (RMSE) on the benchmarks. According to our calculation, the RMSE of linear model was ±10.7 mm, while the rheological model was ±5.0 mm, with an improvement of about 53%.

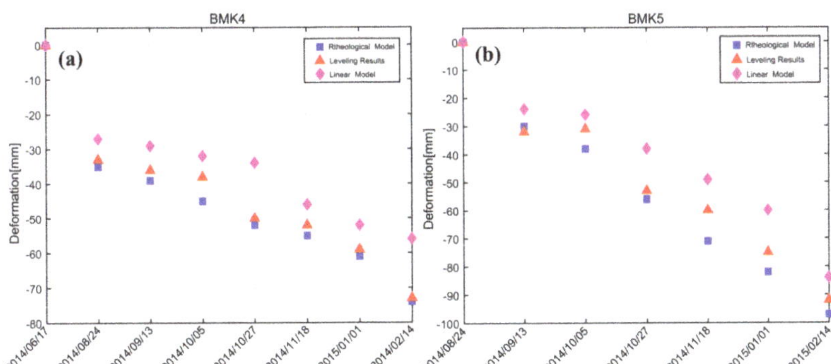

Figure 13. Time-series deformation results compared with leveling measurements on benchmarks: (a) BMK 4 in Figure 3a, and (b) BMK 5.

Table 3. RMSE comparison on benchmarks (mm).

	BMK4	**BMK5**	**RMSE**
Linear velocity model	±10.3	±11.0	±10.7
Rheological model	±3.4	±6.5	±5.0

6. Conclusions

In this paper, a time-series deformation model considering rheological parameters was proposed, and the rheological parameters of elastic modulus and viscosity were introduced into a traditional empirical functional model. Based on the functional relationship between strain and time in the Kelvin rheological model, the function between LOS deformation and rheological parameters was established and the original linear deformation model was improved. The genetic algorithm method was used to solve for the initial values of the model parameters, and the simplex method was used for subsequent optimization. In order to verify the feasibility and reliability of the model and the parameter estimation algorithm, a simulated experiment was designed to obtain the RMSE of four unknown parameters in the model (viscosity, elasticity modulus, linear velocity, and height correction). In the real data experiment, a stretch of highway in Fuoshan, Guangdong province was selected as the test area. The SBAS algorithm was used to process 16 TerraSAR X high-resolution images. Four unknown parameters of the measured area were estimated, and the time-series deformation results of the measured area were inverted eventually. Through an analysis of the results, we found that the higher the elastic modulus and viscosity were, the lower the deformation was. It can also be concluded that the overall temporal characteristics of the time-series deformation showed a non-linear trend of variation, with a gradual decrease of subsidence velocity in the early stage and a small recovery in the later stage. In order to verify the reliability of the results, the HP-deformation component of the interferogram was analyzed. Compared with the pure linear velocity model, the HP-deformation component of the rheological model was reduced by 34%, indicating that the modeling effect was effectively improved after adding the non-linear rheological component into the model. The external accuracy was evaluated by ground-level measurements, with an RMSE of ±5.0 mm for the proposed model, an improvement of 53% compared with the pure linear velocity model.

In our processing, it was very time consuming to carry out a point-wise genetic search algorithm, so the parameter results for the whole area of Lungui Road and its surroundings were not obtained. Our future study will be focused on the parameter estimation algorithm, in order to improve efficiency and accuracy.

Author Contributions: X.X. designed the experiments and produced the results; X.X. and L.C. analyzed the precipitation data; X.X., Z.Y., and Z.S. helped to collect and analyze the leveling measurement in the real data experiment; Z.Y. and L.C. contributed to the discussion of the results; and X.X. drafted the manuscript. All authors contributed to the study and reviewed and approved the manuscript.

Funding: This work was supported by the National Natural Science Foundation (No: 41701536, 61701047, 41674040, 41201468) of China, the Natural Science Foundation of Hunan Province (No. 2017JJ3322, 2019JJ50639), the Key Project of Education Department of Hunan Province (No. 18A148, 16B004), and the Key Laboratory of Special Environment Road Engineering of Hunan Province (No. kfj130405).

Acknowledgments: The TerraSAR dataset used in this study was provided by DLR (Deutsches Zentrum für Luftund Raumfahrt: DLR. No: MTH3393).

Conflicts of Interest: The authors declare no conflict of interest.

References

1. Xue, X.; Song, L.; Jia, L.; Le, Y.; Ge, H. New prediction method for postconstruction settlement of soft-soil roadbed of expressway. *Chin. J. Geotech. Eng.* **2011**, *33*, 125–130.
2. Zhang, J.; Peng, J.; Zheng, J.; Yao, Y. Characterisation of stress and moisture-dependent resilient behaviour for compacted clays in south china. *Road Mater. Pavement Des.* **2018**, 1–14. [CrossRef]
3. Hanssen, R.F. *Radar Interferometry: Data Interpretation and Error Analysis*; Kluwer Academic: New York, NY, USA, 2001.
4. Zebker, H.; Villasenor, J. Decorrelation in interferometric radar echoes. *IEEE Trans. Geosci. Remote Sens.* **1992**, *30*, 950–959. [CrossRef]
5. Ferretti, A.; Prati, C.; Rocca, F. Permanent scatterers in SAR Interferometry. *IEEE Trans. Geosci. Remote Sens.* **2001**, *39*, 8–20. [CrossRef]

6. Berardino, P.; Fornaro, G.; Lanari, R.; Sansosti, E. A new algorithm for surface deformation monitoring based on small baseline differential SAR interferograms. *IEEE Trans. Geosci. Remote Sens.* **2002**, *40*, 2375–2383. [CrossRef]

7. Lanari, R.; Mora, O.; Manunta, M.; Mallorqui, J.J. A small-baseline approach for investigating deformations on full-resolution differential SAR interferograms. *IEEE Trans. Geosci. Remote Sens.* **2004**, *42*, 1377–1386. [CrossRef]

8. Zhang, L.; Lu, Z.; Ding, X.; Jung, H.S.; Feng, G.; Lee, C.W. Mapping ground surface deformation using temporally coherent point SAR interferometry: Application to Los Angeles Basin. *Remote Sens. Environ.* **2012**, *117*, 429–439. [CrossRef]

9. Hu, X.; Oommen, T.; Lu, Z.; Wang, T.; Kim, J.W. Consolidation settlement of Salt Lake County tailings impoundment revealed by time-series InSAR observations from multiple radar satellites. *Remote Sens. Environ.* **2017**, *202*, 199–209. [CrossRef]

10. Zhu, J.; Li, Z.; Hu, J. Research progress and methods of InSAR for deformation monitoring. *Acta Geod. Cartogr. Sin.* **2017**, *46*, 1717–1733.

11. Lin, H.; Ma, P.; Wang, W. Urban infrastructure health monitoring with spaceborne aperture radar interferometry. *Acta Geod. Cartogr. Sin.* **2017**, *46*, 1421–1433.

12. Qin, X.; Yang, M.; Wang, H.; Yang, T.; Lin, J.; Liao, M. Application of high-resolution PS-InSAR in deformation characteristics probe of urban rail transit. *Acta Geod. Cartogr. Sin.* **2016**, *45*, 713–721.

13. Tapete, D.; Morelli, S.; Fanti, R.; Casagli, N. Localizing deformation along the elevation of linear structures: An experiment with space-borne InSAR and RTK GPS on the Roman aqueducts in Rome, Italy. *Appl. Geogr.* **2015**, *58*, 65–83. [CrossRef]

14. Lazecky, M.; Perissin, D.; Bakon, M.; De Sousa, J.M.; Hlavacova, I.; Real, N. Potential of satellite InSAR techniques for monitoring of bridge deformations. In Proceedings of the 2015 IEEE Joint Urban Remote Sensing Event (JURSE), Lausanne, Switzerland, 30 March–1 April 2015; pp. 1–4.

15. Dong, S.; Samsonov, S.; Yin, H.; Ye, S.; Cao, Y. Time-series analysis of subsidence associated with rapid urbanization in Shanghai, China measured with SBAS InSAR method. *Environ. Earth Sci.* **2014**, *72*, 677–691. [CrossRef]

16. Yu, B.; Liu, G.; Li, Z.; Zhang, R.; Jia, H.; Wang, X.; Cai, G. Subsidence detection by TerraSAR-X interferometry on a network of natural persistent scatterers and artificial corner reflectors. *Comput. Geosci.* **2013**, *58*, 126–136. [CrossRef]

17. Xing, X.M.; Wen, D.; Chang, H.C.; Chen, L.F.; Yuan, Z.H. Highway deformation monitoring based on an integrated CRInSAR algorithm—Simulation and real data validation. *Int. J. Pattern Recognit. Artif. Intell.* **2018**, *32*, 1850036. [CrossRef]

18. Dai, K.; Liu, G.; Li, Z.; Ma, D.; Wang, X.; Zhang, B.; Tang, J.; Li, G. Monitoring highway stability in permafrost regions with x-band temporary scatterers stacking InSAR. *Sensors* **2018**, *18*, 1871. [CrossRef] [PubMed]

19. Qin, X.; Yang, T.; Yang, M.; Zhang, L.; Liao, M. Health diagnosis of major transportation infrastructures in Shanghai metropolis using high-resolution persistent scatterer interferometry. *Sensors* **2017**, *17*, 2770. [CrossRef] [PubMed]

20. Zhang, Y.; Liu, Y.; Jin, M.; Jing, Y.; Liu, Y.; Wei, S.; Wei, J.; Chen, Y. Monitoring land subsidence in Wuhan city (China) using the SBAS-InSAR method with radarsat-2 imagery data. *Sensors* **2019**, *19*, 743. [CrossRef] [PubMed]

21. Cusson, D.; Trischuk, K.; Hébert, D.; Hewus, G.; Gara, M.; Ghuman, P. Satellite-based InSAR monitoring of highway bridges—Validation case study on the north-channel bridge in Ontario, Canada. *Transp. Res. Rec. J. Trans. Res. Board* **2018**, *2672*, 76–86. [CrossRef]

22. Hooper, A.; Bekaert, D.; Spaans, K.; Arıkan, M. Recent advances in SAR interferometry time series analysis for measuring crustal deformation. *Tectonophysics* **2012**, *514*, 1–13. [CrossRef]

23. Li, S.S.; Li, Z.W.; Hu, J.; Sun, Q.; Yu, X.Y. Investigation of the seasonal oscillation of the permafrost over Qinghai-Tibet plateau with SBAS-InSAR algorithm. *Chin. J. Geophys.* **2013**, *56*, 1476–1486.

24. Xing, X.M.; Zhu, J.; Wang, Y.; Yang, Y. Time series ground subsidence inversion in mining area based on CRInSAR and PSInSAR integration. *J. Cent. South Univ.* **2013**, *20*, 2498–2509. [CrossRef]

25. Zhang, Y.; Wu, H.a.; Sun, G. Deformation model of time series interferometric SAR techniques. *Acta Geod. Cartogr. Sin.* **2012**, *41*, 864–869.

26. Hetland, E.; Musé, P.; Simons, M.; Lin, Y.; Agram, P.; DiCaprio, C. Multiscale InSAR time series (MINTS) analysis of surface deformation. *J. Geophys. Res. Solid Earth* **2012**, *117*. [CrossRef]

27. Huang, Y. *Research on the Settlement of Embankment and Mechanism of Pavement for Freeway*; Central South University: Changsha, China, 2010.

28. Zhao, W.; Shi, J. *Consolidation and Rtheology for Soft Clay*; Hohao University Press: Nanjing, China, 1996.

29. Huang, M. *Highway Engineering Materials Rheology*; Southwest Jiaotong University Press: Chengdu, China, 2010.

30. You, S.; Zhang, Z.; Hongguang, J.I. A thermodynamic constitutive model for creep behavior of rocks and its application. *J. Univ. Min. Technol.* **2016**, *45*, 507–513.

31. Zhao, R.; Li, Z.; Feng, G.; Wang, Q.; Hu, J. Monitoring surface deformation over permafrost with an improved SBAS-InSAR algorithm: With emphasis on climatic factors modeling. *Remote Sens. Environ.* **2016**, *184*, 276–287. [CrossRef]

32. Liu, L.; Yan, Q.; Sun, H. Study on model of rheological property of soft clay. *Rock Soil Mech.* **2006**, *27*, 214–217.

33. Tian, Y. *Genetic Algorithms Research Based on Nonlinear Least Squares Estimation*; Wuhan University: Wuhan, China, 2003.

34. Yang, Z.; Li, Z.; Zhu, J.; Yi, H.; Hu, J.; Feng, G. Deriving dynamic subsidence of coal mining areas using InSAR and logistic model. *Remote Sens.* **2017**, *9*, 125. [CrossRef]

35. Li, T. *Deformation Monitoring by Multi-Temporal InSAR with Both Point and Distributed Scatterers*; Southwest Jiaotong University: Chengdu, China, 2014.

36. Costantini, M.; Rosen, P.A. A Generalized Phase Unwrapping Approach for Sparse Data. In Proceedings of the IEEE International Geoscience & Remote Sensing Symposium, Hamburg, Germany, 28 June–2 July 1999.

37. Shi, S. Shear strength, modulus of rigity and young's modulus of concrete. *China Civil Engin. J.* **1999**, *32*, 47–52.

38. Xiangjun, F. Viscosity (rigidity coefficient) of high concentration turbid water. *J. Hydraul. Eng.* **1982**, *3*, 59–65.

39. Main Climatic Characteristics and Typical Weather Event of Fuoshan. Available online: www.fs121.com/foreportinfo.aspx?Id=132832 (accessed on 1 December 2016).

Article

Aspect Entropy Extraction Using Circular SAR Data and Scattering Anisotropy Analysis

Fei Teng [1,2,3], Wen Hong [1,2] and Yun Lin [4,*]

[1] Key Laboratory of Technology in Geospatial Information Processing and Application System,
Chinese Academy of Sciences, Beijing 100190, China; scutengfei@163.com (F.T.); whong@mail.ie.ac.cn (W.H.)

[2] Institute of Electronics, Chinese Academy of Sciences, Beijing 100190, China

[3] School of Electronic, Electrical and Communication Engineering, University of Chinese Academy
of Sciences, Beijing 101408, China

[4] School of Electronic Information Engineering, North China University of Technology, Beijing 100144, China

* Correspondence: ylin@ncut.edu.cn; Tel.: + 86-135-8154-3995

Received: 27 November 2018; Accepted: 15 January 2019; Published: 16 January 2019

Abstract: In conventional synthetic aperture radar (SAR) working modes, targets are assumed isotropic because the viewing angle is small. However, most man-made targets are anisotropic. Therefore, anisotropy should be considered when the viewing angle is large. From another perspective, anisotropy is also a useful feature. Circular SAR (CSAR) can detect the scattering variation under different azimuthal look angles by a 360-degree observation. Different targets usually have varying degrees of anisotropy, which aids in target discrimination. However, there is no effective method to quantify the degree of anisotropy. In this paper, aspect entropy is presented as a descriptor of the scattering anisotropy. The range of aspect entropy is from 0 to 1, which corresponds to anisotropic to isotropic. First, the method proposed extracts aspect entropy at the pixel level. Since the aspect entropy of pixels can discriminate isotropic and anisotropic scattering, the method prescreens the target from the isotropic clutters. Next, the method extracts aspect entropy at the target level. The aspect entropy of targets can discriminate between different types of targets. Then, the effect of noise on aspect entropy extraction is analyzed and a denoising method is proposed. The Gotcha public release dataset, an X-band circular SAR data, is used to validate the method and the discrimination capability of aspect entropy.

Keywords: CSAR; anisotropy; aspect entropy; discrimination

---

1. Introduction

Synthetic aperture radar (SAR) is a high-resolution imaging radar that works all-weather and all-day [1]. SAR is widely used in military and civil fields because it assists in target analysis [2–4]. Scattering of a target is aspect-dependent. Therefore, targets are divided into two categories according to their scattering characteristic across azimuth: isotropic targets and anisotropic targets. In conventional SAR working modes, such as strip-map mode and spotlight mode, targets are assumed isotropic because the view angle is small. However, scattering variation cannot be ignored in some new SAR working modes. For example, circular SAR (CSAR) has a circular trajectory in order to observe the target at 360 degrees [5–7]. The aperture is so long that scattering differences in azimuth must be considered. The anisotropic scattering behavior is also a useful feature that can be obtained via CSAR. It can be used for target discrimination as man-made targets are usually anisotropic, while natural targets are usually isotropic. Additionally, different types of targets usually have varying degrees of anisotropy.

Because anisotropic behavior has many uses, it has been widely studied in recent years [8–10]. Most of the studies are based on a polarimetric SAR system. Ferro-Famil et al. analyze the responses of anisotropic targets under different azimuth look angles [11,12]. Some researchers use polarimetric CSAR to obtain complete scattering information of targets. Xue et al. use polarimetric scattering entropy to analyze the anisotropic scattering [13]. Li et al. propose an anisotropic scattering detection method to characterize targets [14]. However, these methods all require the use of full-polarization data. In addition to the polarization characteristic, the scattering intensity is also aspect-dependent. Therefore, we can obtain the scattering behavior by using single-polarization CSAR data. Stojanovic et al. use the sub-aperture method to extract the curve of the radar cross section (RCS) amplitude of pixels versus aspect angles using single-polarization CSAR data [15]. The curve intuitively shows whether a target is anisotropic or isotropic. However, the curve is a high-dimensional feature that is not easy to use and the degree of anisotropy cannot be quantified by the curve.

In this paper, we define aspect entropy as a descriptor of scattering anisotropy. Aspect entropy ranges from 0 to 1, which corresponds to anisotropic to isotropic. Our simulation results show the effectiveness of aspect entropy in quantifying the degree of anisotropy. As a result, we propose the extraction method of aspect entropy using real CSAR data. First, we propose the aspect entropy extraction method at the pixel level based on the sub-aperture method. Using the result of pixel-wise aspect entropy extraction, anisotropic pixels that belong to targets can be discriminated from isotropic clutters by thresholding. Next, we propose the aspect entropy extraction method at the target level. Thus, aspect entropy of targets can be extracted. The result can be used to analyze the scattering anisotropy of different targets and it has the capability of discrimination. During the aspect entropy extraction by using the real data, the RCS curve will have noise. Therefore, the effect of noise on aspect entropy extraction is studied. The simulation result shows that the aspect entropy is more accurate in high signal-to-noise ratio. Only high scatterings in the RCS curve is important on anisotropic target discrimination. Therefore, we proposed a RCS curve denoising method and it is shown effective by the simulation.

The Gotcha public release dataset is used to verify our aspect entropy extraction methods at the pixel and target levels. The result shows that the aspect entropy of pixels and targets can be extracted from CSAR data. Aspect entropy of pixels can be used to discriminate between isotropic and anisotropic scattering. The proposed RCS curve denoising method can remove the noise from the RCS curve extracted from the real data. It makes the result of aspect entropy extraction more accurate. Since the aspect entropy of different types of targets falls into different ranges, targets can be discriminated from each other by the aspect entropy value.

2. Concept of Aspect Entropy

Because the scattering of a target is aspect dependent, CSAR is helpful in detecting the anisotropic scattering behavior of a target. Radar cross section (RCS) is a measure representing the scattering ability of the incident electromagnetic wave [16]. RCS is related to the physical characteristics of the target and the parameters of the electromagnetic wave. Therefore, if the curve of the RCS amplitude versus the aspect angle (hereafter referred to as RCS curve) is known, we can see the scattering behavior of the target and judge whether a target is isotropic or anisotropic.

We can simulate the RCS curve for the selected shapes. The simulation is conducted at 10 GHz using vertical polarization with a 45° look-down angle. Four canonical shapes of the same size were used: a square plate dihedral set horizontally (named dihedral A), a square plate dihedral set vertically (named dihedral B), a triangular trihedral and a top-hat. The size is approximately 10 times longer than wavelength. They are all made of both perfect electric conductor and perfect magnetic conductor. The simulation is aimed at showing the ideal scattering mechanisms of these shapes. Therefore, these models are placed in the free space, whose relative permittivity is 1 and dielectric loss tangent is 0. The condition of the simulation is a more ideal than an anechoic chamber experiment. Figure 1 shows the models of the four canonical shapes. The results of the simulation are shown in Figure 2. The results show that the curve of the top-hat is smooth because it is isotropic. Since the dihedral and trihedral are both anisotropic, the high scatterings are concentrated in a limited range. However, we can discriminate between them by the degree of anisotropy, suggesting that different targets usually have different degrees of anisotropy. The scattering mechanism of dihedral A is very different from the trihedral. The scattering mechanism of dihedral B and trihedral are close but still discrepant. Targets can be discriminated if we find a method to quantify the anisotropy by a calculation. The calculation should be concerned with the scattering mechanism from different angles of view and not the RCS amplitude.

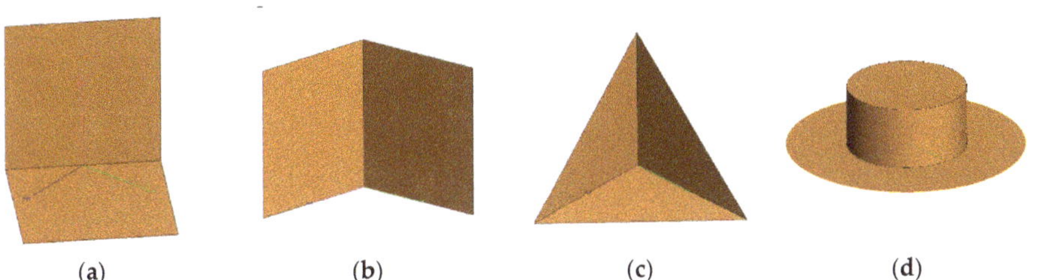

Figure 1. Models of canonical shapes. (**a**) Dihedral A. (**b**) Dihedral B. (**c**) Trihedral. (**d**) Top-hat.

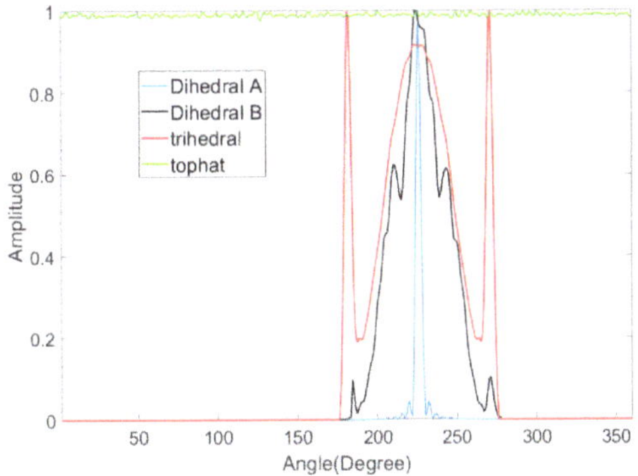

Figure 2. Radar cross section (RCS) curves of canonical shapes.

Early physicists defined entropy as a measure of disorder. Entropy was introduced later in many other fields according to its application in physics. Shannon presented information entropy to describe the uncertainty of the information source [17]. Electromagnetism is the concept of polarization entropy, which quantifies the disorder of scattering [18]. Polarization entropy ranges from 0 to 1, which corresponds to zero scattering to perfect depolarizing. It is widely used in SAR image analysis [19,20]. The scattering in different aspect angles is similar to the scattering in different polarization types. Therefore, we present aspect entropy as a descriptor of scattering anisotropy. We can obtain the RCS amplitudes $R(k)$ follow the angle $\theta(k)$ from curves as shown in Figure 2, where $k = 1, 2, \cdots, n$. The pseudo-probability $P(k)$ of scattering in $\theta(k)$ can be calculated by

$$P(k) = \frac{R(k)}{\sum_{k=1}^{n} R(k)}. \tag{1}$$

Then, aspect entropy is defined as

$$H_a = -\sum_{k=1}^{n} P(k) \log_n P(k). \tag{2}$$

H_a ranges from 0 to 1. As shown in equation (2), H_a is normalized by the sum of $P(k)$ and is not concerned with RCS amplitude. Aspect entropy is inversely proportional to the pseudo-probability $P(k)$. If the scattering is strong in some angles, the aspect entropy will be lower. Consequently, the aspect entropy of an anisotropic target is lower because the scattering of an anisotropic target at certain angles is much stronger than at other angles. The aspect entropy for an isotropic target is higher because the scattering is stable in all azimuth angles. We calculate the aspect entropy of the four models mentioned above and the results are listed in Table 1. The results show that aspect entropy can indicate the anisotropy diversity of the four shapes, and therefore can be used as a descriptor of anisotropy. The aspect entropy value of dihedral A and the top-hat have a large difference between these shapes. The scattering mechanisms of dihedral B and the trihedral are similar. It seems that the values are closed but the difference is big enough to discriminate them. The aspect entropy can still discriminate these two kinds of shapes by using the real data and the result is shown in Section 4.

Table 1. The aspect entropy of canonical shapes.

Shape	Aspect Entropy
Dihedral A	0.3823
Dihedral B	0.7131
Trihedral	0.7625
Top-hat	0.9999

3. Aspect Entropy Extraction

In this section, aspect entropy extraction methods at the pixel level and the target level are proposed respectively. Since the smallest unit of a CSAR image is a pixel, it is convenient to obtain the RCS curve of a pixel by using the sub-aperture method. Extracting the aspect entropy from pixels can help us analyze the scattering characteristics for the structure of a target and offer an overview of full-scene anisotropy. At the application level, targets are usually the object of analysis. Therefore, we propose the aspect entropy extraction method at the target level based on the aspect entropy of the pixels. The RCS curve of the target must be defined to calculate the aspect entropy. Targets of the same type are supposed to have similar aspect entropy, while the aspect entropy of different types of targets are diverse. Thus, targets can be discriminated from each other by using the aspect entropy value.

In addition, we study the effects of noise on aspect entropy extraction. Aspect entropy is more accurate in a high signal-to-noise ratio (SNR). Only high scatterings are interested in the RCS curve of the anisotropic target. Therefore, we can remove the noise out of the RCS curve. Then we propose a denoising method for the RCS curve. The result of simulations shows that aspect entropy is more accurate after denoising.

3.1. Aspect Entropy Extraction Method at the Pixel Level

In the SAR image, a pixel is the smallest unit, so it is reasonable to analyze the scattering anisotropy at the pixel level. The aspect entropy of pixels can be extracted by the method described below. A flowchart of the process is shown in Figure 3. First, the full-aperture is divided into sub-apertures. Second, the coherent complex image of each sub-aperture is obtained. Third, the absolute value of each pixel as the RCS amplitude is used to obtain the RCS curve of each pixel. Finally, the aspect entropy of each pixel is calculated. Details of the procedure are explained in the text below.

The first and second step establish the process for the sub-aperture method. The full-aperture is divided into $k(k = 1, 2, \cdots, n.)$ sub-apertures with the same width θ_w. There is a trade-off when we choose the width θ_w. The width must be large enough to obtain a high azimuth resolution. However, if the width is too large, the RCS amplitude will be inaccurate because the sub-aperture method uses the mean value of RCS amplitudes in θ_w as the RCS amplitude in the central angle $\theta(k)$. The coherent complex image of each sub-aperture is obtained by using the back projection (BP) algorithm [21,22].

When using the BP algorithm, the pixel size should be small enough to ensure that the scattering characteristic of each pixel is accurate. If the pixel size is too small, it will cause a substantial amount of computation. The pixel size can be set according to the theoretical resolution of the system. Then, we obtain the RCS curve through the 360° observation of each pixel. The coherent complex image $I_n(i,j)$ of each sub-aperture is the imaging result by using the BP algorithm. For each sub-aperture, the absolute pixel value of pixel (i,j) is used as the RCS amplitude $R_{ij}(k)$ in angle $\theta(k)$. For each pixel (i,j), the pseudo-probability $P_{ij}(k)$ of scattering in $\theta(k)$ can be obtained by Equation (1). The aspect entropy $H_a(i,j)$ of each pixel (i,j) can be calculated by Equation (2).

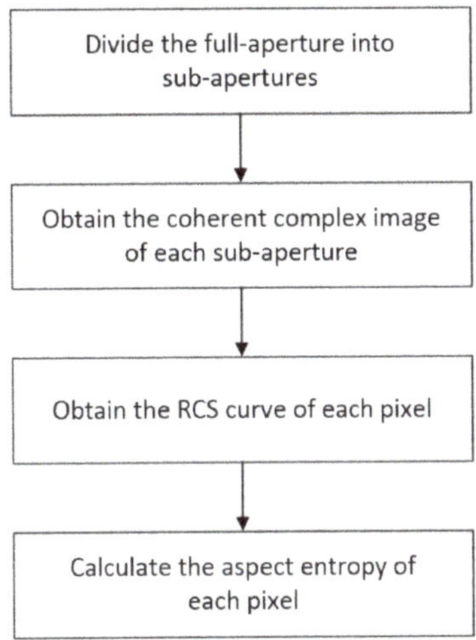

Figure 3. Flowchart of aspect entropy extraction at the pixel level.

3.2. Aspect Entropy Extraction Method at the Target Level

Target discrimination and classification are important applications of SAR. The capability of aspect entropy on target discrimination is shown in Section 2. The aspect entropy of the target must be extracted to enable it to have a broader range of application. Usually, targets of interest are anisotropic because they contain many dihedral and trihedral structures. Natural clutters and man-made clutters such as lawns, trees, and roads are usually isotropic. Therefore, pixels from these clutters have a lower aspect entropy while pixels from the targets have a higher aspect entropy. Anisotropic pixels can be discriminated from isotropic pixels according to the aspect entropy value. We analyzed the scattering anisotropy of the targets by using the anisotropy of the pixels. To accomplish this, we propose the extraction method of aspect entropy at the target level. The process contains four steps and is described in Figure 4. First, the images of targets are extracted and the aspect entropy of the pixels is obtained. Second, the binary image is obtained by thresholding. Third, the RCS curves of the targets are obtained. Finally, the aspect entropy of each target is calculated.

The constant false-alarm rate (CFAR) or generalized likelihood ratio test (GLRT) [23,24] methods can be used to extract the targets from the image. For our purposes, the image of the target is extracted manually in this paper. The aspect entropy can be obtained using the same method described in subsection A. After thresholding according to the aspect entropy value, we can obtain a binary image. The value 1 represents anisotropic pixels, and 0 represents isotropic pixels. In the Circular SAR image, the pixel size is much smaller than the target size. A whole target or a structure of a target consists of many pixels. In the binary image, the targets consist of anisotropic pixels. Different types of targets are expected to have different aspect entropy. Therefore, we can analyze the scattering anisotropy of different types of targets using aspect entropy. Aspect entropy is calculated using the RCS curve as mentioned above. In subsection A, the RCS curve is associated with the pixel. Therefore, it is necessary to obtain the RCS curve of the target. If a target contains N anisotropic pixels after discrimination and the RCS curves of the pixels are $R_m(k)$ respectively, where $m = 1, 2, \cdots, N$, we define the RCS curve of the target as

$$R(k) = \sum_{m=1}^{N} R_m(k). \tag{3}$$

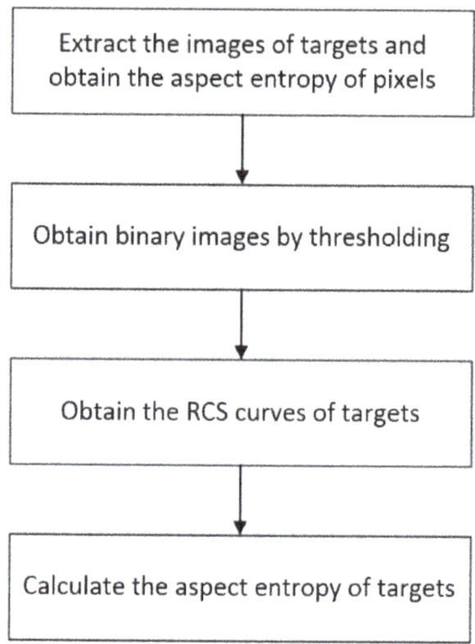

Figure 4. Flowchart of aspect entropy extraction at the target level.

Scattering of the target at a certain angle is accomplished by the scattering of each pixel at this angle. Thus, the RCS curve of a target can be obtained by accumulation. This is equivalent to using a single pixel to represent the whole target. Next, we can calculate the aspect entropy of the target using (1) and (2). The aspect entropy is now applicable to the scattering anisotropy analysis at the target level.

3.3. Denoising of the RCS curve

When we extract the RCS curve by using the real data, the amplitudes are not 0 in angles which do not scatter the waves. These RCS amplitudes are regarded as noise in the RCS curves of targets. The noise mainly comes from the side lobes of clutters. During the coherent imaging process of a CSAR image, the value of a pixel from the image is actually affected by the side lobes of pixels around it. For example, a target pixel scatters the wave around θ and some clutters around this pixel scatter the wave in other angles. The RCS amplitude of this pixel is combined with its main lobe and the side lobes of the clutters. Consequently, isotropic targets are not influenced by the noise and clutters because these are isotropic too. While for an anisotropic target, the RCS amplitudes in other angles are not 0 in the RCS curve extracted by our method. The aspect entropy will become higher and inaccurate. Therefore, we have to denoise for the RCS curve of the anisotropic target.

Here, we define the SNR of the RCS curve as the ratio between the maximum power of the target and the power of the noise. To study the effect of the noise on the result of aspect entropy, we use the models mentioned in Section 2. Varying levels of noise are added into the RCS curve for the three anisotropic models: dihedral A, dihedral B, and the trihedral. The SNR ranges from 10 dB to 40 dB. As shown in Figure 5a, we choose SNR = 20 dB to illustrate the result of adding noise to the RCS curve. The percentage error for aspect entropy calculation in different SNR is shown in Figure 5b. For a certain anisotropic target, the higher the SNR is, the less the error is. With the same

SNR, the percentage error is greater for targets that have higher degree of anisotropy. When we judge whether a target is anisotropic or not, we only interested in whether there are any high scatterings in some angles. Therefore, for anisotropic targets, the noise in other angles can be removed from the RCS curve. The easiest way to wipe out the noise is by setting a proper threshold value and eliminating as much noise as possible. The threshold value cannot be set directly. If we simply reset the amplitudes under a value to 0, lower scatterings belonging to targets would be removed too. Although these low scatterings are not the main lobe, they reflect the scattering characteristics of the target. Therefore, they should be protected during the denoising. The threshold value should be as close as possible to the ceiling of the noise.

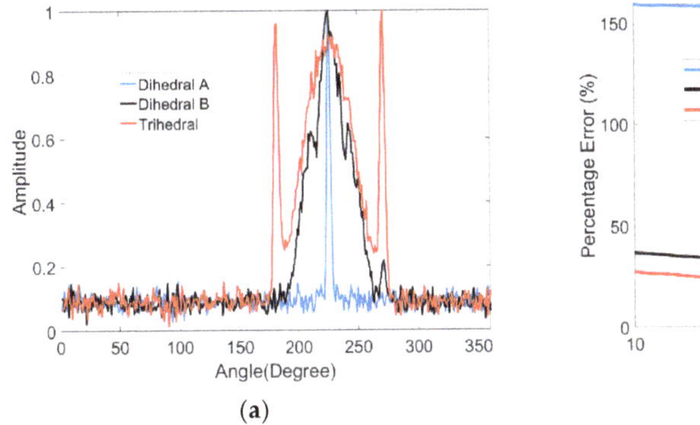
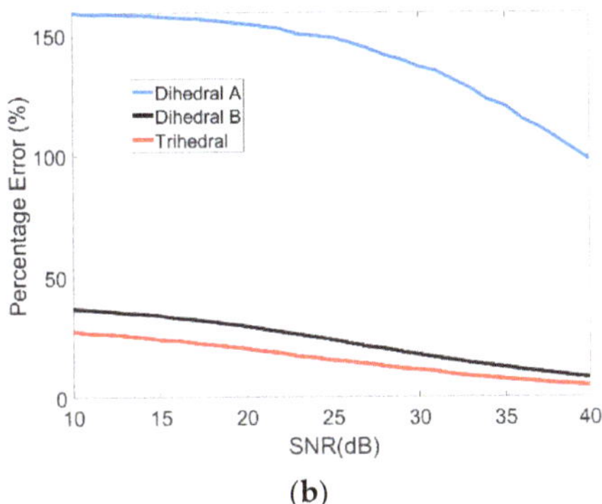

(a) **(b)**

Figure 5. (**a**) RCS curves with noise (SNR = 20 dB). (**b**) Percentage error for aspect entropy calculation in different SNR.

We proposed a denoising method for the RCS curve and Figure 6 is the diagram of the whole procedure. The RCS curve of dihedral B with $SNR = 20\,\text{dB}$ is used as an example. To estimate the noise level of the RCS curve, the high scatterings should be removed. The energy concentration parameter W is defined as an approximation for the target high scattering persistence angle by Zhao et al. [25]. It is calculated by

$$W = \frac{\sum\limits_{k=1}^{n} R(k)}{R(k)_{max}}. \tag{4}$$

First, calculate the energy concentration parameter of the RCS curve and round it up to an integer W. Sort the RCS amplitudes in descending order and remove the top W amplitudes. Figure 6a shows the result of sorting and the red line represents $W = 67$. Second, calculate the mean value μ and the standard deviation σ for all remaining amplitudes, which are the noise. The red line shown in Figure 6b is $\mu = 0.0954$. Third, reset the RCS amplitudes under $T = \mu + 2\sigma$ to 0. Here, we use double the standard deviation σ to ensure that the threshold value is higher than most of the noise. The red line shown in Figure 6c represents $T = \mu + 2\sigma = 0.1362$. The result of RCS curve denoising is shown in Figure 6d.

Figure 7 shows the comparison of the percentage error for aspect entropy calculation in different SNR between with and without denoising. It can be found that the denoising method significantly improved the accuracy of aspect entropy in low SNR. With the increase of the SNR, the percentage error after denoising has a small rebound, especially for the dihedral A. This is because the threshold value we set is higher than the ceiling of the noise to ensure that as much noise as possible can be removed. The level of the noise is very low when the SNR is high, so the threshold value can be higher than the low scatterings of the target. Therefore, the aspect entropy will be lower than the truth value. Although there is a rebound, the percentage error is still lower than it without denoising when the SNR is high. The simulation result shows that our denoising method is effective.

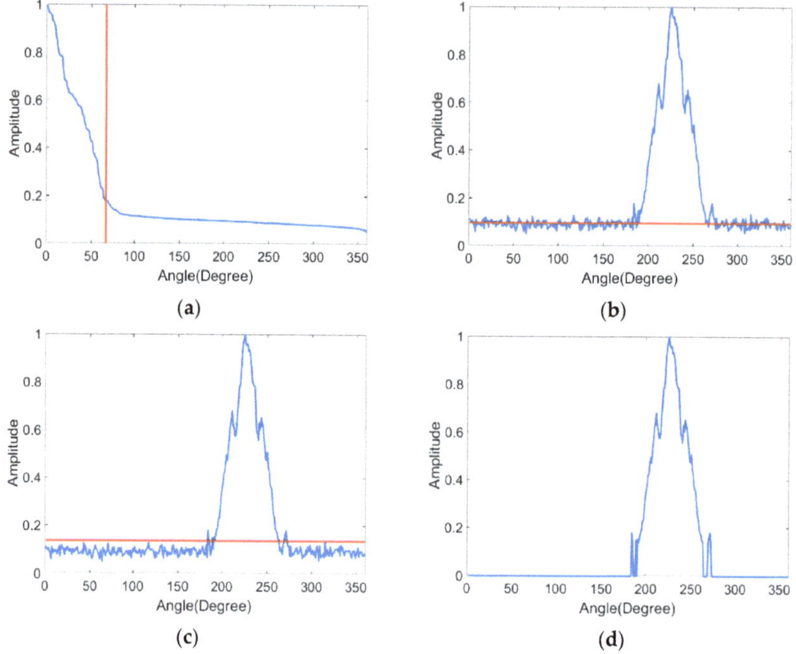

Figure 6. Diagram of the denoising procedure. (**a**) The result of sorting the RCS amplitudes into descending order and calculation of the energy concentration. (**b**) The result of mean value calculation for the noise. (**c**) The result of threshold value calculation. (**d**) The result of RCS curve denoising.

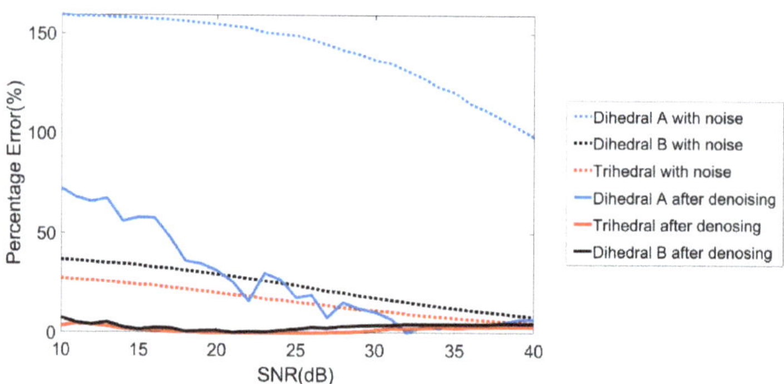

Figure 7. Comparison of the percentage error for aspect entropy calculation.

4. Experiment Results and Analysis

The Gotcha public release dataset was used to illustrate our method. Gotcha data consists of SAR phase history data collected at X-band with a 640 MHz bandwidth with full 360-degree azimuth coverage and full polarization [26]. There are many targets of interest in the imaging scene, including civilian vehicles, a top-hat, trihedrals, and dihedrals.

4.1. Aspect Entropy Extraction at the Pixel Level

Our method uses only one polarization, and thus we select one pass and its' HH polarization from the Gotcha data. Figure 8 is the coherent complex image of the full scene obtained by using the BP algorithm, and the pixel size is 0.025 m.

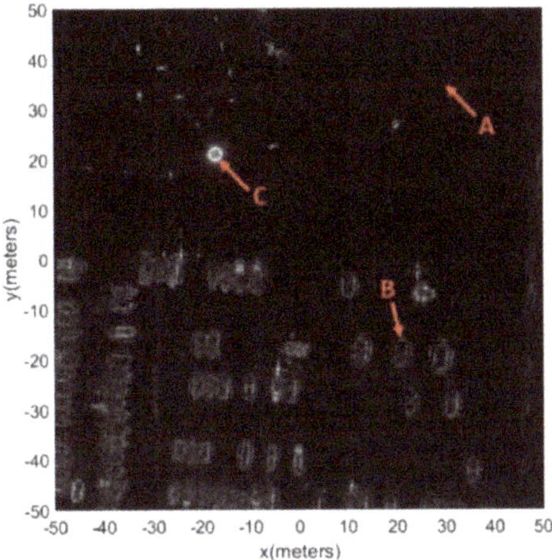

Figure 8. Coherent complex image of the full scene.

When using the sub-aperture method to obtain the RCS curves, we found that using 360 sub-apertures offered a good compromise between the azimuth resolution and the preciseness of the RCS amplitude. As shown in Figure 8, three pixels were selected to show the result of the RCS curve extraction. Pixel A represents a pixel from the lawn. Pixel B represents a pixel from the frame of the vehicle. Pixel C represents a pixel from the edge of the top-hat. Figure 9 shows the optical images of the three targets. Figure 10 shows the RCS curves of the three pixels. Pixel A from the lawn is shown as isotropic scattering. Pixel B and pixel C are anisotropic because both are man-made metal structures. After obtaining the RCS curves of the pixels, the aspect entropy can be calculated. Figure 11 is the aspect entropy image of the full scene. The color bar indicates that darker colors denote a lower aspect entropy which means the pixels are more anisotropic, while the lighter color denotes a higher entropy and more isotropic. The result is as we expected. Pixels from vehicles and calibration targets show up as a dark color because they scatter the wave near a certain angle, which leads to a lower aspect entropy. Pixels from the lawn and roads show up in light color because they scatter the wave in all azimuth angles with similar intensity. Therefore, the aspect entropy can quantify the scattering anisotropy of pixels. Anisotropic scattering and isotropic scattering can be discriminated from each other in the aspect entropy image.

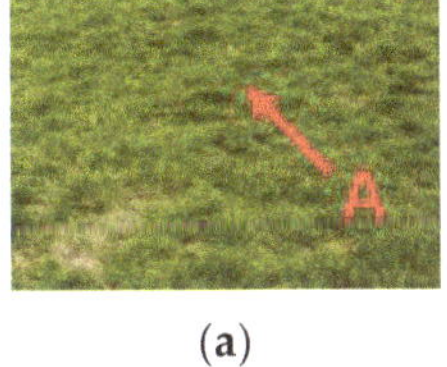
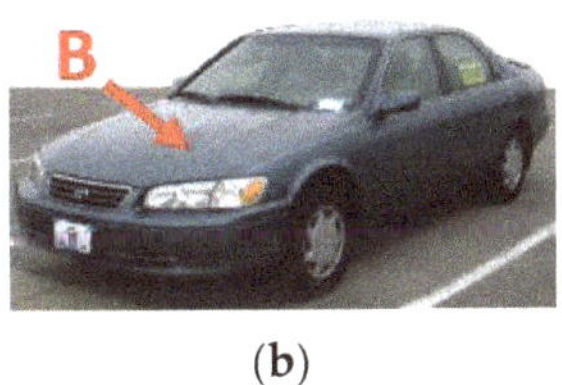
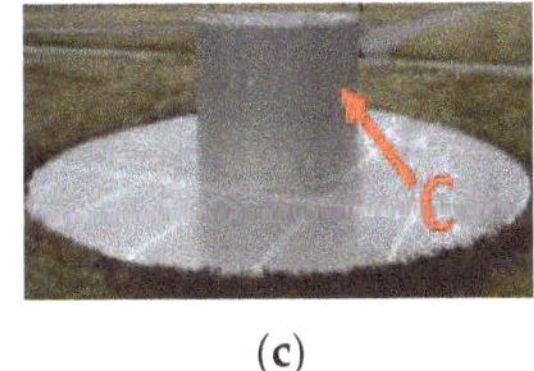

(**a**)　　　　　　　　　　(**b**)　　　　　　　　　　(**c**)

Figure 9. Optical images of three pixels. (**a**) Pixel A of the lawn. (**b**) Pixel B of a vehicle. (**c**) Pixel C of the top-hat.

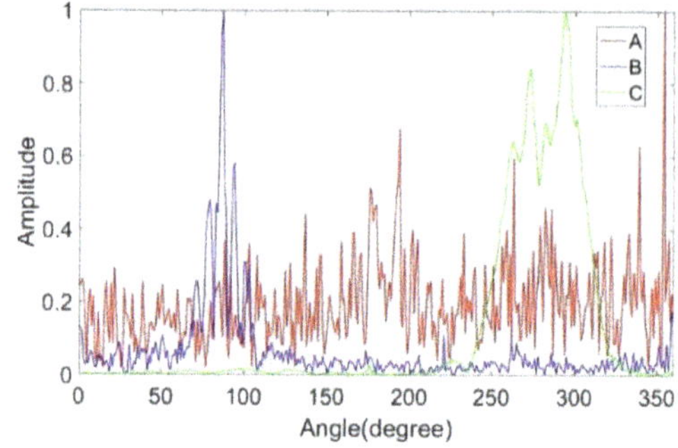

Figure 10. RCS curves of three pixels.

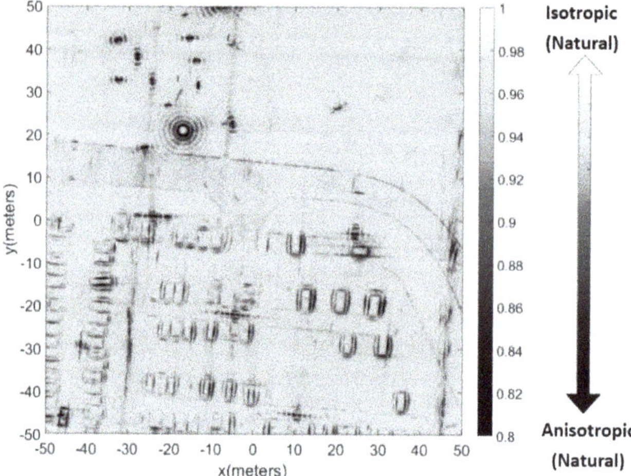

Figure 11. Aspect entropy image of the full scene.

4.2. Aspect Entropy Extraction at the Target Level

We choose a dihedral, a trihedral, a vehicle, and a top-hat from the scene as examples to illustrate the procedure of target aspect entropy extraction. Figure 12 shows the aspect entropy images of the four targets. The results indicate that 0.91 is a suitable threshold value. Figure 13 shows the binary images of the four targets. According to Figure 13, anisotropic pixels from the targets can be segmented from isotropic clutters by thresholding. RCS curves of the four targets are obtained by using Equation (3), and the result is shown in Figure 14a. The RCS curves of the dihedral, trihedral, and the top-hat are similar to the result of the simulation discussed in Section 2. In addition, we can judge that the dihedral used in Gotcha experiment is set vertically because the RCS curve of it is the same as the curve of dihedral B in the simulation. The RCS curve of the vehicle is as expected. The four sides of the vehicle cause substantial scattering in four directions and barely any scattering in other directions. Unlike the RCS curves obtained by the simulation in Section 2, these RCS curves have the noise. Therefore, we use the method mentioned above to denoise the RCS curve. The parameters are $T_1 = 0.3, k = 2$. Figure 14b shows the RCS curves after denoising. It can be seen that most of the noise can be moved out and high scatterings are preserved well. The results show that our proposed aspect entropy extraction method can obtain the RCS curve of the target and extract the aspect entropy of the target. Our proposed RCS curve denoising method can preserve the useful information and obtain a good denoising effect.

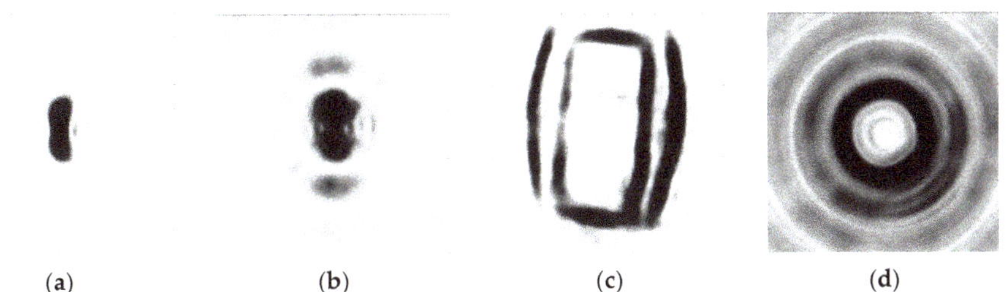

Figure 12. Aspect entropy images of targets. (a) Dihedral. (b) Trihedral. (c) Vehicle. (d) Top-hat.

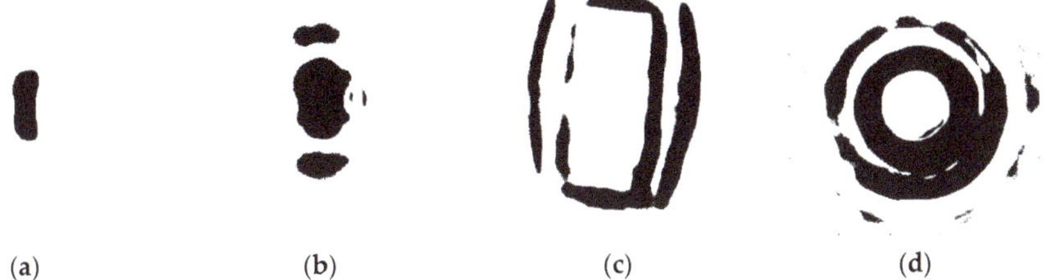

Figure 13. Binary images of the targets. (a) Dihedral. (b) Trihedral. (c) Vehicle. (d) Top-hat.

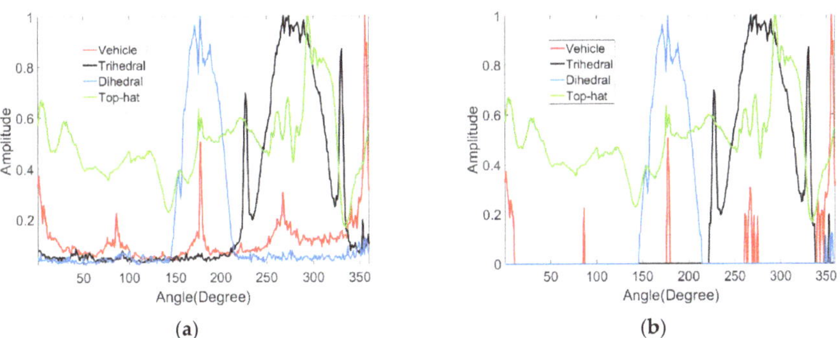

Figure 14. (a) RCS curves of the four targets. (b) RCS curves of the four targets after denoising.

After obtaining the RCS curves of the targets, aspect entropy can be calculated using Equations (1) and (2). The full scene had a limited number of trihedrals and dihedrals that imaged clearly. To analyze the scattering anisotropy of the targets, six vehicles, six trihedrals, three dihedrals, and a top-hat were selected manually, and their aspect entropies were calculated. The locations and aspect entropy with and without denoising of the targets are listed in Table 2. The lower aspect entropy value represents the scattering of a target is more concentrated. Targets of the same type have similar aspect entropy values because they have the same scattering mechanism. The aspect entropy of dihedrals and trihedrals extracted from the real data without denoising is close to 1, which is much higher than the simulation results in Section 2. The aspect entropy after denoising is closer to the result of simulation and it has a greater ability of discrimination and clustering. The aspect entropy of targets of the same type is concentrated in a limited range. The aspect entropy of vehicles is in the range [0.5567, 0.6418]. The aspect entropy of trihedrals is in the range [0.7878, 0.7918]. The aspect entropy of dihedrals is in the range [0.6805, 0.7019]. The aspect entropy of the top-hat is 0.9927. In Figure 15, it is clear that vehicles, dihedrals, trihedrals, and the top-hat are clustered in different ranges. The ranges of different types are not coincident. The aspect entropy values of dihedral B and the trihedral are close in the simulation result so it is unclear whether the aspect entropy value can discriminate these two kinds of shapes in the real experiment. We set all the dihedrals as dihedral B in the Gotcha experiment. As can be seen in Figures 2 and 14, the RCS curves of these two kinds of shapes are similar in both the simulation and the real experiment. The true scattering mechanisms of the targets can be restored well by the RCS curves after denoising. Thus, the aspect entropy values are close to the true value. In the experimental results, the aspect entropy values of dihedral are around 0.69 and the aspect entropy values of trihedral is around 0.79. Each kind of target was clustered well. Therefore, aspect entropy can discriminate between dihedral B and the trihedral despite their scattering mechanisms being similar. Therefore, different targets can be discriminated from each other according to the value of the aspect entropy. Beyond that, the RCS curve denoising greatly enhances the discrimination capabilities of aspect entropy.

Table 2. The aspect entropy of targets.

Target	Location	Aspect Entropy without Denoising	Aspect Entropy after Denoising
Chevy Malibu	(9.97, −5.22)	0.9469	0.5733
Ford Taurus Wag	(12.43, −18.21)	0.9468	0.5567
Toyota Camry	(20.66, −18.71)	0.9627	0.6118
Nissan Sentra	(31.42, −28.87)	0.9460	0.5753
Hyundai SantaFe	(22.68, −28.30)	0.9579	0.6024
Chevy Prizm	(35.44, −41.72)	0.9536	0.5639
Trihedral 1	(−24.39, 32.96)	0.8916	0.7918
Trihedral 2	(−32.50, 33.41)	0.8887	0.7895
Trihedral 3	(−32.14, 42.54)	0.8883	0.7915
Trihedral 4	(−28.09, 38.67)	0.8868	0.7878
Trihedral 5	(−13.86, 37.70)	0.8885	0.7918
Trihedral 6	(−5.12, 22.98)	0.9001	0.7907
Dihedral 1	(−15.55, 42.96)	0.8583	0.6805
Dihedral 2	(−18.58, 33.53)	0.8587	0.7019
Dihedral 3	(−26.15, 17.50)	0.8735	0.6841
Top-hat	(−17.00, 21.00)	0.9927	0.9927

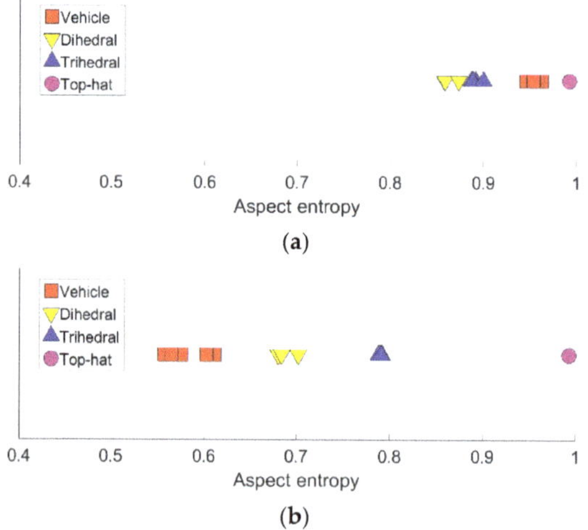

Figure 15. Visual result of discrimination. (**a**) Without denoising. (**b**) After denoising.

The aspect entropy of the targets is affected by the frequency and polarization of the microwave. The aspect entropy of the targets is also significantly affected by the posture and size. Thus, the experimental results prove the capability of target discrimination in this X-band CSAR data and scene.

5. Conclusions

Scattering anisotropy analysis is important and useful in the field of SAR. Previous studies have mostly used polarimetric SAR data. In this paper, we use the CSAR data to analyze anisotropic scattering behavior across the azimuth. Aspect entropy is presented as a descriptor of scattering anisotropy, ranging from 0 to 1, which corresponds to anisotropic to isotropic. We verify that the aspect entropy can be the descriptor of scattering anisotropy by simulation. In addition, the effects of noise on the aspect entropy result is studied and a RCS curve denoising method is proposed. Aspect entropy extraction methods at the pixel and target level are respectively proposed using single-polarization CSAR data. The Gotcha public release dataset is used to illustrate our aspect entropy extraction methods. The results show that the aspect entropy of the pixel and the target can be successfully extracted by our methods. The value of the aspect entropy helps us to analyze the scattering anisotropy of the CSAR image. At the pixel level, aspect entropy can discriminate isotropic and anisotropic scattering. At the target level, it can discriminate different types of targets from each other. Further research will focus on the practical application of aspect entropy. It can be combined with other features and used for target detection or classification.

Author Contributions: Conceptualization, F.T.; methodology, F.T. and Y.L.; validation, F.T.; writing—original draft preparation, F.T.; writing—review and editing, W.H and Y.L.; supervision, W.H and Y.L.

Funding: This work was supported by the National Natural Science Foundation of China under Grant 61571421, Grant 61501210, Grant 61431018 and Grant 61331017.

Conflicts of Interest: The authors declare no conflict of interest.

References

1. Moreira, A.; Prats-Iraola, P.; Younis, M.; Krieger, G.; Hajnsek, I.; Papathanassiou, K.P. A tutorial on synthetic aperture radar. *IEEE Geosci. Remote Sens. Mag.* **2013**, *1*, 6–43. [CrossRef]
2. Fornaro, G.; Reale, D.; Serafino, F. Four-dimensional SAR imaging for height estimation and monitoring of single and double scatterers. *IEEE Trans. Geosci. Remote Sens.* **2009**, *47*, 224–237. [CrossRef]
3. Migliaccio, M.; Gambardella, A.; Tranfaglia, M. SAR polarimetry to observe oil spills. *IEEE Trans. Geosci. Remote Sens.* **2007**, *45*, 506–511. [CrossRef]
4. Banerjee, A.; Burlina, P.; Chellappa, R. Adaptive target detection in foliage-penetrating SAR images using alpha-stable models. *IEEE Trans. Image Process.* **1999**, *8*, 1823–1831. [CrossRef] [PubMed]
5. Lin, Y.; Hong, W.; Tan, W.; Wang, Y.; Xiang, M. Airborne circular SAR imaging: Results at P-band. In Proceedings of the 2012 IEEE International Geoscience and Remote Sensing Symposium, Munich, Germany, 22–27 July 2012; pp. 5594–5597.
6. Dupuis, X.; Martineau, P. Very high resolution circular SAR imaging at X band. In Proceedings of the 2014 IEEE Geoscience and Remote Sensing Symposium, Quebec City, QC, Canada, 13–18 July 2014; pp. 930–933.
7. Chan, T.; Kuga, Y.; Ishimaru, A. Experimental studies on circular SAR imaging in clutter using angular correlation function technique. *IEEE Trans. Geosci. Remote Sens.* **1999**, *37*, 2192–2197. [CrossRef]
8. Ponce, O.; Prats, P.; Pinheiro, M.; Rodriguez-Cassola, M.; Scheiber, R.; Reigber, A.; Moreira, A. Fully polarimetric high-resolution 3-D imaging with circular SAR at L-band. *IEEE Trans. Geosci. Remote Sens.* **2014**, *52*, 3074–3090. [CrossRef]
9. Moses, R.; Potter, L.; Cetin, M. Wide-angle SAR imaging. *Proc. SPIE* **2004**, *5427*, 164–175.
10. Zhao, Y.; Lin, Y.; Hong, W.; Yu, L. Adaptive imaging of anisotropic target based on circular-SAR. *Electron. Lett.* **2016**, *52*, 1406–1408. [CrossRef]
11. Ferro-Famil, L.; Reigber, A.; Pottier, E.; Boerner, W.M. Scene characterization using subaperture polarimetric SAR data. *IEEE Trans. Geosci. Remote Sens.* **2003**, *41*, 2264–2276. [CrossRef]
12. Ferro-Famil, L.; Pottier, E. Urban area remote sensing from L-band PolSAR data using time-frequency techniques. In Proceedings of the 2007 Urban Remote Sensing Joint Event, Paris, France, 11–13 April 2007; pp. 1–6.
13. Xue, F.; Lin, Y.; Hong, W.; Chen, S.; Shen, W. An Improved H/α Unsupervised Classification Method for Circular PolSAR Images. *IEEE Access* **2018**, *6*, 34296–34306. [CrossRef]

14. Li, Y.; Yin, Q.; Lin, Y.; Hong, W. Anisotropy Scattering Detection from Multiaspect Signatures of Circular Polarimetric SAR. *IEEE Geosci. Remote Sens. Lett.* **2018**, *15*, 1575–1579. [CrossRef]

15. Stojanovic, I.; Cetin, M.; Karl, W.C. Joint space aspect reconstruction of wide-angle SAR exploiting sparsity. *Proc. SPIE* **2008**, *6970*. [CrossRef]

16. Odendaal, J.W.; Joubert, J. Radar cross section measurements using near-field radar imaging. *IEEE Trans. Instrum. Meas.* **1996**, *45*, 948–954. [CrossRef]

17. Shannon, C.E. A mathematical theory of communication. *Bell Syst. Tech. J.* **1948**, *27*, 623–656. [CrossRef]

18. Cloude, S.R. Concept of polarization entropy in optical scattering. *Opt. Eng.* **1995**, *34*, 1599–1610. [CrossRef]

19. Shan, Z.; Wang, C.; Zhang, H.; An, W. Improved four-component model-based target decomposition for polarimetric SAR data. *IEEE Geosci. Remote Sens. Lett.* **2012**, *9*, 75–79. [CrossRef]

20. Wakabayashi, H.; Matsuoka, T.; Nakamura, K.; Nishio, F. Polarimetric characteristics of sea ice in the sea of Okhotsk observed by airborne L-band SAR. *IEEE Trans. Geosci. Remote Sens.* **2004**, *42*, 2412–2425. [CrossRef]

21. Ulander, L.M.H.; Hellsten, H.; Stenstrom, G. Sythetic-aperture radar processing using fast factorized back-projection. *IEEE Trans. Aerosp. Electron. Syst.* **2003**, *39*, 769–776. [CrossRef]

22. Ponce, O.; Prats, P.; Rodriguez-Cassola, M.; Scheiber, R.; Reigber, A. Processing of circular SAR trajectories with fast factorized back-projection. In Proceedings of the 2011 IEEE International Geoscience and Remote Sensing Symposium, Vancouver, BC, Canada, 24–29 July 2011; pp. 3692–3695.

23. Kuttikkad, S.; Chellappa, R. Non-Gaussian CFAR techniques for target detection in high resolution SAR images. In Proceedings of the 1st International Conference on Image Processing, Austin, TX, USA, 13–16 November 1994; pp. 910–914.

24. Lombardo, P.; Sciotti, M.; Kaplan, L.M. SAR prescreening using both target and shadow information. In Proceedings of the 2001 IEEE Radar Conference, Atlanta, GA, USA, 1–3 May 2001; pp. 147–152.

25. Zhao, Y.; Lin, Y.; Hong, W.; Shen, W.; Xue, F. Target aspect feature extraction and application from multi-aspect high resolution SAR. In Proceedings of the 2017 IEEE International Geoscience and Remote Sensing Symposium (IGARSS), Fort Worth, TX, USA, 23–28 July 2017; pp. 1784–1787.

26. Ertin, E.; Austin, C.; Sharma, S.; Moses, R.; Potter, L. GOTCHA experience report: Three-dimensional SAR imaging with complete circular apertures. *Proc. SPIE* **2007**, *6568*, 656–802.

Article

Improving the Accuracy of Two-Color Multiview (2CMV) Advanced Geospatial Information (AGI) Products Using Unsupervised Feature Learning and Optical Flow

Berkay Kanberoglu [1,*] **and David Frakes** [2]

[1] School of Electrical, Computer and Energy Engineering, Arizona State University, Tempe, AZ 85281, USA

[2] School of Biological and Health Systems Engineering, Arizona State University, Tempe, AZ 85281, USA; dfrakes@asu.edu

* Correspondence: bkanbero@asu.edu

Received: 5 May 2019; Accepted: 3 June 2019; Published: 8 June 2019

Abstract: In two-color multiview (2CMV) advanced geospatial information (AGI) products, temporal changes in synthetic aperture radar (SAR) images acquired at different times are detected, colorized, and overlaid on an initial image such that new features are represented in cyan, and features that have disappeared are represented in red. Accurate detection of temporal changes in 2CMV AGI products can be challenging because of 'speckle noise' susceptibility and false positives that result from small orientation differences between objects imaged at different times. Accordingly, 2CMV products are often dominated by colored pixels when changes are detected via simple pixel-wise cross-correlation. The state-of-the-art in SAR image processing demonstrates that generating efficient 2CMV products, while accounting for the aforementioned problem cases, has not been well addressed. We propose a methodology to address the aforementioned two problem cases. Before detecting temporal changes, speckle and smoothing filters mitigate the effects of speckle noise. To detect temporal changes, we propose using unsupervised feature learning algorithms in conjunction with optical flow algorithms that track the motion of objects across time in small regions of interest. The proposed framework for distinguishing between actual motion and misregistration can lead to more accurate and meaningful change detection and improve object extraction from an SAR AGI product.

Keywords: SAR; 2CMV; change detection; optical flow; k-means; K-SVD

---

1. Introduction

One important use of synthetic aperture radar (SAR) imagery is in detecting changes between datasets from different imaging passes. Target and coherent change detection in SAR images have been extensively researched [1–4]. In two-color multiview (2CMV) advanced geospatial information (AGI) products, the changes are colorized and overlaid on an initial image such that new features are represented in cyan, and features that have disappeared are represented in red. In order to create the change maps, images are cross-correlated pixel-by-pixel to detect the changes. 2CMV products show changes at the pixel level and are often misleadingly dominated with red and cyan colors. Figure 1 shows a portion of a sample 2CMV image. In the sample images, there is an airplane visibly parked next to a building near the bottom center. It can be seen that many of the pixels in the 2CMV image are colored either red or cyan even if there is no change in the area.

Useful interpretation of temporal changes represented in 2CMV AGI products can be challenging because of speckle noise susceptibility and false positives that result from small orientation differences between objects imaged at different times. When every small intensity change creates a colored pixel, it becomes more difficult for operators and/or algorithms to detect meaningful changes and identify corresponding objects of interest.

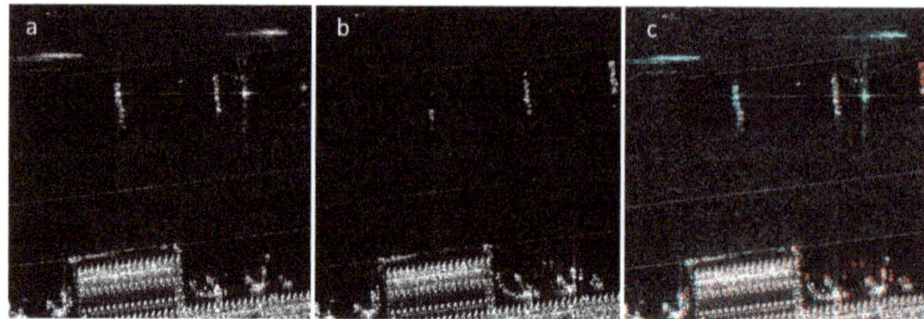

Figure 1. (**a**) reference image; (**b**) mission image; (**c**) two-color multiview (2CMV) image. In both images, there is an airplane visibly parked next to an airport building near the bottom center. In the second image (**b**), the airplane seems rotated by a small degree. The sharp edges of the building are slightly misregistered in the images and these registration errors are false positives in the 2CMV image.

In this work, we introduce a new framework of image processing methods for the efficient generation of 2CMV products toward extraction of advanced geospatial intelligence. Before false positive and object detection algorithms are performed, speckle and smoothing filters are used to mitigate the effects of speckle noise. Then, the number of false positive detections is reduced by applying: (1) unsupervised feature learning algorithms and (2) optical flow algorithms that track the motion of objects across time in small regions of interest.

There have been a number of change detection studies using thresholding [5–8], extreme learning machine [9,10], Markov random fields [11,12] and combinations of feature learning and clustering algorithms [13–19]. Optical flow fields can be used to distinguish between objects that have actually moved between frames and those that are in the same location but are slightly misregistered. Both cases of apparent motion can result in 2CMV detection, but they obviously differ greatly in terms of meaning. Investigation of the state-of-the-art in SAR image processing indicates that differentiating between these two general cases is a problem that has not been well addressed. Algorithms that mitigate speckle noise effects well and distinguishing between actual motion and misregistration can lead to better change detection. There is a lack of published methods for efficient generation of 2CMV products from SAR images, which serves as another motivating factor for this work.

The paper is organized in four sections. Following this introduction, Section 2 gives a brief background on the filtering, unsupervised feature learning, and optical flow techniques that were used and describes the stages of the proposed framework. Section 3 presents simulation results. Section 4 discusses the results and the contributions of the proposed methods.

2. Materials and Methods

In this section, we describe the key methods and steps of our image processing approach for generating change maps that drive the 2CMV representation and eliminating false positives in those maps.

2.1. Speckle Noise Filtering

Speckle noise is an inherent problem in SAR images [20] and causes difficulties for image interpretation by increasing the mean grey level of a local region. In order to mitigate speckle noise effects, we tested different speckle filter designs. Filters that were included in the testing were Frost [21], Enhanced Frost [22], Lee [23], Gamma-MAP [24], SRAD [25] and Non-Local Means [26]. In the end, Enhanced Frost filter was used in the algorithm due to its relatively straightforward implementation and comparable performance.

In [22], it was proposed to divide images into areas of three classes. The first class is comprised of homogeneous areas. The second class is comprised of heterogeneous areas wherein speckle noise is to be reduced, while preserving texture. The third class is comprised of areas containing isolated point targets that filtering should preserve. The Enhanced Frost filter output can be given as:

$$\hat{I}(t_o) = \begin{cases} \bar{I}, & \text{for } C_l(t_o) < C_u, \\ \bar{I}K_1 \exp\left[-K(C_l(t_o) - C_u)/(C_{max} - C_l(t_o))|t|\right], & \text{for } C_u \leq C_l(t_o) \leq C_{max}, \\ I & \text{for } C_l(t_o) \geq C_{max}, \end{cases} \tag{1}$$

where $t_o = (x_o, y_o)$ is the spatial coordinate, $\bar{I}$ is the mean intensity value inside the kernel, K is the filter parameter, K_1 is a normalizing constant, and $|t|$ is the absolute value of the pixel distance from the center of the kernel at t_o. The rest of the parameters are

$$C_u = \sqrt{\frac{1}{L}},$$

$$C_l(t_o) = \sigma/\bar{I}, \text{ and}$$

$$C_{max} = \sqrt{1 + \frac{2}{L}},$$

where C_u is the speckle coefficient of variation of the image, $C_l(t_o)$ is the local coefficient of variation of the filter kernel centered at t_o, C_{max} is the upper speckle coefficient of variation of the image, and L is the number of looks. In our implementation, instead of L, we used "equivalent number of looks" (ENL). It can be defined as $ENL = \mu^2/\sigma^2$, where μ is the mean and σ is the standard deviation.

2.2. k-Means Clustering

The k-means clustering algorithm attempts to partition p observations into k clusters such that each observation belongs to the nearest cluster mean (centroid) [27]. The k-means algorithm iteratively tries to find k centroids for each cluster, while minimizing a within-cluster sum of squares

$$argmin \sum_{i=1}^{k} \sum_{x_j \epsilon S} \|x_j - \mu_j\|^2,$$

where x_j is the jth observation and μ_j is the mean point (centroid) in the cluster. The basic steps of the algorithm are given in Algorithm 1:

Algorithm 1 k-means clustering algorithm

---

1. Initialize the centroids: Assign k points as the initial group centroids.

2. Calculate the distance of each point to the centroids and assign the point to the cluster that has the closest centroid.

3. After the assignment of all the points, recalculate the new values of the centroids.

4. Repeat Steps 2 and 3 until the centroid locations converge to a fixed value.

2.3. K-SVD

K-SVD is a dictionary learning algorithm that is used for training overcomplete dictionaries for sparse representations of signals [28,29]. It is an iterative method that is a generalization of the k-means clustering algorithm. The K-SVD algorithm alternates between two stages: (1) sparse coding stage, and (2) dictionary update stage. In the first stage, a pursuit algorithm is used to sparsely code the input data based on the current dictionary. Based on Ref. [29], the Batch Orthogonal Matching Pursuit (Batch-OMP) algorithm can be used in this step. In the second stage, the dictionary atoms are updated to better fit the data via a singular value decomposition (SVD) approach. The basic steps of the K-SVD algorithm are given in Algorithm 2.

Algorithm 2 K-SVD algorithm

Task: Find the best dictionary to represent the data samples $\{y_i\}_{i=1}^{N}$, $y_i \epsilon R^N$ as sparse compositions by solving:

$$min_{D,X}\{\|Y - DX\|_F^2\} \text{ subject to } \forall_i, \|x_i\|_0 \leq T_0.$$

Initialization: Set the dictionary matrix $D^{(0)} \epsilon R^{n \times K}$ with l^2 normalized columns. Set $J = 1$.

Iterations: Repeat until convergence:

- Sparse coding stage: Use any pursuit algorithm to compute the representation vectors x_i for each sample y_i by approximating the solution of

$$i = 1, 2, ..., N, \quad min_{x_i}\{\|y_i - Dx_i\|_2^2\} \text{ subject to } \|x_i\|_0 \leq T_0.$$

- Dictionary update stage: For each column $k = 1, 2, ..., K$ in D^{J-1},

    - Define the group of samples that use this atom, $w_k = \{i \,|1 \leq i \leq N, x_T^k(i) \neq 0\}$

    - Compute the overall representation error matrix, E_k, by

$$E_k = Y - \sum_{j \neq k} d_j x_T^j$$

    - Restrict E_k by choosing only the columns corresponding to w_k, and obtain E_k^R.

    - Apply SVD decomposition $E_k^R = U\Delta V^T$. Choose the updated dictionary column $\tilde{d}_k$ to be the first column of U. Update the coefficient vector x_R^k to be the first column of V multiplied by $\Delta(1,1)$.
- Set $J = J + 1$.

2.4. Optical Flow

Optical flow is the apparent motion of objects in image sequences that results from relative motion between the objects and the imaging perspective. In one canonical optical flow paper [30], two kinds of constraints are introduced in order to estimate the optical flow: the smoothness constraint and the brightness constancy constraint. In this section, we give a brief overview of the optical flow algorithm we employ in the proposed methodology.

Optical flow methods estimate the motion between two consecutive image frames that were acquired at times t and $t + \delta t$. A flow vector for every pixel is calculated. The vectors represent approximations of image motion that are based in large part on local spatial derivatives. Since the flow velocity has two components, two constraints are needed to solve for it. The brightness constancy constraint assumes that the brightness of a small area in the image remains constant as the area moves from image to image. Image brightness at the point (x,y) in the image at time t is denoted here as $I(x, y, t)$. If the point moves by δx and δy in time δt, then, according to the brightness constancy constraint:

$$\frac{dI}{dt} = 0. \tag{2}$$

This can also be stated as:

$$I(\mathbf{r} + \delta\mathbf{r}, t + \delta t) = I(\mathbf{r}, t), \tag{3}$$

where $\mathbf{r} = (x, y, 1)^T$ and $\mathbf{r} + \delta\mathbf{r} = (x + \delta x, y + \delta y, 1)^T$. However, the brightness constancy constraint is restrictive. A less restrictive brightness constraint was chosen to address the intensity changes in SAR images. In Reference [31], it is proposed that the brightness constancy constraint can be replaced with a more general constraint that allows a linear transformation between the pixel brightness values. This way, the brightness change can be non-zero, or:

$$\frac{dI}{dt} \neq 0.$$

The formulation that allows a linear transformation between the pixel brightness values is less restrictive, and can be written as:

$$I(\mathbf{r} + \delta\mathbf{r}, t + \delta t) = M(\mathbf{r}, t)I(\mathbf{r}, t) + C(\mathbf{r}, t). \tag{4}$$

After using the Taylor series, the revised constraint equation can be obtained:

$$I_t + I_\mathbf{r} \cdot \mathbf{r}_t - Im_t - c_t = 0, \tag{5}$$

where $m_t = \lim_{\delta t \to 0} \frac{\delta m}{\delta t}$ and $c_t = \lim_{\delta t \to 0} \frac{\delta c}{\delta t}$.

The relaxed brightness constraint error is:

$$\epsilon_I = \iint (I_t + I_\mathbf{r} \cdot \mathbf{r}_t - Im_t - c_t)^2 \, dx \, dy. \tag{6}$$

Equation (6) can be combined with the other constraint errors to produce the final functional to be minimized:

$$\epsilon_{total} = \epsilon_I + \lambda_s \epsilon_s + \lambda_m \epsilon_m + \lambda_c \epsilon_c, \tag{7}$$

where λ_s, λ_m, and λ_c are error weighting coefficients. The remaining errors are given as:

$$\epsilon_s = \iint ||\nabla\mathbf{r}_t||_2^2 dxdy,$$

$$\epsilon_m = \iint ||\nabla m_t||_2^2 dxdy,$$

$$\epsilon_c = \iint ||\nabla c_t||_2^2 dxdy.$$

Substituting the approximated Laplacians into the Euler–Lagrange equations, a single matrix equation can be derived:

$$\mathbf{A}\mathbf{f} = \mathbf{g}(\bar{\mathbf{f}}), \tag{8}$$

where

$$\mathbf{A} = \begin{pmatrix} I_x^2 + \lambda_s & I_x I_y & -I_x I & -I_x \\ I_x I_y & I_y^2 + \lambda_s & -I_y I & -I_y \\ -I_x I & -I_y I & I^2 + \lambda_m & I \\ -I_x & -I_y & I & 1 + \lambda_c \end{pmatrix}, \mathbf{f} = \begin{pmatrix} u \\ v \\ m_t \\ c_t \end{pmatrix}, \mathbf{g}(\bar{\mathbf{f}}) = \begin{pmatrix} \lambda_s \bar{u} - I_x I_t \\ \lambda_s \bar{v} - I_y I_t \\ \lambda_m \bar{m}_t + I_t I \\ \lambda_c \bar{c}_t + I_t \end{pmatrix}.$$

These equations have to be solved iteratively. The solution is given by:

$$\mathbf{f} = \mathbf{A}^{-1}\mathbf{g}(\bar{\mathbf{f}}), \tag{9}$$

where

$$\mathbf{A}^{-1} = \frac{1}{\alpha} \begin{pmatrix} \lambda_c \lambda_m \lambda_s + \lambda_m \lambda_s + I^2 \lambda_c \lambda_s + I_y^2 \lambda_c \lambda_m & -I_x I_y \lambda_c \lambda_m & I_x I \lambda_c \lambda_s & I_x \lambda_m \lambda_s \\ -I_x I_y \lambda_c \lambda_m & \lambda_c \lambda_m \lambda_s + \lambda_m \lambda_s + I^2 \lambda_c \lambda_s + I_y^2 \lambda_c \lambda_m & I_y I \lambda_c \lambda_s & I_y \lambda_m \lambda_s \\ -I_x I \lambda_c \lambda_s & I_y I \lambda_c \lambda_s & (I_x^2 + I_y^2)\lambda_c \lambda_s + \lambda_c \lambda_s^2 + \lambda_s^2 & -I \lambda_s^2 \\ I_x \lambda_m \lambda_s & I_y \lambda_m \lambda_s & I \lambda_s^2 & (I_x^2 + I_y^2)\lambda_m \lambda_s + \lambda_m \lambda_s^2 + I^2 \lambda_s^2 \end{pmatrix}$$

and

$$\alpha = \lambda_m \lambda_s^2 + I^2 \lambda_c \lambda_s^2 + (I_x^2 + I_y^2 + \lambda_s)\lambda_c \lambda_m \lambda_s.$$

The equations can then be solved iteratively for other pixels with:

$$\mathbf{f}^{k+1} = \mathbf{A}^{-1}\mathbf{g}(\bar{\mathbf{f}}^{k}), \tag{10}$$

where k is the iteration number. This way the matrix $\mathbf{A}^{-1}$ need only be computed once. More details about this optical flow algorithm can be found in Ref. [31].

2.5. Image Processing Steps

In this section, we describe the image processing approach for extracting change maps. The inputs are two registered SAR images of the same field of view that were taken at different times, i.e., "reference" image and "mission" image. Due to the large size of the images, images were divided into subimages for processing.

In the denoising step, an Enhanced Frost filter, as described in Section 2.1, with a 5×5 window size was first used to mitigate the speckle noise effects. Then, a 9×9 low pass filter was used to smooth the test areas in order to obtain more uniform flow fields in the optical flow processing step. The remaining steps are grouped in three stages and described in the following subsections. The detailed flow diagram shown in Figure 2 can be used as a guide for the following descriptions.

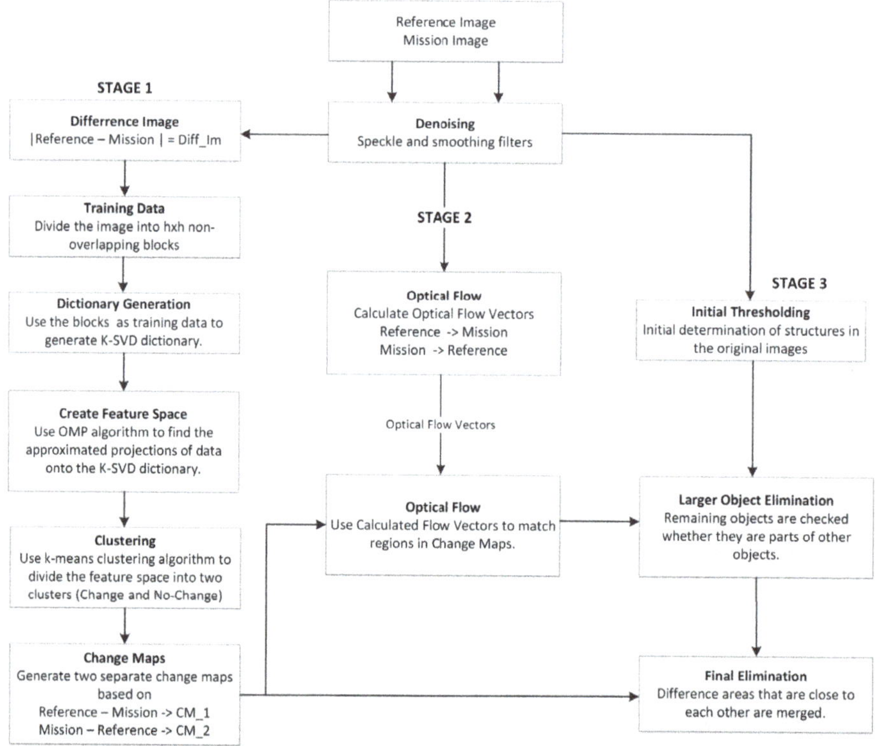

Figure 2. Flow diagram of the proposed framework.

2.5.1. First Stage: Generation of Change Maps Using Unsupervised Feature Learning

Two change maps are needed for a 2CMV representation of an SAR image pair. Each change map represents the changes that exist in the corresponding SAR image. In this stage, we generate a combined change map and separate it into two change maps. In order to generate the combined change map, we used an approach similar to that was used in [13]. In the original approach, an eigenvector

space is created by performing principle component analysis (PCA) on the difference image and
k-means algorithm classifies the projections onto the eigenvector space into two classes: e.g., change
and no-change. The basic steps are given in Algorithm 3. It should be noted that, in our framework,
PCA was replaced with K-SVD because one can adjust the dictionary size and the sparsity constraint
to obtain change maps with different levels of details. Figure 3 shows two change map results with
different dictionary sizes.

Algorithm 3 Generating change maps

- **Difference Image:**

$$X_{dif} = |Reference - Mission|$$

- **Training Data:** Divide X_{dif} into $h x h$ non-overlapping blocks.

- **Dictionary Generation:** Use the K-SVD algorithm to generate an overcomplete dictionary.

- **Create Feature Space:**

    - Generate $h x h$ blocks for each pixel in X_{dif} where the pixel is in the center of the block.

    - Use OMP algorithm to generate the projections of the data onto the dictionary.

- **Clustering:** Use the k-means algorithm to classify the feature space into two classes, e.g. change and no-change.

- **Change maps:** Use the two classes to generate the combined change map. Divide the combined change map into two
  separate change maps based on the changes that occur in the images.

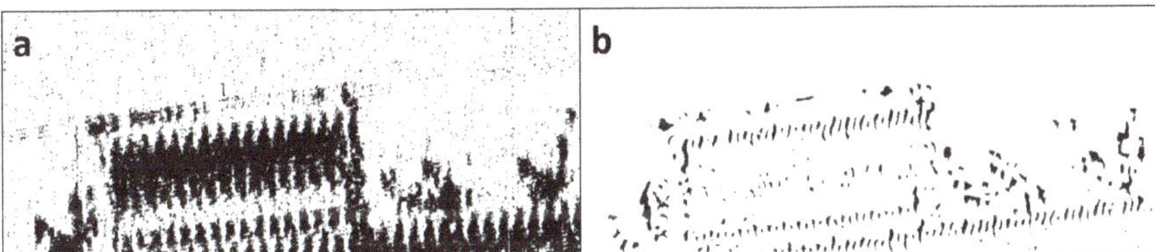

Figure 3. (**a**) change map with dictionary size = 30 atoms with 30 non-zero coefficients; (**b**) change map
with dictionary size = 15 with three non-zero coefficients. Note that a larger dictionary size with more
non-zero coefficients captures more changes.

After the change maps are generated, object properties such as area and location are calculated
and, based on a user-defined area threshold, insignificant change areas are excluded from the change
maps. The remaining change areas are then overlaid onto the reference image. In the 2CMV image,
the areas that exist only in the reference image are colored in cyan and the areas that exist only in the
mission image are colored in red. A sample 2CMV image after this stage is shown in Figure 4.

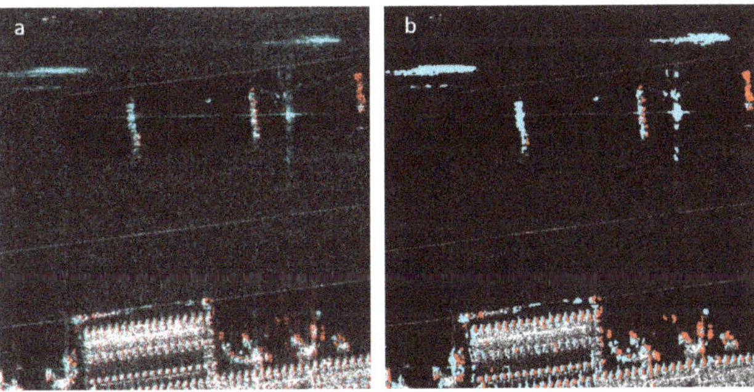

Figure 4. (**a**) original 2CMV image; (**b**) 2CMV image after Stage 1. Note that there are several false
positives around the ridges of the building. In the second image, change colors (red and cyan) were
made more pronounced to highlight the false positives.

In a previous work, this stage was replaced by adaptive thresholding [32].

2.5.2. Second Stage: Optical Flow

Figure 4 displays a 2CMV image after the first stage wherein it is clear that additional processing is needed to improve results because the ridges of the building in both images are slightly misregistered and they are shown as changes in both images. The primary improvement that is targeted with additional processing is reducing the number of false positives in the image. This goal can be accomplished with the use of the optical flow (OF) method described in Section 2.4. To manage computational complexity, the optical flow algorithm is performed on 256×256 pixel image blocks. Note that optical flow is calculated based on the original reference and mission images.

After obtaining the flow vectors, the direction of the majority of flow vectors is determined. The flow vectors that are in this direction are applied to the two first stage change maps to find matches. In the reference image, OF vectors are used to move the detected change areas in the flow direction. The destination of an area is then compared with the same location in the mission image. If there is a matching area based on location and size, then the two change areas are excluded from the change maps. The same process is performed in the opposite direction to match mission image change areas in the reference image. Figure 5 illustrates this step.

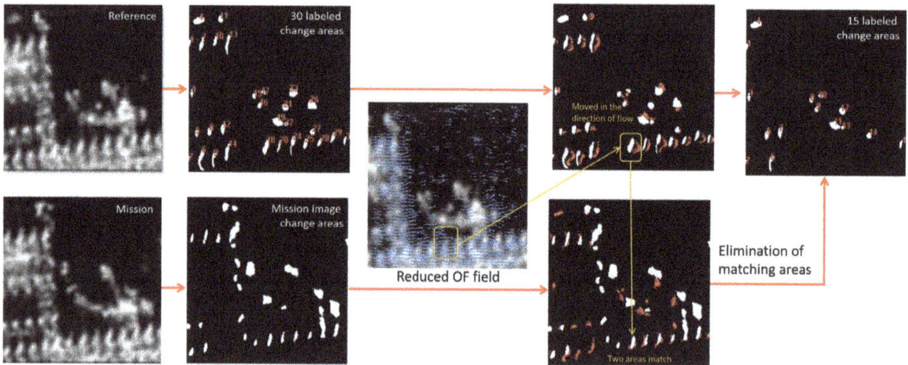

Figure 5. Elimination of false positives using optical flow. Change areas are moved along the flow direction in the reference image change map. Moved areas (shown in red) from the reference image are overlaid onto the mission image change map. The overlapped areas are then removed.

2.5.3. Third Stage: OF Assisted Object Extraction

This stage has two main parts: extraction and elimination. Extraction is performed by an adaptive thresholding method that is similar to the one used in [32]. In this stage, the thresholding is performed on the original images to extract/label objects. The resulting two thresholded images are processed in two ways. First, OF vectors are used on the images to match the objects. The main difference from the second stage is that the flow vectors are used on the original thresholded images, not on the change maps. Change maps do not necessarily contain objects, and the goal is to find objects that moved between the two images. Objects with possibility of movement are labeled and compared against the areas in the change maps. It should also be noted that only some parts of an object can be detected as a change, and these detected changes can be used as a guide to extract the full object.

After this process, the labeled areas in the change maps are overlaid on the reference image and checked whether they are a part of a larger object in the image. If the labeled area is found to be a part of a larger object, then the same location in the mission image is checked for the same object. In the case of two similar objects around the same location, it can be assumed that the detected object is a false negative and excluded from the difference map. After these two methods are performed, the output of this stage is generated by simply taking the intersection of the two results. Figure 6 shows how this process converts the reference image in (a) to the final output in (e).

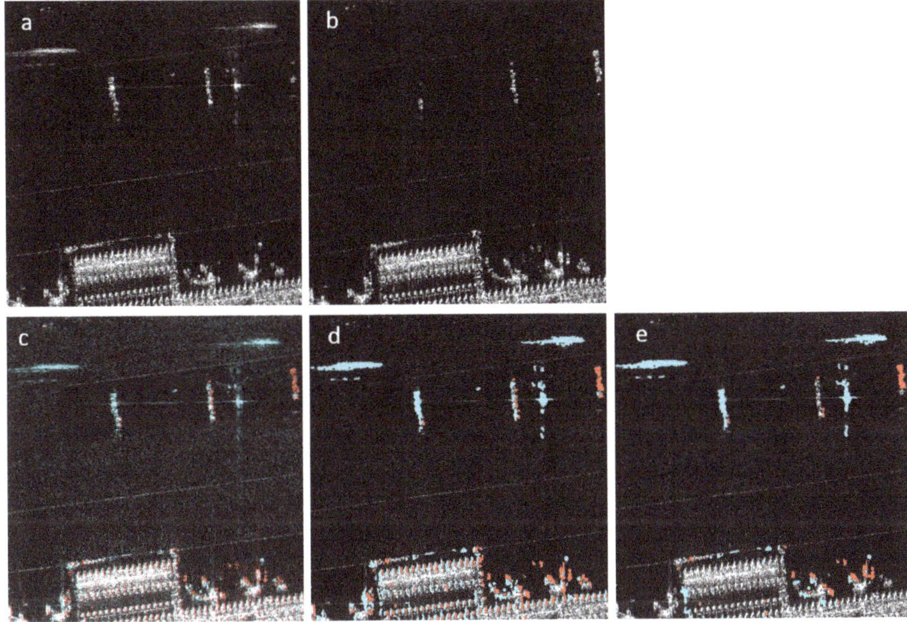

Figure 6. (**a**) reference image; (**b**) mission image; (**c**) original 2CMV image; (**d**) 2CMV image after using dictionary learning and clustering (Stage 1); (**e**) final 2CMV image. False positives are reduced.

3. Results

The proposed algorithm was compared against three change detection methods: PCAKM [13], GaborTLC [18], and NR-ELM [10]. All three methods are implemented with their default parameters by using the publicly available code provided by the authors. The first dataset consisted of 1024 × 1024 regions from an SAR image pair provided by Lockheed Martin (Bethesda, MD, USA). The data were acquired with various Lockheed Martin SAR units, one example of which is an airborne long range, all weather, day/night, X-band SAR unit with a resolution of 1 m. The selected regions contained speckle noise and false positives that resulted from registration and perspective problems. 2CMV images were generated for each method. The visual results are shown in Figure 7. NR-ELM was more susceptible to noise compared to the other methods. It was noted that unsupervised dictionary learning and clustering algorithms were effective at removing false positives that did not match object profiles. Optical flow was effective for removing difficult false positives that resulted from registration and perspective problems.

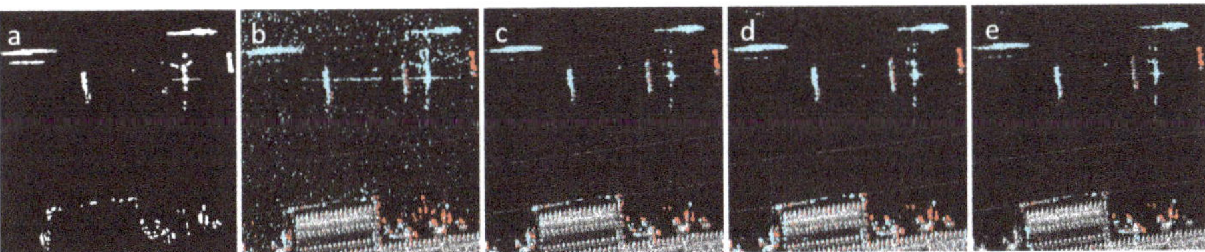

Figure 7. Results by (**a**) manual ground truth; (**b**) NR-ELM; (**c**) GaborTLC; (**d**) PCAKM; (**e**) proposed method.

From the ground truth map, the actual number of pixels belonging to the unchanged class and changed class are calculated, denoted as N_u and N_c, respectively. With this information, five objective metrics are adopted for quantitative evaluation. False positive (FP) is the number of pixels belonging to the unchanged class but falsely classified as changed class. False negative (FN) is the number of pixels belonging to the changed class but falsely classified as unchanged class. The overall error (OE) is calculated by FP + FN. Percentage correct classification (PCC) and Kappa coefficient (KC) are as follows:

$$PCC = \frac{(N_c - FN) + (N_u - FP)}{N_u + N_c} \times 100\%,$$

$$KC = \frac{PCC - PRE}{1 - PRE},$$

where proportional reduction in error (PRE) is defined as

$$PRE = \frac{(N_c - FN + FP) \cdot N_c + (N_u - FP + FN) \cdot N_u}{(N_u + N_c)^2}.$$

The results of the quantitative metrics are given in Table 1.

Table 1. Results for the SAR dataset.

Methods	FP	FN	OE	PCC (%)	KC (%)	Time (s)
NR-ELM	97377	4276	101653	0.9031	0.2993	202.1
GaborTLC	20160	8449	28609	0.9727	0.5809	74.6
PCAKM	35135	6251	41386	0.9605	0.51	6.4
Proposed method	3865	12852	16717	0.9841	0.6569	199.1

In addition to these results, the proposed framework was tested on an ensemble of 1024×1024 regions from the same SAR dataset. In many representative image regions where registration errors were prevalent, false positive detections were reduced by over 60%. Filtering of speckle noise and adaptive thresholds improved the quality of the object extraction and helped identify false positives. Establishing false positive motion/error thresholds, in accordance with initial image registration, can be key for continued improvement. It is also a challenge to extract only regions with intensity value changes. It is possible that wavelet based methods might be more successful with such a task.

For the second test, a more standard dataset was used. The San Francisco dataset has been used in change detection studies and its ground truth change map was provided in [33]. It consists of two SAR images over the city of San Francisco that were acquired by ERS-2 C-band SAR sensor with VV polarization. The images were provided by the European Space Agency with a resolution of 25-m. These two images were captured in August 2003 and May 2004, respectively. The size of the images were 256×256 for this test. The change maps of the methods can be seen in Figure 8.

The results of the quantitative metrics are given in Table 2. The proposed framework performed comparable to PCAKM as a change detection algorithm. The San Francisco dataset doesn't contain registration and perspective errors with speckle noise.

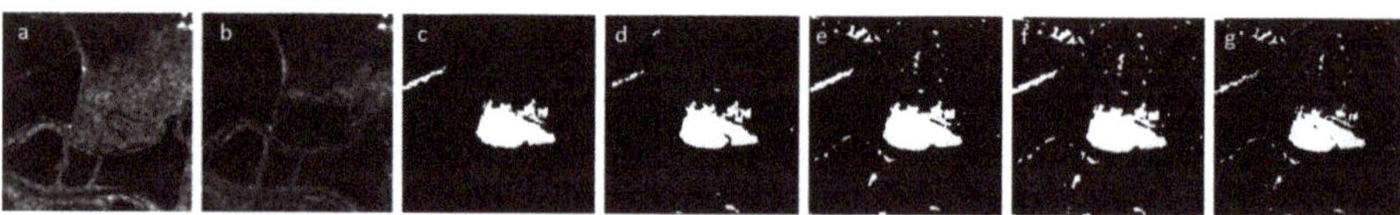

Figure 8. (**a,b**) San Francisco dataset; (**c**) ground truth; (**d**) NR-ELM; (**e**) GaborTLC; (**f**) PCAKM; (**g**) proposed method.

Table 2. Results for the San Francisco dataset.

Methods	FP	FN	OE	PCC (%)	KC (%)
NR-ELM	328	440	768	0.9883	0.9107
GaborTLC	1376	60	1436	0.9781	0.8539
PCAKM	1855	73	1928	0.9706	0.8115
Proposed method	836	685	1521	0.9768	0.8277

It should be noted that the proposed framework provided better results compared to the other methods when the datasets contain registration and perspective errors with speckle noise. Otherwise, the performance of the proposed method is comparable to PCAKM as a change detection algorithm since the optical flow processing stage cannot provide matching regions in the images.

Even though the computational complexity was not an issue during the course of this work, the speckle filtering, optical flow processing and merging are computationally expensive processes. On a dual core computer (Intel Core i7 6500U, Santa Clara, CA, USA) with 16 GB of memory, it takes slightly less than 3.5 min to process one region. There are many factors that are contributing to this time. Code was written in the MATLAB environment (R2016a, MathWorks, Natick, MA, USA) and not optimized for performance.

4. Conclusions

It was shown that unsupervised feature learning algorithms can be effectively used in conjunction with optical flow methods to generate 2CMV AGI products. Other image processing methods like noise reduction and adaptive thresholding were used to improve object extraction in the proposed methodology. Results demonstrated the ability of the techniques to reduce false positives by up to 60% in the provided SAR image pairs. However, there is still room for further improvement. For example, it was noticed that optical flow object matches close to image block borders can be overlooked due to the inaccuracy of flow vectors near the block borders. This problem can be addressed with a multigrid approach that leverages overlapping image blocks. Using this approach, if an object pair is close to the border in one block, then it will be near the center of an overlapping block. It has also been noted that only some parts of an object can be detected as a change, and the detected parts can be used as a guide to segment the full object. Objects that are close to one another can be merged to provide a more holistic analysis of the scene and further reduce the number of false positive object detections. However, it must be concurrently ensured that false positive reduction is not overly aggressive to the point that false negatives are generated. More recent optical flow or motion estimation algorithms can be investigated as an alternative to the one utilized in this work. The chosen optical flow method is suitable for the tested dataset and performs adequately as expected since it takes into account the intensity changes between images. The choice of K-SVD over PCA increased the computational complexity while allowing flexibility over the details of the change maps by changing the dictionary size and the number of non-zero coefficients. Dictionaries with higher number of non-zero coefficients provided more detailed change maps. For future work, investigating the correlation between the quantitative metrics and the parameters in the framework (e.g., dictionary size, etc.) can provide insight into tuning the framework for different types of datasets. Other methods can be researched as alternatives to the K-SVD method in the framework.

Author Contributions: Conceptualization, B.K. and D.F.; methodology, B.K.; software, B.K.; validation, B.K. and D.F.; investigation, B.K.; data curation, D.F.; writing—original draft preparation, B.K.; writing—review and editing, D.F.; supervision, D.F.; project administration, D.F.; funding acquisition, D.F.

Funding: This work was supported in part by Lockheed Martin.

Acknowledgments: The authors would like to acknowledge SenSIP for the center's valuable contributions to this work.

Conflicts of Interest: The authors declare no conflict of interest.

References

1. El-Darymli, K.; McGuire, P.; Power, D.; Moloney, C. Target Detection in Synthetic Aperture Radar Imagery: A State-of-the-Art Survey. *J. Appl. Remote Sens.* **2013**, *7*, 071598. [CrossRef]
2. El-Darymli, K.; Gill, E.W.; McGuire, P.; Power, D.; Moloney, C. Automatic Target Detection in Synthetic Aperture Radar Imagery: A State-of-the-Art Review. *IEEE Access* **2016**, *4*, 6014–6058. [CrossRef]
3. Ashok, H.G.; Patil, D.R. Survey on Change Detection in SAR Images. In Proceedings of the IJCA Proceedings on National Conference on Emerging Trends in Computer Technology, Shirpur, India, 28–29 March 2014; pp. 4–7.
4. Ren, W.; Song, J.; Tian, S.; Wu, W. Survey on Unsupervised Change Detection Techniques in SAR Images1. In Proceedings of the 2014 IEEE China Summit International Conference on Signal and Information Processing (ChinaSIP), Xi'an, China, 9–13 July 2014; pp. 143–147.
5. Bazi, Y.; Bruzzone, L.; Melgani, F. An Unsupervised Approach Based on the Generalized Gaussian Model to Automatic Change Detection in Multitemporal SAR Images. *IEEE Trans. Geosci. Remote Sens.* **2005**, *43*, 874–887. [CrossRef]
6. Bovolo, F.; Bruzzone, L. A Detail-Preserving Scale-Driven Approach to Change Detection in Multitemporal SAR Images. *IEEE Trans. Geosci. Remote Sens.* **2005**, *43*, 2963–2972. [CrossRef]
7. Moser, G.; Serpico, S.B. Generalized Minimum-Error Thresholding for Unsupervised Change Detection from SAR Amplitude Imagery. *IEEE Trans. Geosci. Remote Sens.* **2006**, *44*, 2972–2982. [CrossRef]
8. Sumaiya, M.N.; Kumari, R.S.S. Logarithmic Mean-Based Thresholding for SAR Image Change Detection. *IEEE Geosci. Remote Sens. Lett.* **2016**, *13*, 1726–1728. [CrossRef]
9. Jia, L.; Li, M.; Zhang, P.; Wu, Y. SAR Image Change Detection Based on Correlation Kernel and Multistage Extreme Learning Machine. *IEEE Trans. Geosci. Remote Sens.* **2016**, *54*, 5993–6006. [CrossRef]
10. Gao, F.; Dong, J.; Li, B.; Xu, Q.; Xie, C. Change Detection from Synthetic Aperture Radar Images Based on Neighborhood-Based Ratio and Extreme Learning Machine. *J. Appl. Remote Sens.* **2016**, *10*, 10–14. [CrossRef]
11. Melgani, F.; Bazi, Y. Markovian Fusion Approach to Robust Unsupervised Change Detection in Remotely Sensed Imagery. *IEEE Geosci. Remote Sens. Lett.* **2006**, *3*, 457–461. [CrossRef]
12. Yousif, O.; Ban, Y. Improving SAR-Based Urban Change Detection by Combining MAP-MRF Classifier and Nonlocal Means Similarity Weights. *IEEE J. Sel. Top. Appl. Earth Observ.* **2014**, *7*, 4288–4300. [CrossRef]
13. Celik, T. Unsupervised Change Detection in Satellite Images Using Principal Component Analysis and k-Means Clustering. *IEEE Geosci. Remote Sens. Lett.* **2009**, *6*, 772–776. [CrossRef]
14. Li, W.; Chen, J.; Yang, P.; Sun, H. Multitemporal SAR Images Change Detection Based on Joint Sparse Representation of Pair Dictionaries. In Proceedings of the 2012 IEEE International Geoscience and Remote Sensing Symposium, Munich, Germany, 22–27 July 2012; pp. 6165–6168.
15. Lu, X.; Yuan, Y.; Zheng, X. Joint Dictionary Learning for Multispectral Change Detection. *IEEE Trans. Cybern.* **2017**, *47*, 884–897. [CrossRef]
16. Ghosh, A.; Mishra, N.; Ghosh, S. Fuzzy Clustering Algorithms for Unsupervised Change Detection in Remote Sensing Images. *Inf. Sci.* **2011**, *181*, 699–715. [CrossRef]
17. Nguyen, L.H.; Tran, T.D. A Sparsity-Driven Joint Image Registration and Change Detection Technique for SAR Imagery. In Proceedings of the 2010 IEEE International Conference on Acoustics, Speech and Signal Processing, Dallas, TX, USA, 14–19 March 2010; pp. 2798–2801.
18. Li, H.; Celik, T.; Longbotham, N.; Emery, W.J. Gabor Feature Based Unsupervised Change Detection of Multitemporal SAR Images Based on Two-Level Clustering. *IEEE Geosci. Remote Sens. Lett.* **2015**, *12*, 2458–2462.
19. Gong, M.; Su, L.; Jia, M.; Chen, W. Fuzzy Clustering With a Modified MRF Energy Function for Change Detection in Synthetic Aperture Radar Images. *IEEE Trans. Fuzzy Syst.* **2014**, *22*, 98–109. [CrossRef]
20. Dekker, R.J. Speckle Filtering in Satellite SAR Change Detection Imagery. *Int. J. Remote Sens.* **1998**, *19*, 1133–1146. [CrossRef]
21. Frost, V.; Stiles, J.A.; Shanmugan, K.S.; Holtzman, J.C. A Model for Radar Images and Its Application to Adaptive Digital Filtering of Multiplicative Noise. *IEEE Trans. Pattern Anal. Mach. Intell.* **1982**, *PAMI-4*, 157–166. [CrossRef]
22. Lopes, A.; Touzi, R.; Nezry, E. Adaptive Speckle Filters and Scene Heterogeneity. *IEEE Trans. Geosci. Remote Sens.* **1990**, *28*, 992–1000. [CrossRef]

23. Lee, J.S. Digital Image Enhancement and Noise Filtering by Use of Local Statistics. *IEEE Trans. Pattern Anal. Mach. Intell.* **1980**, *PAMI-2*, 165–168. [CrossRef]
24. Lopes, A.; Nezry, E.; Touzi, R.; Laur, H. Maximum a Posteriori Filtering and First Order Texture Models in SAR Images. In Proceedings of the 10th Annual International Symposium on Geoscience and Remote Sensing, College Park, MD, USA, 20–24 May 1990; pp. 2409–2412.
25. Yu, Y.; Acton, S. Speckle Reducing Anisotropic Diffusion. *IEEE Trans. Image Process.* **2002**, *11*, 1260–1270.
26. Coupe, P.; Hellier, P.; Kervrann, C.; Barillot, C. NonLocal Means-based Speckle Filtering for Ultrasound Images. *IEEE Trans. Image Process.* **2009**, *18*, 2221–2229. [CrossRef]
27. Gonzalez, R.; Woods, R. *Digital Image Processing*, 3rd ed.; Prentice-Hall: Upper Saddle River, NJ, USA, 2006.
28. Aharon, M.; Elad, M.; Bruckstein, A. K-SVD: An Algorithm for Designing Overcomplete Dictionaries for Sparse Representation. *IEEE Trans. Signal Process.* **2006**, *54*, 4311–4322. [CrossRef]
29. Rubinstein, R.; Zibulevsky, M.; Elad, M. *Efficient Implementation of the K-SVD Algorithm Using Batch Orthogonal Matching Pursuit*; Technical Report; Computer Science Department, Technion: Haifa, Israel, 2008.
30. Horn, B.; Schunck, B. Determining Optical Flow. *Artif. Intell.* **1980**, *17*, 185–203. [CrossRef]
31. Gennert, M.; Negahdaripour, S. *Relaxing the Brightness Constancy Assumption in Computing Optical Flow*; A.I. Lab Memo 975; Massachusetts Institute of Technology: Cambridge, MA, USA, 1987.
32. Kanberoglu, B.; Frakes, D. Extraction of Advanced Geospatial Intelligence (AGI) from Commercial Synthetic Aperture Radar Imagery. In Proceedings of the Algorithms for Synthetic Aperture Radar Imagery XXIV 2017, Anaheim, CA, USA, 9–13 April 2017; Volume 10201, p. 1020106.
33. Gao, F.; Liu, X.; Dong, J.; Zhong, G.; Jian, M. Change Detection in SAR Images Based on Deep Semi-NMF and SVD Networks. *Remote Sens.* **2017**, *9*, 435. [CrossRef]

Article

Refocusing Moving Ship Targets in SAR Images Based on Fast Minimum Entropy Phase Compensation

Xiangli Huang [1,2], Kefeng Ji [1,2,*], Xiangguang Leng [2], Ganggang Dong [3] and Xiangwei Xing [4]

[1]   State Key Laboratory of Complex Electromagnetic Environment Effects on Electronics and Information System, National University of Defense Technology, Changsha 410073, China; huangxiangli_2008@126.com

[2]   School of Electronic Science, National University of Defense Technology, Changsha 410073, China; luckight@163.com

[3]   National Laboratory of Radar Signal Processing, Xidian University, Xi'an 710071, China; dongganggang@xidian.edu.cn

[4]   Beijing Institute of Remote Sensing Information, Beijing 100192, China; xingxiangwei@nudt.edu.cn

*   Correspondence: jikefeng@nudt.edu.cn; Tel.: +86-731-8457-6384

Received: 15 January 2019; Accepted: 4 March 2019; Published: 7 March 2019

Abstract: Moving ship targets appear blurred and defocused in synthetic aperture radar (SAR) images due to the translation motion during the coherent processing. Motion compensation is required for refocusing moving ship targets in SAR scenes. A novel refocusing method for moving ship is developed in this paper. The method is exploiting inverse synthetic aperture radar (ISAR) technique to refocus the ship target in SAR image. Generally, most cases of refocusing are for raw echo data, not for SAR image. Taking into account the advantages of processing in SAR image, the processing data are SAR image rather than raw echo data in this paper. The ISAR processing is based on fast minimum entropy phase compensation method, an iterative approach to obtain the phase error. The proposed method has been tested using Spaceborne TerraSAR-X, Gaofeng-3 images and airborne SAR images of maritime targets.

Keywords: synthetic aperture radar (SAR); inverse synthetic aperture radar (ISAR); moving ship; refocusing; fast minimum entropy

---

1. Introduction

Synthetic aperture radar (SAR) is widely employed in military surveillance, geography mapping and resource surveying. High-resolution SAR image is of great significance for homeland and military security [1–4]. A moving ship is not a static target during image formation, and the SAR imaging results are blurred and defocused [5–7]. It is necessary to refocus the defocused ship for ship recognition, and precise motion compensation becomes a key element of refocusing.

The motion between the moving target and the radar contains translation and rotation motions [8,9]. Translation motion is the main cause of image defocusing. Motion compensation can eliminate the translation motion affection on image. It is a significant step of inverse synthetic aperture radar (ISAR) processing for refocusing moving target. Motion compensation commonly includes two steps. The first is range alignment which is coarse compensation [10,11]. The second is phase compensation which compensates the Doppler frequency shift caused by movement [1,5]. Range alignment and phase compensation are also referred to as autofocusing. Due to imperfection of coarse compensation, this paper focuses on phase compensation.

Methods for phase compensation may be divided into three categories. The first category is scatter-based algorithms, such as dominant scatter processing (DSP) method [12,13], phase gradient autofocus (PGA) method [14–16]. The DSP method is intuitive in concept and easy to implement, but it needs high-quality prominent point in echo, otherwise the image is inferior [13]. The PGA method estimates phase error with part of data selected by isolating defocused targets via center shifting and window operations. The disadvantage of PGA is that the imaging is sensitive to the selection of the dominant scatters, window length, and the iteration times [15]. The second category is optimization algorithms, including Doppler centroid tracking (DCT) method [17,18], maximum contrast (MC) method [19,20], maximum likelihood (ML) method [21,22] and minimum entropy (ME) method [23,24]. DCT is a classical autofocus method with good robustness and a small amount of computation. However, only the rotational motion is ignored, the DCT approach can provide the maximum likelihood estimation of the phase error. When one scatter or multiple scatters are located in a range bin, the motion compensation accuracy will decrease [18]. The MC method assumes a mathematical model for the received signal. The model parameters (target radial velocity, acceleration) are achieved by the maximum contrast criterion of the image. It should be pointed out that the method requires two-dimensional search parameters at the same time and the computation is large [19]. The ME method which is based on the overall information of image takes the entropy as the cost function. The focused image can be achieved by numerical iterative until minimum entropy are acquired. It has well performance under low signal noise ratio (SNR), while the efficiency is lower and the computation time is large [23]. The last category is other algorithms, such as spare representation (SR) method [25]. A sparse metric is defined to iteratively estimate the sparse scatterer coefficients and phase errors, while the SR method is proposed to only deal with autofocus issues and cannot simultaneously obtain high-resolution images. The scatter-based algorithms are based on the processing of dominant scatter center and pay attention to the phase history of isolated scatter center. The optimization algorithms obtain phase error via image quality evaluation. The image contrast and entropy are maximal and minimum respectively when the image are well-focused. The scatter-based algorithms are higher computational efficiency than optimization algorithms, but the image quality constructed with former algorithms are worse than latter algorithms results. The previous methods all have some drawbacks and are mainly applied in raw echo data.

There are two issues involved in processing raw echo data. One is that the aforementioned methods are often applied to raw echo data, while the moving ship's position is hard to ascertain in raw data [6]. Hence those methods may have to cope with the entire raw dataset. The invalid data would occupy a massive amount of computation time when processing all the raw data. The other is that there may be a few ship targets with different motions in the raw data. Raw data is repeatedly processed for specific parameters of different moving ship, and the computation cost greatly increases [6]. The processing data are sub-images selected from ordinary SAR images, not raw data, and the sub-images are converted into the raw echo data domain by an inversion algorithm. The sub-images in the raw echo data domain are refocused with the ISAR technique, and the moving ships can then be well-focused. This thought has the advantage of easily locating moving ships in SAR images and the data size of sub-image is also smaller than the entire raw data.

To deal with these problems, a refocusing method for moving ships based on fast minimum entropy phase compensation is proposed in this paper. The processing data are sub-images containing moving ships rather than the raw echo data, and the ISAR technique is based on fast minimum entropy phase compensation. The refocusing method has three advantages, the first is that the computational burden is low, due to the smaller data size of the sub-image than the raw echo data. The second is the procedures of the inversion algorithm and image reconstruction are simple. The last is that the ISAR technique based on fast minimum entropy phase compensation has good image quality and computational efficiency.

The remainder of this paper is organized as follows: in Section 2, basic procedures of moving ship refocusing in SAR images is presented; ISAR processing based on fast minimum entropy phase compensation is elaborated in Section 3. In Section 4, experiments based on real SAR images are performed and the conclusions are presented in Section 5.

2. Basic Procedures of Moving Ship Refocusing in SAR Images

Moving ships are regularly blurred and defocused in SAR images due to the translation motion with respect to the scene center. ISAR technology is a common method to form well-focused images, but it has two basic problems. One is that SAR images contain a very large number of ships, each with its own motion. Hence, it needs sub-images which only contain a single ship target. The other is the input of the ISAR processor is raw data, not SAR images, so it needs to get input data for the ISAR system which can serve as raw data. The basic procedures of refocusing moving ship in SAR image are represented in Figure 1.

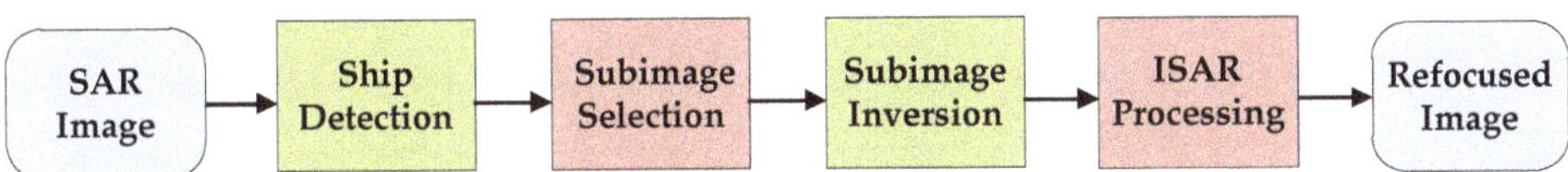

Figure 1. Block scheme of defocused target refocusing in SAR image.

The detailed procedures are as follows:

(a) Input a single look complex (SLC) image;
(b) Implement ship detection with software;
(c) Select sub-images, where each sub-image includes only a single defocused ship and has the same spatial resolution as the original image;
(d) Invert the sub-image to the equivalent raw data domain via an inversion method [26–29];
(e) Exploit ISAR processing to generate a focused image of the ship.

The sub-image needs to be inverted to equivalent raw data-like data containing only the target echo, background and residual clutter. The common inversion method is known as the range Doppler inversion [26].

When the angle variation is not too large, and the rotation vector is sufficiently stable during radar imaging, the range Doppler (RD) algorithm is applied for reforming SAR or ISAR images with high accuracy after motion compensation. The polar grid in the spatial frequency domain can be taken as a nearly regularly sampled rectangular grid with no need for interpolations. The main advantage of this method is the low amount of computation and it is the reason why it has been employed in many references [26,29]. The disadvantage is that the RD algorithm can only be used on low resolution SAR images when the spatial resolution is of the order of meters. The RD reformation algorithm is used when the SAR image is StripMap data. In this paper, the inverse range Doppler (IRD) algorithm is applied via a two-dimensional fast Fourier transform (FFT). The flowcharts of the RD algorithm and inverse RD algorithm are shown in Figure 2.

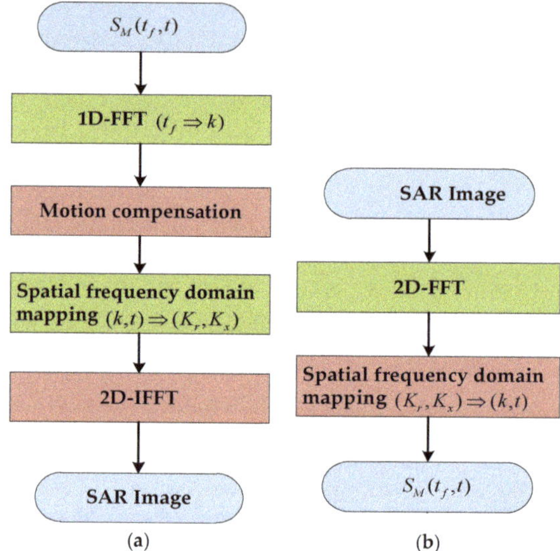

(a)   (b)

Figure 2. RD and IRD algorithm flow charts. (**a**) RD algorithm flowchart; (**b**) IRD algorithm flowchart.

3. ISAR Processing Based on Fast Minimum Entropy Phase Compensation

We propose an ISAR processing method based on fast minimum entropy phase compensation to refocus sub-images of moving ship targets. The approach is a RD algorithm exploiting a Fourier transform to create the image. The fast minimum entropy phase compensation method is a non-parametric method and can be applied to arbitrary targets, even high-order polynomial models.

3.1. Signal Model

The geometry of the ISAR system is depicted in Figure 3, where the radar is located at $r(0, 0, h)$ in the (X, Y, Z) coordinates system. The reference coordinates system (z_1, z_2, z_3) is set at the target point $p(x0, y0, 0)$. The distance between the radar and the target is $R_0(t)$. The back-scattering property of the target is represented by $\xi(z)$ and z is the vector that locates a generic scatter point on the reference coordinate system. The received signal from the moving target can then be written as follows [30,31]:

$$S_R(f,t) = rect(\frac{t}{T_{obs}})rect(\frac{f-f_0}{B})e^{-j\frac{4\pi f}{c}R_0(t)}\int_V \xi(z)e^{-j\frac{4\pi f}{c}(z^T \bullet i_{R_0(t)})}dz \qquad (1)$$

where f_0 is the carrier frequency, B the signal bandwidth and T_{obs} the observation time. $i_{R_0(t)}$ is the unit vector of $R_0(t)$ and V is the spatial domain where the reflectivity function z^T is defined.

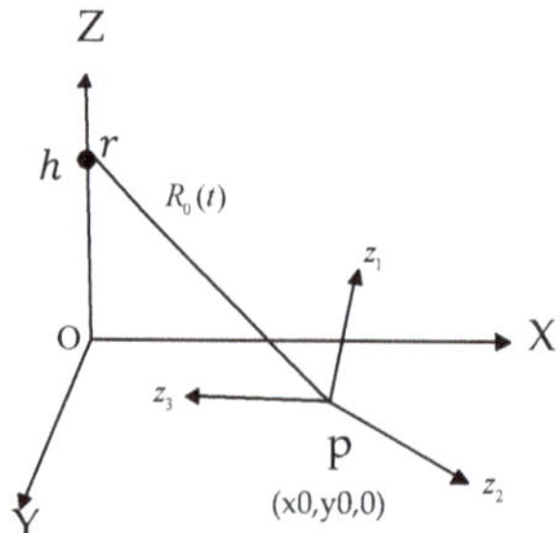

Figure 3. Geometry of the ISAR system.

The radial motion compensation can be achieved by removing the phase term $e^{-j\frac{4\pi f}{c}R_0(t)}$ and the received signal after motion compensation can be expressed as follows:

$$S_R(f,t) = rect(\frac{t}{T_{obs}})rect(\frac{f-f_0}{B})\int_V \xi(z)e^{-j\frac{4\pi f}{c}\left(z^T\bullet i_{R_0(t)}\right)}dz \tag{2}$$

The distance $R_0(t)$ can be approximated by a polynomial. This can be written as follows:

$$R_0(t) \approx \alpha + \beta t + \frac{1}{2}\gamma t^2 \tag{3}$$

where $\alpha = R_0(0)$, $\beta = \dot{R}_0(0)$ and $\gamma = \ddot{R}_0(0)/2$, $\alpha = R_0(0)$ cannot cause the defocusing in the image, the β and γ represent the target radial velocity and acceleration which also are related to the Doppler frequency parameter in (4):

$$\begin{cases} f_{dc} = \frac{2f}{c}\beta \\ f_{dr} = \frac{4f}{c}\gamma \end{cases} \tag{4}$$

The quadratic term $\frac{1}{2}\gamma t^2$ is the main cause of defocus in SAR images [3,32].

3.2. Phase Compensation Based on Fast Minimum Entropy Method

The features of ship, image entropy and image contrast (IC) are regarded as the evaluation criteria of image focusing. If the image is well-focused, the entropy and IC attain their minimum and maximum values, respectively [25]. Because there are no ground truths of ship features compared with refocused ships, we assume that the geometrical features of the ship are compact for a well-focused ship. The well-focused ships have smaller geometrical features than defocused ships. The geometrical features contain the length, width and the areas of ship.

An ISAR image $I(k,n)$ is a two-dimensional complex image, where k is the range sample number, n is the cross-range number. The entropy of the two-dimensional image is written as follows [33]:

$$E(I) = \sum_{k=0}^{K-1}\sum_{n=0}^{N-1} \frac{|I(k,n)|^2}{S} \ln \frac{S}{|I(k,n)|^2} \tag{5}$$

where S is the total energy of the image:

$$S = \sum_{k=0}^{K-1}\sum_{n=0}^{N-1} |I(k,n)|^2 \tag{6}$$

The entropy is relatively small when the image is well-focused, and image refocusing is assessed via Equation (5). The IC denotes the normalized effective power of the image intensity and gives a measure of the image focus. If the image is well-focused, the IC value of the image is large. IC definition is considered as the ratio of the standard deviation to the mean of the amplitude. The IC is written as follows [34]:

$$IC(I) = \frac{\sqrt{E\left\{[I(k,n) - E\{I(k,n)\}]^2\right\}}}{E\{I(k,n)\}} \tag{7}$$

where E represents the spatial mean operator.

The motion compensation mainly compensates for the translation motion between the target and the radar. It contains two steps, one is range alignment, and the other is phase compensation [8,11]. The raw data-like data $R(m,n)$ is inverted from SLC SAR images and range alignment is a coarse compensation, the raw data-like data $R(m,n)$ no longer implement range alignment and the phase compensation is the main step of ISAR processing in this paper.

Phase compensation and ISAR imaging can be written as follows [23,32,35]:

$$I(k,n) = IFFT_{2D}\{R(k,n) \cdot \exp(-j\phi(m))\} \tag{8}$$

where $I(k,n)$ is the SAR image, $\phi(m)$ represents the phase error in Equation (2). m is the number of samples in the range direction, n is the cross-range samples number. The key step to phase compensation is estimation of $\phi(m)$. The phase error estimation $\hat{\phi}(m)$ is obtained by minimizing the entropy of ISAR image:

$$\hat{\phi}(m) = \mathrm{argmin} E(I) \tag{9}$$

It means that the Equation (9) should be satisfied with (10):

$$\frac{\partial E}{\partial \phi(m)} = 0 \tag{10}$$

Since the total energy S is constant, the cost function E can be redefined as:

$$E'(I) = -\sum_{k=0}^{K-1}\sum_{n=0}^{N-1} |I(k,n)|^2 \ln|I(k,n)|^2 \tag{11}$$

Therefore Equation (10) is equivalent to:

$$\frac{\partial E'}{\partial \phi(m)} = 0 \tag{12}$$

The derivation function of the entropy with respect to $\phi(m)$ is obtained from (13):

$$\frac{\partial E'}{\partial \phi(m)} = -\sum_{k=0}^{K-1}\sum_{n=0}^{N-1} [1 + \ln|I(k,n)|^2]\frac{\partial |I(k,n)|^2}{\partial \phi(m)} \tag{13}$$

Since $|I(k,n)|^2 = I(k,n)I^*(k,n)$, there are:

$$\frac{\partial |I(k,n)|^2}{\partial \phi(m)} = 2\mathrm{Re}(I^*(k,n)\frac{\partial I(k,n)}{\partial \phi(m)}) \tag{14}$$

Then substituting Equation (14) into Equation (13), we have:

$$\frac{\partial E'}{\partial \phi(m)} = -\mathrm{Re}\sum_{k=0}^{K-1}\sum_{n=0}^{N-1} [1 + \ln|I(k,n)|^2]I^*(k,n)\frac{\partial |I(k,n)|}{\partial \phi(m)} \tag{15}$$

The derivative of $I(k,n)$ with respect to $\partial \phi(m)$ is acquired as follows:

$$\frac{\partial I(k,n)}{\partial \phi(m)} = -jR(k,n)\exp(-j\phi(m))\exp(-j\frac{2\pi}{M}km) \tag{16}$$

Substituting Equation (16) into (15), one obtains:

$$\frac{\partial E'}{\partial \phi(m)} = -2M\mathrm{Im}\{\exp[-j\phi(m)]w^*(m)\} \tag{17}$$

where:

$$w(m) = \sum_{k=0}^{K-1} R^*(k,n)R_l \tag{18}$$

$$R_l = \frac{1}{M} \sum_{m=0}^{M-1} [1 + \ln\left|I(k,n)^2\right|]I(k,n) \exp(j\frac{2\pi}{M}km) \tag{19}$$

Finally, Equation (17) is equal to zero, and $\hat{\phi}(m)$ is obtained:

$$\exp(-j\hat{\phi}(m)) = \frac{w(m)}{|w(m)|} \tag{20}$$

3.3. The Procedures of Phase Compensation Method

The flowchart of the proposed phase compensation method is shown in Figure 4. The steps of phase compensation method based on fast minimum-entropy are described as follows:

Step 1: Input the raw data-like data and utilize DCT method [17] to obtain the initial phase error $\hat{\phi}(m)$, l is the number of iterations.

Step 2: Compensate phase error by substituting $\phi(m)$ with $\hat{\phi}(m)$. Two-dimensional inverse fast Fourier transform is performed to generate the ISAR image.

Step 3: Calculate entropy of the ISAR image and set tolerance T which is used to stop the iterations. If $E_l(I) - E_{l-1}(I)$ is greater than the T, the iteration continues; otherwise, the process terminates.

Step 4: Obtain R_l by Fourier transform $\ln(|I(k,n)|) \cdot I^*(k,n)$ and calculate the $w(m)$.

Step 5: Update phase error estimation $\hat{\phi}(m)$ and $l = l + 1$, go back to Step 2.

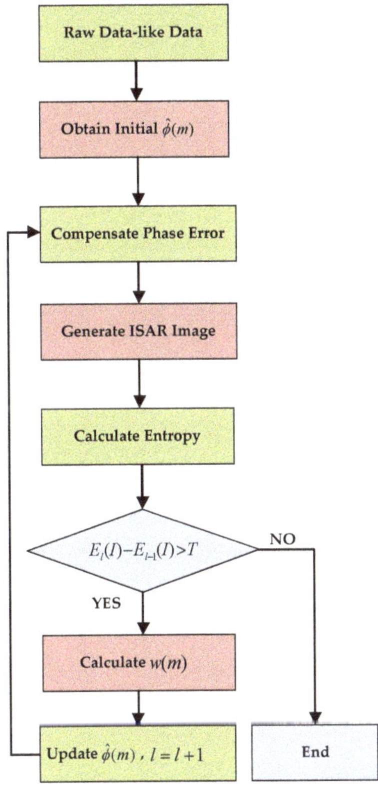

Figure 4. The basic flowchart of phase compensation method based on fast minimum-entropy.

4. Experimental Results and Discussions

Experimental results based on real SAR images are presented in this section to quantitatively evaluate the validity of the proposed method. The experimental data include spaceborne and airborne SAR data. The parameters of the SAR images are presented in Table 1.

Table 1. The parameters of four SAR images.

Image	Image01	Image02	Image03	Image04
Product	TerraSAR-X	Gaofeng-3	Gaofeng-3	Airborne
Mode	Strip	UFS	UFS	Strip
Resolution (M)	3	3	3	0.5
PRF (Hz)	3472.134984	2014.078491	1977.984863	500.0000
Band (MHz)	120.00	80.00	80.00	150.00
Polarization	VV	HH	HH	HH
Wave Length (m)	0.031040	0.055517	0.055517	0.056564
Slant-Range (km)	629.17	7127.22	7137.52	4.62
Velocity (m/s)	7088.636524	7563.162316	7568.372931	55.599743

The SAR image perform ships detection with software [36] and sub-images containing defocused ship are selected in advance. The raw data-like data is derived from sub-images with the IRD algorithm. The number of iterations l is set at 300. The results of the proposed method are compared with the results of the DCT [17] and PGA [14] methods.

4.1. Spaceborne SAR Data

The spaceborne SAR data contains data from two kinds of satellite system. One is TerraSAR-X system images, the other is Gaofeng-3 system images. The defocused ship targets of TerraSAR-X images are depicted in Figure 5. The two ships are numbered as ship01 and ship02. The ships are still blurred and defocused due to their motion, and need motion compensation.

Figure 5. Two defocused ships detected in a TerraSAR-X image (sub-images of ship01 and ship02). The red rectangles are the zoomed sub-images.

The refocused results of ship01 with the three methods are shown in Figure 6. The original sub-image, DCT and PGA results are respectively depicted in Figure 6a–c. The result with the proposed method is shown in Figure 6d. The image quality of the proposed method result is better than the original image, DCT and PGA results. The cross-range of the image is well-focused and blurred regions are reduced in Figure 6d. The refocused results of ship02 are shown in Figure 7. The image quality of the proposed method result is also superior to the other results.

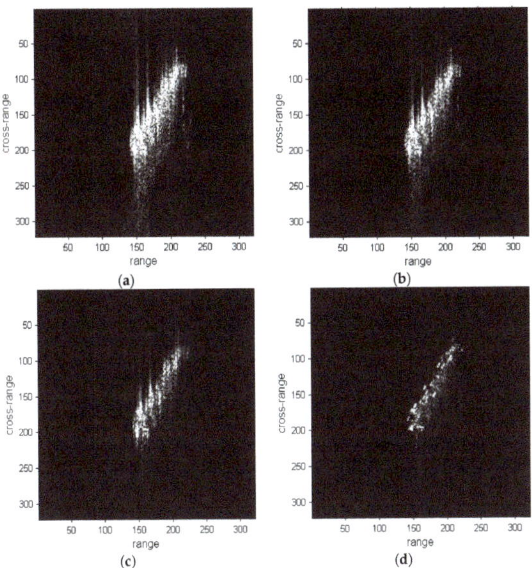

Figure 6. Refocused images of the ship01 sub-image. (**a**) Sub-image of ship01; (**b**) Refocused image with the DCT method; (**c**) Refocused image with the PGA method; (**d**) Refocused image with the proposed method.

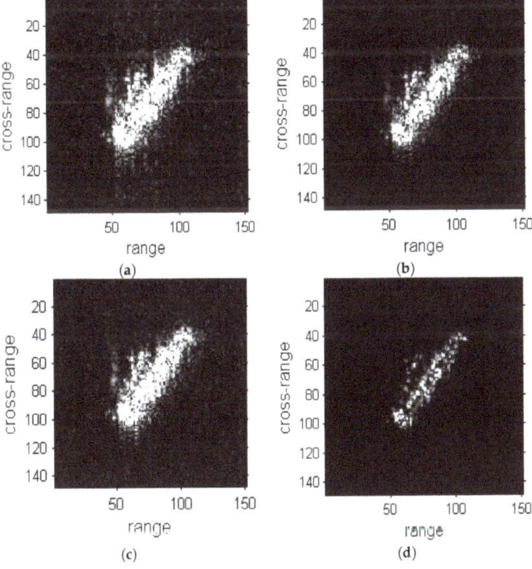

Figure 7. Refocused images of ship02 sub-image. (**a**) Sub-image of ship02; (**b**) Refocused image with the DCT method; (**c**) Refocused image with the PGA method; (**d**) Refocused image with the proposed method.

Image entropy and IC value are the two criteria to assess the image quality. The convergences of the two criteria versus the number of iterations are depicted in Figure 8. As the number of iterations increases, there has been decrease in entropy and increase in IC. Both criteria of ship01 have been greatly improved in Figure 8, and the focusing of the ship01 image is enhanced, while the criteria change of ship02 are not as obvious as for ship01 and the enhancement in refocusing of ship02 is not as good as for ship01.

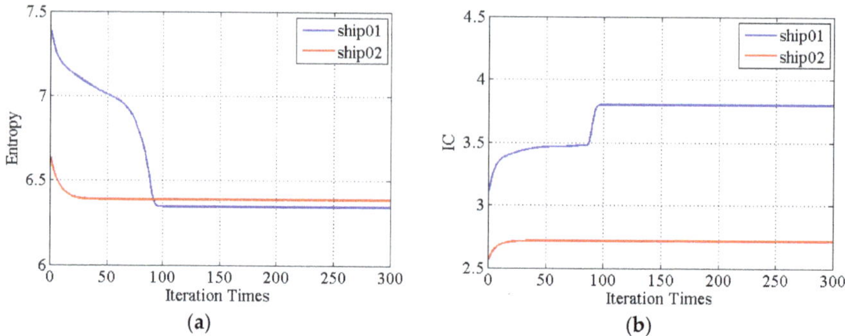

Figure 8. The entropies and IC values convergence of two sub-images versus the number of iterations. (**a**) Entropies change of the two sub-images during iteration; (**b**) IC values change of the two sub-images during iteration.

The geometrical features of the ships are extracted for further assessment. The features include the length, width and area. Since there are no ground truths for the ship features, it is impossible to compare extracted geometrical features with ground truths. The geometrical features of ships in focused images should be smaller than the defocused images.

The extracted features of ship01 and ship02 are illustrated in Figure 9. It is evident that the ship lengths extracted from the proposed method results are smaller than other results in Figure 9a. The widths and areas of proposed method results are also the least of the four results in Figure 9b,c. These results indicate that the refocusing result of the proposed method is better than that of the other methods.

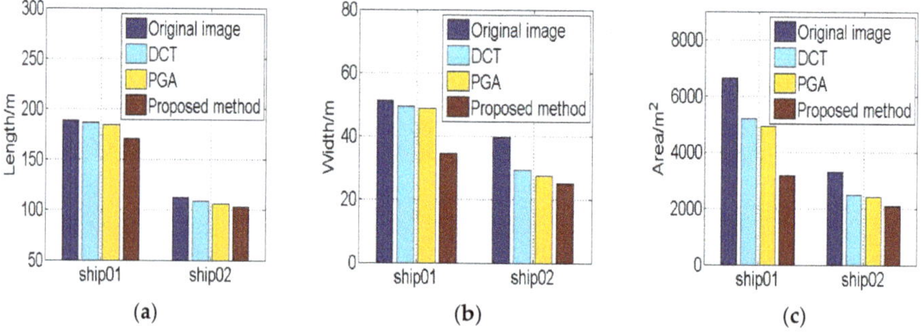

Figure 9. The geometrical features of refocusing ship01 and ship02. (**a**) Lengths of ships; (**b**) Widths of ships. (**c**) Areas of ships.

The defocused ships detected in two Gaofeng-3 SAR images are illustrated in Figure 10. The two ships are numbered as ship03 and ship04, respectively. It is obvious that the two ships are blurred and defocused in the SAR images.

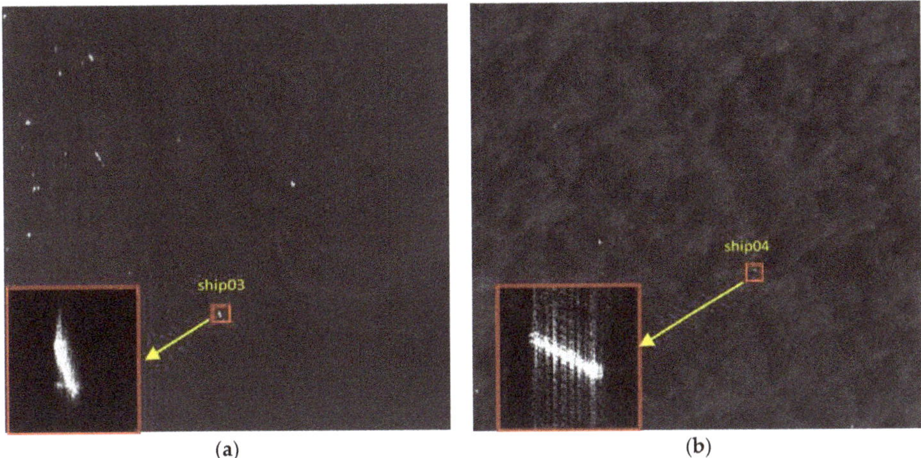

(a)　　　　　　　　　　(b)

Figure 10. Two defocused ships are detected in Gaofeng-3 images. (**a**) Sub-image of ship03 located in the Gaofeng-3 image; (**b**) Sub-image of ship 04 located in the Gaofeng-3 image. The red rectangles are the zoomed sub-images.

The refocusing results of ship03 are presented in Figure 11. The result of the proposed method is better than the DCT and PGA method results in Figure 11d. The left side regions of the ship are obviously well-focused and become clear in Figure 11d.

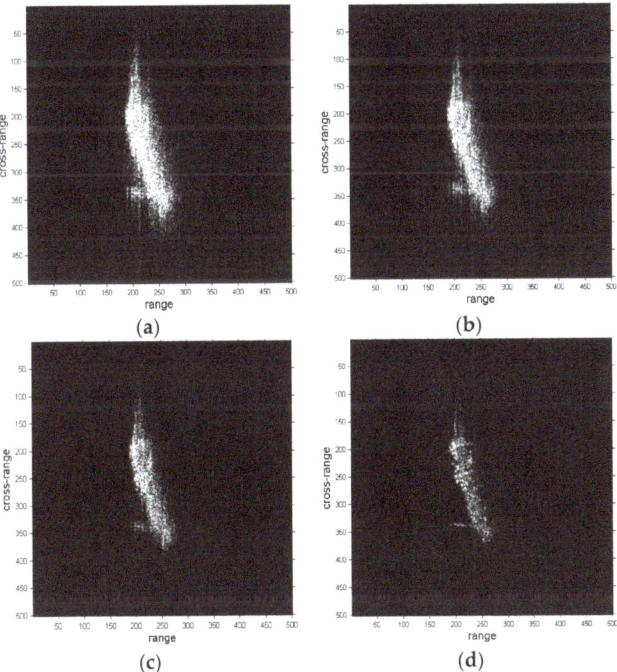

(a)　　　　　　　　　　(b)

(c)　　　　　　　　　　(d)

Figure 11. Refocused images of ship03 sub-image. (**a**) Sub-image of ship03; (**b**) Refocused image with DCT method; (**c**) Refocused image with PGA method; (**d**) Refocused image with proposed method.

The results of ship04 are shown in Figure 12, and the results of DCT, PGA and the proposed method are presented in Figure 12b–d, respectively. The focusing of DCT and PGA results produce a certain improvement compared with the original images, but are not well-focused. The blurring in the cross-range direction is greatly removed in Figure 12d. As stated previously, the image quality of the proposed method results are superior to the original images, DCT and PGA results from a visual point of view.

The entropies and IC values during iteration are represented in Figure 13. The entropies are getting smaller and IC values larger with iteration, showing that the quality of the image is improved. The entropies and IC values of ship04 are changing greatly, demonstrating that the image quality of ship04 sub-image processing with the proposed method is good. The geometrical features of ship03 and ship04 are shown in Figure 14. The refocusing feature results with the proposed method are also smaller than the other method results.

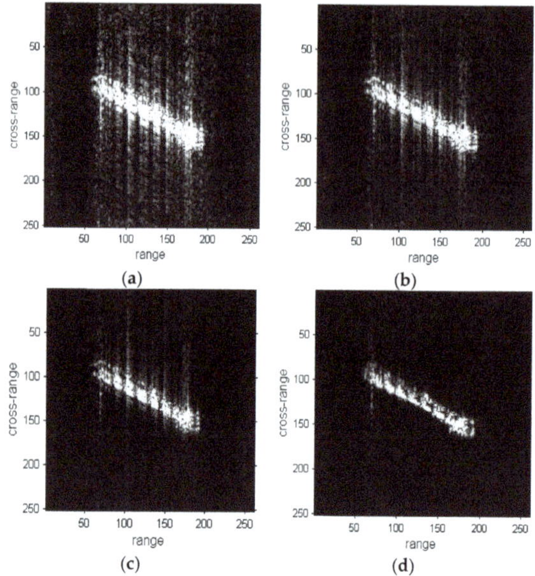

Figure 12. Refocused images of ship04 sub-image. (**a**) Sub-image of ship04; (**b**) Refocused image with DCT method; (**c**) Refocused image with PGA method; (**d**) Refocused image with proposed method.

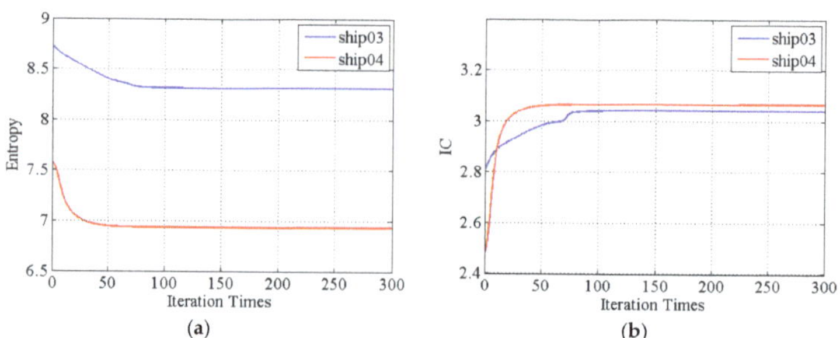

Figure 13. The entropies and IC values convergence of two sub-images versus the number of iterations. (**a**) Entropy change of the two sub-images during iteration. (**b**) IC values change of the two sub-images during iteration.

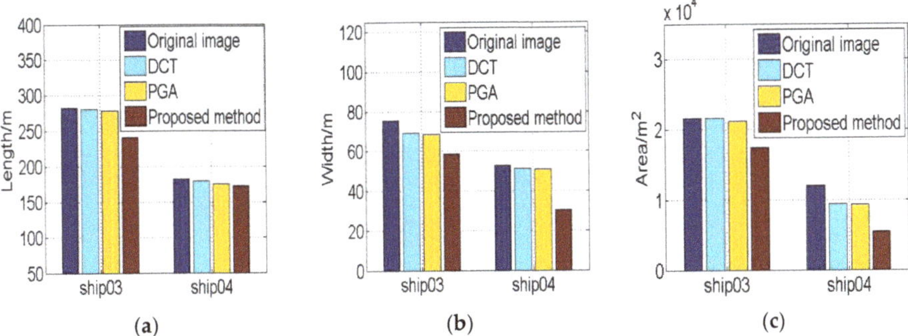

Figure 14. The geometrical features of refocusing ship03 and ship04. (**a**) Lengths of ships; (**b**) Widths of ships. (**c**) Areas of ships.

4.2. Airborne SAR Data

In addition to spaceborne SAR data, an airborne SAR dataset is exploited to further validate the procedure. The defocused ships are marked in the airborne SAR image in Figure 15. The two ships are numbered as ship05 and ship06. The resolution of the airborne SAR image is higher than that of the spaceborne SAR images.

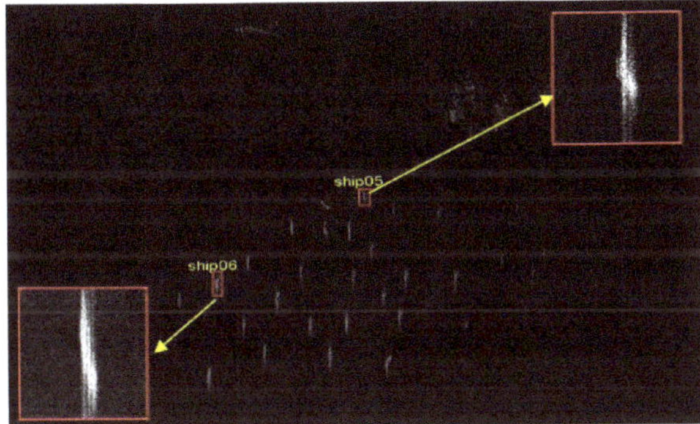

Figure 15. Two defocused ships are detected in airborne SAR images. Sub-images of ship05 and ship06 are located in airborne SAR image. The red rectangles are the zoomed sub-images.

The defocusing of moving ship in airborne SAR image is more evident. The Doppler rate shift induced by target motion in the airborne SAR system is greater than in the spaceborne SAR system. The Doppler rate has great influence on the image quality and a little error will result in image defocusing and decreased resolution. The degree of defocusing is directly related to the Doppler rate shift [5].

The experimental results of ship05 are shown in Figure 16. As can be noted by observing Figure 16d that the refocused result with the proposed method has some improvement in focusing compared with the original images, DCT and PGA results. The results of ship06 are illustrated in Figure 17. The images with ISAR processing all have certain improvement in focusing compared to the original image. The result of the proposed method is still better than the DCT and PGA method results.

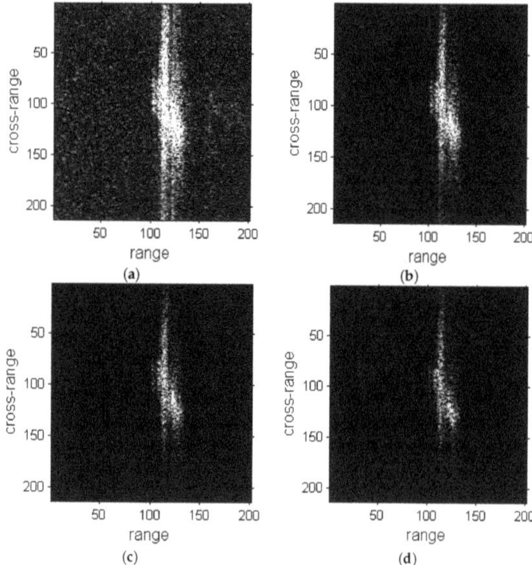

Figure 16. Refocused images of ship05 sub-image. (**a**) Sub-image of ship05; (**b**) Refocused image with the DCT method; (**c**) Refocused image with the PGA method; (**d**) Refocused image with the proposed method.

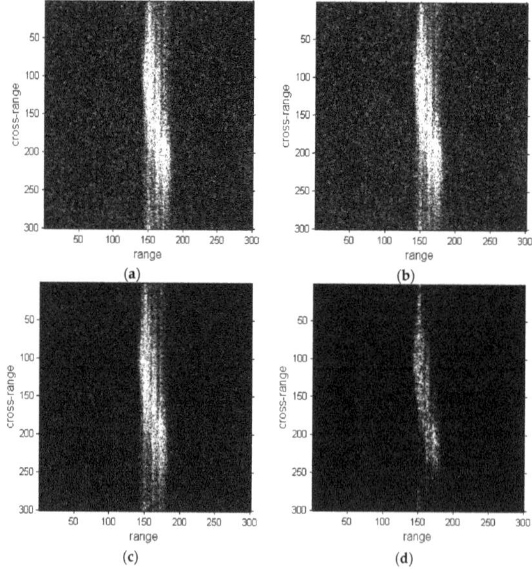

Figure 17. Refocused images of ship06 sub-image. (**a**) Sub-image of ship06; (**b**) Refocused image with the DCT method; (**c**) Refocused image with the PGA method; (**d**) Refocused image with the proposed method.

The ISAR processing is not able to produce a well-focused target from airborne SAR images and the refocusing are not as good as for spaceborne SAR sub-images. The reason is that the IRD inversion is applied in SAR images for which the resolution is of the order of meters, while the resolution of airborne SAR images is higher than one meter. The inversion of high-resolution SAR images (higher than 1 m) would be need a more accurate inversion algorithm.

The entropies and IC values change are illustrated in Figure 18. The entropy and IC value of the proposed method results have some changes with iteration. This means the image quality shows some improvement. The features of ship05 and ship06 are shown in Figure 19. The features of the proposed method result are the least yet.

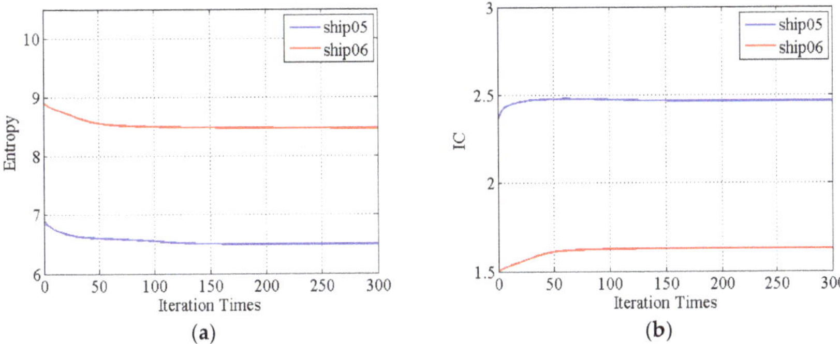

(a)   (b)

Figure 18. The entropies and IC values convergence of two sub-images versus the number of iterations. (a) Entropies change of the two sub-images during iteration; (b) IC values change of the two sub-images during iteration.

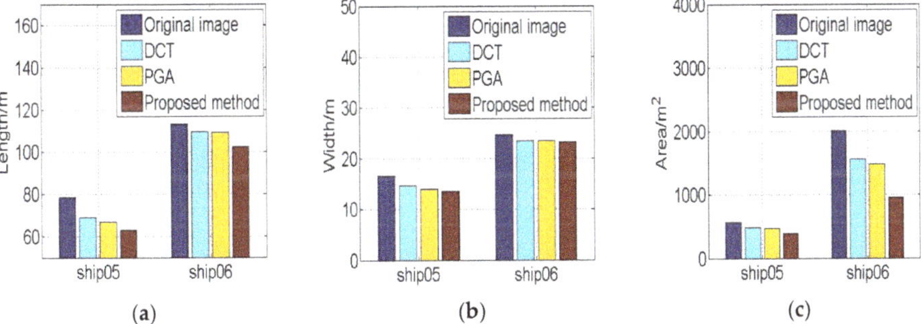

(a)   (b)   (c)

Figure 19. The geometrical features of refocusing ship05 and ship06. (a) Lengths of ships; (b) Widths of ships. (c) Areas of ships.

The entropies and IC values of the original and refocused images are calculated for quantitative evaluation of the focusing. The performance of the proposed method is compared with the DCT, and PGA methods and the original images. The entropies of sub-images with different method processing are presented in Table 2. All entropies of the proposed method results are lower than others in the table. This means that the focusing of the proposed method result could outperform the other method results. The IC values of the sub-images are presented in Table 3. The IC values of the proposed method results are quite larger than the other results, particularly for spaceborne sub-images.

Table 2. The entropies of six sub-images.

Sub-Image	ship01	ship02	ship03	ship04	ship05	ship06
Original Image	7.39	6.59	8.72	7.61	6.92	8.92
DCT	7.26	6.54	8.69	7.58	6.89	8.89
PGA	7.05	6.51	8.64	7.50	6.81	8.76
Proposed Method	6.35	6.30	8.32	6.94	6.50	8.45

Table 3. The IC values of six sub-images.

Sub-Image	ship01	ship02	ship03	ship04	ship05	ship06
Original Image	2.97	2.42	2.81	2.44	2.33	1.49
DCT	3.11	2.50	2.85	2.48	2.37	1.50
PGA	3.25	2.54	2.88	2.56	2.38	1.53
Proposed Method	3.80	2.67	3.04	3.07	2.47	1.63

5. Conclusions

By exploiting the ISAR technique based on fast minimum entropy phase compensation, a refocusing method for moving ships in SAR images is proposed in this paper. The basic procedures of refocusing in SAR images are built. The processing data are SAR images rather than raw echo data. The algorithms inverting images to the raw data domain are described. This makes the processing flow simple and reduces the computational burden. The ISAR processing based on the fast minimum entropy phase compensation method iteratively obtains the phase error estimation by constructing a cost function of entropy. The experimental results based on spaceborne and airborne SAR data verify the effectiveness of the proposed method.

Though the experimental results are good, the airborne SAR sub-images (resolution higher than 1 m) are not well focused via ISAR processing. The high-resolution SAR image inversion needs to be investigated in future work.

Author Contributions: X.H. and X.L. conceived and designed the experiments; X.H. performed the experiments and analyzed the data; K.J., G.D. and X.X. contributed materials; X.H. and K.J. wrote the paper; G.D. revised and polished the paper.

Funding: This work is supported by the National Natural Science Foundation of China, grant numbers 61372163, 61331015 and 61601035.

Acknowledgments: The authors would like to thank the anonymous reviewers for their competent comments and helpful suggestions.

Conflicts of Interest: The authors declare no conflict of interest.

References

1. Martorella, M.; Pastina, D.; Berizzi, F.; Lombardo, P. Spaceborne radar imaging of maritime moving targets with the Cosmo-Skymed SAR system. *IEEE J. Sel. Top. Appl. Earth Observ. Remote Sens.* **2014**, *7*, 2797–2810. [CrossRef]

2. Crisp, D.J. *The State-of-the-Art in Ship Detection in Synthetic Aperture Radar Imagery*; DSTO Information Sciences Laboratory: Edinburgh, Australia, 2004.

3. Dong, G.; Kuang, G.; Wang, N.; Wang, W. Classification via sparse representation of steerable wavelet frames on Grossmann manifold: Application to target recognition in SAR image. *IEEE Tran. Image Process.* **2017**, *26*, 2892–2904. [CrossRef] [PubMed]

4. Kang, M.; Ji, K.; Leng, X.; Xing, X.; Zou, H. Synthetic aperture radar target recognition with feature fusion based on a stacked autoencoder. *Sensors* **2017**, *17*, 192. [CrossRef] [PubMed]

5. Noviello, C.; Fornaro, G.; Mertorella, M. Focused SAR image formation of moving targets based on Doppler parameter estimation. *IEEE Trans. Geosci. Remote Sens.* **2015**, *53*, 3460–3470. [CrossRef]

6. Zhang, Y.; Sun, J.; Lei, P.; Li, G.; Hong, W. High-resolution SAR-based ground moving target imaging with defocused ROI data. *IEEE Trans. Geosci. Remote Sens.* **2016**, *54*, 1062–1073. [CrossRef]

7. Wang, B.; Xu, S.; Wu, W.; Hu, P.; Chem, Z. Adaptive ISAR imaging of maneuvering targets based on a modified Fourier transform. *Sensors* **2018**, *18*, 1370. [CrossRef] [PubMed]

8. Lv, X.; Xing, M.; Wang, C.; Zhang, S. ISAR imaging of maneuvering targets based on the range centroid Doppler technique. *IEEE Tran. Image Process.* **2010**, *19*, 141–153.

9. Zheng, J.; Liu, H.; Liu, Z.; Liu, Q. ISAR imaging of ship targets based on an integrated cubic phase bilinear autocorrelation function. *Sensors* **2017**, *17*, 498. [CrossRef] [PubMed]

10. Wang, J.; Liu, X. Improved global range alignment for ISAR. *IEEE Trans. Aerosp. Electron. Syst.* **2007**, *43*, 1070–1075. [CrossRef]

11. Wang, J.; Kasilingam, D. Global range alignment for ISAR. *IEEE Trans. Aerosp. Electron. Syst.* **2003**, *39*, 351–357. [CrossRef]

12. Steinberg, B. Radar imaging from a distorted array: The radio camera algorithm and experiment. *IEEE Trans. Antennas Propag.* **1981**, *29*, 740–748. [CrossRef]

13. Zheng, B.; Wei, Y. Improvements of autofocusing techniques for ISAR motion compensation. *Acta Electron. Sin.* **1996**, *24*, 74–79.

14. Wahl, D.E. Phase gradient autofocus—A robust tool for high resolution SAR phase correction. *IEEE Trans. Aerosp. Electron. Syst.* **1994**, *30*, 827–835. [CrossRef]

15. Qing, J.; Xu, H.; Liang, X.; Li, Y. An improved phase gradient autofocus algorithm used in real-time processing. *J. Radars* **2015**, *4*, 600–607.

16. Zhu, D.; Jiang, R.; Mao, X.; Zhu, Z. Multi-Subaperture PGA for SAR autofocusing. *IEEE Trans. Aerosp. Electron. Syst.* **2013**, *49*, 468–488. [CrossRef]

17. Itoh, T.; Sueda, H.; Watanabe, Y. Motion compensation for ISAR via centroid tracking. *IEEE Trans. Aerosp. Electron. Syst.* **1996**, *32*, 1191–1197. [CrossRef]

18. Zhu, Z.; Qiu, X.; She, Z. ISAR motion compensation using modified Doppler centroid tracking method. *Trans. Nanjing Univ. Aeronaut. Astronaut.* **1995**, *12*, 195–201.

19. Martorella, M.; Berizzi, F.; Haywood, B. Contrast maximization based technique for 2-D ISAR autofocusing. *IEE Proc. Radar Sonar Navig.* **2005**, *152*, 253–262. [CrossRef]

20. Berizzi, F.; Corsini, G. Autofocusing of inverse synthetic radar images using contrast optimization. *IEEE Trans. Aerosp. Electron. Syst.* **1996**, *32*, 1185–1191. [CrossRef]

21. Simmons, S.; Evans, R. Maximum likelihood autofocusing of radar images. In Proceedings of the IEEE 1995 International Radar Conference, Alexandria, VA, USA, 8–11 May 1995; pp. 410–415.

22. Berizzi, F.; Pinelli, G. Maximum-likelihood ISAR image autofocusing technique based on instantaneous frequency estimation. *IEE Proc. Radar Sonar Navig.* **1997**, *144*, 284–292. [CrossRef]

23. Li, X.; Liu, G.; Ni, J. Autofocusing of ISAR images based on entropy minimization. *IEEE Trans. Aerosp. Electron. Syst.* **1999**, *35*, 1240–1251. [CrossRef]

24. Yang, L.; Xing, M.; Zhang, L.; Sheng, J.; Bao, Z. Entropy-based motion error correction for high-resolution spotlight SAR imagery. *IET Radar Sonar Navig.* **2012**, *6*, 627–637. [CrossRef]

25. Du, X.; Duan, C.; Hu, W. Sparse representation based autofocusing technique for ISAR images. *IEEE Geosci. Remote Sens. Lett.* **2013**, *51*, 1826–1835. [CrossRef]

26. Martorella, M.; Giusti, E.; Berizzi, F.; Bacci, A.; Mese, E.D. ISAR based technique for refocusing non-cooperative targets in SAR images. *IET Radar Sonar Navig.* **2012**, *6*, 332–340. [CrossRef]

27. Khwaja, A.; Ferro-Famil, L.; Pottier, E. Efficient SAR raw data generation for anisotropic urban scenes based on inverse processing. *IEEE Geosci. Remote Sens. Lett.* **2009**, *6*, 757–761. [CrossRef]

28. Khwaja, A.; Ferro-Famil, L.; Pottier, E. SAR raw data generation using inverse SAR image formation algorithms. In Proceedings of the 2006 IEEE International Symposium on Geoscience and Remote Sensing (IGARSS), Denver, CO, USA, 31 July–4 August 2006. [CrossRef]

29. Martorella, M.; Berizzi, F.; Giusti, E.; Bacci, A. Refocusing of moving targets in SAR images based on inversion mapping and ISAR processing. In Proceedings of the 2011 IEEE Radar Conference, Kansas, MO, USA, 23–27 May 2011; pp. 68–72.

30. Martorella, M.; Acito, N.; Berizzi, F. Statistical CLEAN technique for ISAR imaging. *IEEE Trans. Geosci. Remote. Sens.* **2007**, *45*, 3552–3560. [CrossRef]

31. Kang, M.; Bae, J.; Lee, S.; Kim, K. Efficient ISAR autofocus via minimization of Tsallis entropy. *IEEE Trans. Aerosp. Electron. Syst.* **2016**, *52*, 2952–2960. [CrossRef]

32. Yang, J.; Zhang, Y. An airborne SAR moving target imaging and motion parameters estimation algorithm with azimuth-dechirping and the second-order keystone transform applied. *IEEE J. Sel. Top. Appl. Earth Observ. Remote Sens.* **2015**, *8*, 3967–3976. [CrossRef]

33. Wang, J.; Liu, X. SAR minimum-entropy autofocus using an adaptive-order polynomial model. *IEEE Geosci. Remote Sens. Lett.* **2006**, *3*, 512–516. [CrossRef]

34. Wu, W.; Hu, P.; Xu, S.; Chen, Z. Image registration for InSAR based on joint translational motion compensation. *IET Radar Sonar Navig.* **2017**, *11*, 1597–1603. [CrossRef]

35. Zeng, T.; Wang, R.; Li, F. SAR image autofocus utilizing minimum-entropy criterion. *IEEE Geosci. Remote Sens. Lett.* **2013**, *10*, 1552–1556. [CrossRef]
36. Leng, X.; Ji, K.; Zhou, S.; Xing, X.; Zou, H. An adaptive ship detection scheme for Spaceborne SAR imagery. *Sensors* **2016**, *16*, 1345. [CrossRef] [PubMed]

Article

Ship Classification in High-Resolution SAR Images via Transfer Learning with Small Training Dataset

Changchong Lu and Weihai Li *

Department of Electronic Engineering and Information Science, University of Science and Technology of China, Hefei 230027, China; ll964183@mail.ustc.edu.cn
* Correspondence: whli@ustc.edu.cn

Received: 19 November 2018; Accepted: 20 December 2018; Published: 24 December 2018

Abstract: Synthetic aperture radar (SAR) as an all-weather method of the remote sensing, now it has been an important tool in oceanographic observations, object tracking, etc. Due to advances in neural networks (NN), researchers started to study SAR ship classification problems with deep learning (DL) in recent years. However, the limited labeled SAR ship data become a bottleneck to train a neural network. In this paper, convolutional neural networks (CNNs) are applied to ship classification by using SAR images with the small datasets. To solve the problem of over-fitting which often appeared in training small dataset, we proposed a new method of data augmentation and combined it with transfer learning. Based on experiments and tests, the performance is evaluated. The results show that the types of the ships can be classified in high accuracies and reveal the effectiveness of our proposed method.

Keywords: synthetic aperture radar (SAR); convolutional neural networks (CNNs); deep learning (DL); ship classification

---

1. Introduction

Synthetic aperture radar (SAR) is an active Earth observation system that can be installed on planes, satellites, spacecraft, etc. It can perform observations on the ground all day and in all weather conditions. Now we can get more high-resolution SAR images by recent development of SAR satellites, e.g., RADARSAT-2, TerraSAR-X [1,2], etc. By using these images, lots of applications can be implemented. Pieralace et at al. [3] presents a new simple and very effective filtering technique, which is able to process full-resolution SAR images. Gambardella at al. [4] presents a methodological approach for a fast and repeatable monitoring, which can be applied to higher resolution data. Ship classification is an important application of SAR images.

Researchers usually used traditional classification methods in ship classification, including image processing, feature extraction and selection, and classification. Feature extraction is a key step in ship classification. Researchers widely used geometric features, scattering features in feature extraction [5]. For geometric features, it contains ship area, ship rectangularity, moment of inertia, fractal dimension, spindle direction angle and ratio of length to width [6], etc. For scattering features, it contains superstructure scattering features [7], three-dimensional scattering feature [8], radar-cross-section (RCS) [9], and symmetric scattering characterization (SSCM) [10], etc. As for classifiers, artificial neural networks (ANNs) [11] can establish a general classification scheme by training, which makes it widely used in ship classification. Support vector machines (SVM) [12] is also a popular model. Researchers also proposed some methods to get high-classification accuracy [13–15], these methods have high requirements for on features and classifiers, so these methods could not applied in other datasets. With the development in neural networks, researchers now focus on processing SAR images with deep neural networks (DNNs) [16–18].

The deep neural network (DNN) is artificial neural network (ANNs)'s promotion, which includes lots of hidden layers between the input and output layers [19,20]. DNN can give good expression of an object by its deep architectures and performs well in modeling complex nonlinear relationships [21]. DNN have many popular models, such as recurrent neural networks (RNNs) [22] and convolutional deep neural networks (CNNs). Nowadays, CNNs [23] are playing an important role in detection and recognition. One of the most remarkable results was its application in the ImageNet data set. The ImageNet dataset includes over 15 million images with 22,000 different categories. By using a model called Alexnet [24], the researchers achieved a remarkable result, which have reduced the error rates of 8.2% in top-one error rates and 8.7% in top-five error rates than the previous work [24,25]. Furthermore, CNNs have achieved many impressive results in computer the vision area, such as handwritten digits recognition [26], traffic sign recognition [27], and face recognition [28].

Deep learning applications in SAR images have been gradually used. Zhang et al. [5] showed an application in ship classification with transfer learning and fine-tuning. Kang et al. [16] used deep neural networks for SAR ship detection and get good results. Chen et al. [29] presented a method for classification of 10 categories of SAR images, and showed application in ship target recognition. By using deep learning methods, SAR images avoid complicated feature extraction, which completed by deep networks. It can obviously improve the performance of classifiers. However, there are still many problems. Compared with computer vision, SAR image interpretation has the same purpose—extracting useful information from images—but the processed SAR image is significantly different from visible light image, mainly reflected in the band, imaging principle, projection direction, angle of view, etc. Furthermore, the small dataset may be a problem, too. Therefore, when we use the method in SAR images, we need to fully consider these problems.

As with ship classification, many issues may arise when training CNNs. One common issue is over-fitting. Over-fitting can be explained as the neural network models the training data too well and perform bad in data which is different from training data. If a model learns the most of detail and noise of training data, it cannot get good performance in new data, then the over-fitting happens. Some useless information such as noise and random fluctuations have been learned while training as parts of the models. Then the models cannot have good generalize ability in new data. CNNs are prone to over-fitting when model have rare dependencies in the training data. To solve this problem, CNNs often require tens of thousands of examples to train adequately. However, in many cases, we cannot get enough training data in SAR applications, which may cause severe over-fitting. A popular way to solve this problem is data augmentation, such as flipping, brightening, and contrast [30]. Another way to solve this problem is transfer learning, which often been applied in natural images [31].

The application which using deep neural networks in SAR ship images worth paying attention in improving the performance in SAR ship detection and classification. When dealing with SAR images, we should take both the data processed method and CNN training method into consideration. To get good performance, we should retain the important information of the images and get enough images when we make data augmentation and use some good models for training to reduce the amount requirement of training set. In this paper, we proposed a new method for SAR ship classification. First, we proposed a new data augmentation method which can keep important information while increasing the amount of data and achieve the requirements of the dataset. Then, by coupling transfer learning with the processed data, we can get good performance in classification. When dealing with a small training dataset, the method can successfully enlarge the training datasets. With the enlarged datasets coupled with transfer learning, CNNs can avoid the over-fitting issue and achieve excellent classification accuracy. Comparison experiments demonstrated the good performance of our method.

The remainder of this paper is organized as follows. Section 2 presents an introduction of the theory for CNNs, transfer learning and components of our method. The details of the experiments are described in Section 3, we also present our discussion in this section. Finally, Section 4 offers the conclusion.

2. Materials and Methods

2.1. CNN Models

A CNN is composed of an input layer, an output layer, and multiple hidden layers. The hidden layers can be convolutional layers, pooling layers, activation layers, and fully connected layers [32,33]. The first multilayer CNN is a simple convolutional network consisting of seven levels called LeNet-5 which was proposed by LeCun et al. in 1998 [23]. Authors used it in classifying digits, and it was applied by several banks to recognize hand-written numbers. The pixels of images were 32×32. With the large developments in GPU, the layers of CNNs had become much deeper. In recent years, many new models had been proposed, such as Alexnet [24], Resnet [34], VGG16 [22], etc. In this paper, we mainly use two different models for ship classification in order to do comparative test. For further study, we also compared our method with other popular models.

- Traditional CNN models

The first model we used is traditional CNN models, as shown in Figure 1a. These models consist of four convolution layers, four max pooling layers and two fully connected layers. We used Leaky ReLU as our activation function. To avoid the problem of over-fitting, dropout has been used [35].

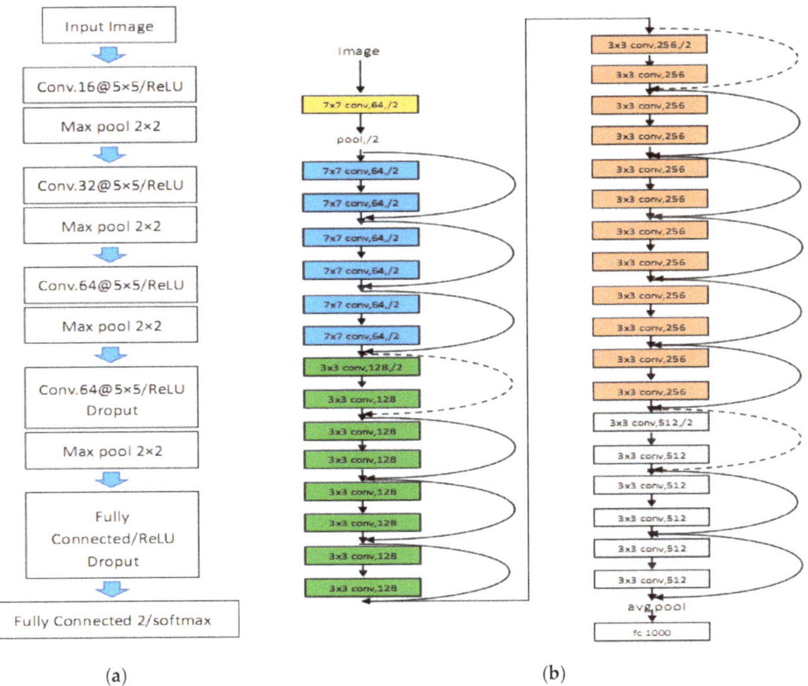

(a)　　　　　　　　　(b)

Figure 1. Learning Models. (**a**) Traditional CNN models. (**b**) Resnet-34 models.

- Resnet models

Another model we used is transfer learning the Resnet-50 [34] model. Resnet models used a connection method named shortcut connection, as shown in Figure 2. The models are stacked by multiple blocks. By using this structure, network can obviously improve its performance.

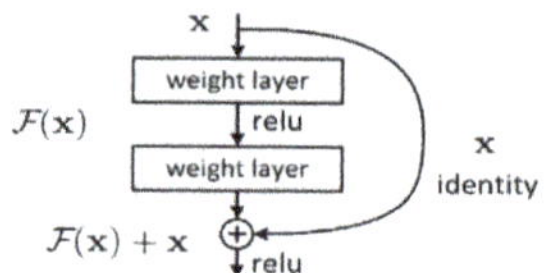

Figure 2. Shortcut connection.

Figure 1b presents Resnet-34 models. Table 1 presents Resnet models with different layers.

Table 1. Resnet models

Layer Name	Output Size	18-Layer	34-Layer	50-Layer	101-Layer	152-Layer
conv1	112×112	7×7, 64, stride 2				
		3×3 max pool, stride 2				
conv2_x	56×56	$\begin{bmatrix} 3 \times 3, & 64 \\ 3 \times 3, & 64 \end{bmatrix} \times 2$	$\begin{bmatrix} 3 \times 3, & 64 \\ 3 \times 3, & 64 \end{bmatrix} \times 3$	$\begin{bmatrix} 1 \times 1, & 64 \\ 3 \times 3, & 64 \\ 1 \times 1, & 256 \end{bmatrix} \times 3$	$\begin{bmatrix} 1 \times 1, & 64 \\ 3 \times 3, & 64 \\ 1 \times 1, & 256 \end{bmatrix} \times 3$	$\begin{bmatrix} 1 \times 1, & 64 \\ 3 \times 3, & 64 \\ 1 \times 1, & 256 \end{bmatrix} \times 3$
conv3_x	28×28	$\begin{bmatrix} 3 \times 3, & 128 \\ 3 \times 3, & 128 \end{bmatrix} \times 2$	$\begin{bmatrix} 3 \times 3, & 128 \\ 3 \times 3, & 128 \end{bmatrix} \times 4$	$\begin{bmatrix} 1 \times 1, & 128 \\ 3 \times 3, & 128 \\ 1 \times 1, & 512 \end{bmatrix} \times 4$	$\begin{bmatrix} 1 \times 1, & 128 \\ 3 \times 3, & 128 \\ 1 \times 1, & 512 \end{bmatrix} \times 4$	$\begin{bmatrix} 1 \times 1, & 128 \\ 3 \times 3, & 128 \\ 1 \times 1, & 512 \end{bmatrix} \times 8$
conv4_x	14×14	$\begin{bmatrix} 3 \times 3, & 256 \\ 3 \times 3, & 256 \end{bmatrix} \times 2$	$\begin{bmatrix} 3 \times 3, & 256 \\ 3 \times 3, & 256 \end{bmatrix} \times 6$	$\begin{bmatrix} 1 \times 1, & 256 \\ 3 \times 3, & 256 \\ 1 \times 1, & 1024 \end{bmatrix} \times 6$	$\begin{bmatrix} 1 \times 1, & 256 \\ 3 \times 3, & 256 \\ 1 \times 1, & 1024 \end{bmatrix} \times 23$	$\begin{bmatrix} 1 \times 1, & 256 \\ 3 \times 3, & 256 \\ 1 \times 1, & 1024 \end{bmatrix} \times 36$
conv5_x	7×7	$\begin{bmatrix} 3 \times 3, & 512 \\ 3 \times 3, & 512 \end{bmatrix} \times 2$	$\begin{bmatrix} 3 \times 3, & 512 \\ 3 \times 3, & 512 \end{bmatrix} \times 3$	$\begin{bmatrix} 1 \times 1, & 512 \\ 3 \times 3, & 512 \\ 1 \times 1, & 2048 \end{bmatrix} \times 3$	$\begin{bmatrix} 1 \times 1, & 512 \\ 3 \times 3, & 512 \\ 1 \times 1, & 2048 \end{bmatrix} \times 3$	$\begin{bmatrix} 1 \times 1, & 512 \\ 3 \times 3, & 512 \\ 1 \times 1, & 2048 \end{bmatrix} \times 3$
	1×1	average pool, 1000-d fc, softmax				
FLOPs		1.8×10^9	3.6×10^9	3.8×10^9	7.6×10^9	11.3×10^9

- Other popular models

In this paper, we also used some popular models to do transfer learning, such as Alexnet, Vgg-16 net, etc. These models are shown in Figure 3.

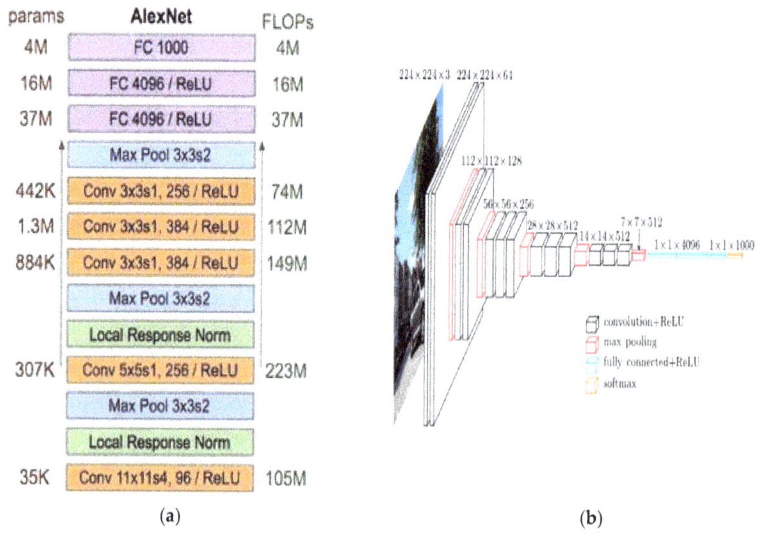

Figure 3. Other popular models. (**a**) Alexnet model. (**b**) VGG-16 model.

2.2. *Learning of CNNs*

2.2.1. Environment

The networks are implemented in Pytorch 0.3.0. All layers were designed to match the size of images. The input has a dimension of $224 \times 224 = 50176$. The output was implemented using the softmax operation and consists of three classes.

2.2.2. Stochastic Gradient Descent with Momentum

Gradient descent is a commonly used optimization algorithm for neural network which can solve many simple problems. However, when the training dataset is very large, we can find that using simple gradient descent method may consume much computing resources, and the convergence process will be slow. At the same time, since all the training data are considered for each calculation in gradient descent method, it may cause over-fitting. To solve this problem, SGD have been proposed.

In stochastic gradient descent (SGD), each training example $x^{(i)}$ and label $y^{(i)}$ with be updated as

$$\theta = \theta - \alpha \cdot \nabla_\theta J\left(\theta; x^{(i)}; y^{(i)}\right) \tag{1}$$

During training, the network only calculates the loss of one sample per iteration, then gradually traverse all samples and complete an epoch calculation. By using this method, although it may produce large fluctuations in a simple, but the result may finally converge successfully. The amount of calculation is greatly reduced, so the speed can also be improved.

Momentum [36] is a commonly used acceleration technique in gradient descent method. Stochastic gradient descent with momentum can be expressed as

$$v = \beta \times v - \alpha \cdot \nabla_\theta J\left(\theta; x^{(i)}; y^{(i)}\right) \tag{2}$$

$$\theta \leftarrow \theta + v \tag{3}$$

β is momentum coefficient. It can be understood as, if the last momentum (i.e., v) is the same as the negative gradient direction of this time, then the magnitude of this decline will increase. By using momentum, we can accelerate the convergence process [29].

2.2.3. Learning Rate

When deal with CNNs training, learning rate is an important parameter. Usually, we expected to get the result as soon as possible, so the learning rate we used will be large. However, in common situations, by using large learning rate may cause concussion, and the result may not behave as expected. The commonly used method is reducing the learning rate during training. The initial learning rate often takes as 0.01 or 0.001, which is set to decrease the loss function quickly. Then users should adjust the learning rate by several epochs or iterations according to the accuracy during training [29]. The learning rate often been reduced by a factor of 0.1 or 0.5.

2.3. *Proposed Method with Transfer Learning*

2.3.1. Our Method

As mentioned in Section 1, CNNs often require a large amount of data to train adequately. However, as with the SAR dataset, we do not have enough images for training. Therefore, we must do something to avoid the problem of over-fitting.

One effective way is data augmentation. There are many methods to achieve our requirement, including adding noise, changing colors, flipping, etc. Figure 4 illustrates the original SAR ship images and Figure 5 shows some common ways of data augmentation.

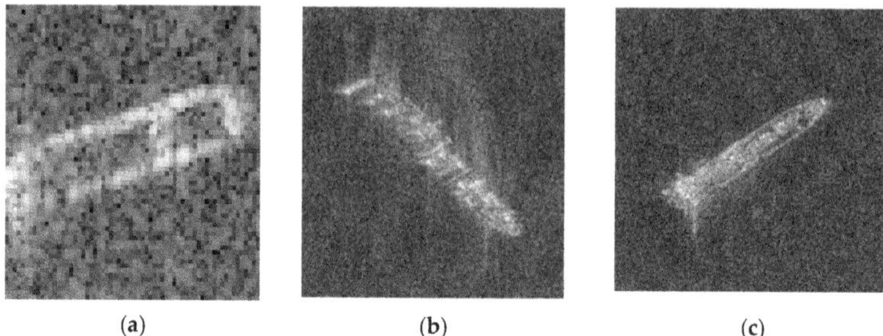

(a) (b) (c)

Figure 4. Illustrates the different target classes extracted from SAR images. (**a**) Bulk Carrier with four times magnification. (**b**) Container Ship. (**c**) Oil Tanker.

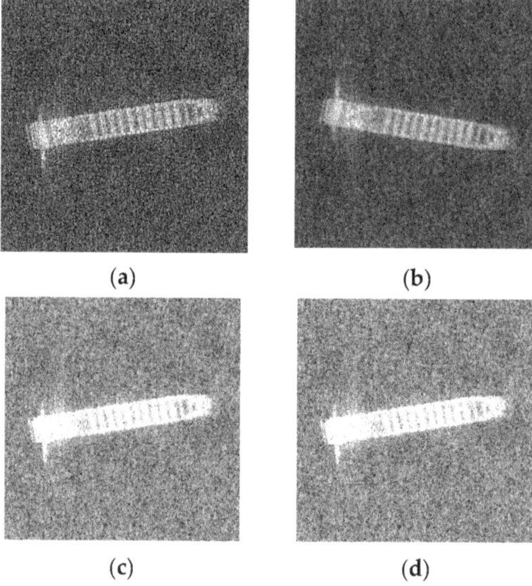

(a) (b)

(c) (d)

Figure 5. Image and image with flipping, brightening, and sharpness. (**a**) An original image. (**b**) Image with flipping. (**c**) Image with brightening. (**d**) Image with sharpness.

One of the most popular methods is random crop. It can significantly increase the amount of data. Recently, researchers usually use some fixed frame to do random crop. For example, using a frame of 224×224 to do random crop in a picture of 256×256. It not only increases the data, but it also retains the most information of the data.

However, not all traditional ways of data processing can be helpful when dealing with SAR images. The operation of data processing may loss original information and amplify noise information. Sometimes, random crop may lose some important information. When the main information of images diverges of the center, using random crop cannot perform well, as shown in Figure 6, the bow and the stern of the ship have been cut. In many traditional cases, such as image classification with a cats and dogs dataset, they usually have tens of thousands images for training and testing. When dealing with these problems, though using random crop may lose some information, but they could still keep the main information of images. This information is sufficient for further work. When concerned with SAR images, this method cannot get good performance as expected.

Figure 6. Image with random crop.

If we control our crop frame to make sure that the main parts of the pictures haven't been cut, as shown in Figure 7, it should be noticed that these pictures can be seen as one image and its translations. This method may have good performance in some simple models, which can extract most of the features of the images by training with that data. However, when concerned with deep neural networks, such as Resnet-50, the repeated images cannot give more contributions for training. Thus the performance cannot have more improvement.

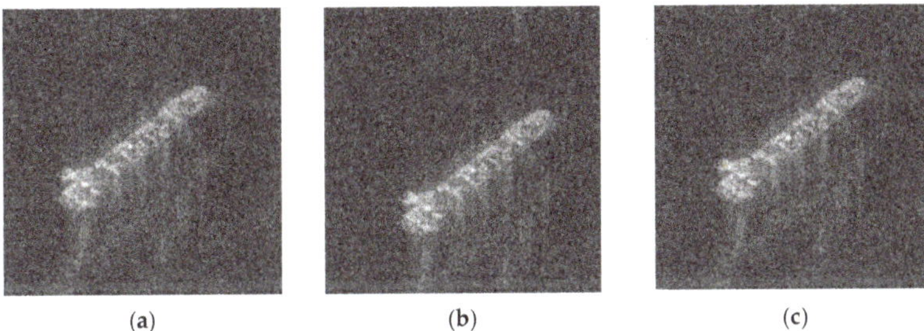

Figure 7. Images with random crop. (**a**–**c**) are three examples of images with random crop.

Rotating is also a popular way for data augmentation, it can keep main information of images. However, people generally rotate images in an integer multiple of 90 degrees: 90, 180, 270. If we rotate the image by a random number of degrees, some problems may appear. As shown in Figure 8, when we rotate image 20 degrees, the black area may introduce a lot of disturbance into CNN training.

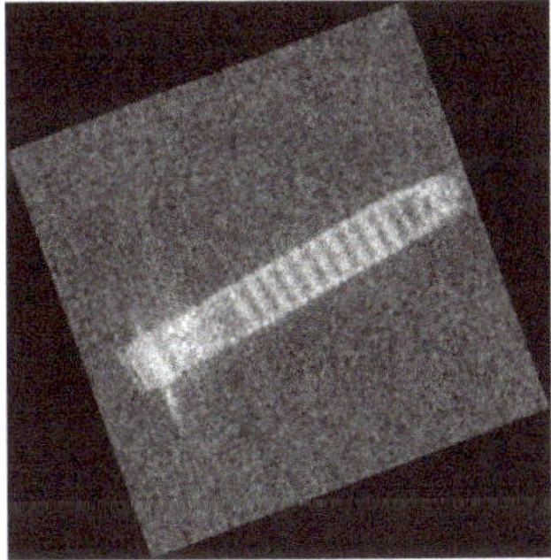

Figure 8. Image rotation of 20 degrees.

To solve this problem, we expansion the images in their edges with pixel pads to remove the black area. The initial image and the image with rotating have been shown in Figure 9.

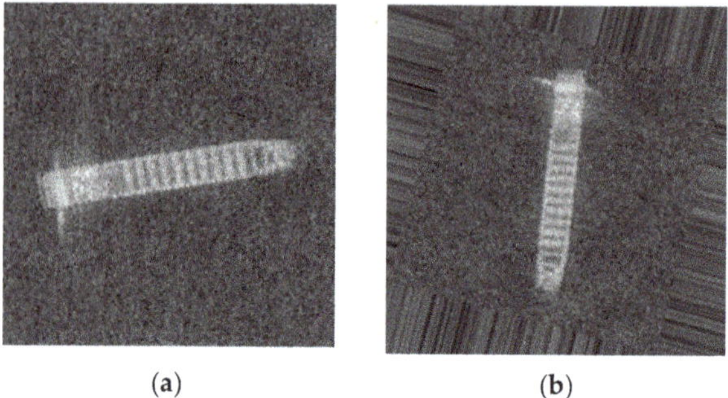

(a)   (b)

Figure 9. Rotate image. (**a**) Original image. (**b**) Image rotated 256 degrees.

By using the proposed method, we not only increase the number of images, but we can also retain the important information of the images. Some noise is still in the background, but concerning SAR images, that noise of the sea surface can be accepted. Because of the characteristics of SAR images, the angle of pictures may cause a lot of difference in CNN training than simple transposition. The data augmentation of all images allows the network to produce better feature representations.

2.3.2. Transfer Learning and Fine-Tuning

Transfer learning is a learning technique that apply the known knowledge from one problem to other related problem [31]. Transfer learning and fine-tune can successfully train a deep model with small datasets. In this paper, we used some pre-trained models for transfer learning, such as Resnet-50, Vgg-16, etc. Depending on structure of different models, we also do some fine-tuning in their layers. The low-level neural layers learned by deep learning models are useful for extracting features such as corners, edges. So, we make the assumption that the lower level neural layers share common features. The methods take the weights of the models in the low-level features as the inputs instead of random weights. Here, we only need to change their fully connected layers and classifiers. For example, the Alexnet model has three linear layers, one dropout layer, and two ReLU layers in its classifier, and the number of its output types are 1000. In our dataset, we need three types for output. So we changed its last layer with output of 3, as shown in Figure 10. The classifier and the fully connected layers should be changed as required.

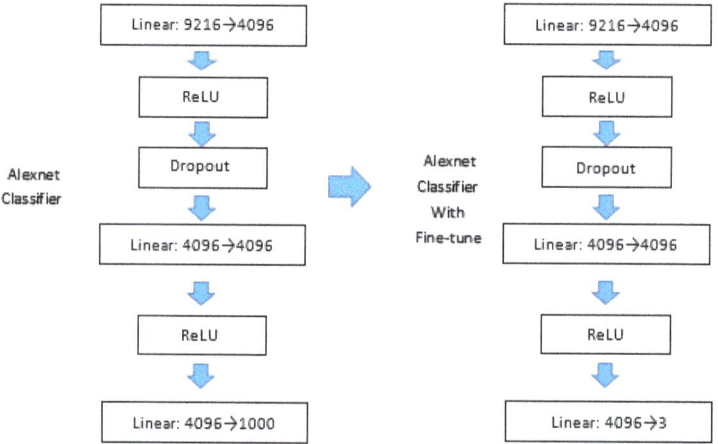

Figure 10. Alexnet with fine-tuning.

3. Results and Discussions

In this section, experiments are carried out to evaluate the performance of the proposed method. Besides, the comparison with other methods indicates the outperformance of the proposed method. We will discuss the results after our experiments.

3.1. Datasets

In this part, we will present our datasets. The SAR ship data set is derived from six full-size SAR images from TerraSAR-X stripmap mode, with a resolution of 2 × 1.5 m in the range and azimuth directions. Slices of SAR ship image are obtained by target detection. And with the aid of the real ground information provided by AIS, all vessels are manually annotated by interpretation experts. The data set includes a total of three types of ships: Bulk Carrier (B), Container Ship (C), and Oil Tanker (OT). A total of 250 stripmap images were used, which were composed of 150 Bulk Carrier images, 50 Container Ship images, and 50 Oil Tanker images. The data size of Bulk Carrier is 64 × 64, and the data size of Container Ship and Oil Tanker is 256 × 256. The chips of ships were shown in Figure 4.

These ship chips are split into training, validation dataset, with percentages of 70% and 30%, respectively.

3.2. Hyperparameters

In this paper, the momentum we set is 0.9. The whole network is trained purely supervised using SGD with a minibatch size [37] of 64 examples, combined with a weight decay parameter of 0.00000001. We rotated images in every 3 degrees for data augmentation. In this paper, the learning rate is initially 0.001 and is reduced by a factor of 0.1 after 100 epochs.

3.3. Experimental Data

By data augmentation, we have three different datasets, named D1, D2, D3. The D1 classification data set is processed with traditional ways of flipping, brightening, sharpness, etc. The operation we did was shown in Table 2. Therefore, we can expand the dataset seven times. The data are divided into two sets: the training data set Dtrain1 (with 1440 samples) and the validation data set Dval1 (with 560 samples). Samples are shown in Figure 11.

Table 2. Traditional augmentation.

Operation	Parameter
Rotate	90,180
Brightening	1.5
Color enhancement	1.5
Contrast	1.5
Sharpness	3.0
Flip	Top bottom

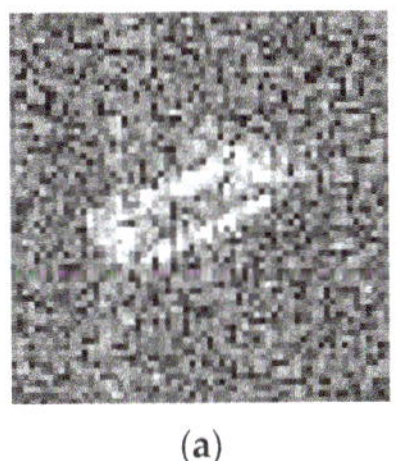
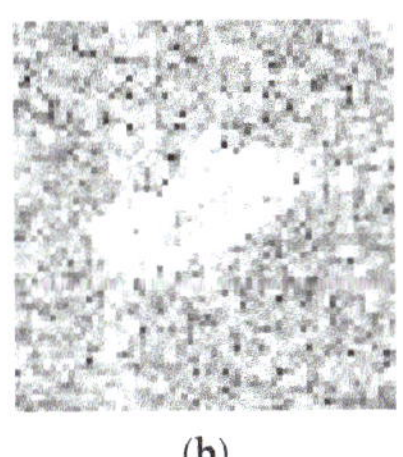
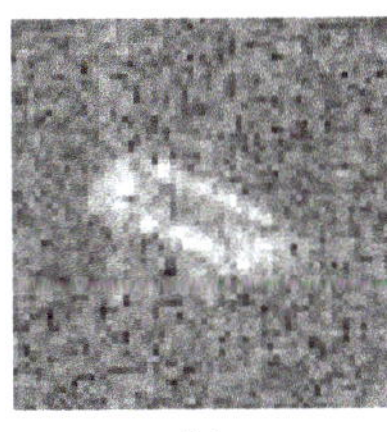

(a)   (b)   (c)

Figure 11. *Cont.*

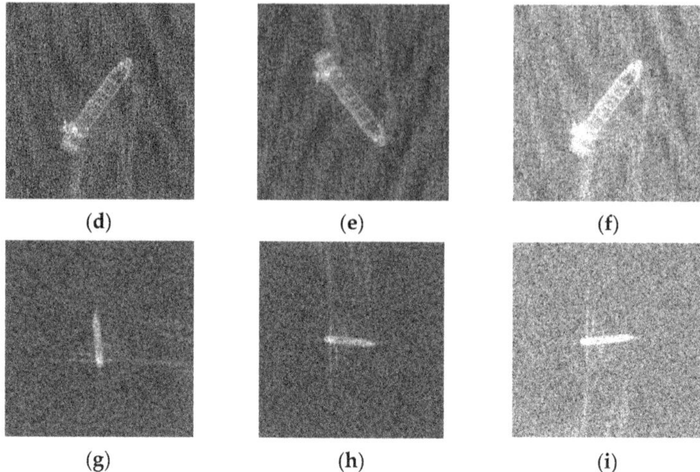

Figure 11. Images with traditional ways. (**a–c**) Bulk Carrier images with processing. (**d,e**) Container Ship images with processing. (**g–i**) Oil Tanker images with processing.

The D2 classification data set is processed with random crop, divided into two sets: the training data set Dtrain2 (with 14270 samples) and the validation data set Dval2 (with 8400 samples), as shown in Figure 12.

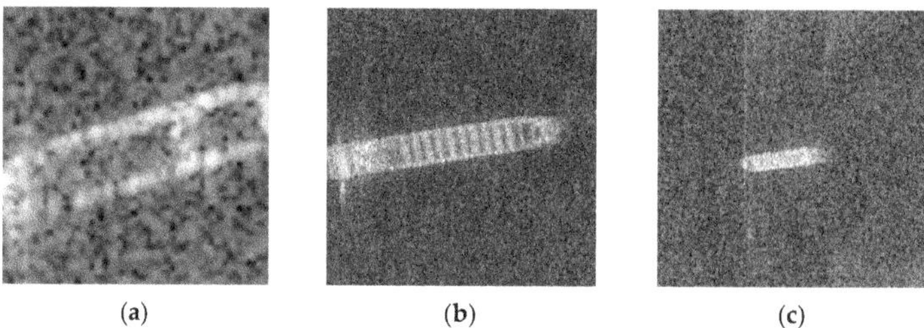

Figure 12. Images with random crop. (**a**) Bulk Carrier image with random crop. (**b**) Container ship image with random crop. (**c**) Oil Tanker image with random crop.

The D3 classification data set is processed with proposed method, divided into two sets: the training data set Dtrain3 (with 13,823 samples) and the validation data set Dval3 (with 8127 samples), as shown in Figure 13.

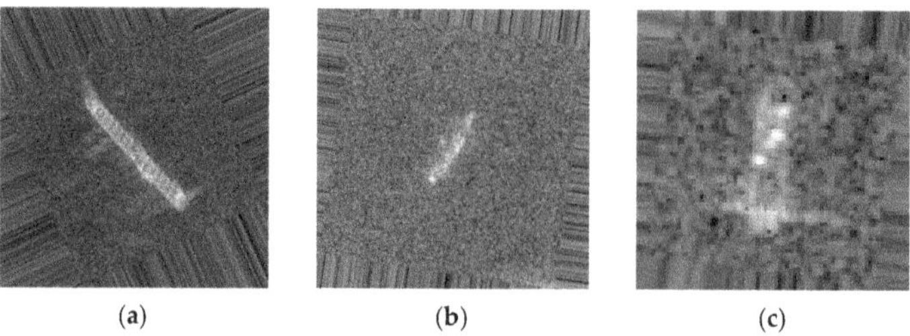

Figure 13. Images with rotate. (**a**) Container Ship image with rotate. (**b**) Oil Tanker image with rotate. (**c**) Bulk Carrier image with rotate.

These three datasets were trained with models as shown in Section 2.1. The image chips of the dataset are listed in Table 3. To confirm the correctness of our methods, we also use the original image in validation dataset as our test dataset. The test datasets are listed in Table 4.

Table 3. D1, D2, and D3 dataset.

D1 Dataset			D2 Dataset			D3 Dataset		
Label	Train	Validation	Label	Train	Validation	Label	Train	Validation
Bulk Carrier	896	304	Bulk Carrier	6110	4560	Bulk Carrier	5595	4255
Container Ship	272	128	Container Ship	4080	1920	Container Ship	4114	1936
Oil Tanker	272	128	Oil Tanker	4080	1920	Oil Tanker	4114	1936

Table 4. Test dataset.

Label	Test
Bulk Carrier	38
Container Ship	16
Oil Tanker	16

All of the experiments have been done several times, and we listed the average results of the experiments.

3.4. First Experiment: Datasets Using Traditional CNN model

In the first experiment, we used the traditional model, as shown in Figure 1a, deal with D1, D2, and D3 datasets.

The results of the first experiment are summarized in Table 5, where accuracies achieved by the traditional CNN models on our validation data set are listed for the considered three classes of maritime targets.

Table 5. Datasets using traditional CNN model.

Dataset	Accuracy (%)
D1	91.43
D2	87.49
D3	88.76

From Table 5, we can see that the traditional CNN models have low accuracy in the D2 and D3 datasets, but perform better with D1 dataset. This is expected, as the traditional CNN model we used have only 10 layers, which is a simple neural network. The D1 dataset has a total of 2000 images for training, so the simple network can produce features pretty well, while when consider with D2 and D3 dataset, the simple network cannot give good expression. Therefore, they had lower accuracy. Furthermore, when using traditional CNN models, it may take more than 1000 epochs when the results become steady.

3.5. Second Experiment: Datasets Using Resnet-50 Model

In the second experiment, we used the Resnet-50 models deal with D1, D2, and D3 datasets. The models was shown in Figure 1b.

The results of the second experiment are summarized in Table 6.

Table 6. Datasets using Resnet-50 model.

Dataset	Accuracy (%)
D1	94.67
D2	95.43
D3	98.52

As shown in Table 6, these results show that the deep networks with transfer learning have better performance than the traditional CNN models when dealing with the three datasets. We can achieve at least 3% higher accuracies for classification when compared with dataset using traditional CNN model. With deep networks, the images with data augmentation can be trained effectively. These results in Table 6 suggest that the proposed method is able to give more target details of the input image for training, it has 3% higher accuracies compared to the method which used random crop and have almost 4% higher accuracies compared to the method which used traditional ways of data augmentation. These results proved that our method is partly corrected.

3.6. Third Experiment: Dataset Using Other Models

In the third experiment, we used other models compared with Resnet-50 models in the D3 dataset. The results are summarized in Table 7.

Table 7. D3 dataset using different models.

Model	Accuracy (%)
Resnet-50	98.52
Alexnet	96.31
VGG-16	98.46
Densenet-121	98.96
Resnet-34	97.24

The results in Table 7 show that with our method, many models can get good performance, and Densenet-121 model have better performance than other popular models, but it may take much time for training. VGG-16 net also performs well. Because of the good performance in Densenet-121, we guess that with deeper networks, further improvement may be attained.

3.7. f1-Score and Misclassified Ships

As shown in Figure 4, we can find that the Bulk Carrier images are bright in its edges and have small length and width. But the Container Ship and Oil Tanker are pertained to be with long length and short width. Therefore, it can be predicted that we could easily classified Bulk Carrier with its size characteristics. However, distinguishing between Container Ships and Oil Tankers may be a challenge.

For further study, we also use precision (the ratio of true positives and predicted positives) and recall (the ratio of true positives and all positives samples) to evaluate our results of D3 dataset, which was trained with Resnet-50 models. And we use f1-score to combine precision and recall into one. The f1-score is then given as

$$f1score = 2 \times \frac{Precision * Recall}{Precision + Recall},\tag{4}$$

The results of the experiment are summarized in Table 8.

Table 8. D3 dataset with Resnet-50 models.

Label	Precision	Recall	f1-Score
Bulk Carrier	1	1	1
Container Ship	0.9763	0.9355	0.9555
Oil Tanker	0.9381	0.9773	0.9573
Avg. total	0.9715	0.9709	0.9709

Some misclassified samples are shown in Figure 14.

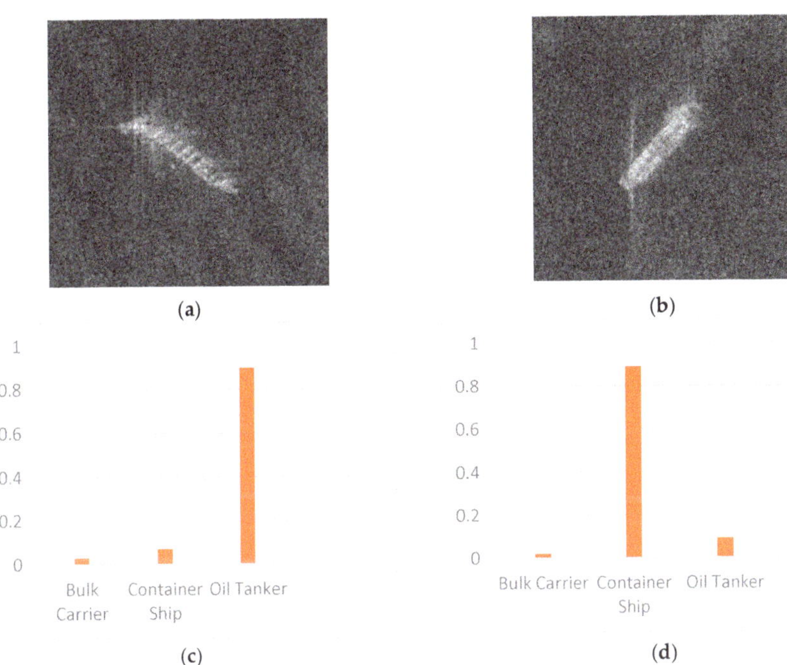

Figure 14. Misclassified ships. (**a**) Container Ship misclassified as Oil Tanker. (**b**) Oil Tanker misclassified as Container Ship. (**c**,**d**) are probabilities of three categories.

The results in Table 8 show that the classifier has a high f1-score for the classes Bulk Carrier. Due to Bulk Carrier's distinct shape and size characteristics, it is easy to classify. As shown in Table 8, we can entirely classify the Bulk Carrier images. Consider Container Ships and Oil Tankers, distinguishing between them is a challenge in ship classification, because they have similar shapes and sizes. However, with our method, we can achieve the scores of 0.9555 and 0.9573, which proves that our method performs well in all of the three classes.

3.8. Comparison with Other Methods

The good performance has been shown from the results in Tables 2–8. To further confirm the importance of combing data processing and training, we did a new experiment. We used the original dataset with no data processing for training and watched the performance with transfer learning and simple CNN models. The results were shown in Table 9.

Table 9. Experiment using original dataset.

Method	Accuracy (%)
Original dataset with simple CNN models	85.71
Original dataset with transfer learning	94.93

When compared the results with Tables 2–8. we can easily find that with no data processing in our dataset, the CNNs cannot perform well either using transfer learning or not. The experimental results confirm that both the data processing and the way in CNN training should be concerned to get better performance.

To confirm our method is effective, we also compare with other methods. In [5], the authors used deep neural networks with transfer learning and fine-tuning for SAR ship classification and they achieved good results. In [17], a multiple input resolution CNN model is proposed and its performance is evaluated. In [38], authors proposed a novel ship classification model combining kernel extreme learning machine and dragonfly algorithm in binary space. The result was shown in Table 9. Because of the dataset and the model, the result may not be exactly correct. When compared our method with these studies, in [5,38], authors did not do data augmentation but training data with fine-tune and a new CNN model. In [17], authors processed data with multiple resolution and used random crop for data augmentation.

The result in Table 10 shows that our method can get better performance compared with other proposed method by taking a compromise between data processing and CNN training, both in accuracy and f1-score.

Table 10. Comparison with other method.

Method	Accuracy (%)	f1-Score
Our Method	98.52	0.9715
Method in [5]	97.62	0.9565
Method in [17]	Unknown	0.9443
Method in [38]	Unknown	0.9404

3.9. Performance in Test Dataset

We also did an experiment with a test dataset. As shown in Table 4, because the test datasets only include 70 images, the data may undulate significantly. The results are summarized in Table 11.

Table 11. Experiment with test dataset.

Accuracy (%)	Classified/Real
98.57	69/70

The result shows that with our method, we can also get good performance with a test dataset.

4. Conclusions

This paper presented a new method for SAR targets classification on TerraSAR-X high-resolution images. To verify our method, we take several experiments to compare. The experimental results reveal that: (1) compared with the dataset trained via random crop, traditional data augmentation, our method achieve the best performance with regard to classification accuracy. (2) With the proposed method in our dataset, the Densenet-121 model scored the best classification performance with an accuracy of 98.96%, other models like VGG-16, Resnet-50 also perform well. (3) When compared with other researchers' work, our method can get at least 1% higher accuracies. We also have advantage in f1-score. (4) Our method also has good generalization ability. Our paper has presented the application of our method in ship classification, the point of ship classification is that we cannot get enough high-resolution SAR ship images. Therefore, if we can solve the data problem, we think we can also get good results when promoting our method to other SAR fields with more images and details. Other procedures in the preprocessing step of the images, such as images with GAN, SRCNN, may be the focus of future work.

Author Contributions: C.L. conceived the manuscript and conducted the whole experiment. W.L. contributed to the organization of the paper and also the experimental analysis.

Funding: This research was supported by the National Natural Science Foundation of China (61331020).

Acknowledgments: We would like to thank the anonymous reviewers for carefully reading the manuscript and providing valuable comments and suggestions.

Conflicts of Interest: The authors declare no conflict of interest.

References

1. Brusch, S.; Lehner, S.; Fritz, T.; Soccorsi, M.; Soloviev, A.; van Schie, B. Ship surveillance with TerraSAR-X. *IEEE Trans. Geosci. Remote Sens.* **2011**, *49*, 1092–1103. [CrossRef]

2. Crisp, D.J. A ship detection system for RADARSAT-2 dual-pol multi-look imagery implemented in the ADSS. In Proceedings of the 2013 International Conference on Radar, Adelaide, Australia, 9–12 September 2013; pp. 318–323.

3. Pieralice, F.; Proietti, R.; La Valle, P.; Giorgi, G.; Mazzolena, M.; Taramelli, A.; Nicoletti, L. An innovative methodological approach in the frame of Marine Strategy Framework Directive: A statistical model based on ship detection SAR data for monitoring programmes. *Mar. Environ. Res.* **2014**, *102*, 18–35. [CrossRef] [PubMed]

4. Gambardella, A.; Nunziata, F.; Migliaccio, M. A physical full-resolution SAR ship detection filter. *IEEE Geosci. Remote Sens. Lett.* **2008**, *5*, 760–763. [CrossRef]

5. Wang, Y.; Wang, C.; Zhang, H. Ship Classification in High-Resolution SAR Images Using Deep Learning of Small Datasets. *Sensors* **2018**, *18*, 2929. [CrossRef] [PubMed]

6. Dong, J.-M.; Li, Y.-Q.; Deng, B. Ship targets recognition by ships' feature in SAR image. *J. Shaanxi Norm. Univ. (Nat. Sci. Ed.)* **2004**, *32*, 203–205.

7. Jiang, M.; Yang, X.; Dong, Z.; Fang, S.; Meng, J. Ship classification based on superstructure scattering features in SAR images. *IEEE Geosci. Remote Sens. Lett.* **2016**, *13*, 616–620. [CrossRef]

8. Touzi, R.; Raney, R.K.; Charbonneau, F. On the use of permanent symmetric scatterers for ship characterization. *IEEE Trans. Geosci. Remote Sens.* **2004**, *42*, 2039–2045. [CrossRef]

9. Liu, C.; Vachon, P.W.; Geling, G.W. Improved ship detection with airborne polarimetric SAR data. *Can. J. Remote Sens.* **2005**, *31*, 122–131. [CrossRef]

10. Touzi, R.; Charbonneau, F. The SSCM for ship characterization using polarimetric SAR. In Proceedings of the 2003 IEEE International Geoscience and Remote Sensing Symposium, IGARSS'03, Toulouse, France, 21–25 July 2003; Volume 1, pp. 194–196.

11. Russell, S.J.; Norvig, P.; Canny, J.F.; Malik, J.M.; Edwards, D.D. *Artificial Intelligence: A Modern Approach*; Prentice Hall: Upper Saddle River, NJ, USA, 2003.

12. Vapnik, V.N. An overview of statistical learning theory. *IEEE Trans. Neural Netw.* **1999**, *10*, 988–999. [CrossRef]

13. Makedonas, A.; Theoharatos, C.; Tsagaris, V.; Anastasopoulos, V.; Costicoglou, S. Vessel classification in cosmo-skymed SAR data using hierarchical feature selection. In Proceedings of the 36th International Symposium on Remote Sensing of Environment, Berlin, Germany, 11–15 May 2015; Schreier, G., Skrovseth, P.E., Staudenrausch, H., Eds.; Copernicus Gesellschaft Mbh: Gottingen, Germany, 2015; Volume 47, pp. 975–982.

14. Lang, H.; Zhang, J.; Zhang, X.; Meng, J. Ship classification in SAR image by joint feature and classifier selection. *IEEE Geosci. Remote Sens. Lett.* **2016**, *13*, 212–216. [CrossRef]

15. Lang, H.; Wu, S. Ship classification in moderate-resolution SAR image by naive geometric features-combined multiple kernel learning. *IEEE Geosci. Remote Sens. Lett.* **2017**, *14*, 1765–1769. [CrossRef]

16. Kang, M.; Ji, K.; Leng, X.; Lin, Z. Contextual region-based convolutional neural network with multilayer fusion for SAR ship detection. *Remote Sens.* **2017**, *9*, 860. [CrossRef]

17. Bentes, C.; Velotto, D.; Tings, B. Ship Classification in TerraSAR-X Images with Convolutional Neural Networks. *IEEE J. Ocean. Eng.* **2018**, *43*, 258–266. [CrossRef]

18. Xing, X.; Ji, K.; Zou, H.; Chen, W.; Sun, J. Ship classification in TerraSAR-X images with feature space based sparse representation. *IEEE Geosci. Remote Sens. Lett.* **2013**, *10*, 1562–1566. [CrossRef]

19. Bengio, Y. *Learning Deep Architectures for AI*; Foundations and Trends®in Machine Learning: Hanover, MA, USA, 2009; Volume 2, pp. 1–127.

20. Schmidhuber, J. Deep learning in neural networks: An overview. *Neural Netw.* **2015**, *61*, 85–117. [CrossRef] [PubMed]

21. Szegedy, C.; Toshev, A.; Erhan, D. Deep neural networks for object detection. In Proceedings of the Advances in neural Information Processing Systems, Lake Tahoe, NV, USA, 5–8 December 2013; pp. 2553–2561.

22. Mikolov, T.; Karafiát, M.; Burget, L.; Černocký, J.; Khudanpur, S. Recurrent neural network based language model. In Proceedings of the Eleventh Annual Conference of the International Speech Communication Association, Makuhari, Japan, 26–30 September 2010.

23. LeCun, Y.; Bottou, L.; Bengio, Y.; Haffner, P. Gradient-based learning applied to document recognition. *Proc. IEEE* **1998**, *86*, 2278–2324. [CrossRef]

24. Krizhevsky, A.; Sutskever, I.; Hinton, G.E. Imagenet classification with deep convolutional neural networks. In Proceedings of the Advances in Neural Information Processing Systems, Lake Tahoe, NV, USA, 3–6 December 2012; pp. 1097–1105.

25. He, K.; Zhang, X.; Ren, S.; Jian, S. Delving deep into rectifiers: Surpassing human-level performance on imagenet classification. In Proceedings of the IEEE International Conference on Computer Vision, Santiago, Chile, 11–18 December 2015; pp. 1026–1034.

26. Chellapilla, K.; Puri, S.; Simard, P. High performance convolutional neural networks for document processing. In Proceedings of the Tenth International Workshop on Frontiers in Handwriting Recognition, La Baule, France, 23–26 October 2006.

27. Aghdam, H.H.; Heravi, E.J. *Guide to Convolutional Neural Networks: A Practical Application to Traffic-Sign Detection and Classification*; Springer: Heidelberg, Germany, 2017.

28. Lawrence, S.; Giles, C.L.; Tsoi, A.C.; Back, A.D. Face recognition: A convolutional neural-network approach. *IEEE Trans. Neural Netw.* **1997**, *8*, 98–113. [CrossRef]

29. Chen, S.; Wang, H.; Xu, F.; Jin, Y.Q. Target classification using the deep convolutional networks for SAR images. *IEEE Trans. Geosci. Remote Sens.* **2016**, *54*, 4806–4817. [CrossRef]

30. Van Dyk, D.A.; Meng, X.L. The art of data augmentation. *J.Comput. Graph. Stat.* **2001**, *10*, 1–50. [CrossRef]

31. West, J.; Ventura, D.; Warnick, S. *Spring Research Presentation: A Theoretical Foundation for Inductive Transfer*; Brigham Young University, College of Physical and Mathematical Sciences: Provo, UT, USA, 2007.

32. Chen, Z.Q.; Li, C.; Sanchez, R.V. Gearbox fault identification and classification with convolutional neural networks. *Shock Vib.* **2015**, *2015*, 390134. [CrossRef]

33. Ciresan, D.C.; Meier, U.; Masci, J.; Gambardella, L.M.; Schmidhuber, J. Flexible, high performance convolutional neural networks for image classification. In Proceedings of the IJCAI Proceedings-International Joint Conference on Artificial Intelligence, Barcelona, Spain, 16–22 July 2011; Volume 22, p. 1237.

34. He, K.; Zhang, X.; Ren, S.; Sun, J. Deep residual learning for image recognition. In Proceedings of the IEEE Conference on Computer Vision and Pattern Recognition, Seattle, WA, USA, 27–30 June 2016; pp. 770–778.

35. Zeiler, M.D.; Fergus, R. Visualizing and understanding convolutional networks. In Proceedings of the European Conference on Computer Vision; Zurich, Switzerland, 6–12 September 2014; pp. 818–833.

36. LeCun, Y.; Bottou, L.; Orr, G.; Müller, K.R. Efficient BackProp. In *Neural Networks: Tricks of the Trade*; Orr, G., Müller, K., Eds.; Lecture Notes in Computer Science: Heidelberg, Germany, 1524.

37. Bottou, L. Large-scale machine learning with stochastic gradient descent. In Proceedings of the COMPSTAT'2010, Paris, France, 22–27 August 2010; pp. 177–186.

38. Wu, J.; Zhu, Y.; Wang, Z.; Song, Z.; Liu, X.; Wang, W.; Zhang, Z.; Yu, Y.; Xu, Z.; Zhang, T.; et al. A novel ship classification approach for high resolution SAR images based on the BDA-KELM classification model. *Int. J. Remote Sens.* **2017**, *38*, 6457–6476. [CrossRef]

Article

Simulation and Analysis of SAR Images of Oceanic Shear-Wave-Generated Eddies

Yuhang Wang [1,2,3], **Min Yang** [1,2] **and Jinsong Chong** [1,2,*]

1   National Key Lab of Microwave Imaging Technology, Beijing 100190, China;
    wangyuhang15@mails.ucas.ac.cn (Y.W.); minyang993@126.com (M.Y.)
2   Institute of Electronics, Chinese Academy of Sciences, Beijing 100190, China
3   School of Electronics, Electrical and Communication Engineering, University of Chinese Academy of
    Sciences, Beijing 101408, China
*   Correspondence: iecas_chong@163.com; Tel.: +86-010-5888-7125

Received: 23 February 2019; Accepted: 26 March 2019; Published: 29 March 2019

Abstract: Synthetic Aperture Radar (SAR) is widely used in oceanic eddies research. High-resolution SAR images should be useful in revealing eddy features and investigating the eddy imaging mechanism. However, SAR imaging is affected by various radar parameters and environmental factors, which makes it quite difficult to learn directly from SAR eddy images. In order to interpret and evaluate eddy images, developing a proper simulation method is necessary. However, seldom has a SAR simulation method for oceanic eddies, especially for shear-wave-generated eddies, been established. As a step forward, we propose a simulation method for oceanic shear-wave-generated eddies. The Burgers-Rott vortex model is used to specify the surface current field of the simulated eddies. Images are then simulated for a range of different radar frequencies, radar look directions, wind speeds, and wind directions. The results show that the simulated images are consistent with actual SAR images. The effects of different radar parameters and wind fields on SAR eddy imaging are analyzed by qualitative and quantitative methods. Overall, the simulated images produce a surface pattern and brightness variations with characteristics resembling actual SAR images of oceanic eddies.

Keywords: oceanic eddies; shear-wave-generated eddies; burgers-Rott vortex model; SAR image simulation

---

1. Introduction

As an important part of ocean dynamics, the formation, motion, and dissipation of eddies are significant research issues for oceanographers. The movement of eddies will agitate seawater and expand the scope of biological organisms, thus affecting the distribution of organics and the transportation of heat and salt in the ocean. Synthetic Aperture Radar (SAR) is capable of acquiring high-resolution images all-day and in all weather, and it is sensitive to minor changes of ocean surface roughness produced by eddies. Therefore, SAR images are advantageous in the identification and study of oceanic eddies. At present, SAR images are widely used for the detection [1–5] and statistical research [6–8] of oceanic eddies.

SAR imaging of eddies is mainly determined by four mechanisms. The first one is the shear-wave mechanism [8–12], which is associated with the wave/current interactions in the zones of current shear, and the eddy manifests itself in the form of regions of modulated normalized radar cross section (NRCS) and twisted into spirals in SAR images. The second one is the film mechanism [8–10,13], which is caused by the suppression of gravity-capillary waves by surface films of natural origin. Such surface films alter the sea surface tension by smoothing ripples and cause the diminishing of NRCS. The third eddy manifestation mechanism is one possible due to variations in the wind field caused by changes in the atmospheric boundary layer across an oceanic temperature front [8,9,13,14]. The fourth mechanism contributes to eddy visualization in SAR images by tracing currents with visible particles, e.g., ice pieces [8–10,15].

Oceanic eddies are captured by SAR, 29% of which are shear-wave-generated; the rest are mostly film-generated [8]. Film generated eddies, owing to the notable smoothing effect on ocean surface roughness, are usually obvious and easy to detect. However, shear-wave-generated eddies, appearing on SAR images through surface roughness modulation from wave-current interaction, are usually less obvious. Image features of shear-wave-generated eddies are affected by various factors, such as wind, current, seafloor topography, SAR parameters, etc. Features of shear-wave-generated eddies are usually the product of several of the above factors, which are sometimes even hard to recognize. Besides, the imaging conditions of actual SAR images is usually unknown and hard to control, which makes it quite difficult to conclude feature patterns of eddies directly from SAR images. Thus, this paper proposes a simulation method whose results may provide guidance for interpreting the features of the shear-wave-generated eddies in SAR images and facilitate detection and visualization of shear-wave-generated eddies in future works.

Shear-wave-generated eddies in SAR images will present brightness variations which correspond to the smoothed and rough region [16]. Ref. [11] presents a schematic diagram of imaging geometry for an idealized anticyclonic eddy, as shown in Figure 1. A dark line appears on the upper left and lower right portions of the eddy, and a bright line on the lower left and upper right of the eddy. This brightness variation of the shear-wave-generated eddy depends on the radar look direction and wind field. By analyzing two airborne SAR images of the same eddy under two orthogonal flight tracks, Lyzenga [11] found that under two orthogonal radar look directions, the eddy spirals show the opposite brightness variation. Johannessen [12] also found that the changes in the radar look direction result in varying NRCS along eddy spirals. In addition, Ivanov [10] mentioned that wind speed has an effect on the eddy feature in SAR images. However, the SAR parameters used to observe the eddies are only a subset of the full parameter set. Thus, the results of these SAR images only represent some of the possible eddy observations.

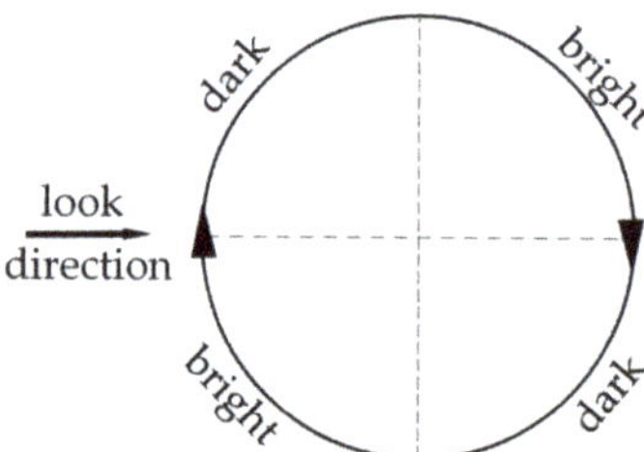

Figure 1. Schematic diagram of imaging geometry for idealized anticyclonic eddy [11].

In order to interpret and evaluate the effects of the radar look direction and wind field conditions on SAR eddy imaging, numerical imaging simulations can serve as an effective tool for the systematical investigation of brightness variations of shear-wave-generated eddies. Cooper et al. [16] used the inertial instability model and the Environmental Research Institute of Michigan (ERIM) Ocean Model to simulate SAR images of film-generated eddies and analyzed the effects of radar parameters and wind fields on eddy imaging. However, SAR image simulation for shear-wave-generated eddies has rarely been performed, and the influence of different radar parameters and wind field conditions on imaging characteristics of these eddies has not been discussed systematically. Therefore, the objective of the present study is to propose a simulation method to simulate radar backscatter images for shear-wave-generated eddies and analyze the influence of the radar look direction and wind field conditions on SAR eddy imaging.

The remainder of this paper is organized as follows: In Section 2, the simulation method and the model used is described in detail. Section 3 verifies the correctness of the proposed method based on ENVISAT-1 ASAR and ERS-2 SAR images. The influence of different radar looks directions, wind directions and wind speeds on SAR eddy imaging are discussed in Section 4. We conclude with a discussion of the applicability of our research for future investigations on shear-wave-generated eddies in Section 5.

2. Simulation Method of SAR Images of Oceanic Shear-Wave-Generated Eddies

The simulation process was divided into two steps: Firstly, the current field of eddy was established on the basis of the Burgers-Rott vortex model. Then, the established current field of the eddy and sea surface wind field were inputted into an oceanic SAR imaging simulation model, and the simulated image was obtained.

2.1. Establishment of Surface Current Field

The features of shear-wave-generated eddies mainly present themselves as spirals in SAR images. Therefore, the hydrodynamic vortex model which presents the eddy spirals should be used to establish the eddy current field. Typical vortex models which are qualified include the Rankine vortex [17], Oseen vortex [18], Burgers-Rott vortex [19–22], and Sullivan vortex [23]. The Rankine vortex model assumes that the vorticity of its core is discontinuous, which is not the case for shear-wave-generated eddies. The Oseen vortex model is hypothesized to be a plane flow; however, shear-wave-generated eddies should be three-dimensional. Therefore, two three-dimensional vortex models with an extra axial velocity have been proposed, i.e., the Burgers-Rott and the Sullivan vortex models. The Burgers-Rott vortex model is a spiral vortex instead of a two-cell vortex like that of the Sullivan vortex model. Due to the fact that the features of shear-wave-generated eddies mainly present as spirals in SAR images, the Burgers-Rott vortex model which can present the eddy spirals is most qualified to establish the current field of the shear-wave-generated eddy.

The Burgers-Rott vortex model is actually an exact solution obtained from the Navier-Stokes equation with the assumption that the vortex is stationary and axisymmetric, and the radial velocity V_r and the circumferential velocity V_θ are independent of the axial coordinate z [19–22], i.e.,

$$V_r = V_r(r), V_\theta = V_\theta(r), V_z = V_z(r, z) \tag{1}$$

The continuity equation of the incompressible flow is

$$\frac{\partial(rV_r)}{\partial r} + \frac{\partial(rV_z)}{\partial z} = 0 \tag{2}$$

The projection of the N-S equation in the circumferential direction can be expressed as

$$V_r \frac{\partial V_\theta}{\partial r} + \frac{V_r V_\theta}{r} = v\left(\frac{\partial^2 V_\theta}{\partial r^2} + \frac{1}{r}\frac{\partial V_\theta}{\partial r} - \frac{V_\theta}{r^2}\right) \tag{3}$$

where v is the kinematic viscosity coefficient of the molecule. Providing the circulation of velocity $\Gamma = 2\pi r V_\theta$, Equation (3) can be rewritten as

$$\frac{d^2\Gamma}{dr^2} = \left(\frac{1}{r} + \frac{V_r}{v}\right)\frac{d\Gamma}{dr} \tag{4}$$

Since the radial velocity V_r is only a function of the radial coordinate r, the general solution of Equation (4) can be expressed as

$$\Gamma = c_1 H(r) + c_2 \tag{5}$$

where c_1 and c_1 are constants, and

$$H(r) = \int_0^r x \exp\left\{\int_0^x [V_r(s)/v]ds\right\}dx \tag{6}$$

Considering $r \to 0$ at the vortex core, $\Gamma \to 0$ should be ensured. Consequently, we know that $c_2 = 0$. At $r \to \infty$, $\Gamma \to \Gamma_0$, we can obtain $c_1 = \Gamma_0/H(\infty)$. Following this, the circumferential velocity V_θ can be computed as

$$V_\theta = \frac{H(r)}{2\pi r H(\infty)}\Gamma_0 \tag{7}$$

To determine the unknown relationship between the axial velocity V_z, the radial velocity V_r and r, it is assumed that the V_z is independent of the radial coordinate r and has a linear relationship with z, i.e.,

$$V_z = \alpha z, \alpha > 0 \tag{8}$$

where α is a constant named suction intensity. Substituting Equation (8) into Equation (2), $d(rV_r)/dr = -\alpha r$ can be obtained, and the general solution is $V_r = -\frac{\alpha}{2}r + c/r$. Considering $r \to 0$ at the vortex core, a finite velocity is ensured. Therefore, the integral constant c is zero, and

$$V_r = -\frac{\alpha}{2}r \tag{9}$$

By substituting Equation (9) into Equation (6), $H(r)$ can be obtained. Following this, the circumferential velocity V_θ can be computed as

$$V_\theta = \frac{\Gamma_0}{2\pi r}\left[1 - \exp(-\frac{\alpha r^2}{4v})\right] \tag{10}$$

According to the above derivation, the three-dimensional velocity field of the vortex in the cylindrical coordinate system is given by Equations (8)–(10). Transform Equations (9) and (10) into the Cartesian rectangular coordinate, and the vortex velocity field can be expressed as

$$\begin{aligned}
V_x &= -\frac{\alpha}{2}x - \frac{\Gamma_0 \alpha}{8\pi v}y \\
V_y &= \frac{\Gamma_0 \alpha}{8\pi v}x - \frac{\alpha}{2}y
\end{aligned} \tag{11}$$

where V_x is the component of the velocity field in the x direction, and V_y is the component of the velocity field in the y direction. Therefore, by setting the parameter values of α and Γ_0, different two-dimensional current fields of vortex can be obtained according to Equation (11).

Figure 2 shows the simulated current fields of vortex. The parameters of the simulation are given in Table 1. Comparing Figure 2a,b, the value of α clearly affects the velocity magnitude of the vortex current field; the velocity magnitude increases with increasing α values. Moreover, the sign of α determines the rotation direction of the vortex current field. The rotation direction is clockwise when α is positive, but anticlockwise when α is negative, as shown in Figure 2b,c. The value of Γ_0/v affects the shape of the vortex; the larger the value of Γ_0/v, the larger the curvature of the vortex, as shown in Figure 2b,d.

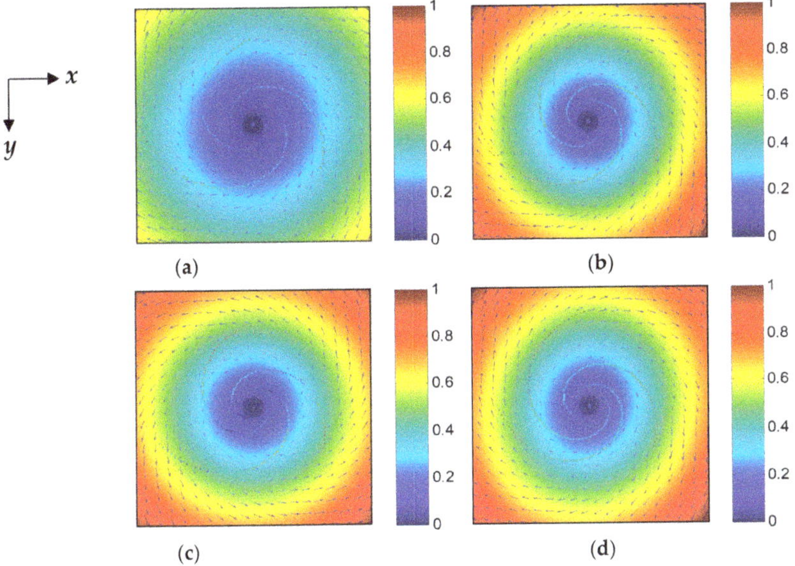

Figure 2. Simulated current fields of vortex based on the Burgers-Rott model. (**a**) $\alpha = 0.0204$, $\Gamma_0/v = 10\pi$, the maximum velocity of the vortex current field $V_{max} = 0.68$ m/s, and the rotation direction of the vortex current field is clockwise. (**b**) $\alpha = 0.0272$, $\Gamma_0/v = 10\pi$, $V_{max} = 1$ m/s, and the current field is clockwise. (**c**) $\alpha = -0.0272$, $\Gamma_0/v = 10\pi$, $V_{max} = 1$ m/s, and the current field is anticlockwise. (**d**) $\alpha = 0.0272$, $\Gamma_0/v = 12\pi$, $V_{max} = 1$ m/s, and the current field is clockwise.

Table 1. The simulation parameters in Figure 2.

Figure	α	Γ_0/v	V_{max}	Rotation Direction
(a)	0.0204	10 π	0.68 m/s	clockwise
(b)	0.0272	10 π	1 m/s	clockwise
(c)	−0.0272	10 π	1 m/s	anticlockwise
(d)	0.0272	12 π	1 m/s	clockwise

2.2. SAR Image Simulation

After the establishment of the eddy current field, an oceanic SAR imaging simulation model was required for simulating SAR eddy images. For this purpose, we selected M4S,f which was developed by Roland Romeiser of the University of Hamburg, Germany. M4S is a software toolkit based on a modified composite surface model for numerical simulations of SAR imaging of oceanic surface current features [24–26], and it can simulate surface wave spectra modulated by spatially varying currents. Different wind fields, radar and platform parameters can be set to investigate their impact on SAR eddy features with the same ocean surface current field induced by eddies.

A flow chart of the simulation process is presented in Figure 3. M4S calculates the NRCS of the sea surface under given radar parameters by reading the sea surface wind field and current field as input data. Among them, the hydrodynamic parameters include α and Γ_0/v. The wind field parameters contain wind speed and direction. The SAR parameters include radar frequency, incidence angle, polarization, and look direction. The platform parameters are flight height and velocity. With the input current field established by the Burgers-Rott vortex model, wind field, and radar parameters, a simulated SAR eddy image is shown in Figure 4. The size of the scene is 20 km × 20 km, and the platform and radar parameters are given in Table 2. The hydrodynamic parameter α and Γ_0/v was set to be −0.001486 and 10π, respectively. The wind speed was 5 m/s, and the wind direction was 225°.

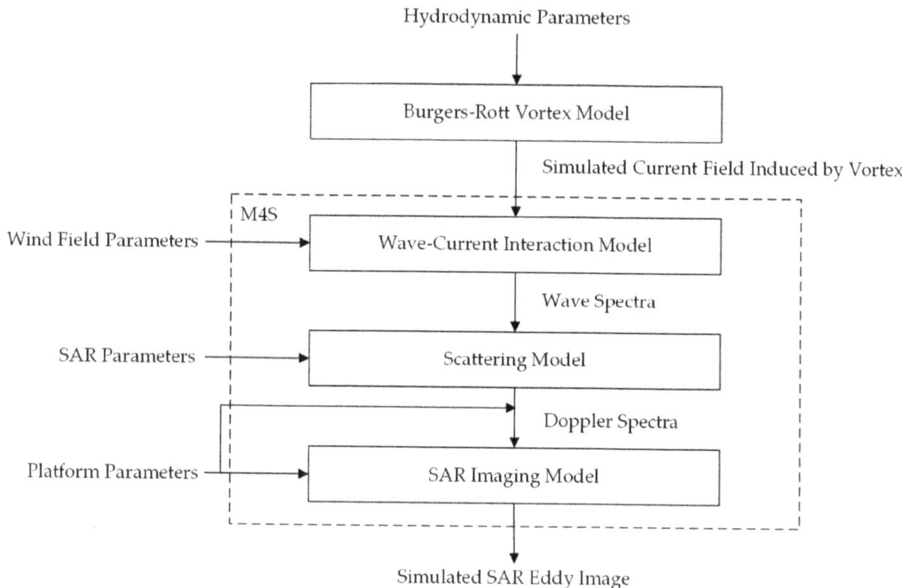

Figure 3. Flow chart of the simulation method.

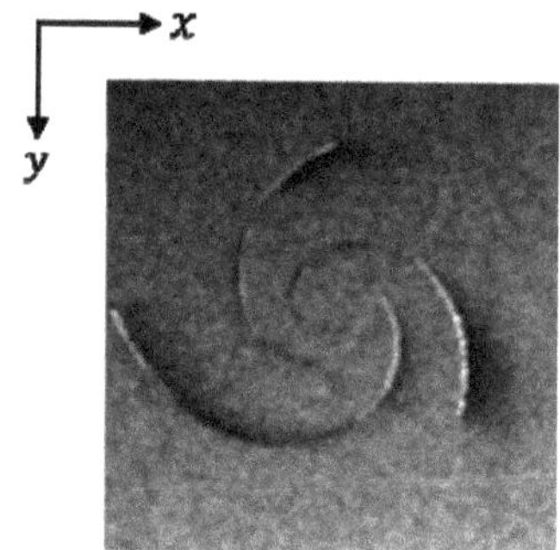

Figure 4. Simulated SAR image of an eddy. The size of the scene is 20 km × 20 km.

Table 2. The simulation parameters of Figure 4.

Parameters	Values
Radar look direction	180°
Incidence angle	23°
Radar frequency	C-band
Polarization	HH
Platform height	800 km
Platform velocity	7455 m/s

3. Validation of the Simulation Method

In this section, simulated eddy images were compared with actual SAR images to validate the rationality of the proposed simulation method.

3.1. Example 1: The Experimental Validation Using ENVISAT-1 ASAR Image

Figure 5 is an ENVISAT-1 ASAR IM image acquired in Luson Strait, on 27 April 2005 at 01:53:42 UTC. The related radar parameters are listed in Table 3. Frame A (20 km × 20 km) highlights a shear-wave-generated eddy with a diameter of about 11.2 km, which is shown in detail in Figure 6.

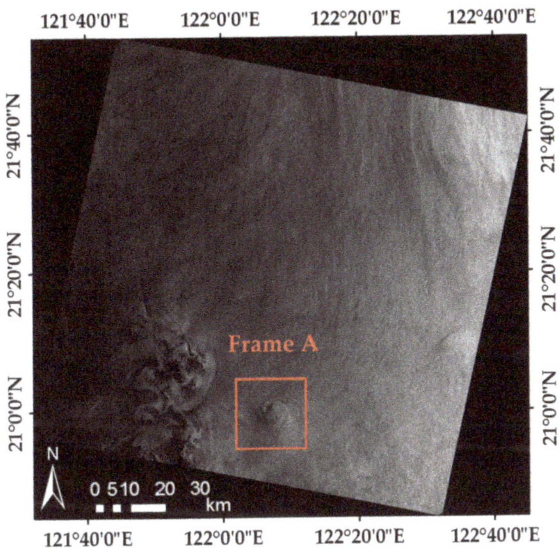

Figure 5. ENVISAT-1 ASAR image of the Luson Strait acquired on 27 April 2005 at 01:53:42 UTC. Frame A (20 km × 20 km) highlights a shear-wave-generated eddy.

Table 3. The related parameters of ENVISAT-1 ASAR.

Platform	Polarization	Band	Boresight Incidence Angle	Platform Height	Platform Velocity
ENVISAT-1	HH	C	23.1°	800 km	7455

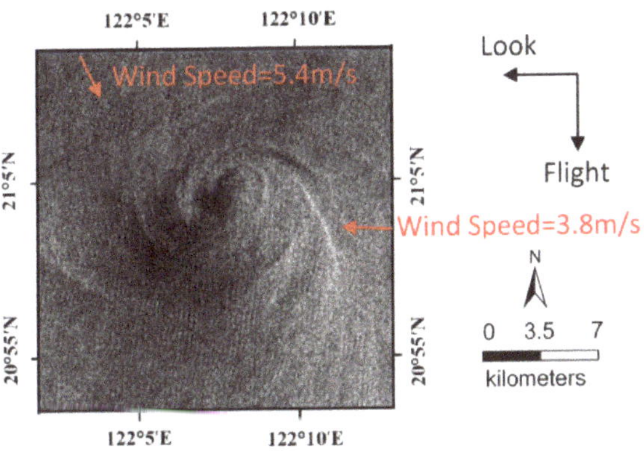

Figure 6. Enlargement of the shear-wave-generated eddy in Frame A. The flight and look direction of ENVISAT-1 ASAR are indicated by black arrows. Two wind vectors are shown as red arrows.

In Figure 6, the flight and look direction of ENVISAT-1 ASAR are indicated by black arrows. Two wind vectors were identified and shown as red arrows. The sea surface wind data was obtained from QuikSCAT on 27 April 2005 [27]. The grid resolution of the wind field is 25 km × 25 km over the ocean surface. The wind speeds and directions from left to right are listed as follows: (a) 5.44 m/s, 159.5° and (b) 3.80 m/s, 273.6°, where the wind direction is defined as the angle clockwise from the north in degrees. In addition, the corresponding current field reanalysis data was obtained from the Global Ocean Data Assimilation System (GODAS) with a spatial resolution of $(1/3)° \times 1°$. A five-day average of the current field data from 25 April 2005 to 29 April 2005 was considered.

According to the above data, the wind speed near the eddy area is 5.21 m/s, the wind direction is 223.6°, and the current velocity is 0.26 m/s. Therefore, α was set to be -0.001486 according to Equation (11). Γ_0/v was set to be 10π, adjusting the simulated eddy spirals to fit well with the actual eddy shape. The radar parameters were set with reference to the ENVISAT-1 ASAR parameters listed in Table 3. The size of the current field is 20 km×20 km, and the spatial resolution is 100 m × 100 m.

The comparison between the simulated and ENVISAT-1 ASAR image is illustrated in Figure 7a,b. The radar look and wind direction are indicated by black arrows. The eddy shape and brightness variations along eddy spirals in the simulated image appear to be consistent with the ENVISAT-1 ASAR image. Under the counterclockwise direction (the cyclonic eddy in the northern hemisphere rotates counterclockwise), the spirals show brightness variations from the outside to inside. The brightness variations along the longest spiral line A follow the order bright-dark-bright-dark, while spiral line B and C follows bright-dark. Such features are explained by changes in the spectral density of Bragg waves responsible for the backscattering of radar signals [28]. In addition, brightness variations along eddy spirals are periodic. A change from bright to dark can be defined as one alternation cycle, and the alternation cycle is related to the scale of the spirals, i.e., the alternation cycle increases as the spirals become longer.

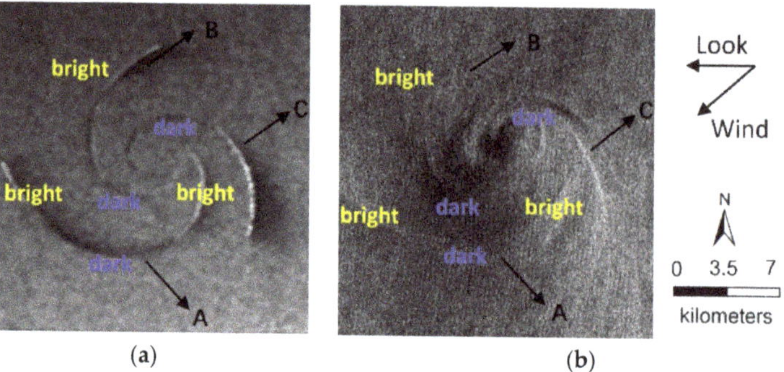

Figure 7. Comparison of (**a**) the simulated SAR eddy image and (**b**) the ENVISAT-1 ASAR image under the same radar parameters and wind field conditions. The radar look and wind direction are indicated by black arrows.

3.2. Example 2: The Experimental Validation Using ERS-2 SAR Image

To further verify the feasibility of the method, we performed a similar simulation experiment of an ERS-2 SAR image. Figure 8 is an ERS-2 SAR image acquired in the Luson Strait, on 11 June 2010 at 01:25:47 UTC. The related radar parameters are listed in Table 4. Frame B (24 km × 24 km) highlights a shear-wave-generated eddy with a diameter of about 24 km, which is shown in detail in Figure 9. In Figure 8, oceanic wakes generated by the island in the upper right corner of the image also exist, but this is beyond the scope of this paper.

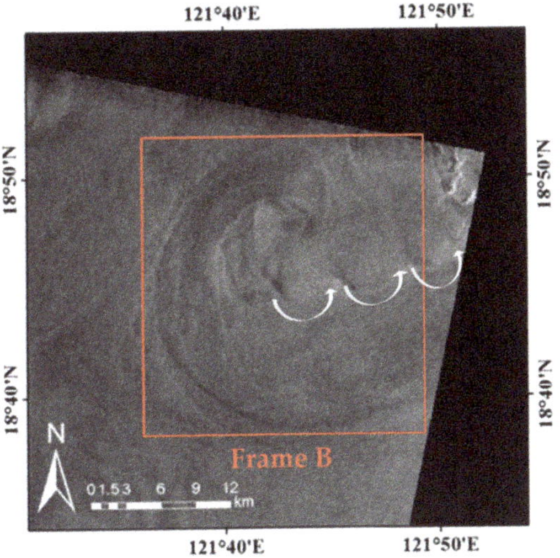

Figure 8. ERS-2 SAR image of the Luson Strait acquired on 11 June 2010 at 01:25:47 UTC. Frame B (24 km × 24 km) highlights a shear-wave-generated eddy.

Table 4. The related parameters of ERS-2 Synthetic Aperture Radar (SAR).

Platform	Polarization	Band	Boresight Incidence Angle	Platform Height	Platform Velocity
ERS-2	VV	C	23°	780 km	7500 m/s

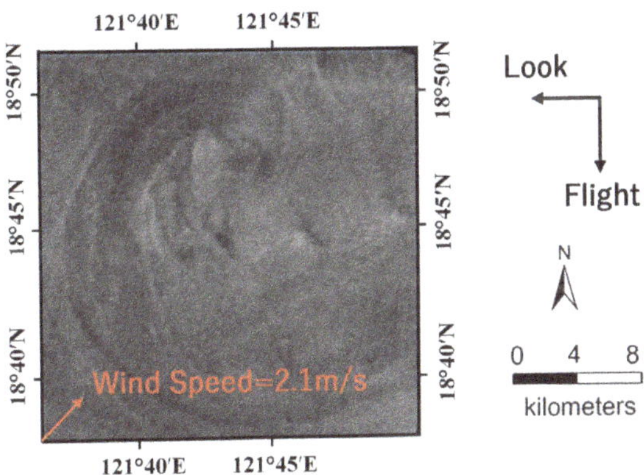

Figure 9. Enlargement of the shear-wave-generated eddy in Frame B. The flight and look direction of ERS-2 SAR are indicated by black arrows. The wind vector is shown as a red arrow.

In Figure 9, the flight and look direction of ERS-2 SAR are indicated by black arrows. The wind vector was identified and shown as a red arrow. The sea surface wind field reanalysis data was obtained from the Europe Centre for Medium-Range Weather Forecasts (ECMWF) on 11 June 2010. The grid resolution of the wind field is 0.125° × 0.125° over the ocean surface. In addition, the corresponding current field reanalysis data was obtained from GODAS with a spatial resolution of $(1/3)° × 1°$. A five-day average of the current field data from 10 June 2010 to 14 June 2010 was considered. According to the above data, the wind speed near the eddy area is 2.1 m/s, the wind direction is 45°, and the current velocity is 0.23 m/s. Therefore, α was set to be 0.000657 according to Equation (11). The radar parameters were set in reference to the ERS-2 SAR parameters listed in Table 4. The size of the current field is 24 km × 24 km, and the spatial resolution is 100 m × 100 m.

The comparison between the simulated and ERS-2 SAR image is illustrated in Figure 10a,b. The radar look and wind direction are indicated by black arrows. The island wakes are omitted during the simulation since we only focused on the eddy spiral. The eddy shape and brightness variations along the eddy spiral in the simulated image appear to be consistent with the ERS-2 SAR image. Under the clockwise direction (the anticyclonic eddy in the northern hemisphere rotates clockwise), the brightness variations along the spiral from the outside to the inside follow the order dark-bright. This result is consistent with the imaging geometry for an idealized anticyclonic eddy in Figure 1. Figure 10a shows the opposite of brightness variations along the eddy spiral, compared to Figure 1, since their radar look directions are opposite.

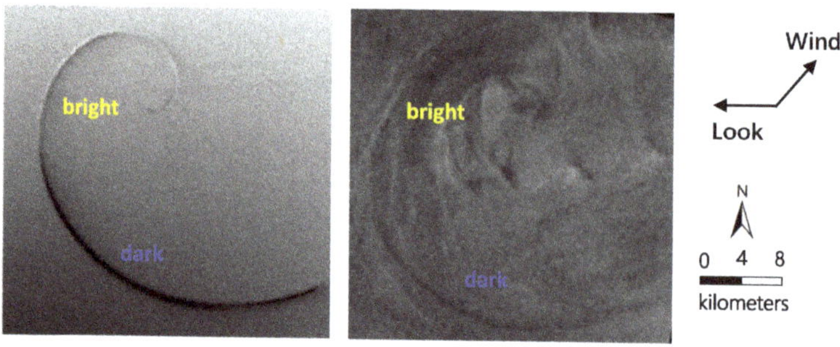

Figure 10. Comparison of (**a**) the simulated SAR eddy image and (**b**) the ERS-2 SAR image under the same radar parameters and wind field conditions. The radar look and wind direction are indicated by black arrows.

According to the above experimental validations, the proposed simulation method can realize an SAR image simulation of shear-wave-generated eddies. Nevertheless, some differences still exist between the simulated and actual SAR images. One of the most distinctive differences is around the eddy cores, where the NRCS of the simulated eddy should be darker. We believe that this is due to the incapability of the M4S model to take a three-dimensional current field as input. The altitude change caused by the eddy will generate a vertical velocity component, and it gets larger near the eddy core. Though the Burgers-Rott vortex model can generate a three-dimensional current field, M4S can only recognize the plane components, thus causing the inadequate modulation of NRCS around the eddy cores.

Some fine structures of the eddies cannot be simulated, although the agreement between the observed and simulated brightness variations of the eddy spirals is generally good. The radar imaging model used for this study is still based on a number of simplifying assumptions. For example, it is not clear whether the actual effect of a spatially varying atmospheric stratification on the surface wave field is always adequately represented by the effect of the proposed equivalent variations of V_x and V_y. Furthermore, our surface wave model does not yet account for effects like wave breaking [29] or feedback between the surface roughness and wind stress [30]. The inclusion of such effects may lead to changes in the simulated radar signatures. Nevertheless, our proposed method is mainly focused on the features along the eddy spirals, since fully simulating the eddy features is too complicated; besides, eddy spirals are usually the only features which are detectable from actual SAR images. Our conclusions of the eddy spiral features may facilitate eddy detection, such as supervised or semi-supervised machine learning of shear-wave-generated eddy detection. We believe that the main results of this study are quite robust and not very sensitive to future model modifications.

4. Influence of Radar Look Direction and Wind Field on SAR Eddy Imaging

In this section, the images under different radar look directions, wind directions and wind speeds were generated by the proposed simulation method to analyze their influence on SAR eddy imaging.

Brightness varies not only along the eddy spirals but also throughout the SAR eddy image. Such brightness variations are, as a matter of fact, a modulation of the normalized radar cross section, which is also referred to as NRCS. The eddy spiral presents itself due to its higher or lower NRCS than the background. Along an eddy spiral, whether it gets brighter or darker, can be quantified as the NRCS contrast between values along the spiral and surrounding it. Therefore, $\Delta\sigma$ is defined as the maximum NRCS contrast, or in other words, it is calculated from the brightest or darkest part of the eddy spiral to represent its visibility from the simulated image. The larger the value of $\Delta\sigma$, the clearer the spirals. To avoid the bias that may result from speckles or thermal noises, an average NRCS of twenty pixels was used for each along or beside the spiral to calculate the difference. In addition, the NRCS dynamic range of the background can also affect the visibility of the eddy, so $\Delta\sigma_r$ is defined as the NRCS contrast of the entire background image; it is also calculated from twenty pixels on average similar to $\Delta\sigma$. The larger the value of $\Delta\sigma_r$, the larger the NRCS contrast of the overall SAR image. Fifty simulations under each radar frequency have been conducted; the eddy spiral features and the calculated $\Delta\sigma$ and $\Delta\sigma_r$ are almost the same.

In general, the influence of the radar look direction, wind direction and wind speed on SAR eddy imaging will be analyzed from four aspects: (1) the brightness variation along eddy spirals and (2) the brightness variation of the SAR image, which can be directly distinguished from simulated images; (3) the visibility of eddy spirals and (4) the brightness contrast of the SAR image, which can be quantified using $\Delta\sigma$ and $\Delta\sigma_r$ respectively.

4.1. Influence of Radar Look Direction

During the simulations, the radar look directions are defined as the angle counterclockwise from the x axis of the current field in degrees, and they were selected as 0°, 90°, 180°, and 270° respectively. The four look directions with respect to the given current field are shown in Figure 11. The red arrows represent radar look directions. The x axis of the current field and wind direction are indicated by black arrows. Simulated SAR images under the four look directions are given in Figure 12. The parameters of the simulations are shown in Table 5.

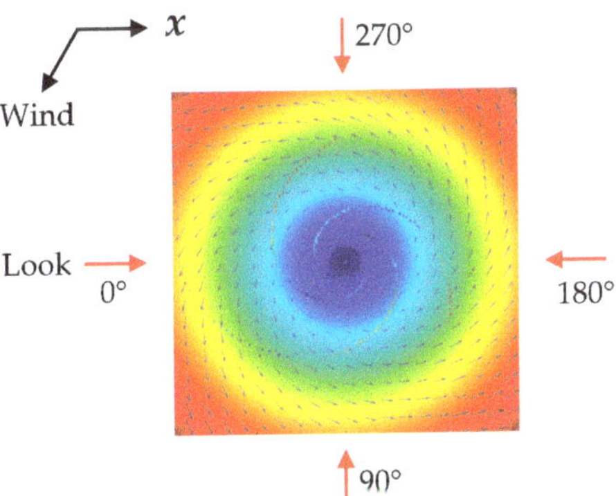

Figure 11. Four radar look directions with respect to the given current field. The red arrows represent the four radar look directions. The x axis of the current field and wind direction are indicated by black arrows.

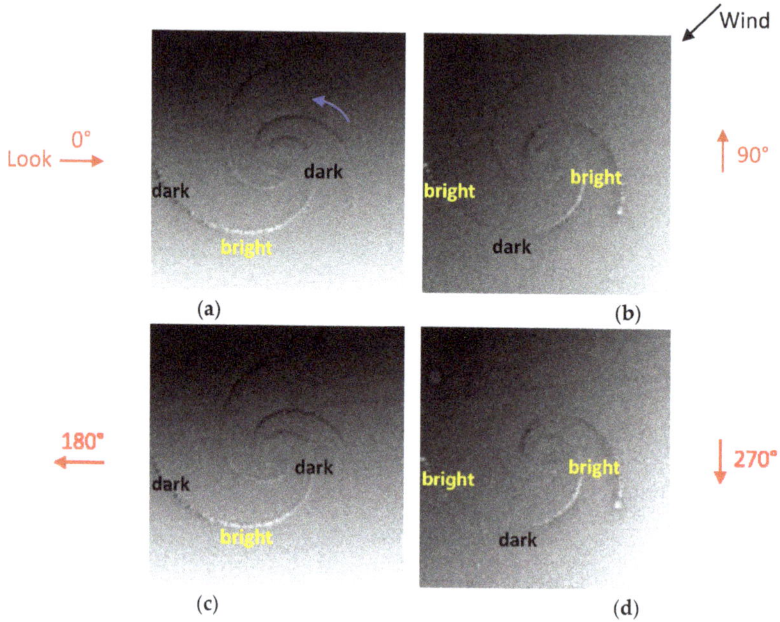

Figure 12. Simulated SAR eddy images under different radar look directions. (**a–d**) correspond to the look directions of 0°, 90°, 180°, and 270° respectively. The black arrow represents the wind direction and the blue arrow indicates the rotation direction of the current field.

Table 5. The simulation parameters of Figure 12.

Parameters	Values
Radar look direction	0°, 90°, 180°, 270°
Incidence angle	25°
Radar frequency	C-band
Polarization	HH
Platform height	800 km
Platform velocity	7455 m/s
Wind direction	225°
Wind speed	6 m/s
Rotation direction of current field	counterclockwise
Maximum Current velocity	1 m/s

In Figure 12, the red arrows represent the four different look directions, the black arrow represents the wind direction and the blue arrow indicates that the rotation direction of the eddy current field is counterclockwise. As shown in Figure 12a–d, the eddy spirals present different brightness variations, which are obviously related to the radar look directions. Under the look directions of 0° and 180°, the brightness variations are both dark-bright-dark from the outside to the inside of the eddy spirals. Meanwhile, under the look directions of 90° and 270°, the brightness variations are both bright-dark-bright. This indicates that the brightness variations along the eddy spirals are the same in two parallel look directions. Meanwhile, under two orthogonal look directions, the brightness variations are opposite. In addition, the results show that the simulated SAR images obtained under each look direction are darker in the upper left portion, but brighter in the lower right portion, which is due to the influence of the wind direction, as described in Section 4.2.

In general, the radar look direction determines the brightness variations along the eddy spirals, but it has a limited effect on the brightness variations of the SAR image. To further analyze the influence of the radar look direction on SAR eddy imaging, $\Delta\sigma$ and $\Delta\sigma_r$ of Figure 12a–d were calculated. The results are shown in Figure 13. To verify the validity of this result under different radar frequencies, L-, S-, and X-band were also considered.

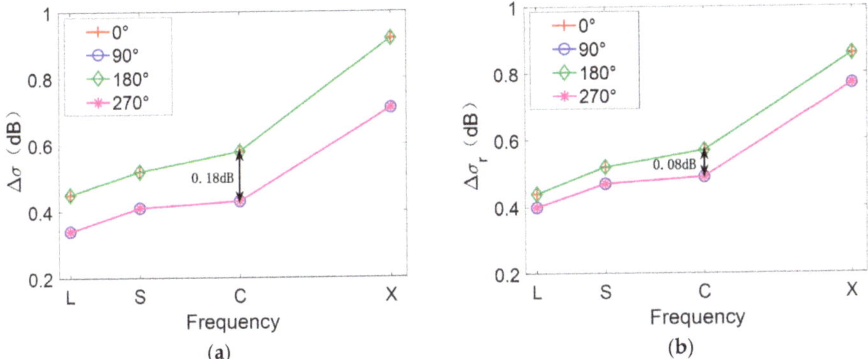

Figure 13. (a) Normalized radar cross section (NRCS) contrast of eddy spirals $\Delta\sigma$ and (b) NRCS contrast of SAR image $\Delta\sigma_r$ under different look directions and radar frequencies. Fifty simulations are averaged to reduce speckle bias.

As shown in Figure 13a, the values of $\Delta\sigma$ are the same in two parallel look directions, such as $0°$ and $180°$, which indicates the same visibility of eddy spirals under two parallel look directions. Moreover, Figure 13a,b shows that there is a considerable value difference of $\Delta\sigma$ between two orthogonal look directions; however, the value difference of $\Delta\sigma_r$ is relatively small. For example, under the look direction of $0°$ and $180°$ at C-band, the value difference of $\Delta\sigma$ is about 0.18 dB, while the value difference of $\Delta\sigma_r$ is only 0.08 dB. This indicates that the radar look direction has more influence on the visibility of eddy spirals than the brightness contrast of the overall SAR image. In addition, when the look direction is $0°$ or $180°$, the values of $\Delta\sigma$ are larger, which means that the eddy is relatively more obvious and conducive to be observed by SAR in this condition. On the other hand, as the radar frequency increases, the values of $\Delta\sigma$ and $\Delta\sigma_r$ increase gradually. It is apparent that the eddy features in the SAR images become clearer at higher radar frequencies.

In summary, the radar look direction mainly affects brightness variations and the visibility of eddy spirals. The simulation results of Figure 12a are consistent with the imaging geometry for the idealized anticyclonic eddy in Figure 1. When the radar look direction is $0°$ and the current field direction is counterclockwise, the longest spiral of the cyclonic eddy changes from dark to bright and then to dark, which is opposite to the brightness variations of the idealized anticyclonic eddy. Furthermore, the conclusion of two orthogonal look directions is in accordance with the analysis of an ERS-1 SAR eddy image in Ref. [11], thus verifying the effectiveness of the simulation.

4.2. Influence of Wind Direction

During the simulations, the wind directions are defined as the angle counterclockwise from the x axis of the current field in degrees and were selected as $45°$, $135°$, $225°$, and $315°$ respectively. The radar look direction was $180°$, and the other parameters were the same as those described in Section 4.1, as shown in Table 6. The simulated SAR images under the four different wind directions are given in Figure 14.

Table 6. The simulation parameters of Figure 14.

Parameters	Values
Radar look direction	180°
Incidence angle	25°
Radar frequency	C-band
Polarization	HH
Platform height	800 km
Platform velocity	7455 m/s
Wind direction	45°, 135°, 225°, 315°
Wind speed	6 m/s
Rotation direction of current field	counterclockwise
Maximum Current velocity	1 m/s

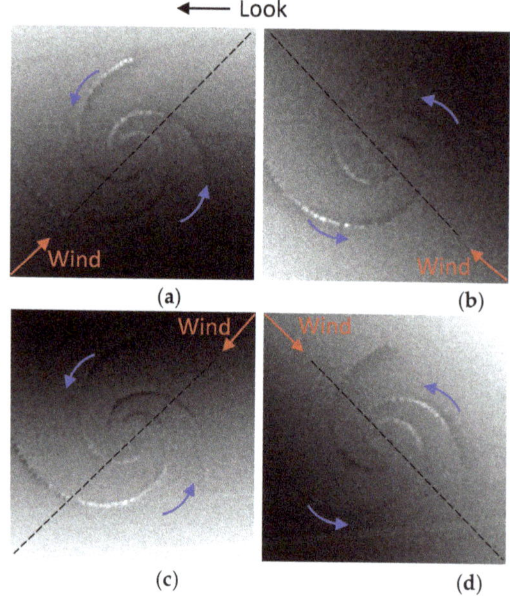

Figure 14. Simulated SAR images under different wind directions. (**a**–**d**) correspond to the wind directions of 45°, 135°, 225°, and 315° respectively. The black arrow represents the radar look direction and the blue arrow indicates the rotation direction of the current field.

In Figure 14, the red arrows represent four different wind directions, the black arrow represents the radar look direction, and the blue arrows indicate that the eddy current field rotates in the counterclockwise direction. Each simulated SAR image is divided into two parts by its diagonal. As shown in Figure 14a–d, the brightness varies across the image and the brightness variation is obviously related to the wind directions, that is, half of the image with the current field direction opposite to the wind direction is brighter, whereas the other half is darker. This phenomenon can also be observed in Figure 7. The lower right portions of Figure 7a,b are brighter than the other area is. This indicates that the wind direction determines the brightness variations of the overall SAR image. Though the brightness variations along the eddy spirals is different under different wind direction, it is merely caused by the brightness variation of the entire SAR image, since when half of the image is brighter, so are the eddy spirals in it. To further analyze the influence of wind direction on SAR eddy imaging, $\Delta\sigma$ and $\Delta\sigma_r$ of Figure 14a–d were calculated. The results are shown in Figure 15. L-, S-, and X-band were also considered.

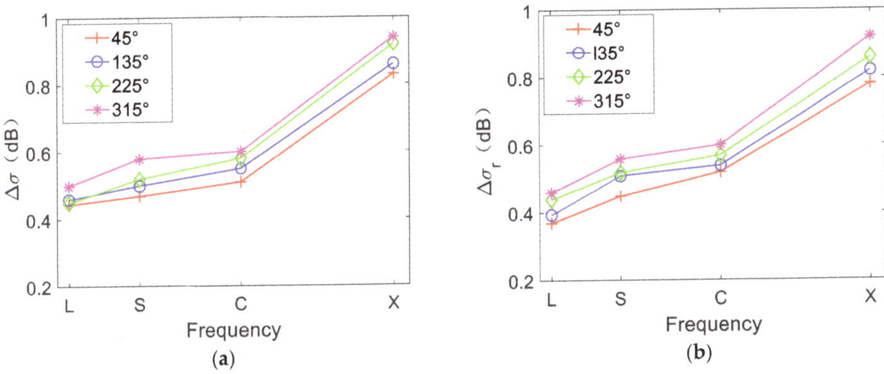

Figure 15. (a) NRCS contrast of eddy spirals $\Delta\sigma$ and (b) NRCS contrast of SAR image $\Delta\sigma_r$ under different wind directions and radar frequencies. Fifty simulations are averaged to reduce speckle bias.

Figure 15a,b shows that under different wind directions, the values of $\Delta\sigma$ and $\Delta\sigma_r$ are slightly different. This implies that wind direction has minor effect on the visibility of eddy spirals and the brightness contrast of the overall SAR image. Therefore, wind direction generally affects the brightness variations of the SAR image, and half of the image with the current field direction opposite to the wind direction is brighter, whereas the other half is darker. In addition, the same conclusion as in Figure 13 can be obtained. The values of $\Delta\sigma$ and $\Delta\sigma_r$ increase as the radar frequency increases.

4.3. Influence of Wind Speed

To analyze the influence of wind speed on SAR eddy imaging, the radar look direction and wind direction were kept constant, and wind speeds were set to be 4 m/s, 7 m/s and 10 m/s respectively. The other parameters of simulations are given in Table 7. The simulated SAR images under different wind speeds are shown in Figure 16.

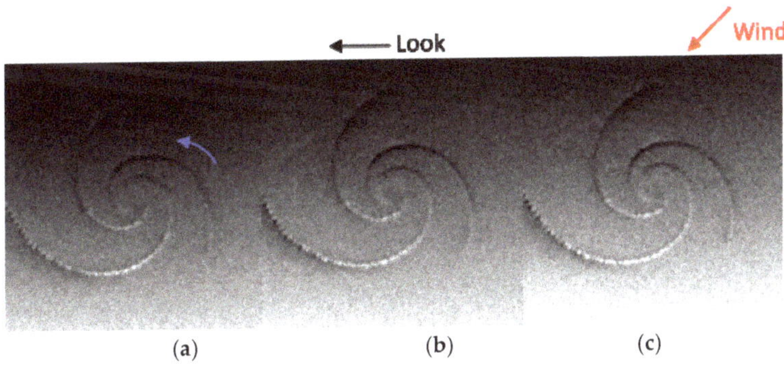

Figure 16. Simulated SAR images under different wind speeds. (a–c) correspond to wind speeds of 4 m/s, 7 m/s, and 10 m/s respectively. The red arrow represents the wind direction, the black arrow represents the radar look direction, and the blue arrow indicates the rotation direction of the current field.

Table 7. The simulation parameters of Figure 16.

Parameters	Values
Radar look direction	180°
Incidence angle	25°
Radar frequency	C-band
Polarization	HH
Platform height	800 km
Platform velocity	7455 m/s
Wind direction	225°
Wind speed	4 m/s, 7 m/s, 10 m/s
Rotation direction of current field	counterclockwise
Maximum Current velocity	1 m/s

In Figure 16, the red arrow represents the wind direction, the black arrow represents the radar look direction, and the blue arrow indicates that the eddy current field rotates in the counterclockwise direction. As shown in Figure 16a–c, wind speed does not change the brightness variations along the eddy spirals, but affects the brightness variations of the overall SAR image. With increases in wind speed, the entire simulated image becomes brighter. To further analyze the influence of wind speed on SAR eddy imaging, $\Delta\sigma$ and $\Delta\sigma_r$ of Figure 16 were calculated. The results are shown in Figure 17. L-, S-, and X-band were also considered.

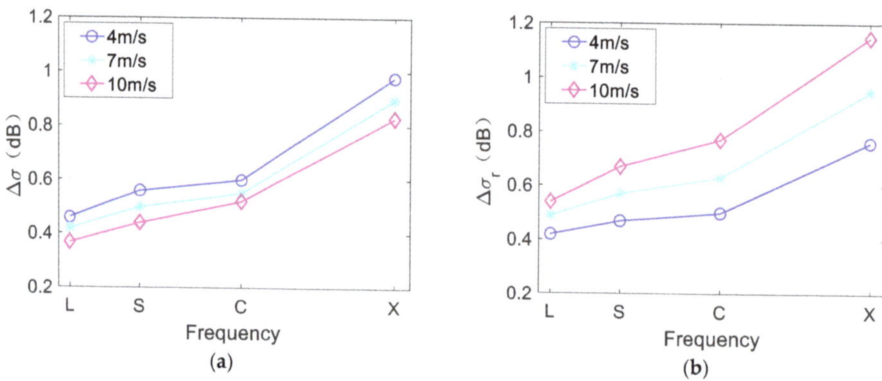

Figure 17. (**a**) NRCS contrast of eddy spirals $\Delta\sigma$ and (**b**) NRCS contrast of SAR image $\Delta\sigma_r$ under different wind speeds and radar frequencies. Fifty simulations are averaged to reduce speckle bias.

In Figure 17a, with increasing wind speed, the value of $\Delta\sigma$ becomes smaller, indicating that the eddy spirals become less obvious. Meanwhile, in Figure 17b, the value of $\Delta\sigma_r$ increases with the increasing wind speed, indicating that the brightness contrast of the entire SAR image increases. In addition, there is a considerable value difference of $\Delta\sigma_r$ between different wind speeds, but the value difference of $\Delta\sigma$ is relatively small. This suggest that the wind speed has more influence on the brightness contrast of the overall SAR image than on the visibility of the eddy spirals.

On the other hand, the results indicate that the values of $\Delta\sigma$ and $\Delta\sigma_r$ are larger at higher radar frequencies, which is the same as the conclusion drawn from Figures 13 and 15. Referring to existing theories, a possible explanation for this conclusion should be the less defocusing effect as the radar frequency gets higher. According to SAR imaging theory, the high resolution along the flight direction is realized by synthesizing a large virtual aperture within the synthetic aperture time. However, the best resolution along the flight direction is restricted by its actual antenna aperture length, which is half of the antenna aperture. To achieve the best resolution, the synthetic aperture time should be [31]:

$$T = 0.886 \frac{cR}{DV_{SAR}f} \tag{12}$$

where f is the radar frequency, c is the speed of light, R is the nearest range between the platform and imaged target, V_{SAR} is the platform velocity, and D is the actual antenna aperture. The synthetic aperture time, as a matter of fact, is the integrating time for the backscattered energy of a target to be well focused. For the same set of antennae, a longer integrating time is needed for a higher radar frequency, according to Equation (12). However, within the integrating time, moving targets will get defocused, and this is inevitable especially when imaging the ocean surface. Therefore, under a higher radar frequency, the SAR image of the eddy suffers less from defocusing due to a shorter integrating time, the spirals are more obvious and the brightness contrast is larger. However, many additional multi-frequency radar images of shear-wave-generated eddies will be needed for a validation of this conclusion, in our opinion. This issue needs to be addressed in more detail in future projects and experiments. We think that an important and solid conclusion can be drawn from our results, despite some unresolved theoretical problems.

5. Conclusions

Based on the Burgers-Rott vortex model, a SAR image simulation method for oceanic shear-wave-generated eddies is proposed in this paper. Furthermore, comparative analyses have proven that the simulated images correspond well to the actual SAR images.

The simulated SAR images indicate that eddy spirals exhibit brightness variations, and the alternation cycles of brightness variations are related to the scale of eddy spirals. However, the quantitative relationship between the alternation cycles and the scale of spirals still needs to be resolved through further statistical comparisons between the simulated and actual SAR images.

SAR images simulated under different radar look directions show that the look direction mainly affects the SAR imaging of eddy spirals. The brightness variations along eddy spirals remain the same under two parallel look directions but show opposite trends under two orthogonal look directions. The visibility of eddy spirals under two parallel look directions is also the same and the spirals are more obvious under radar look directions of $0°$ or $180°$. SAR images simulated under different wind fields show that wind direction and wind speed mainly affect the SAR imaging of the whole eddy area. Wind direction affects the brightness variations throughout the SAR image, and half of the image with the current field direction opposite to the wind direction is brighter, whereas the other half is darker. Wind speed affects the brightness variations and the brightness contrast of the SAR image. With an increased wind speed, the image is brighter and its brightness contrast is higher. Therefore, in future SAR observations of eddies, brightness features due to different radar look directions and wind field should be considered while interpreting eddy images. Moreover, radars at higher frequencies also facilitate the observation of eddy features. To our knowledge, a comparable agreement between observed and simulated radar signatures of the shear-wave-generated eddies at more than one frequency, look direction and wind field has never been conducted in previous studies.

SAR imaging of oceanic eddies is affected by radar parameters and environmental factors, and the simulation method proposed in this paper can facilitate research on eddy features by changing radar parameters and environmental conditions. The simulation results can interpret and evaluate the effects of radar look direction and wind field conditions on SAR eddy imaging and provide guidance for interpreting eddy features in SAR images. Nevertheless, the proposed simulation method and results in this context are mainly focused on shear-wave-generated eddies. Other SAR imaging mechanisms of oceanic eddies, including film mechanisms, thermal mechanisms, and ice mechanisms, need to be resolved through further research.

Sensors **2019**, *19*, 1529

Author Contributions: Conceptualization, Y.W.; Methodology, Y.W. and M.Y.; Validation, Y.W. and M.Y.; Project Administration, J.C.; Supervision, J.C.; Writing—Original Draft, Y.W.; Writing—Review and Editing, J.C.

Funding: This research received no external funding.

Acknowledgments: The authors would like to thank Antony K. Liu of NASA Goddard Space Flight Center and Ming-Kuang Hsu of Technology and Science Institute of Northern Taiwan for their ENVISAT ASAR image data. The authors also thank to the anonymous reviewers for their constructive comments and recommendations.

Conflicts of Interest: The authors declare no conflict of interest.

References

1. Karimova, S. An Approach to Automated Spiral Eddy Detection in Sar Images. In Proceedings of the Igarss IEEE International Geoscience & Remote Sensing Symposium, Fort Worth, TX, USA, 23–28 July 2017.
2. Dreschler-Fischer, L.; Lavrova, O.; Seppke, B.; Gade, M.; Bocharova, T.; Serebryany, A.; Bestmann, O. Detecting and Tracking Small Scale Eddies in the Black Sea and the Baltic Sea using High-Resolution Radarsat-2 and Terrasar-x Imagery (dteddie). In Proceedings of the Geoscience and Remote Sensing Symposium, Quebec City, QC, Canada, 13–18 July 2014; pp. 1214–1217.
3. Johannessen, J.A.; RØed, L.P.; Wahl, T. Eddies detected in ERS-1 SAR images and simulated in reduced gravity model. *Int. J. Remote Sens.* **2012**, *14*, 2203–2213. [CrossRef]
4. Schuler, D.L.; Lee, J.S.; Grandi, G.D. Spiral Eddy Detection Using Surfactant Slick Patterns and Polarimetric Sar Image Decomposition Techniques. In Proceedings of the IEEE International Geoscience & Remote Sensing Symposium, Anchorage, AK, USA, 20–24 September 2004.
5. Liu, A.K.; Peng, C.Y.; Schumacher, J.D. Wave-current interaction study in the gulf of alaska for detection of eddies by synthetic aperture radar. *J. Geophys. Res. Oceans* **1994**, *99*, 10075–10085. [CrossRef]
6. Karimova, S.; Gade, M. Improved statistics of sub-mesoscale eddies in the baltic sea retrieved from SAR imagery. *Int. J. Remote Sens.* **2016**, *37*, 2394–2414. [CrossRef]
7. Xu, G.; Yang, J.; Dong, C.; Chen, D.; Wang, J. Statistical study of submesoscale eddies identified from synthetic aperture radar images in the luzon strait and adjacent seas. *Int. J. Remote Sens.* **2015**, *36*, 4621–4631. [CrossRef]
8. Karimova, S. Spiral eddies in the baltic, black and caspian seas as seen by satellite radar data. *Adv. Space Res.* **2012**, *50*, 1107–1124. [CrossRef]
9. Karimova, S.S.; Lavrova, O.Y.; Solov'Ev, D.M. Observation of eddy structures in the baltic sea with the use of radiolocation and radiometric satellite data. *Izv. Atmos. Ocean. Phys.* **2012**, *48*, 1006–1013. [CrossRef]
10. Ivanov, A.Y.; Ginzburg, A.I. Oceanic eddies in synthetic aperture radar images. *J. Earth Syst. Sci.* **2002**, *111*, 281–295. [CrossRef]
11. Lyzenga, D.; Wackerman, C. Detection and classification of ocean eddies using ERS-1 and aircraft SAR images. In Proceedings of the Third ERS Symposium on Space at the service of our Environment, Florence, Italy, 14–21 March 1997.
12. Johannessen, J.A.; Shuchman, R.A.; Digranes, G.; Lyzenga, D.R.; Wackerman, C.; Johannessen, O.M.; Vachon, P.W. Coastal ocean fronts and eddies imaged with ERS 1 synthetic aperture radar. *J. Geophys. Res. Oceans* **1996**, *101*, 6651–6667. [CrossRef]
13. Alpers, W.; Brandt, P.; Lazar, A.; Dagorne, D.; Sow, B.; Faye, S.; Hansen, M.; Rubino, A.; Poulain, P.M.; Bremer, P. A small-scale oceanic eddy off the coast of West Africa studied by multi-sensor satellite and surface drifter data, and by a numerical model. *Remote Sens. Environ.* **2013**, *129*, 132–143. [CrossRef]
14. Friedman, K.S.; Li, X.; Pichel, W.G.; Clemente-Colon, P. Eddy detection using radarsat-1 synthetic aperture radar. In Proceedings of the Geoscience and Remote Sensing Symposium (IGARSS '04), Anchorage, AK, USA, 20–24 September 2004; pp. 4707–4710.
15. Mitnik, L.M.; Dubina, V.A.; Shevchenko, G.V. Ers sar and envisat asar observations of oceanic dynamic phenomena in the southwestern okhotsk sea. In Proceedings of the 004 Envisat & ERS Symposium, Salzburg, Austria, 6–10 September 2004.
16. Cooper, A.L.; Shen, C.Y.; Marmorino, G.O.; Evans, T. Simulated radar imagery of an ocean "spiral eddy". *IEEE Trans. Geosci. Remote Sens.* **2005**, *43*, 2325–2331. [CrossRef]
17. Rankine, W.J.M. *A Manual of Applied Mechanics*, 3rd ed.; Charles Griffin and Company: London, UK, 1872.

18. Oseen, C.W. *Ber die Stoke'sche Formel und Über Eine Verwandte Aufgabe in der Hydrodynamik*; Almqvist & Wiksell: Stockholm, Sweden, 1911.

19. Burgers, J. *Application of a Model System to Illustrate Some Points of the Statistical Theory of Free Turbulence*; Springer: Dordrecht, Netherlands, 1940.

20. Burgers, J.M. A mathematical model illustrating the theory of turbulence. *Adv. Appl. Mech.* **1948**, *1*, 171–199.

21. Rott, N. On the viscous core of a line vortex. *Z. Angew. Math. Phys. ZAMP* **1958**, *9*, 543–553. [CrossRef]

22. Rott, N. On the viscous core of a line vortex ii. *Zeitschrift für Angewandte Mathematik und Physik* **1959**, *10*, 73–81. [CrossRef]

23. Sullivan, R.D. A two-cell vortex solution of the navier-stokes equations. *J. Aerosp. Sci.* **1959**, *26*, 767–768. [CrossRef]

24. Romeiser, R.; Alpers, W.; Wismann, V. An improved composite surface model for the radar backscattering cross section of the ocean surface: 1. Theory of the model and optimization/validation by scatterometer data. *J. Geophys. Res. Oceans* **1997**, *102*, 25237–25250. [CrossRef]

25. Romeiser, R.; Alpers, W. An improved composite surface model for the radar backscattering cross section of the ocean surface: 2. Model response to surface roughness variations and the radar imaging of underwater bottom topography. *J. Geophys. Res. Oceans* **1997**, *102*, 25251–25267. [CrossRef]

26. Romeiser, R.; Thompson, D.R. Numerical study on the along-track interferometric radar imaging mechanism of oceanic surface currents. *IEEE Trans. Geosci. Remote Sens.* **2000**, *38*, 446–458. [CrossRef]

27. Liu, A.K.; Hsu, M.K. Deriving ocean surface drift using multiple SAR sensors. *Remote Sens.* **2009**, *1*, 266–277. [CrossRef]

28. Mitnik, L.; Dubina, V.; Lobanov, V. Cold Season Features of the Japan Sea Coastal Zone Revealed by Ers Sar. In Proceedings of the IEEE International Geoscience & Remote Sensing Symposium, Boston, MA, USA, 8–11 July 2008.

29. Lyzenga, D.R. Effects of Wave Breaking on Sar Signatures Observed Near the Edge of the Gulf Stream. In Proceedings of the International Geoscience & Remote Sensing Symposium, Lincoln, NE, USA, 31 May 1996.

30. Romeiser, R.; Schmidt, A.; Alpers, W. A three-scale composite surface model for the ocean wave-radar modulation transfer function. *J. Geophys. Res. Oceans* **1994**, *99*, 9785–9801. [CrossRef]

31. Gumming, I.G.; Wong, F.H. *Digital Processing of Synthetic Aperture Radar Data: Algorithms and Implementation*; Artech House: Norwood, MA, USA, 2005; p. 625.

MDPI

St. Alban-Anlage 66

4052 Basel

Switzerland

Tel. +41 61 683 77 34

Fax +41 61 302 89 18

www.mdpi.com

Sensors Editorial Office

E-mail: sensors@mdpi.com

www.mdpi.com/journal/sensors